BIOTIC STRESS MANAGEMENT IN RICE

Molecular Approaches

BIOTIC STRESS MANAGEMENT IN RICE

Molecular Approaches

Edited by
Md. Shamim, PhD
K. N. Singh, PhD

APPLE
ACADEMIC
PRESS

Apple Academic Press Inc.
3333 Mistwell Crescent
Oakville, ON L6L 0A2 Canada

Apple Academic Press Inc.
9 Spinnaker Way
Waretown, NJ 08758 USA

© 2017 by Apple Academic Press, Inc.

First issued in paperback 2021

Exclusive worldwide distribution by CRC Press, a member of Taylor & Francis Group
No claim to original U.S. Government works

ISBN-13: 978-1-77463-675-6 (pbk)
ISBN-13: 978-1-77188-525-6 (hbk)

Library and Archives Canada Cataloguing in Publication

Biotic stress management in rice : molecular approaches / edited by Md. Shamim, PhD, K.N. Singh, PhD.

Includes bibliographical references and index.
Issued in print and electronic formats.
ISBN 978-1-77188-525-6 (hardcover).--ISBN 978-1-315-36553-4 (PDF)

1. Rice--Disease and pest resistance--Genetic aspects. 2. Rice--Molecular aspects. 3. Rice-- Breeding. I. Shamim, Md., 1985-, author, editor II. Singh, K. N. (Kapildeo Narayan), 1956-, author, editor

SB191.R5B56 2017   633.1'89   C2017-900922-2   C2017-900923-0

Library of Congress Cataloging-in-Publication Data

Names: Shamim, Md., 1985- editor. | Singh, K. N. (Kapildeo Narayan), 1956-editor.
Title: Biotic stress management in rice : molecular approaches / editors: Md. Shamim, PhD, K.N. Singh, PhD.
Description: Waretown, NJ : Apple Academic Press, 2017. | Includes bibliographical references and index.
Identifiers: LCCN 2017003499 (print) | LCCN 2017008871 (ebook) | ISBN 9781771885256 (hardcover : alk. paper) | ISBN 9781315365534 (ebook)
Subjects: LCSH: Rice--Effect of stress on. | Plant molecular biology. | Crop improvement.
Classification: LCC SB112.5 .B565 2017 (print) | LCC SB112.5 (ebook) | DDC 633.1/8--dc23
LC record available at https://lccn.loc.gov/2017003499

Apple Academic Press also publishes its books in a variety of electronic formats. Some content that appears in print may not be available in electronic format. For information about Apple Academic Press products, visit our website at **www.appleacademicpress.com** and the CRC Press website at **www.crcpress.com**

ABOUT THE EDITORS

Md. Shamim, PhD

Dr. Md. Shamim is working as Assistant Professor cum Jr. Scientist in the Department of Molecular Biology and Genetic Engineering at Bihar Agricultural University, Sabour, India. He is the author or coauthor of 25 peer-reviewed journal articles, eight book chapters, and two conference papers. He has one authored book and one practical book to his credit. He is an editorial board member of several national and international journals. Recently, Dr. Shamim received the Young Scientist Award 2016 for his research work on Biotechnology by the Bioved Research Institute of Agriculture, Technology and Sciences, Allahabad, India.

Dr. Shamim acquired his BSc (Biology) degree from Dr. Ram Manohar Lohia Avadh University, Faizabad, India, and his MSc (Biotechnology) and PhD (Agricultural Biotechnology) degrees from Narendra Deva University of Agriculture and Technology, Kumarganj, Faizabad, India, with specialization in biotic stress management in rice through molecular and proteomics tools. Dr. Shamim was the awarded Maulana Azad National Fellowship Award from the University Grants Commission, New Delhi, India, during his PhD degree program.

Before joining Bihar Agricultural University, Sabour, Dr. Shamim worked at the Indian Agricultural Research Institute, New Delhi, where he was engaged in heat-responsive gene regulation in wheat. Dr. Shamim also has working experience at the Indian Institute of Pulses Research, Kanpur, India, on molecular and phylogeny analysis of several *Fusarium* fungus of pulses and has also done research at the Biochemistry Department of Dr. Ram Manohar Lohia Institute on plant protease inhibitor isolation and their characterization.

He is a member of the soil microbiology core research group at Bihar Agricultural University (BAU), where he helps with providing appropriate direction and assisting with prioritizing the research work on PGPRs. He has proved himself as an active scientist in the area of biotic stress management in rice, especially in yellow stem borer management by isolating protease inhibitor from jackfruit seeds and sheath blight resistance mechanism in wild rice, cultivated rice, and other hosts.

Dr. K. N. Singh, PhD

Dr. K. N. Singh is working as Professor and Head in the Department of Plant Molecular Biology and Genetic Engineering at Narendra Deva University of Agriculture and Technology, Kumaranj, Faizabad, India. He is the author or coauthor of 50 peer-reviewed journal articles, 10 book chapters, and four conference papers. He has one authored book and one practical book. He is an editorial board member of many journals.

Professor K. N. Singh matriculated from the Bihar School Examination Board with a national scholarship. He earned his BSc (Hons.) and MSc from Science College, Patna and then received a MPhil from Jawaharlal Nehru University, New Delhi, in life sciences. He earne his PhD from Cambridge University (UK) on a Govt. of India overseas fellowship program. He then joined The Energy and Resources Institute, TERI, New Delhi, for a year before joining Tamil Nadu Agricultural University as Assistant Professor and then Associate Professor in the Centre for Plant Molecular Biology (CPMB). He joined Narendra Deva University of Agriculture and Technology as a Professor and has been heading the biotechnology department for the last 10 years. He was a visiting scientist under the Rockefeller Program at IRRI, Philippines. He is a fellow of the Indian Society of Agricultural Biochemist and a life member of many national and international societies. Dr. Singh has proved himself as an active scientist in the area of biotic stress management in rice, pigeon pea, and sesamum.

CONTENTS

LIST OF CONTRIBUTORS

Kahkashan Arzoo
Department of Plant Pathology, G. B. Pant University of Agriculture and Technology, Pantnagar 263145, Uttarakhand, India

Deepa Bisht
National JALMA Institute for Leprosy and Other Mycobacterial Diseases, Tajganj, Agra 282001, Uttar Pradesh, India

Anirudha Chattopadhyay
C. P. College of Agriculture, S. D. Agricultural University, S. K. Nagar, Palanpur 385506, Gujarat, India

Indranil Dasgupta
Department of Plant Molecular Biology, University of Delhi, South Campus, New Delhi 110021, India. E-mail: indranil58@yahoo.co.in

D. K. Dwivedi
Department of Plant Molecular Biology and Genetic Engineering, N. D. University of Agriculture and Technology, Kumarganj, Faizabad 224229, Uttar Pradesh, India

Erayya
Department of Plant Pathology, Bihar Agricultural University, Sabour, Bhagalpur 813210, Bihar, India. E-mail: erayyapath@gmail.com

Geeta
Govt. P. G. College Noida, Gautam Buddhnagar, Sec-39, Noida, Uttar Pradesh, India

Raja Hussain
Department of Plant Molecular Biology and Genetic Engineering, N. D. University of Agriculture and Technology, Kumarganj, Faizabad 224229, Uttar Pradesh, India

Juhi
Department of Ethnobiology, Jiwaji University, Gwalior 474011, Madhya Pradesh, India

Ravi Kesari
Department of Molecular Biology and Genetic Engineering, Bihar Agricultural University, Sabour, Bhagalpur 813210, Bihar, India

N. A. Khan
Department of Plant Molecular Biology and Genetic Engineering, N. D. University of Agriculture and Technology, Kumarganj, Faizabad 224229, Uttar Pradesh, India

B. S. Kharayat
Division of Plant Pathology, ICAR-Indian Agricultural Research Institute, Pusa, New Delhi 110012, India

Anand Kumar
Department of Plant Breeding and Genetics, Bihar Agricultural University, Sabour, Bhagalpur 813210, Bihar, India

Deepak Kumar
Research and Development Unit, Shri Ram Solvent Extractions Pvt. Ltd., U. S. Nagar Jaspur 244712, Uttarakhand, India

Dharmendra Kumar
Department of Plant Pathology, N. D. University of Agriculture and Technology, Kumarganj, Faizabad 224229, Uttar Pradesh, India. E-mail: dkumar_nduat@yahoo.in

Gaurav Kumar
Department of Plant Molecular Biology, University of Delhi, South Campus, New Delhi 110021, India

Mahesh Kumar
Department of Molecular Biology and Genetic Engineering, Bihar Agricultural University, Sabour, Bhagalpur 813210, Bihar, India. E-mail: maheshkumara2z@gmail.com

Pintoo Kumar
Department of Nematology, N. D. University of Agriculture and Technology, Kumarganj, Faizabad 224229, Uttar Pradesh, India

Ravi Ranjan Kumar
Department of Molecular Biology and Genetic Engineering, Bihar Agricultural University, Sabour, Bhagalpur 813210, Bihar, India

Sanjeev Kumar
Department of Plant Pathology, Bihar Agricultural University, Sabour, Bhagalpur 813210, Bihar, India

John M. Lima
National Research Centre on Plant Biotechnology, Pusa Campus, 110012 New Delhi, India

Neelam Maurya
Department of Plant Pathology, N. D. University of Agriculture and Technology, Kumarganj, Faizabad 224229, Uttar Pradesh, India

Deepti Nagaich
Banasthali Vidyapeeth, Jaipur 304022, Rajasthan, India

Nagateja Natra
Department of Plant Pathology, Irrigated Agriculture Research and Extension Center, Washington State University, 24106 N. Bunn Road, Prosser, WA 99350, USA

Narendra Shankar Pandey
Department of Environmental Sciences, Babu Banarsi Das University, Lucknow 226028, Uttar Pradesh, India

Pramila Pandey
Department of Plant Molecular Biology and Genetic Engineering, N. D. University of Agriculture and Technology, Kumarganj, Faizabad 224229, Uttar Pradesh, India. E-mail: pramila28@gmail.com

Poonam
Department of Plant Molecular Biology and Genetic Engineering, N. D. University of Agriculture and Technology, Kumarganj, Faizabad 224229, Uttar Pradesh, India

Nishant Prakash
Krishi Vigyan Kendra, Arwal, Bihar Agricultural University, Sabour, Bhagalpur 813210, Bihar, India

Tushar Ranjan
Department of Molecular Biology and Genetic Engineering, Bihar Agricultural University, Sabour, Bhagalpur 813210, Bihar, India

Sarita
Plant Molecular Biology and Genetic Engineering Division, CSIR-National Botanical Research Institute, Lucknow 226001, India; Academy of Scientific and Innovative Research, New Delhi 110022, India. E-mail: ssarita07@gmail.com

Md. Shamim

Department of Plant Molecular Biology and Genetic Engineering, N. D. University of Agriculture and Technology, Kumarganj, Faizabad 224229, Uttar Pradesh, India. Present address: Department of Molecular Biology and Genetic Engineering, Bihar Agricultural University, Sabour, Bhagalpur 813210, Bihar, India. E-mail: shamimnduat@gmail.com

Divakar Sharma

National JALMA Institute for Leprosy and Other Mycobacterial Diseases, Tajganj, Agra 282001, Uttar Pradesh, India

Mohammed Wasim Siddiqui

Department of Food Science and Technology, Bihar Agricultural University, Sabour, Bhagalpur, 813219, Bihar, India

K. N. Singh

Department of Plant Molecular Biology and Genetic Engineering, N. D. University of Agriculture and Technology, Kumarganj, Faizabad 224229, Uttar Pradesh, India. E-mail: kapildeos@hotmail.com

U. M. Singh

International Rice Research Institute, ICRISAT, Patancheru 502324, Tealangana State, India. E-mail: uma.singh@irri.org

Deepti Srivastava

Department of Plant Molecular Biology and Genetic Engineering, N. D. University of Agriculture and Technology, Kumarganj, Faizabad 224229, Uttar Pradesh, India. Present address: CSIR-National Botanical Research Institute, 226001 Lucknow, Uttar Pradesh, India. E-mail: deeptifzd@gmail.com

R. Srivastava

Division of Molecular and Life Sciences, College of Science and Technology, Hanyang University, Ansan, Republic of Korea

Kapil Kumar Tiwari

C. P. College of Agriculture, S. D. Agricultural University, S. K. Nagar, Palanpur 385506, Gujarat, India. E-mail: kapil21282@gmail.com

Amit Verma

College of Basic Science and Humanities, S. D. Agricultural University, S. K. Nagar, Palanpur 385506, Gujarat, India

Prashant Yadav

Department of Plant Molecular Biology and Genetic Engineering, N. D. University of Agriculture and Technology, Kumarganj, Faizabad 224229, Uttar Pradesh, India; Division of Crop Improvement, National Research Centre on Rapeseed-Mustard, Sewar, Bharatpur 321303, Rajasthan, India

S. Yadav

Plant Biotechnology Division, Forest Research Institute, Dehradun, 248 006, Uttarakhand, India

Sunil Kumar Yadav

Plant Molecular Biology and Genetic Engineering Division, CSIR-National Botanical Research Institute, Lucknow 226001, India; Academy of Scientific and Innovative Research, New Delhi 110022, India

LIST OF ABBREVIATIONS

AGs	anastomosis groups
ALL	actual lesion length
BBMV	brush border membrane vesicles
BLB	bacterial leaf blight
BPB	bacterial panicle blight
BPH	brown plant hopper
BSA	bulked segregant analysis
BTH	benzothiadiazole-s-methyl ester
CAPS	cleaved amplified polymorphic sequence
CP	coat protein
CRISPR/Cas9	clustered regularly interspaced short palindromic repeats/ CRISPR-associated protein)
DH	dead heart
dsRNAs	double-stranded RNAs
ELISA	enzyme-linked immunosorbent assay
ESI Q-TOF MS	electrophoresis and electrospray ionization quadrupole-time of flight mass spectrometry
ETI	effector-triggered immunity
FS	false smut
GLH	green leafhopper
GM	genetically modified
Hph	hygromycin phosphotransferase
HPR	host plant resistance
HR	homologous recombination
HYVs	high-yielding varieties
ISR	induced systemic resistance
ISSR	inter simple sequence repeat
JA	jasmonic acid
LH	lesion height
MABB	marker-assisted backcross breeding
MAGP	marker-assisted gene pyramiding
MARS	marker-assisted recurrent selection
MAS	marker-assisted selection
NILs	near-isogenic lines

ORFs	open reading frames
PAMPs	pattern-associated molecular patterns
PCR	polymerase chain reaction
PMA	phenylemercuric acetate
PR	pathogenesis-related
PRRs	pattern recognition receptors
PSB	pink stem borer
PTI	patterns-triggered immunity
QPCR	quantitative polymerase chain reaction
QTLs	quantitative trait loci
RAPD	randomly amplified polymorphic DNA
RB	resistant bulk
rDNA	ribosomal DNA
RFLP	restriction fragment length polymorphism
RILs	recombinant inbred lines
RISC	RNA-induced silencing complex
RNAi	RNA-interference
ROSs	reactive oxygen species
RP	recurrent parent
RTBV	rice tungro bacilliform virus
RTD	rice tungro disease
RTSV	rice tungro spherical virus
RVD	repeat-variable di-residue
SA	salicylic acid
SAR	systemic acquired resistance
SB	susceptible bulk
ShB	sheath blight
siRNA	small-interfering RNAs
SSB	striped stem borer
SSR	single sequence repeats
SVMs	support vector machines
T3SS	type III secretion system
TA	tiller angle
TALEN	transcription activator-like effector nucleus
TALEs	transcriptional activator-like effectors
vsiRNAs	virus-mediated siRNAs
WE	white ear
YSB	yellow stem borer
ZFNs	zinc-finger nucleases

PREFACE

Ninety percent of rice is produced and consumed in the Asia-Pacific region. With the rapid nurturing of prosperity and urbanization, per capita rice consumption has started declining in the middle- and high-income Asian countries. But nearly a fourth of the Asian population is still poor, and population is growing at 1.8% per year. So far, the annual growth rate for rice consumption in the Asia-Pacific region has kept pace with the demand, more through yield increase rather than area expansion. Nowadays, remarkable efforts are being made by scientists and breeders to raise rice productivity by modifying and manipulating rice plants to survive under different types of stresses.

This book is focused on biotic stress of rice and its management. Rice serves as a host for a number of diseases and insect-pests: 54 in the temperate zone, and about 500 in tropical countries. Of the major diseases of rice, 45 are different types fungal diseases, 10 different bacterial, 15 different viral, and 75 are insect-pests and nematodes that cause economic losses. The major diseases are bacterial blight, bacterial leaf streak, and bacterial sheath rot. Many of the serious rice diseases are caused by several fungi. Some of the diseases, such as blast, sheath blight, brown spot, narrow brown leaf spot, sheath rot, and leaf scald, are of economic significance in major rice-growing areas of the world. Ten major bacterial diseases have been reported in rice. Twelve virus diseases of rice have been identified, and the important ones are tungro, grassy stunt, ragged stunt, stripe, and dwarf virus (in temperate Asia). Brown plant hoppers, stem borers, and gall midges are among the major insect-pests, which cause loss in rice production. Efforts have been made to understand the genetic basis of resistance and susceptibility. The studies are directed to understand the host–plant interaction in rice that have given rise to specialized breeding programs for resistance to diseases and insect-pests. In addition to these advancements in molecular breeding, marker-assisted selection and transgenic approaches open new ways to increase the resistance in rice for better production under several biotic stresses.

This book, *Biotic Stress Management in Rice: Molecular Approaches,* covers the above-mentioned advancements in marker-assisted selection, molecular breeding, gene pyramiding, and transgenic approaches of biotic stress management in rice. The book is comprised of 12 chapters. Chapter 1 describes the recent understandings of the molecular biology of rice tungro disease and its management. This chapter provides an in-depth study on the different approaches against the rice tungro virus management through transgenics. Chapter 2 illustrates the most important bacterial disease of rice. The bacterial blight resistance gene pyramiding in different improved rice variety is discussed. Different examples of resistance genes, that is, *Xa,* etc. are cited in the chapter for better understanding. Chapters 3 and 4 elaborately discuss the two major fungal pathogens (blast and sheath blight). Different resistance rice cultivars, their resistance loci, and quantitative trait loci mapping in the important rice cultivars have been also discussed. Chapter 5 deals with false smut, a fungal pathogen, which is affecting rice production heavily in certain area. The chapter deals with the different approaches for the false-smut pathogen detection and management through different molecular and other methods. Chapter 6 describes advances in brown spot resistance breeding and management in rice. An in-depth discussion on brown spot disease management technologies and lacuna has been also included. Several nematodes affect the rice growth and production. The nematodes infestation begins right from the roots, and infestation remains a problem up to the completion of life of rice plant and this has been discussed in Chapter 7. Chapter 8 discusses molecular approaches for yellow stem borer management in rice through some moderately resistance rice cultivars. Different transgenic technologies and other insecticidal proteins have been also discussed in this chapter. Chapter 9 deals with leaf folder, another important insect of rice that affects the rice production worldwide. In this chapter, different *cry* proteins, their construct, and transformation process have been discussed elaborately. In Chapter 10, detailed information on postharvest losses by different insect-pest and by disease is listed and their toxins production has been also discussed, including the handling process right from the harvesting stage to final marketing. Chapter 11 describes the different major gene pyramiding techniques in rice and their importance. The different major gene pyramided in the improved rice varieties have been listed and discussed. Chapter 12 deals with the recent advancement made for the tackling of several biotic stresses causing agents, that is, for insect-pest, nematodes, bacteria, fungi, and viruses through RNA interference.

We believe that this book will become a standard reference work for the molecular approaches in rice disease management for the higher production through molecular breeding and transgenic approaches. The editors would appreciate receiving useful comments from readers that may assist in the development and advancement of future editions of the book.

—**K. N. Singh**

CHAPTER 1

MOLECULAR BIOLOGY OF RICE TUNGRO VIRUSES AND STRATEGIES FOR THEIR CONTROL

GAURAV KUMAR and INDRANIL DASGUPTA*

Department of Plant Molecular Biology, University of Delhi, South Campus, New Delhi 110021, India

Corresponding author. E-mail: indranil58@yahoo.co.in

CONTENTS

ABSTRACT

Rice tungro disease is the most important viral disease of rice in south and southeast Asia. Two viruses, *Rice tungro bacilliform virus* (RTBV), a DNA virus, and *Rice tungro spherical virus* (RTSV), an RNA virus, are found to be present in rice plants affected by tungro. Concerted efforts by several groups of researchers in India and abroad have built up a rich knowledge resource on the above viruses. A defense strategy targeting one or both the viruses promises to deliver tungro resistance in rice. Multiple approaches toward resistance against RTBV and RTSV have resulted in a number of reports indicating success in achieving this goal. Constitutive expression of viral RNAi-inducing DNA constructs strengthens the antiviral RNAi response and results in viral resistance. Similarly, expression of coat protein gene also results in resistance. Finally, manipulating expression factors required for the viral promoters have also been used to produce viral resistance in rice.

1.1   INTRODUCTION

Rice tungro disease (RTD) is a viral disease of rice, prevalent in south and southeast Asia. RTD is responsible for an annual loss in excess of 10^9 US dollars in the affected countries (Herdt, 1991) and about 2% loss of production in India (Muralidharan et al., 2003). RTD appeared in India in the late 1960s (Raychaudhury et al., 1967a,b), soon after the first description of the disease from the International Rice Research Institute, Philippines (Rivera & Ou, 1965). Since then, studies on RTD have resulted in the accumulation of significant knowledge regarding the causative viruses, pathological and biochemical changes, vector transmission, and resistance genes (Azzam & Chancellor, 2002). In the last two decades, several attempts have been made to engineer resistance against the disease and to further characterize resistance genes. In this chapter, the important information related to development of viral resistance in rice have been described, and new developments aimed at obtaining novel resistance strategies have been discussed.

1.2   REVIEW OF RECENT ADVANCES IN RESEARCH ON RTD

RTD is known by various names in countries of Asia, although its characteristic symptoms remain the same, namely yellow-orange foliar discoloration and stunting. However, there is a strong influence of the genetic background of the rice variety used; highly susceptible varieties such as Taichung Native-1 show severe symptoms, whereas varieties such as Pusa Basmati-1 show mild symptoms. RTD is spread by an insect vectorfound in abundance in rice fields, the green leafhopper (GLH), *Nephotettix virescens* (Ling, 1974). Viral particles showing icosahedral morphology as well as showing bacilliform morphology are found to be associated with RTD. The bacilliform-shaped particles comprise *Rice tungro bacilliform virus* (RTBV) and the isometric particles *Rice tungro spherical virus* (RTSV), having DNA and RNA as the genetic materials, respectively (Jones et al., 1991). GLH shows an interesting transmission pattern with respect to the two viruses. Studies have clearly revealed that GLH can transmit RTSV independent of RTBV, but the latter is always transmitted with RTSV (Cabauatan & Hibino, 1985). Hence, RTSV can be thought of functioning as a helper virus for the transmission of RTBV. RTSV, on its own, causes mild stunting, whereas RTBV causes the typical yellow-orange foliar symptoms in affected plants. In the presence of both the viral components, the symptoms are accentuated. Hence, it was concluded quite early on that RTSV provides the transmission functions for the viral complex, but the symptoms are caused mainly by RTBV (Cabauatan & Hibino, 1988; Hibino & Cabauatan, 1987). This hypothesis was proven when it was shown that a cloned RTBV DNA was capable of infecting rice plants, independent of GLH, if introduced through *Agrobacterium* in a binary plasmid (Dasgupta et al., 1991). The *Agrobacterium* cells containing the cloned RTBV DNA were injected into the meristematic region of 15–20-day-old rice plants, which in turn showed symptoms of RTD, albeit in a mild form, indicating strongly that RTBV causes most of the symptoms of RTD. The precise mechanism by which RTSV provides the transmission functions are yet unknown.

1.3   GENE FUNCTIONS AND VARIABILITY OF RTBV

RTBV belongs to the virus family *Caulimoviridae*, genus *Tungrovirus* and is its sole member. All caulimovirids have a circular, double-stranded DNA genome of approximately 7–8 kbp, with genes arranged only in one direction. All of them are also described as pararetroviruses, because they use reverse transcriptase to convert their terminally redundant full-length transcripts into DNA. In addition, a conserved sequence representing the binding site for a tRNA primer is present in all of them, which primes cDNA synthesis on a full-length RNA template. All RTBV genomes analyzed are about 8 kbps in size and have four open reading frames (ORFs), ORFs I–IV. Based on biochemical analysis of the expressed proteins, only ORF III has been assigned functions. ORF III encodes a large polyprotein, which has domains exhibiting functions of a movement protein, coat protein (CP), protease, and reverse transcriptase-RNAse-H (Laco & Beachy, 1994; Marmey et al., 2005). This arrangement of genesis is conserved in all pararetroviruses (Hay et al., 1991). The protein encoded by ORF II interacts with RTBV CP, both in vitro as well as in vivo, the interacting domains being essential for infectivity of the virus. This observation points toward the possible essential nature of such interactions and suggests a role of assisting the encapsidation (Herzog et al., 2000). Comparison of the DNA sequences of the full-length RTBV genomes determined almost a decade and a half ago clearly showed that they could be grouped into two types; the south Asian and the southeast Asian groups. The isolates from India and Bangladesh fall in the former group, and those from the countries of Myanmar to Philippines in the latter group (Nath et al., 2002). ORF IV, which is present only in the genus *Tungrovirus*, shows the highest sequence heterogeneity between the two groups of RTBV, as compared to the rest of the ORFs. ORF IV from an isolate from Philippines has been shown to function as of suppressor of gene silencing (Rajeswaran et al., 2014), a function related to viral virulence. The division of RTBV in two groups has been reinforced by the analysis of newer full-length (Banerjee et al., 2011b; Ganesan et al., 2009; Sharma & Dasgupta, 2012; Valarmathi et al., 2016) and partial RTBV sequences (Mangrauthia et al., 2012a), which have become available from new regions in India.

1.3.1 GENE FUNCTIONS OF RTSV

RTSV is a member of the family *Secoviridae*, whose members have single-stranded positive-sense RNAs as their genome. The genome of RTSV encodes a large polyprotein, which can potentially give rise to three CPs and a replicase (rep), in addition to several other proteins (Shen et al., 1993). Two additional small ORFs in the 3' end of the RTSV RNA was initially reported, but was not confirmed in subsequent studies (Verma & Dasgupta, 2007). The roles of the rest of the proteins encoded by the RTSV genome are still obscure. Interestingly, RTSV sequences from across south and southeast Asia show high sequence conservation (Verma & Dasgupta, 2007), which is not the observation for RTBV sequences.

1.3.2 ROLE OF THE VIRUSES IN RTD

Although RTBV and RTSV were known to be associated with RTD since the late 1960s, Koch's postulate has been demonstrated only for RTBV. This means that of the two viruses, only RTBV has been shown to be able to cause RTD when introduced back into rice, albeit in a cloned form (Dasgupta et al., 1991). This is because of the lack of mechanical transmission of RTBV and RTSV, meaning that they cannot be introduced into rice plants by simple rubbing or abrasion of leaves; a process fairly common for many viruses infecting solanaceous crop plants. GLH is absolutely required for transmission of RTBV and RTSV to rice, and there is an absolute requirement of RTSV for RTBV transmission, thereby making it technically challenging to introduce individual viruses into the rice plant. The demonstration of the role of RTBV in RTD could only be achieved using "agroinoculation," a method in which the viral DNA is cloned in a binary vector and introduced to the plant *via Agrobacterium*. Rice plants agroinoculated with cloned RTBV showed mild RTD symptoms and the virus could also be transmitted further by GLH (Dasgupta et al., 1991). The agroinoculated plants showed a recovery later. This suggested that RTBV fulfilled of Koch's postulates for RTD and can be thought to be one of the causative viruses. A similar demonstration for RTSV is still awaited.

1.3.3 CONVENTIONAL METHODS TO ATTAIN RTD RESISTANCE

Using classical breeding methods, RTD management was focused on the availability of resistance genes in the rice germplasm and their transfer to popular cultivated rice varieties. Extensive screening of the available rice germplasm revealed a moderately large number of rice varieties, which showed resistance or tolerance to RTSV, but very few against RTBV (reviewed in Azzam & Chancellor, 2002). The nature of the resistance genes or their modes of action were unknown, and only in very few cases was their chromosomal locations mapped by the help of molecular markers. In addition, in the absence of effective methods to discriminate between the resistance to RTBV and RTSV, and resistance to GLH, the resistance could not be distinguished as being against the viruses or against GLH. The resistance to GLH is believed to be transient in nature and is generally based on the presence of an antifeedant. Such resistance may be easily broken under heavy GLH pressure (Manwan et al., 1985).

1.4   NONCONVENTIONAL APPROACHES TO OBTAIN TUNGRO RESISTANCE

1.4.1   RESISTANCE AGAINST RTSV

The earliest report at engineering RTD resistance used the classical CP-mediated resistance strategy, in which, the viral CP is expressed in plants using a transgene and the expressed CP results in virus resistance. The three CPs of RTSV, either singly or in various combinations, were expressed in rice under constitutive promoters. Expression of the various CPs was detectable using Northern blot to look for the accumulation of the corresponding transcripts. When the transgenic rice plants were challenged with viruliferous GLH, some lines escaped infection, and there was a significant delay in the buildup of RTSV (Sivamani et al., 1999). This was then followed by a study where the rep gene of RTSV was used to engineer resistance by inserting it in sense as well as in antisense orientation. Upon challenge with viruliferous GLH, a significant proportion of the plants showed resistance to RTSV, even against multiple isolates, in the lines containing the rep gene in antisense and in those containing in sense orientation. The investigators also reported a low transmission of RTBV from the transgenic lines, indicating that these transgenic plants would form an effective barrier to the spread of RTD (Huet et al., 1999).

1.4.2   RESISTANCE AGAINST RTBV

It was reported that attempts to achieve resistance against RTBV by expressing various viral proteins were unsuccessful, inappropriate expression levels of the transgenes being cited as a reason (Azzam & Chancellor, 2002). However, Ganesan et al. (2009) reported CP-mediated resistance against RTBV. These transgenic rice lines were raised using the CP of an Indian isolate of RTBV. The transgenic rice lines accumulated the viral CP and displayed low levels of RTBV upon challenge inoculation with viruliferous GLH. The inoculated plants developed mild symptoms of RTD. There was a rough correlation between the levels of CP expression and the degree of resistance.

RNA-interference (RNAi), a conserved defense mechanism in many eukaryotes against invading nucleic acids, such as from viruses (Ratcliffe et al., 1999) can be used also to obtain virus resistance. The principle involves expressing viral double-stranded RNA in plants, usually using transgenes, to trigger the RNAi machinery against viruses. Tyagi et al. (2008) generated transgenic rice lines expressing double-stranded RNA against RTBV ORF IV. These plants, upon challenge inoculation with viruliferous GLH, accumulated RTBV DNA much slowly, as compared to the non-transgenic control lines. The small RNA corresponding to RTBV ORF IV (transgene) accumulated in these plants and had very low levels of the transcripts for the transgene. These properties are a hallmark of RNAi, which is triggered by double-stranded RNA. These plants showed mild stunting and no yellowing, unlike the non-transgenic control plants inoculated in the same manner. The evidence indicated strongly that RNAi against RTBV could reduce its level of accumulation, which results in RTD resistance.

To see whether the transgene can work in variable genetic backgrounds, and in those rice varieties which are grown in regions where RTD is common, selected lines of transgenic plants were used to introgress the transgene into a number of popular rice varieties by back-cross breeding, using molecular markers. Testing of two such lines, parents of which are widely grown in eastern India, at the BC_2 stage has showed the transfer of the resistant trait to the progeny, at the same time, the good agronomical properties of the parental lines were retained (Roy et al., 2012). This represents the first-targeted engineering of RTBV resistance in rice, which needs to be tested under field conditions in the areas endemic to RTD and also against the southeast Asian group of RTBV. The same transgene was also introgressed into rice varieties popular in southern India and the analysis at the genetic level to show the presence of the transgene and the reaction of the lines to RTD, indicated significant resistance to the viruses in the plant lines containing the transgene at the BC_3 stage (Jyothsna et al., 2013, 2014). Some of the above plant lines were further characterized by Valarmathi et al. (2016) to study the accumulation of RTBV DNA and the agronomic traits under controlled glasshouse conditions after inoculation with viruliferous GLH. Certain lines showed 10,000-fold reduction in RTBV titers in the transgenic plants, compared to non-transgenic parent plants and an almost twofold amelioration of stunting and reduction of foliar chlorophyll contents.

Dai et al. (2008) used a different approach to achieve RTBV resistance. The group had earlier reported that rice transcription factors, RF2a and RF2b, which are required for the normal growth and maturation of rice plants, bind strongly to the RTBV promoter (Dai et al., 2004). Hence, upon RTBV infection, these two factors are sequestered by the RTBV promoter resulting in a fall in its availability required for the normal expression of genes for the growth and development of the plant. This leads to the appearance of symptoms of tungro. Using the same logic, the authors overexpressed RF2a and RF2b in rice plants. These lines, apart from showing normal growth and

development, remained largely symptom-free upon challenging with viruliferous GLH containing an isolate of RTBV from Philippines. This novel
approach, although shown to work under laboratory conditions, needs to be
further tested, especially using the south Asian group of RTBV.

1.4.3   CHARACTERIZING RESISTANCE GENES

RTSV resistance genes are fairly widespread in the rice germplasm. In the
cultivar Utri Merah, which is resistant to RTD, the genetic basis of the resistance was investigated (Encabo et al., 2009). By using near isogenic lines
derived from this line and the susceptible cultivar TN-1, the RTD resistance
was linked to the suppression of interacting RTBV and RTSV, but the suppression trait was inherited separately. This indicated genetic independence of
the RTBV and RTSV suppression traits. Later, using a map-based cloning
approach, RTSV resistance gene in UtriMerah was cloned and sequenced.
This was later followed up by the comparison of the homologous fragment
from a number of rice lines showing differential resistance toward RTSV.
Analysis of DNA sequence suggested that the RTSV resistance gene allele
(*tsv1*) corresponded to a eukaryotic translation initiation factor (eIF4G). The
results suggested that *tsv1*, upon being mutated, prevented the accumulation
of RTSV in inoculated plants (Lee et al., 2010). It can be hoped that this
information and the associated genetic markers would be widely used in the
future to introgress this allele into cultivated rice varieties.

1.5   NEW DEVELOPMENTS TOWARD MANAGEMENT OF RTD

1.5.1   GENOME ANALYSIS OF NEW ISOLATES OF RTBV AND RTSV

RTBV has a single promoter driving the expression of the four ORFs. The
RTBV promoter has been the subject of intense investigation in the past,
using the isolates from Philippines, wherein, several regulatory elements
were characterized and some of their protein binding partners identified (Dai
et al., 2006). An Indian RTBV isolate was shown by Mathur & Dasgupta
(2007) to have a very different architecture regarding the functional domains
of the promoters. Not surprisingly, they reported novel expression patterns
of the Indian RTBV promoters and their derivatives, especially negative
expression elements in the downstream regions. A deletion variant of the

Indian RTBV promoter, in which the negative element was removed, showed constitutive expression, compared to the tissue- and development-specific expression pattern of the full-length (native) promoter. The work has been expanded further to show the orientation and position independence of the negative element and its capacity to silence heterologous promoters (Purkayastha et al., 2010). Hence, this negative element is capable of working independent of its position and orientation and can silence strong promoters such as Cauliflower mosaic virus 35S promoter. The importance of these expression elements on the pathology of RTD is an area waiting to be explored.

A recent report on an isolate from Tamil Nadu southern India shows that the RTBV genome is likely to have undergone recombination (Sharma & Dasgupta, 2012). Certain isolates of RTSV were also shown to be the product of recombination (Sailaja et al., 2013). This probably reflects the general principles of natural population dynamics of viral genomes. Although the isolate shows no change in its basic properties, but the evidence of recombination indicates the potential for virulent strains to emerge, which might be products of recombination. Banerjee et al. (2011a) by analyzing the sequence of a new RTBV isolate from West Bengal (named Chinsurah isolate) has concluded that the evolution of RTBV strains is strongly influenced by the geographical region. These results were based on the analysis of conserved motifs of the large intergenic region (Banerjee et al., 2011b). Mangrauthia et al. (2012a) analyzed the sequences of ORF I, ORF II, and ORF IV from isolates of RTBV from Cuttack and Puducherry with those already available in the database and concluded that the sequence elements and motifs are conserved to a high degree. Another recent study from the same group has studied the CP sequences of Cuttack and Puducherry isolates with those already available in the database and concluded that the Indian RTSV isolates could fall into two groups. However, the differences at the protein level were insignificant, thereby strengthening the earlier view of tight conservation of RTSV sequences (Mangrauthia et al., 2012b).

1.5.2 GENE EXPRESSION CHANGES OF RICE PLANT TO INFECTION WITH VIRUSES CAUSING RTD

Plants infected with RTD shows a significant change at the level of overall gene expression providing an insight into the molecular level of symptoms development. Encabo et al. (2009) reported a more than twofold change in the expression level of about 100 genes in an RTSV-susceptible rice plant by RTSV infection, and about 150 genes being significantly regulated by

RTSV infection in an RTSV-resistant plant in which RTSV accumulated at a level much lower than in an RTSV-susceptible plant. Satoh et al. (2012) examined the gene expression responses of the host to RTSV infection in two near-isogenic rice plants (a backcrossed plant and its recurrent parental plant) by microarray (Fig. 1.1). They found many stress responsive genes such as those of jasmonic acid (JA) synthesis as well as of JA-mediated signaling pathways, transcription factors such as WRKY, NAC, and AP2/EREBP (APETALA2/ethylene responsive element binding protein) family genes encoding transcription regulators, pathogenesis-related genes, gluta-thione S-transferases, and genes for HSPs to be differentially regulated. In addition, the genes involved in developmental processes, for example, the homebox gene family, genes involved in synthesis and signaling of plant hormones (auxin, gibberellins, etc.) and the genes for photosynthesis-related processes also showed change in expression level upon RTSV infection. Major cell wall-related genes such as *cellulose synthase (A5 and A6), cellulose synthase-like (A9 and C7)* as well as the precursor genes of α-expansin are also reported to be differentially regulated upon both RTBV and RTSV infection (Budot et al., 2014; Satoh et al., 2012). The above knowledge can form the basis of novel strategies to control RTD, because, in principle, the gene products can be manipulated to change the viral levels and even symp-toms. However, this requires deeper investigation on the role of the plant proteins in pathogenesis.

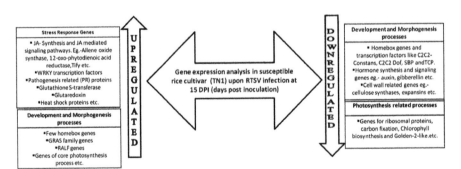

FIGURE 1.1   Changes of gene expression in rice plants in response to RTSV infection (based on Budot et al., 2014; Satoh et al., 2012).

1.5.3 IMPROVED DESIGN OF RNAi VECTORS

RTBV and RTSV depend upon each other as causative agents for RTD, and hence it is generally believed that a resistance strategy against one would be sufficient to control both. However, a resistance approach targeting both the viruses is expected to result in enhanced resistance against RTD. With a view to combine RNAi against RTBV and RTSV, hairpin loop constructs were designed, combining fragments derived from both RTBV and RTSV. In addition, to facilitate removal of the antibiotic selection marker, the well-known *Cre-lox* system has been incorporated (Sharma et al., unpublished). The antibiotic selection marker, *hygromycin-phosphotransferase* (*hpt*) is flanked by two *lox* sites, which will result in the excision of *hpt* upon crossing with a *cre*-expressing line. This will go a long way to improve the consumer acceptability of the rice lines. Keeping this strategy in mind, two sets of gene combinations—RTBV RT-RNAse H, and RTSV Rep and RTBV promoter-RTSV Rep, have been obtained in the *lox*-containing binary vector. The hairpin constructs have been placed downstream to the constitutive CaMV 35S promoter. The testing of the resistance of the plants to RTD is still awaited.

1.5.4 NEW OPPORTUNITIES TO DEVELOP RTD RESISTANCE

An exciting new approach toward developing RTD resistance could target the process of transmission of the viral complex by GLH. Early evidences pointed toward a nonviral factor associated with RTSV (Cabauatan & Hibino, 1985). The unraveling of the nature of the helper factor of RTBV transmission will depend upon the development of membrane feeding methods for GLH, where the acquisition of the viral particles could be analyzed at the biochemical level. Other approaches should include the characterization of new resistance genes against both RTBV and RTSV, which might be revealed by further molecular studies of varieties showing resistance. Using the vast and varied rice germplasm existing in India, such an exercise should yield useful genes, if planned and executed in a time-bound, interdisciplinary, and mission-mode manner, using tightly linked molecular markers. Lastly, the interrelationship between RTBV and RTSV should be analyzed in detail: how they interact within the rice plant and within the GLH vector to cause this unique and important disease.

ACKNOWLEDGMENTS

The work on Rice Tungro is supported by the Department of Biotechnology, Government of India and DST-PURSE Grant to University of Delhi.

KEYWORDS

- tungro
- virus
- resistance
- RNAi
- transgenic

REFERENCES

Azzam, O.; Chancellor, T. C. B. The Biology, Epidemiology and Management of Rice Tungro Disease in Asia. *Plant Dis.* **2002,** *86,* 88–100.

Banerjee, S.; Roy, S.; Tarafdar, J. The Large Intergenic Region of *Rice Tungro Bacilliform Virus* Evolved Differentially among Geographically Distinguished Isolates. *Virus Gen.* **2011a,** *43,* 398.

Banerjee, S.; Roy, S.; Tarafdar, J. Phylogenetic Analysis of Rice Tungro Bacilliform Virus ORFs Revealed Strong Correlation Between Evolution and Geographical Distribution. *Virus Genes* **2011b,** *43,* 398–408.

Budot, B. O.; Encabo, J. R.; Ambita, I. D. V.; Atienza-Grande, G. A.; Satoh, K.; Kondoh, H.; Ulat, V. J.; Mauleon, R.; Kikuchi, S.; Choi, I. R. Suppression of Cell Wall-related Genes Associated with Stunting *Oryza glaberrima* Infected with *Rice Tungro Spherical Virus*. *Front. Microbiol.* **2014,** *5,* 26.

Cabauatan, P. Q.; Hibino, H. Transmission of Rice Tungro Bacilliform and Spherical Viruses by *Nephotettix virescens* Distant. *Philipp. Phytopathol.* **1985,** *21,* 103–109.

Cabauatan, P. Q.; Hibino, H. Isolation, Purification and Serology of Rice Tungro Bacilliform and Rice Tungro Spherical Viruses. *Plant Dis.* **1988,** *72,* 526–528.

Dai, S.; Zhang, Z.; Chen, S.; Beachy, R. N. RF2b a Rice bZIP Transcription Activator, Interacts with RF2a and is Involved in Symptom Development of Rice Tungro Disease. *Proc. Natl. Acad. Sci. U.S.A.* **2004,** *101,* 687–692.

Dai, S.; Wei, X.; Alfonso, A. A.; Pei, L.; Duque, U. G.; Zhang, Z.; Babb, G. M.; Beachy, R. N. Transgenic Rice Plants that Over Express Transcription Factors RF2a and RF2b are Tolerant to Rice Tungrovirus Replication and Disease. *Proc. Natl. Acad. Sci. U.S.A.* **2008,** *105,* 21012–21016.

Dai, S.; Zhang, Z.; Bick, J.; Beachy, R. N. Essential Role of the Box II Cis Element and Cognate Host Factors in Regulating the Promoter of Rice Tungro Bacilliform Virus. *J. Gen. Virol.* **2006,** *87,* 715–722.

Dasgupta, I.; Hull, R.; Eastop, S.; Poggi-pollini, C.; Blakebrough, M.; Boulton, M. I.; Davies, J. W. Rice Tungro Bacilliform Virus DNA Independently Infects Rice after *Agrobacterium*-mediated transfer. *J. Gen. Virol.* **1991,** *72,* 1215–1221.

Encabo, J. R.; Cabauatan, P. Q.; Cabunagan, R. C.; Satoh, K.; Lee, J. H.; Kwak, D. Y.; De Leon, T. B.; Macalalad, R. J.; Kondoh, H.; Kikuchi, S.; Choi, I. R. Suppression of Two

Tungro Viruses in Rice by Separable Traits Originating from Cultivar Utri Merah. *Mol. Plant Microb. Interct.* **2009**, *22*, 1268–1281.

Ganesan, U.; Suri, S. S.; Rajasubramaniam, S.; Rajam, M. V.; Dasgupta, I. Transgenic Expression of Coat Protein Gene of Rice Tungro Bacilliform Virus in Rice Reduces the Accumulation of Viral DNA in Inoculated Plants. *Virus Gen.* **2009**, *39*, 113–119.

Hay, J. M.; Jones, M. C.; Blakebrough, M. L.; Dasgupta, I.; Davies, J. W.; Hull, R. An Analysis of the Sequence of an Infectious Clone of Rice Tungro Bacilliform Virus, A Plant Pararetrovirus. *Nucl. Acids Res.* **1991**, *19*, 2615–2621.

Herdt, R. W. Research Priorities for Rice Biotechnology. In *Rice Biotechnology, Biotechnology in Agriculture*; Khush, G. S., Toenissen, G. H., Eds.; International Rice Research Institute, Cab International: UK 1991; pp 19–54.

Herzog, E.; Guerra-Peraza, O.; Hohn, T. The Rice Tungro Bacilliform Virus Gene II Product Interacts with the Coat Protein Domain of the Viral Gene III Polyprotein. *J. Virol.* **2000**, *74*, 2073–2083.

Hibino, H.; Cabauatan, P. Q. Infectivity Neutralization of Rice Tungro-associated Viruses Acquired by Vector Leafhoppers. *Phytopathology* **1987**, *77*, 473–476.

Huet, H.; Mahendra, S.; Wang, J.; Sivamani, E.; Ong, C. A.; Chen, L.; Kochko, A. D.; Beachy, R. N.; Fauquet, C. Near Immunity to *Rice Tungro Spherical Virus* Achieved in rice by a Replicase-mediated Resistance Strategy. *Phytopathology* **1999**, *89*, 1022–1027.

Jones, M. C.; Gough, K.; Dasgupta, I.; Subba Rao, B. L.; Cliffe, J.; Qu, R.; Shen, P.; Kaniewska, M.; Blakebrough, M.; Davies, J. W.; Beachy, R. N.; Hull, R. Rice Tungro Disease is Caused by a RNA and a DNA Virus. *J. Gen. Virol.* **1991**, *72*, 757–761.

Jyothsna, M.; Manonmani, S.; Rabindran, R.; Dasgupta, I.; Robin, S. Introgression of Transgenic Resistance for Rice Tungro Disease into Mega Variety, ASD 16 of Tamil Nadu through Marker Assisted Backcross Breeding. *Madras Agric. J.* **2013**, *100*, 70–74.

Jyothsna, M.; Manonmani, S.; Pradeep, M. Phenotypic Screening of Rice Tungro Disease Resistant Transgenic ASD 16 Rice Lines Obtained Through Marker Assisted Backcross Breeding. *Trends Biosci.* **2014**, *7*, 149–152.

Laco, G. S.; Beachy, R. N. Rice Tungro Bacilliform Virus Encodes Reverse Transcriptase, DNA Polymerase and Ribonuclease H Activities. *Proc. Natl. Acad. Sci. U.S.A.* **1994**, *91*, 2654–2658.

Lee, J. H.; Muhsin, M.; Atienza, G. A.; Kwak, D. Y.; Kim, S. M.; De Leon, T. B.; Angeles, E. R.; Coloquio, E.; Kondoh, H.; Satoh, K.; Cabunagan, R. C.; Cabauatan, P. Q.; Kikuchi, S.; Leung Hand Choi, I. R. Single Nucleotide Polymorphisms in a Gene for Translation Initiation Factor eIF4G of Rice *Oryza sativa* Associated with Resistance to Rice Tungro Spherical Virus. *Mol. Plant Microb. Interact.* **2010**, *23*, 29–38.

Ling, K. C. Capacity of *Nephotettix virescens* to Infect Rice Seedlings with Tungro. *Philipp. Phytopathol.* **1974**, *10*, 42–49.

Mangrauthia, S. K.; Malathi, P.; Agarwal, S.; Sailaja, B.; Singh, J.; Ramkumar, G.; Krishnaveni, D.; Balachandran, S. M. The Molecular Diversity and Evolution of Rice Tungro Bacilliform Virus from Indian Perspective. *Virus Gen.* **2012a**, *44*, 482–487.

Mangrauthia, S. K.; Malathi, P.; Agarwal, S.; Ramkumar, G.; Krishnaveni, D.; Neeraja, C. N.; Madhav, M. S.; Ladhalakshmi, D.; Balachandran, S. M.; Viraktamath, B. C. Genetic Variation of Coat Protein Gene among the Isolates of Ricetungro Spherical Virus from Tungro-endemic States of the India. *Virus Gen.* **2012b**, *44*, 482–487.

Manwan, I.; Sama, S.; Rizvi, S. A. Use of Varietal Rotation in the Management of Rice Tungro Disease in Indonesia. *Indonesian Agric. Res. Develop. J.* **1985**, *7*, 43–48.

Marmey, P.; Rojas-Mendoza, A.; de Kochko, A.; Beachy, R. N.; Fauquet, C. M. Charac-terization of the Protease Domain of Rice Tungro Bacilliform Virus Responsible for the Processing of the Capsid Protein from the Polyprotein. *Virol. J.* **2005**, *2*, 33.

Mathur, S.; Dasgupta, I. Downstream Promoter Sequence of an Indian Isolate of Rice Tungro Bacilliform Virus Alters Tissue-specific Expression in Host Rice and Acts Differentially in Heterologous System. *Plant Mol. Biol.* **2007**, 65, 259–275.

Muralidharan, K.; Krishnaveni, D.; Rajarajeshwari, N. V. L.; Prasad, A. S. R. Tungro Epidemic and Yield Losses in Paddy Fields in India. *Curr. Sci.* **2003**, *85*, 1143–1147.

Nath, N.; Mathur, S.; Dasgupta, I. Molecular Analysis of Two Complete Rice Tungro Bacil-liform Virus Sequences from India. *Arch. Virol.* **2002**, *147*, 1173–1187.

Purkayastha, A.; Sharma, S.; Dasgupta, I. A Negative Element in the Downstream Region of the Rice Tungro Bacilliform Virus Promoter is Orientation- and Position-independent and is Active with Heterologous Promoters. *Virus Res.* **2010**, *153*, 166–171.

Rajeswaran, R.; Golyaev, V.; Seguin, Zvereva, A. S.; Farinelli, L.; Pooggin, M. M. Inter-actions of Rice Tungro Bacilliform Pararetrovirus and Its Protein P4 with Plant RNA-silencing Machinery. *Mol. Plant. Microbe Int.* **2014**, *27*, 1370–1378.

Ratcliffe, F. G.; MacFarlane, S. A.; Baulcombe, D. C. Gene Silencing without DNA: RNA-mediated Cross Protection between Viruses. *Plant Cell* **1999**, *11*, 1207–1215.

Raychaudhury, S. P.; Mishra, M. D.; Ghosh, A. Preliminary note on Transmission of Virus, a Disease Resembling Tungro of Rice in India and Other virus-like Symptoms. *Plant Dis. Rep.* **1967a**, *51*, 300–301.

Raychaudhury, S. P.; Mishra, M. D.; Ghosh, A. Virus Disease that Resembled Tungro. *Indian Farm.* **1967b**, *173*, 29–33.

Roy, S.; Banerjee, A.; Tarafdar, J.; Senapati, B. K.; Dasgupta, I. Transfer of Transgenes for Resistance to Rice Tungro Disease into High Yielding Rice Cultivars through Gene Based Marker-assisted Selection. *J. Agric. Sci.* **2011**, *158*(3), 1363–1379.

Rivera, C. T.; Ou, S. H. Leafhopper Transmission of Tungro Disease of rice. *Plant Dis. Rep.* **1965**, *49*, 127–131.

Sailaja, B.; Anjum, N.; Patil, Y. K.; Agarwal, S.; Malathi, P.; Krishnaveni, D.; Balachandran, S. M.; Viraktamath, B. C.; Mangrauthia, S. K. The Complete Genome Sequence of a South Indian Isolate of Rice Tungrospherical Virus Reveals Evidence of Genetic Recombination between Distinct Isolates. *Virus Gen. Dec.* **2013**, *47*, 515–523.

Satoh, K.; Kondoh, H.; De Leon, T. B.; Macalalad, R. J.; Cabunagan, R. C.; Cabauatan, P. Q.; Mauleon, R.; Kikuchi, S.; Choi, I. R. Gene Expression Responses to *Rice tungro spherical virus* in Susceptible and Resistant Near-isogenic Rice Plants. *Virus Res.* **2012**, *171*(1), 111–120.

Sharma, S.; Dasgupta, I. Development of SYBR Green I Based Real Time PCR Assays for Quantitative Detection of Rice Tungro Bacilliform Virus and Rice Tungro Spherical Virus. *J. Virol. Met.* **2012**, *181*, 86–92.

Shen, P.; Kaniewska, M.; Smith, C.; Beachy, R. N. Nucleotide Sequence and Genomic Orga-nization of Rice Tungro Spherical Virus. *Virology* **1993**, *193*, 621–630.

Sivamani, E.; Huet, H.; Shen, P.; Ong, C. A.; de Kochko, A.; Fauquet, C.; Beachy, R. N. Rice plant *Oryza sativa* L. Containing Rice Tungro Spherical Virus RTSV Coat Protein Trans-genes are Resistant to Virus Infection. *Mol. Breed.* **1990**, *5*, 177–185.

Tyagi, H.; Rajasubramaniam, S.; Rajam, M. V.; Dasgupta, I. RNA-interference in Rice against Rice Tungro Bacilliform Virus Results in its Decreased Accumulation in Inoculated Rice Plants. *Transgenic Res.* **2008**, *17*, 897–904.

Valarmathi, P.; Kumar, G.; Robin, S.; Manonmani, S.; Dasgupta, I.; Rabindran, R. Evaluation of Virus Resistance and Agronomic Performance of Rice Cultivar ASD 16 after Transfer of Transgene against Rice Tungro Bacilliform Virus by Backcross Breeding. *Virus Gen.* **2016**, *52*, 521–529.

Verma, V.; Dasgupta, I. Sequence Analysis of the Complete Genomes of Two *Rice tungro spherical virus* Isolates from India. *Arch. Virol.* **2007**, *152*, 645–648.

CHAPTER 2

MOLECULAR ASPECTS OF BACTERIAL BLIGHT RESISTANCE IN RICE: RECENT ADVANCEMENT

ANIRUDHA CHATTOPADHYAY[1], DEEPTI NAGAICH[2], JOHN M. LIMA[3], AMIT VERMA[4], and KAPIL KUMAR TIWARI[1*]

[1]C. P. College of Agriculture, S. D. Agricultural University, S. K. Nagar, Palanpur 385506, Gujarat, India

[2]Banasthali Vidyapeeth, Jaipur 304022, Rajasthan, India

[3]National Research Centre on Plant Biotechnology, Pusa Campus, 110012 New Delhi, India

[4]College of Basic Science and Humanities, S. D. Agricultural University, S. K. Nagar, Palanpur 385506, Gujarat, India

[*]Corresponding author. E-mail: kapil21282@gmail.com

CONTENTS

ABSTRACT

Bacterial leaf blight (BLB) is one of the main constraints to the worldwide rice production system. Various management strategies are advocated to mitigate this problem. Among these, use of resistant cultivars is the cheapest and environmentally safest way. From long back, conventional breeding approaches have been mostly adopted to generate a number of resistant cultivars in different countries. And these cultivars are gradually outdated due to the breakdown of resistance and low yield response. The host resistance is overcome by mutant strains of the pathogen with changing population dynamics of *Xanthomonas oryzae* pv. *oryzae* (*Xoo*). This bacterial blight resistance in rice is qualitative and quantitative in nature, where both contribute significantly to the defense response against *Xoo*. So far about 40 resistance genes have been identified in rice against various *Xoo* pathotypes and a few have been cloned and characterized to decipher the host–pathogen interaction. Various molecular approaches have been adopted to deploy them for getting a long-term durable resistance. Gene pyramiding of several resistance genes through marker assisted selection is found to be the most effective strategy. However, with changing dynamics of pathogen population in this intensified agro ecosystem, our prime aim would be to safeguard resistant durability. Continued efforts should also be paid to look forward upon the modern genomics approaches like the development of transgenic and application of genome editing tools for the crop improvement. Therefore, a thorough understanding of molecular mechanisms worked behind BLB resistance is needed.

2.1 INTRODUCTION

Rice is a popular cereal crop grown in more than hundreds of countries, approximately 163 million hectares, producing 486 million tons of milled rice (730 million tons of paddy) with the average of 2.98 t/ha (FAOSTAT, 2012). It is a staple food for more than half of the world's population and a rich source of carbohydrates and energy, provide 20% of world calories requirement (Khush, 2005). Only Asia is a region that produces around 90% of the world's rice: its consumption is very high, which has exceeded 100 kg/capita annually, and it provides more than 50% of the calorie requirement within this area. At present, rice productivity has been increased four times from the past five decades that made possible by the adoption of semi-dwarf high-yielding varieties (HYV) and modern farming technologies; however, from the last few years, rice productivity is near constant. In order to provide food demand of a growing population, estimated rice requirement by 2020 should reach 124 million tons (Kumar et al., 1995). Due to slow improvement in HYV and reduction in natural resources along with biotic and abiotic stresses, increasing rice production for the fulfillment of future food demands is quite challenging.

Rice yield is disassembled by a number of diseases that globally decrease 10–20% of rice production. Rice bacterial leaf blight (BLB) is a one of the major diseases that caused by Gram-negative bacterium *X. oryzae* pv. *oryzae* (*Xoo*) (Ishiyama, 1922). Yield losses caused by *X. oryzae* range from 20% to 30% and can be as high as 80% in some of the cultivated area under severe condition. BLB is a seed-borne disease that was first identified in Japan by a farmer in 1884 (Tagami & Mizukami, 1962). All the developmental stages are susceptible to BLB infection under favorable environmental conditions. Moreover, extreme wind and rains exaggerate the epidemic of BLB. Development of BLB resistant variety is the most economic, effective, and sustainable approach to control disease. Although identification and selection of resistant source through screening under high pressure of BLB have been successfully utilized for the development of resistant varieties, the new virulent mutant strains of pathogen have always been a challenge for BLB resistance rice breeders. In addition to the economic importance of the disease, the rice bacterial blight system serves as a model system for host–pathogen interaction study owing to genome sequence availability of both the interacting partners and the availability of sufficient amount of variation in host and pathogen. Utilizing advanced biotechnological tools, about 40 genes or alleles have been characterized in rice conferred to BLB disease, and few of them were also cloned and frequently used for improvement of rice cultivars. Instead of single gene introgression, pyramiding of multiple genes through marker-assisted selection strategy is quite effective for sustainable control of disease. Nevertheless, some of the advanced transgenic strategies like overexpression and silencing of genes, genome editing tools like transcription activator-like effector nucleus (TALEN) and CRISPR/Cas9 (clustered regularly interspaced short palindromic repeats/CRISPR-associated protein) are also being used in the recent past. Therefore, the purpose of this section is to provide consolidated information about the molecular strategies and approaches of BLB resistance rice improvement program. In broad sense, molecular approaches used in crop improvement of rice bacterial blight disease resistance can be divided in two sections:

    A. Molecular basis of host resistance in rice.
    B. Molecular tools used for crop improvement.

2.2 MOLECULAR BASIS OF HOST RESISTANCE IN RICE

The genetics of disease resistance also varies according to the nature of the host plant resistance and virulence of pathogens. In the language of genetics, host plant resistance can be categorized as qualitative resistance and quantitative resistance. The qualitative resistance is governed by a single gene or oligogenes and is mostly race specific and provides a higher degree of resistance, whereas quantitative host resistance is governed by either multiple genes or quantitative gene loci. It is race nonspecific and provides a generalized resistance to all the races of a pathogen, but due to the complexity of its genetic inheritance, there are certain limitations for its exploitation in resistant breeding program. On the other hand, qualitative resistance is widely used in crop improvement program for its easy manipulation. This qualitative resistance regulated in two ways, through patterns-triggered immunity (PTI) and effector-triggered immunity (ETI)-based defense system (Zhang & Wang, 2013). In PTI, rice plant usually perceives some signals, either pathogen (*Xoo*)-associated molecular patterns (PAMPs) that are highly conserved across the genera or plant-derived damage-associated molecular patterns molecule that are produced during *Xoo* infection (Jones & Dangl, 2006; Takai et al., 2008). After recognition by plasma membrane, localized pattern recognition receptors (PRRs), the downstream defense signaling pathway is activated and confers a broad spectrum resistance to all kinds of pathogenic races of *Xoo*, thus called as basal resistance. In rice, *XA21* gene-encoded protein, that is, serine–threonine protein kinase acts like PRRs which recognizes *XA21* protein secreted by *Xoo* through type I secretion system and induce a higher level of resistance (Bahar et al., 2014; Lee et al., 2009; Song et al., 1995). This *XA21* gene of rice confers a broad spectrum resistance to almost all the races of *Xoo*, whereas other resistance genes (R-genes) like *Xa1, Xa3/Xa26, Xa5, Xa13, Xa25, Xa27*, etc. provide race-specific resistance to various *Xoo* pathotypes. The functioning of these R genes differs from *XA21* and works in "gene for gene" interaction manner (Flor, 1942). The R-gene-encoded receptors can recognize the specific signature proteins (effectors) encoded by avirulence (*Avr*) genes of *Xoo* either directly or indirectly and initiate ETI. ETI is pathogen-race specific and confers a higher level of resistance (Thomma et al., 2011). Thus, qualitative resistance is the outcome compatible interaction between specific R and *Avr*. Therefore, the characteristic of R and Avr gene is important for better understanding of the molecular mechanisms of host resistance.

Besides plant innate immunity, induced plant defense including systemic acquired resistance (SAR) and induced systemic resistance (ISR) is a well-established phenomenon. The SAR and ISR follow the different approach for similar outcomes. In SAR, plant defense is induced by either infection of mild pathogenic agents or treatment with salicylic acid (SA) or its functional analogs, namely, probenazole and benzothiadiazole-s-methyl ester (BTH) which are associated with the transcriptional activation of defense-related genes including pathogenesis-related (PR) proteins (Ryals et al., 1996; Wang et al., 2006). Several studies reveal that SAR in rice branches into *OsNPR1*- and *OsWRKY45*-dependent pathways (Nakayama et al., 2013). *OsNPR1* and *OsWRKY45* gene of rice is upregulated. *OsNPR1* is a rice counterpart of *NPR1*, a transcriptional coactivator that plays a central role in the SA pathway in *Arabidopsis*, and *OsWRKY45* is a BTH inducible rice-specific transcription factor (Takatsuji et al., 2010). In response to the SA/BTH signal, upregulation of *OsNPR1* results in the suppression of photosynthesis activity and reduces the production of reactive oxygen species (ROSs) and, thereby, protects the plants from ROS-induced cellular damage during the defense, whereas *OsWRKY45* activates the biosynthesis of defense metabolites and suppress the cell death *via* ROS scavenging. In this way, SAR-mediated-induced defense provides a broad spectrum resistance against BLB pathogen. In contrast, ISR is induced when plant interacts with either biocontrol agents or plant growth promoting rhizobacteria (PGPR). It is an SA independent signaling pathway regulated by NPR1 gene and activates jasmonic acid and ethylene signaling defense related genes upon pathogen infection (Cartieaux et al., 2008; Verhagen et al., 2004). Rice root colonization with PGPR strain of *Pseudomonas* spp. induces host defense mechanism against foliar pathogens (De Vleesschauwer et al., 2006, 2008). In this way, induced immunity as a whole contributes significantly to achieve a higher level of resistance against bacterial blight pathogen in rice.

2.2.1 MOLECULAR CHARACTERIZATION OF PATHOGENICITY GENES OF XANTHOMONAS ORYZAE PV. ORYZAE

X. oryzae pv. *oryzae* (*Xoo*) is one of the most damaging pathogens of rice due to its epidemic occurrence in various rice growing areas of the world. Its greater pathogenic variation is the main concern. Hence, their identification

and characterization is important for defining host resistance, as high degree
of race-cultivar specificity determining the host resistance. There are over
30 races reported from several countries which are identified based on their
incompatible interaction with different resistant gene (Adhikari et al., 1999;
Noda et al., 1996, 2001). This race profiling is important for the resistance-
breeding program and for further deployment of R genes also. Hence,
molecular detection about different pathogenicity genes of *Xoo* and their
function is important for clear understanding underlying behind durable host
resistance.

First of all, the availability of whole genome sequence information
of three *Xoo* races made it quite easier to enlighten the role of *hrp* genes,
encoding the type III secretion system (T3SS) and the *avr* genes encoding
Avr proteins in the pathogenicity. Further comparative genomics and functional genomics approaches help to understand their mechanism in a better
way. Total nine *hrp* (hypersensitive response [HR] and pathogenicity) genes,
nine *hrc* (HR and conserved) genes and eight *hpa* (*hrp*-associated) genes
were identified for their importance to form T3SSs (Salzberg et al., 2008).
T3SS of *Xoo* is an essential determinant of bacterial pathogenicity (Zhang et
al., 2013). The transcriptional regulation of *hrp-genes* leads to the formation
of a suitable transport system which helps in the translocation of effector
proteins encoded by *Avr*-genes. The effector protein alone, or in combination with *hrp*-gene-encoded harpins, causes disease in susceptible hosts and
induces HR in resistant host and nonhost plants. The comparative genomic
studies indicated different *Xoo* pathotypes and many pathogenicity related
genes, namely, the genes for controlling chemotaxis (Kamoun & Kado,
1990), flagella synthesis (Shen et al., 2001), extracellular enzymes, T3SS
(Ray et al., 2000), and adhesion-related EPS (Dharmapuri & Sonti, 1999),
etc. are highly conserved but the differences in pathogenicity among *Xoo*
races are mainly determined by *avr*-genes (Yang & White, 2004). The variability within *Avr* genes of various *Xoo* races is the key factor to determine
race-cultivar specificity. *AvrXa* series genes of *X. oryzae* pv. *oryzae* are
the member of the *avBs3/pthA* gene family. Usually, most of *AvrXa* series
gene-encoded proteins in *Xoo* have a variable central domain with highly
conserved 5′ and 3′ end terminal. The central domain contains a nearly
identical repeat sequence of 34 amino acids, which varies in repeat number
and composition of amino acids at 12th and 13th position with respect to
each repeat, called the repeat-variable di-residue (RVD) (Leach & White,
1996). The combination of the number of repeats and composition of RVD
in the central domain is an important determinant for host specificity (Yang
& Gabriel, 1995; Yang & White, 2005), whereas the carboxy terminal end

is highly conserved containing three nuclear localization signals (NLS) domain and a transcription activation domain and is essential for avirulence gene function (van den Ackerveken et al., 1996; Zhu et al., 1998). Therefore, structural and functional genomics of *Avr* genes is important for elucidating the virulence or avirulence property in the different host.

AvrBs3/PthA-encoded proteins are mostly transcriptional activator-like effectors (TALEs), which are transported into the plant cell cytoplasm *via* type-III secretion system (Boch & Bonas, 2010). NLS domains are responsible for their translocation into the nucleus, where they act like a transcription factor and regulate the gene expression. Some TALEs contribute to virulence by activation of the host susceptible (S) genes that enhance pathogenicity and subsequently promoting pathogen infection and disease development (Boch et al., 2014). Whereas, some act as avirulence factor by activating resistance genes and induce genotype-specific HR (Bogdanove et al., 2010). To encounter TALEs, sometimes rice plants have evolved R genes like *Xa27* that can recognize TALEs through promoter sequence and activate defense. Identification and characterization of such R genes are important for resistance-breeding program. It will also provide a scientific basis for understanding the molecular events of host resistance in rice against *Xoo*.

2.2.2 MOLECULAR CHARACTERIZATION OF RESISTANCE (R) GENES

In the last few years, several new genomics tools such as molecular markers (Tiwari et al., 2014), gene-expression-based techniques (microarray and transcriptome sequencing analysis), mutation based approaches (TILLING, Insertional Mutagenesis, etc.), and whole genome sequencing approach have been developed that make feasible for identification, mapping, and cloning of genes. For the last few decades, thousands of molecular markers have been developed and utilized worldwide in rice. Molecular markers are essentially required for the investigation of classical method of mapping and map-based cloning of genes. Markers are also essential in marker-assisted selection (MAS), gene pyramiding, and marker-based genotype screening program.

Among all the above discussed strategies, map-based cloning is the most popular and robust method which are often used for discovery of novel genes/alleles. In this method, molecular markers are being used to trace genomic region where the gene of interest is located. After initial screening, small (150–200) mapping population (F_2, DH, RILs, etc.) are often being used for the identification of gene-linked molecular markers. In an efficient approach named bulked segregant analysis (BSA) to identify the linked polymorphic markers for both resistance and susceptibility by pooling equal amount of DNA from extreme resistant and susceptible lines (Giovannoni et al., 1991; Lima et al., 2007). However, BSA is not always useful for the identification of quantitative resistance. Distance between molecular markers and gene is measured in cM. In preliminary mapping, distance between molecular markers may be huge, but the distance can be narrowed down using the next steps of mapping with increased size of mapping population (1000–3000 individuals) and a higher number of polymorphic markers. After narrowing down the gene position, the region present in genomic libraries of bacterial artificial chromosome and yeast artificial chromosome gets screened using identified flanking molecular markers to land on the gene. Now, the predicted gene again can be confirmed by complementation test. Using this method, several "R" genes were identified against bacterial blight.

Disease management based on R genes is a cost-effective, sustainable method and has been used for bacterial blight management in rice. Till date, 40 "R" genes have been identified and named (*Xa1–Xa40*), hence due to duplication in independent discovery of genes actual number got reduced (Table 2.1). Most of the resistance genes were identified from *indica* and *japonica* subspecies, while some genes were identified from wild species of rice. Generally a resistance mechanism generated by interaction between R genes with a pathogen that initiate defense signal transduction lead to disease resistance in most of the biotic stresses. Out of these genes, some of them which were fine mapped and cloned that are discussed in detail.

2.2.2.1   XA1

This gene was first identified from *japonica* rice cultivar Kogyoku on the long arm of chromosome 4 (Sakaguchi, 1967) and have been cloned using map-based strategy for its functional genomics study. *Xa1* is a member of the NBS-LRR class of plant disease-resistance genes. It encodes a cytoplasmic receptor-like protein with nucleotide binding sites (NBS) and Leucine-rich repeats (LRR) domains and quite different from *XA21* and another BB-resistance gene was isolated from rice (Yoshimura et al., 1998). The *Xa1* gene in rice confers resistance to Japanese race 1 of *Xanthomonas oryzae* pv. *oryzae*.

TABLE 2.1 List of R Genes of Rice Conferring Resistance to *Xanthomonas oryzae* pv. *Oryzae*.

Gene	Original source/Accession	Ch	Type of resistance	References
Xa1	*Oryza sativa* ssp. *japonica* (Temperate) (cv. Kogyoku)	4L	Race specific dominant resistance to Japanese race 1	Yoshimura et al. (1998)
Xa2	*Oryza sativa* ssp. *indica* (cv. Tetep)	4L	Race specific dominant resistance to Japanese race 2	He et al. (2006)
Xa3/Xa6/ Xa26	*Oryza sativa* ssp. *japonica* (cv. Wase Aikoku 3	11	Broad spectrum dominant resistance to six Philippine *Xoo* races (race 1, 2, 3, 4, 5, and 9) at the booting stage	Xiang et al. (2006), Hur et al. (2013)
Xa4	*Oryza sativa* var. *indica* (TKM6; IR20, IR22, IR64)	11	Broad spectrum dominant resistance, confers resistance to Philippine race 1, 4, 5, 7, 8, and 10; effective at low temperature	Yoshimura et al. (1995)
xa5	*Oryza sativa* ssp. *indica* Aus-boro line (DZ192), DV85; DV86; DV78	5S	Recessive; race-specific resistance to Phillipine race 1, 4; but susceptible to race 6	Blair et al. (2003), Iyer and McCouch (2004)
Xa7	*Oryza sativa* ssp. *indica* Aus-boro line (DZ78)	6	Dominant resistance, effective at higher temperature	Porter et al. (2003)
xa8	PI231129 (American cultivar)	7	Recessive race-specific resistance to Philippine *Xoo* race 5, 6, & 8; complete to partial resistance against north Indian *Xoo* races at seedling and adult plant stage	Vikal et al. (2014)
Xa10	Cas 209	11L	Dominant race-specific resistance to Philippine races PXO86 (R2), PXO112 (R5), and PXO145 (R7) at all developmental stages	Yoshimura et al. (1983), Gu et al. (2008)
Xa11	IR8	3L	Dominant race-specific resistance to Japanese race IB, II, IIIA, and V	Goto et al. (2009)
Xa12	*Oryza sativa* ssp. *japonica* (cv. Kogyoku)	4	Dominant race-specific resistance to Japanese race V and Philippine race 5	Taura et al. (1992b)
xa13	*Oryza sativa* ssp. *indica* Aus-*Boro* line (cv. BJ1)	8L	Race specific resistance to Philippine race 6 (PXO99)	Sanchez et al. (1999), Chu et al. (2006)

TABLE 2.1 *(Continued)*

Gene	Original source/Accession	Ch	Type of resistance	References
Xa14	*Oryza sativa* ssp. *indica* (cv. TN 1)	4L	Highly resistant to Philippine race 5	Bao et al. (2010), Yuan et al. (2010)
xa15	*Oryzae sativa* ssp. *japonica* (cv. M41, a mutant line of Harebare)	ND	Broad spectrum resistance to Japanese race	Gnanamanickam et al. (1999)
Xa16/ XA16	*Oryzae sativa* ssp. *indica* (cv. Tetep)	ND	Dominant resistant gene to Japanese isolates J8581 and H8584	Noda and Ohuchi (1989), Oryzabase (2011)
Xa17	*Oryzae sativa* ssp. *japonica* (cv. Asominori)	ND	Dominant resistant gene to Japanese isolates J8513	Ogawa et al. (1989), Oryzabase (2011)
Xa18	*Oryzae sativa* ssp. *japonica* (cv. Toyonishiki)	ND	Effective against African *Xoo* strains (raceA3), Burmese isolate BM8417 and BM8429, but not effective against Asian *Xoo* strains	Noda et al. (1996), Gonzalez et al. (2007), Oryzabase (2011)
xa19	*Oryzae sativa* ssp. *indica* cv. XM5, (induced *mutant* line of IR-24)	ND	Resistant to all Phillipine races 1–6	Taura et al. (1991)
xa20	*Oryzae sativa* ssp. *indica* cv. XM6, (induced *mutant* line of IR-24)	ND	Resistant to all Phillipine races 1–6	Taura et al. (1992a)
XA21	Wild Rice (*O. longistaminata*)	11L	Broad spectrum resistance to all *Xoo* races from Philippines and India at post-seedling growth stages	Song et al. (1995)
Xa22(t)	*Oryzae sativa* ssp. *japonica* (cv. Zhachanglong)	11	Broad spectrum resistance to 16 *Xoo* strains from China, Japan and Philippines	Lin et al. (1996), Wang et al. (2003)
Xa23	Wild rice (*Oryza rufipogon*)	11L	Complete dominant and broadest resistance to all most all *Xoo* races from Philippine, Japanese, and Chinese at all growth stages of rice	Wang et al. (2014)
xa24	*Oryza sativa* ssp. *indica* Aus-boro line (cv. DV86)	2L	Dominant resistance to several races of *Xoo* from Philippine, Japanese, and Chinese throughout all growth stages	Mir and Khush (1990), Wu et al. (2008b), Khush and Angeles (1999)

TABLE 2.1 *(Continued)*

Gene	Original source/Accession	Ch	Type of resistance	References
xa25a(t)	Oryza sativa ssp. indica (cv. HX-3, somaclonal mutant)	4L	Resistance to many Philippine, Chinese, and Japanese races	Gao et al. (2001, 2005)
xa25b(t)	Oryza sativa ssp. indica (cv. Minghui 63)	12	Race-specific resistance to Philippine Xoo race 9 in both seedling and adult stages	Chen et al. (2002), Liu et al. (2011)
Xa26	Oryza sativa ssp. indica (cv. Minghui 63)	11L	Broad spectrum resistance to Philippine and Chinese races, both in seedling and adult stage	Yang et al. (2003), Sun et al. (2004)
Xa27(t)	wild rice (Oryza minuta)	6L	Wide spectrum higher resistance to 27 Xoo strain and moderately resistance to three strains	Gu et al. (2004), Wang et al. (1996)
xa28	Oryza sativa ssp. indica (cv. Lota Sail)	ND	Recessive gene for resistance to race 2 and 5	Lee et al. (2003)
Xa29(t)	Oryza officinalis	1	Dominant resistance	Tan et al. (2004)
Xa30(t)	Oryza rufipogon germplasm (Y238)	11L	—	Jin et al. (2007)
Xa31(t)	Zhachanglong	4L	Resistance against Xoo strain OS105 and susceptible to Px061	Wang et al. (2009)
Xa32	Oryzae ustraliensis	11L	Resistant to strains P1 (PXO61), P4 (PXO71), P5 (PXO112), P6 (PXO99), P7 (PXO145), P8 (PXO280), P9 (PXO339), and KX085, but susceptible to P2 (PXO86) and P3 (PXO79)	Zheng et al. (2009)
Xa33	Wild rice (Oryza nivara); accession number 105710	7	Wider spectrum resistance	Kumar et al. (2012)
xa34(t)	Oryza sativa ssp. indica (cv. BG1222)/aus rice cultivar (cv. G1222)	1	High resistant to the Chinese Xoo race V	Chen et al. (2011)
Xa35(t)	Oryza minuta(Acc. No. 101133)	11L	Dominant resistance gene, to PXO61, PXO112, and PXO339	Guo et al. (2010)

TABLE 2.1 *(Continued)*

Gene	Original source/Accession	Ch	Type of resistance	References
Xa36(t)	C4059	11L	–	Miao et al. (2010)
Xa38(t)	*Oryza nivara* (acc. IRGC 81825)	4L	Resistant to all the seven *Xoo* pathotypes prevalent in northern states of India	Bhasin et al. (2012); Cheema et al. (2008), Vikal et al. (2007)
Xa39(t)	FF329	11	Broad-spectrum BB resistance	Zhang et al. (2015)
Xa40(t)	*Oryza sativa* ssp. *indica* line IR65482-7-216-1-2 (background cultivar Junam)	11	High levels of resistance to all Korean *Xoo* races, including new race K3a	Kim et al. (2015)

Ch: Chromosome number.

2.2.2.2 XA3/XA26

This gene was isolated by map-based cloning strategy from *indica* rice cultivar Minghui 63 that confirmed their position on the long arm of chromosome 11 (Sun et al., 2004). Both *Xa3* and *Xa26* are the same gene with identical sequences in the coding region and only one nucleotide substitution occurring at 475 bp upstream of the translation initiation site. Thus, at present it can be represented as *Xa3/Xa26* (Xiang et al., 2006). But their expression level varies with genetic background. This gene encodes a plasma LRR receptor kinase-type protein and provides a broad spectrum resistance to Philippine and Chinese races, both in seedling and adult stage.

2.2.2.3 XA5

This is a recessive resistant gene of rice, which confers specific resistance to some Japanese and Philippine races of *X. oryzae* pv. *oryzae*. It was first identified in *indica* Aus-Boro line (DZ192) in 1977 (Petpisit et al., 1977). Later, this gene has been mapped in the short arm of chromosome 5 (Yoshimura et al., 1984) and positionally cloned by map-based cloning approach (Iyer & McCouch, 2004). This gene encodes gamma subunit of transcription factor IIA (TFIIAγ). TFIIAγ is a general eukaryotic transcription factor; however, previously its role in disease resistance was not confirmed and does not possess a typical structure of a resistance gene class. Sequencing of this gene revealed two nucleotide substitutions resulting an amino acid change between resistant and susceptible cultivars. This association was conserved across 27 resistant and nine susceptible rice lines in the *Aus-Boro* group.

2.2.2.4 XA10

This gene was originally identified from *rice* cultivar Cas 209 (Yoshimura et al., 1983). It is mapped in the long arm of chromosome 11 and cloned by map based cloning approach. The *Xa10* gene encodes an ER membrane protein (*Xa10*) which triggers calcium depletion in the ER and induces host cell death (Tian et al., 2014). It is a dominant gene which confers narrow-spectrum, race-specific resistance to Philippine races PXO86 (R2), PXO112 (R5), and PXO145 (R7) of *Xoo* at all the developmental stages.

2.2.2.5   XA13

The rice gene *xa13* is a recessive resistance allele of the disease-suscepti-bility gene *Os-8N3* (also named *Xa13* or *OsSWEET11*). The Os-*8N3* is a member of the *NODULIN3* (*N3*) gene family. The *xa13* gene was first iden-tified in *indica* Aus-Boro line (cv. BJ1) (Ogawa et al., 1987) and mapped on the long arm of rice chromosome 8 (Sanchez et al., 1999). In fully reces-sive state, it confers resistance to the *Xoo* strains encoding *PthXo1* effector defeated by strains of the pathogen producing any one of the type III effec-tors, namely, *AvrXa7*, *PthXo2*, or *PthXo3*, which all are TALEs (Antony et al., 2010), whereas its dominant counterpart *Xa13* is a susceptibility gene. The expression of both is controlled by designer TALEs to modulate host defense.

2.2.2.6   XA21

The gene *xa21* provides a higher level of resistance to pathogens. This gene was cloned from wild rice *Oryza longistaninata* by map-based strategy and its location was confirmed on the long arm of chromosome 11. The sequence of *XA21* gene found to have leucine-rich repeat motif and serine–threonine kinase-like domain which was involved in cell surface recognition of pathogen-encoding proteins and activate intracellular defense response (Song et al., 1995). This protein acts as a PRR and recognizes an evolu-tionarily conserved pathogen-associated molecular pattern (Lee et al., 2009) and subsequently induced basal defense system in plants. The transgenic plant carrying *Xa21* gene confers resistance to multiple pathogen isolates of *X. oryzae* pv. *oryzae*. Both the engineered lines as well as donor lines showed resistance to 29 isolates of BLB indicating single gene is sufficient to provide multi-isolate resistance.

2.2.2.7   XA25

The recessive gene, *xa25*, was first identified in a somaclonal mutant, HX-3 (Gao et al., 2005). This gene was mapped in the centromeric region of chro-mosome 12 (Chen et al., 2002) and was isolated from *indica* rice cultivar Minghui 63 by a map-based cloning strategy (Liu et al., 2011). Similar to *xa13*, it also encodes a plasma membrane protein of the MtN3/saliva family and confers race-specific resistance to Philippine *Xoo* race 9 at both seedling and adult stages. The proteins encoded by recessive allele, *xa25* and its dominant allele, *Xa25* are differed in eight amino acid sequence. The expression of dominant *Xa25*, but not recessive *xa25*, was rapidly induced by PXO339 but no other *Xoo* strain infections.

2.2.2.8   XA27

This is a dominant resistant gene identified in wild rice (*Oryza minuta*). *Xa27* is mapped on the long arm of chromosome 6 and isolated from *indica* rice line IRBB27 by map-based cloning (Gu et al., 2004). *Xa27* encodes an apoplast protein of 113 amino acids that has no distinguishable sequence similarity to proteins from organisms other than rice (Wu et al., 2008a). It provides a wide spectrum higher resistance to 27 *Xoo* strain and moderately resistance to three strains.

2.3   MOLECULAR TOOLS USED FOR CROP IMPROVEMENT

Management of BLB of rice has been always a herculean task. Various management strategies like cultural, biological, and chemical practices are being proposed to the farmers. But not a single one is fully effective to give the complete solution. Due to its seed-borne nature, and its survivability in wild rice or other grassy weeds, make this disease very complex and continuous supply of inoculum, especially in low-land rice growing areas is a very challenging issue. Therefore, an integrated disease management system including all different approaches should be adopted, where host plant resistance would be considered as a major component. Exploitation of host resistance has been always the safest approach. But identification of resistance source and the development of resistant cultivar is the main problem. Conventional breeding strategies are generally used for the development of cultivar. But these are very slow and time consuming as well as labor-intensive process. It also requires extensive phenotypic evaluation of breeding material, thus takes several years to release a variety. The recent technological advancement in molecular breeding and functional genomics helps to improvise our resistance breeding system and to shape the plant defense system in new ways through following approaches.

2.3.1 MARKER-ASSISTED BREEDING APPROACH

Marker-assisted selection is an approach for precision plant breeding. It enhances the efficiency of conventional plant breeding steps by reducing the time and accurate selection of genetic materials. Abundant information on DNA markers is great promise. These markers which are tightly linked to the resistance traits are commonly used to track the segregation of resistance genes. It helps in marker assisted selection of resistant germplasm and pyramiding R genes within as a single germplasm. Using conventional breeding, gene pyramiding is also very difficult due to the dominance and epistasis gene interaction. Pyramided lines carrying two or more numbers of bacterial blight resistance genes provide broad spectrum resistance in comparison to the nongenic resistant lines (Jeung et al., 2006; Singh et al., 2001; Suh et al., 2009). Nowadays, a combine approach including conventional breeding along with marker-assisted selection becomes very popular in the bacterial blight resistance hybrid rice breeding program (Table 2.2).

TABLE 2.2   Commercially Released MAS Rice Cultivars in Asia.

Combination of gene	Variety/Genotype	Country	Release year/Reference
xa5, xa13, & *XA21*	PR106	India	Singh et al. (2001)
xa5, xa13, & *XA21*	Samba Mahsuri	India	Sundaram et al. (2008)
XA21, xa13, sd-1	Type 3 Basmati	India	Rajpurohit et al. (2011)
Xa4, xa5, xa13, & *XA21*	Mahsuri	India	Guvvala et al. (2013)
Xa39(t)	*O. rufipogon*	India	Sundaram et al. (2014)
Xa38, xa13, XA21	PAU 201	India	Sundaram et al. (2014)
XA21, xa13, xa5, & *Xa4*	Lalat and Tapaswini	India	Sundaram et al. (2014)
XA21, xa13, & *xa5*	Swarna and IR64	India	Sundaram et al. (2014)
XA21 & *xa13*	Pusa Basmati 1	India	Sundaram et al. (2014)
XA21	Zhongyou 1176	China	Cao et al. (2003)
XA21	Zhongyou 6	China	Cao et al. (2003)
xa5/Blast R	RD6	Thailand	Pinta et al. (2013)

Already some rice cultivars were developed and some elite cultivars were genetically improved by using conventional breeding in combination with marker-assisted backcross breeding (MABB) approach. It helps in pyramiding different of resistance genes. This can be exemplified as improved

Basmati 1 (Pusa 1460) which is an improved version of Pusa Basmati 1 carrying resistance genes (*xa13* and *XA21*) for bacterial blight resistance (Singh et al., 2011). In marker-assisted backcross breeding, availability of tightly linked markers to each resistance gene helps in rapid incorporation of R genes into desire germplasm and their subsequent accurate selection (Sundaram et al., 2008). Thus, MABB methodology is now widely used in many countries for resistance gene pyramiding (Huang et al., 1997; Singh et al., 2001). Of many characterized R genes, *XA21* is most widely exploited in many rice-breeding programs throughout the world. *XA21* in combination with other R genes like (*xa5* and *xa13*) was incorporated in different rice cultivars like PR106, Samba Mahsuri, Swarna, IR64, etc. (Singh et al., 2001; Sundaram et al., 2008, 2014). *Xa4* is another resistance gene that is widely used for pyramiding in many commercial rice cultivars to get durable resistance (Mew et al., 1992; Sun et al., 2003). Pyramiding of resistance genes would be the most effective strategy for improving *indica* and *japonica* rice cultivars for bacterial blight resistance.

2.3.2 TRANSGENIC APPROACH

Since long time, conventional breeding and marker assisted breeding were the main methods to breeding bacterial blight resistant rice varieties, but resistant sources available for these methods are extremely limited and time consuming, investment of so many years to develop a new resistant variety. At present, genetic engineering is found to be a very powerful technology for incorporating beneficial trait into different plant species. The Genetically engineered crops cover around 181 million hectares of the global arable land (IASSS, 2014). Genetic engineering is used in various staple food crops like rice, maize, banana, soybean, etc. for biotic and abiotic stress tolerance and is expected to substantially alleviate poverty, hunger and malnutrition (Demont & Stein, 2013). In transgenic approach, rice plants are transformed with one or more transgene(s) by using genetic engineering tools to control of the disease. Transformation technology makes possible for the introduction of a foreign gene/transgene into rice from different source within very short time. The nature and origin of transgenes are varying with the target for pathogens (Singh et al., 2016). Based on the nature and origin of transgenes that can be utilized, the current transgenic approaches can be categorized as two ways, that is, approaches that exploit transgenes of plant origin, and that of transgenes of microbial/nonplant origin.

The transgenic approach has been quite beneficial in rice for exploiting R genes from different wild source. The dominant *XA21* resistant gene was transferred from the wild species *Oryza longistaminata* to *Oryza sativa* cv. IR24 by Khush et al. (1990). Zhai et al. (2004) generated transgenic rice lines containing transgene *XA21* introduced through Agrobacterium-mediated transformation system and this transgene was stably inherited and expressed resistance against BLB in all the transgenic lines. Wei et al. (2008) successfully bred transgenic restorer lines for multiple resistances against BLB, striped stem borer and herbicide. In this way, limits of transferring R genes are minimized. But, sometimes monogenic expression of R-genes is not stable in the long run due to the development new strains of *Xoo* that can overcome the host resistance. This can be exemplified in rice cultivars carrying *Xa4* and *Xa3* genes which lost their resistance to BLB in some areas within a few years (Zhang, 2007). For that, we have to explore new resistance genes from other sources too, instead of *Oryza sativa* and wild rice. Transgenes isolated from other plant host were also tested to develop BLB resistant transgenic. Transfer of seven genes from *Nicotiana benthamiana* encoding a calreticulin protein, an ERF transcriptional factor, a novel solanaceous protein, a hydrolase, a peroxidase, and two proteins with unknown function provide a broad spectrum resistance in rice to *X. oryzae pv. oryzae* (Li et al., 2012). Similarly, transgenic rice plants harboring a resistance gene analog FZ14 derived from *Zizania latifolia* confer bacterial blight resistance (Shen et al., 2011). Several transgenes from microbial/nonplant origin including ROS inducing *GOX* gene (encoding glucose oxidase enzyme) from *Aspergillus niger* (Kachroo et al., 2003), immune response-related genes such as *flagellin* (Takakura et al., 2008), and antimicrobial peptide genes such as *cecropin* B encoding genes from *Bombyx mori* (Sharma et al., 2000) have been also transformed into the rice genome to see their expression and resistance response. These genes, both under constitutive and pathogen-induced expression have shown broad spectrum disease resistance and are significantly effective against various bacterial and fungal diseases of rice.

2.3.3 EXPLOITATION OF GENOME EDITING TOOLS

In the past few years, scientists working on genome modification gene or single nucleotide level by using various methods such as chemical mutagenesis (Eeken & Sobels, 1983; Solnica-Krezel et al., 1994) and transposon-mediated mutagenesis (Marx, 1982; Rubin & Spradling, 1982) but these methods produced large number of phenotypic variants random mutation.

Therefore scientist developed reverse genetic technologies that can be used to make precise genetic manipulations, including homologous recombination (HR) based gene targeting (Dui et al., 2012; Thomas & Capecchi, 1987), genome editing starts with the efficient double strand break generation in the target DNA and these breaks are repaired either by HR or via nonhomologous end joining. This is a convenient method for gene knockout.

Recently, artificially engineered hybrid enzymes such as zinc-finger nucleases (ZFNs), TALENs, TALEs, CRISPR, Cas systems have been developed in various organisms including plants. ZFNs are constructed by fusion of DNA-binding zinc-finger motifs with the nuclease domain of the *FokI* endonuclease (Voytas, 2013) and used for gene targeting in various plants, including *Arabidopsis*, tobacco, maize, and soybean for efficient and heritable mutagenesis (Curtin et al., 2011; Zhang et al., 2010). Moreover, the versatile usage of ZFNs in plants is limited by prohibitive licensing fees, which restricts access to the required design tools developed by the company, Sangamo Bioscience.

The DNA-binding domain of the Xanthomonas TALE protein has also been fused with the *FokI* endonuclease domain to obtain TALENs (Feng et al., 2014; Shan et al., 2013). TALEs from *Xanthomonas* bacteria activate the association of host targeted genes, leading to disease susceptibility or resistance. Several TALEs have been found to be essential virulence factors of *Xoo* in susceptible rice, inducing host S genes and subsequently promoting pathogen infection and disease development (Yang & White, 2004; Yang et al., 2006). For example, some *Xoo* strains use TALE *PthXo1* to induce the S gene *Os8N3* (*Xa13* or *OsSWEET11*) for disease in susceptible rice. Intriguingly, some genetic variations in the promoter of *Os8N3* are nonresponsive to PthXo1 and confer disease resistance to *PthXo1*-dependent Xoo strains, and those S gene alleles are collectively named as the recessive resistance gene *xa13* (Chen et al., 2010; Chu et al., 2006). The "resistance" is not due to the active defense but to a lack of S gene induction and loss of disease susceptibility (Yang et al., 2006).

To encounter TALEs, plants have evolved R genes, such as *Xa27*, that recognize TALEs through the promoter sequence and activate the defense process once induced. *Xa27*, recognizing the cognate TALE *AvrXa27*, is the representative of an unusual class of dominant R genes in plants (Gu et al., 2004, 2005). When gene fused to the nonspecific pathogen-inducible *OsPR1* promoter, the induced *Xa27* conferred resistance to *Xoo* strains due to presence of *AvrXa27* (Gu et al., 2005). Two additional, as-yet uncharacterized rice R genes (*Xa10* for *AvrXa10* and *Xa7* for *AvrXa7*) have similar requirements for the transcription activation domain and nuclear localization motifs of their cognate TALEs as required for *AvrXa27*-mediated induction of *Xa27* (Yang et al., 2000; Zhu et al., 1998). Gene activation of three alleles (*xa27*, *xa13*, and *OsSWEET12*) led to phenotypic changes in disease resistance or susceptibility in response to *Xoo* infection suggesting the feasibility of this approach to the functional analysis of gene. Bacterial blight of rice involves molecular interactions of genes or gene products from both pathogen and rice, and gene regulation in host through pathogen effectors appears to play an important role in the pathogenesis of disease (White & Yang, 2009). In future, these approaches will be used for discovery of genes as well as genome improvement against abiotic stress.

2.4   CONCLUSION AND FUTURE DIRECTION

BLB of rice is one of the main constraints to achieve expected yield. So far, various strategies had been advocated for the management of this disease, but use of resistant cultivars is found to be the most economical. A lot of research has already been carried out, both in the conventional and molecular breeding aspects. The huge genomic diversity within the available germ-plasms of rice is already exploited for getting durable resistance. For that, a large number of resistance genes have been identified, characterized and deployed. But pathogen has been always very clever and adopt new tools to break down the host defense when govern by single gene only. Therefore, screening of more and more number of germplasm, including land races and the wild species of *Oryza*, is needed to be screened against BLB to find out new resistant genes. Development of rice mini-core carrying diverse resistant genes against BLB and other diseases is also a noble approach to bring whole information within an umbrella that will support our resistant breeding program. At present, many mini-cores have been developed in different countries that representative of worldwide collection. So these genotypes can be a valuable source for mining of new genes/allele against BLB.

BLB of rice mostly appears in epidemic form in low land, waterlogged rice growing areas, where problem of rice blast as well as frequent waterlogging is also very serious. To combat such problems, an integrated approach should be adopted. This can only be possible when pyramiding of genes for resistance to bacterial blight and blast and that of waterlogged tolerance will be done. Although the marker assisted breeding may enrich the process by reducing the time, many newer efforts are still required to fine-tune these approaches to be readily usable and applicable. Nowadays, the integration of modern genetic tools, like transgenic approach, with conventional breeding makes it very easier to improve genetic background of our existing cultivar. The benefit of newly popular genome editing tools like TALEN, CRISPR can also be a choice for the development of rice resistance and substantial improvement of existing rice cultivars. Therefore, adoption of interdisciplinary breeding strategies for the enrichment of genetic resource is encouraging and may improve the disease management strategies in sustainable and durable manner.

KEYWORDS

- **gene pyramiding**
- **molecular approaches**
- **durable resistance**
- **biotic stress**
- **marker assisted backcrossing**

REFERENCES

Adhikari, T. B.; Mew, T. W.; Leach, J. E. Genotypic and Pathotypic Diversity in *Xanthomonas oryzae* pv. *oryzae* in Nepal. *Phytopathology* **1999**, *89*, 687–694.

Antony, G.; Zhou, J.; Huang, S.; Li, T.; Liu, B.; White, F.; Yang, B. Rice *xa13* Recessive Resistance to Bacterial Blight is Defeated by Induction of the Disease Susceptibility Gene *Os-11N3*. *Plant Cell* **2010**, *22*, 3864–3876.

Bahar, O.; Pruitt, R.; Luu, D. D.; Schwessinger, B.; Daudi, A.; Liu, F.; Ruan, R.; Fontaine-Bodin, L.; Koebnik, R.; Ronald, P. The *Xanthomonas Ax21* Protein is Processed by the General Secretory System and Is Secreted in Association with Outer Membrane Vesicles. *Peer J.* **2014**, *2*, e242.

Bao, S. Y.; Tan, M. P.; Lin, X. H. Genetic Mapping of a Bacterial Blight Resistance Gene *Xa14* in Rice. *Acta Agron. Sin.* **2010**, *36*, 422–427.

Bhasin, H.; Bhatia, D.; Raghuvanshi, S.; Lore, J. S.; Sahi, G. K.; Kaur, B.; Vikal, Y.; Singh, K. New PCR-based Sequence-tagged Site Marker for Bacterial Blight Resistance Gene *Xa38* of Rice. *Mol. Breed.* **2012**, *30*, 607–611.

Blair, M. W.; Garris, A. J.; Iyer, A. S.; Chapman, B.; Kresovich, S.; McCouch, S. R. High Resolution Genetic Mapping and Candidate Gene Identification at the *xa5* Locus for Bacterial Blight Resistance in Rice (*Oryza sativa* L.). *Theor. Appl. Genet.* **2003**, *107*, 62–73.

Boch, J.; Bonas, U. Xanthomonas *AvrBs3* Family-type III Effectors: Discovery and Function. *Ann. Rev. Phytopath.* **2010**, *48*, 419–436.

Boch, J.; Bonas, U.; Lahaye, T. TAL Effectors-pathogen Strategies and Plant Resistance Engineering. *New Phytol.* **2014**, *204*, 823–832.

Bogdanove, A. J.; Schornack, S.; Lahaye, T. TAL Effectors: Finding Plant Genes for Disease and Defense. *Curr. Opin. Plant Biol.* **2010**, *13*, 394–401.

Cao, L. Y.; Zhuang, J. Y.; Yuan, S. J.; Zhan, X. D.; Zheng, K. L.; Cheng, S. H. Hybrid Rice Resistant to Bacterial Leaf Blight Developed by Marker Assisted Selection. *Rice Sci.* **2003**, *11*, 68–70.

Cartieaux, F.; Contesto, C.; Gallou, A.; Desbrosses, G.; Kopka, J.; Taconnat, L.; Renou, J. P.; Touraine, B. Simultaneous Interaction of *Arabidopsis thaliana* with *Bradyrhizobium* sp. Strain *ORS278* and *Pseudomonas syringae* pv. *tomato* DC3000 Leads to Complex Transcriptome Changes. *Mol. Plant Microb. Int.* **2008**, *21*, 244–259.

Cheema, K. K.; Grewal, N. K.; Vikal, Y.; Sharma, R.; Lore, J. S.; Das, A.; Bhatia, D.; Mahajan, R.; Gupta, V.; Bharaj, T. S.; Singh, K. A Novel Bacterial Blight Resistance Gene from *Oryza nivara* Mapped to 38 kb Region on Chromosome 4L and Transferred to *Oryza sativa* L. *Gen. Res.* **2008**, *90*, 397–407.

Chen, H.; Wang, S.; Zhang, Q. New Gene for Bacterial Blight Resistance in Rice Located on Chromosome 12 Identified from Minghui 63, an Elite Restorer Line. *Phytopathology* **2002**, *92*, 750–754.

Chen, S.; Liu, X.; Zeng, L.; Ouyang, D.; Yang, J.; Zhu, X. Genetic Analysis and Molecular Mapping of a Novel Recessive Gene *xa34*(t) for Resistance against *Xanthomonas oryzae* pv. *oryzae*. *Theor. Appl. Genet.* **2011**, *7*, 1331–1338.

Chen, Y.; Dui, W.; Yu, Z.; Li, C.; Ma, J.; Jiao, R. Drosophila *RecQ5* is Required for Efficient SSA Repair and Suppression of LOH *In Vivo*. *Protein Cell* **2010**, *1*, 478–490.

Chu, Z.; Yuan, M.; Yao, J.; Ge, X.; Yuan, B.; Xu, C.; Li, X.; Fu, B.; Li, Z.; Bennetzen, J. L.; Zhang, Q.; Wang, S. Promoter Mutations of an Essential Gene for Pollen Development Result in Disease Resistance in Rice. *Genes Dev.* **2006**, *20*, 1250–1255.

Curtin, S. J.; Zhang, F.; Sander, J. D.; Haun, W. J.; Starker, C.; Baltes, N. J.; Reyon, D.; Dahlborg, E. J.; Goodwin, M. J.; Coffman, A. P.; Dobbs, D.; Joung, J. K.; Voytas, D. F.; Stupar, R. M. Targeted Mutagenesis of Duplicated Genes in Soybean with Zinc-finger Nucleases. *Plant Physiol.* **2011**, *156*, 466–473.

De Vleesschauwer, D.; Cornelis, P.; Hofte, M. Redox-active Pyocyanin Secreted by *Pseudomonas aeruginosa* 7NSK2 Triggers Systemic Resistance to *Magnaporthe grisea* but Enhances *Rhizoctonia solani* Susceptibility in Rice. *Mol. Plant Microbe Int.* **2006**, *19*, 1406–1419.

De Vleesschauwer, D.; Djavaheri, M.; Bakker, P. A. H. M.; Hofte, M. *Pseudomonas fluorescens* WCS374r-induced Systemic Resistance in Rice against *Magnaporthe oryzae* is Based on Pseudobactin-mediated Priming for a Salicylic Acid-repressible Multifaceted Defense Response. *Plant Physiol.* **2008**, *148*, 1996–2012.

Demont, M.; Stein, A. J. Global Value of GM Rice: A Review of Expected Agronomic and Consumer Benefits. *New Biotech.* **2013**, *5*, 426–436.

Dharmapuri, S.; Sonti, R. V. A. Transposon Insertion in the *gumG* Homologue of *Xanthomonas oryzae* pv. *oryzae* Causes Loss of Extracellular Polysaccharide Production and Virulence. *FEMS Microbiol. Lett.* **1999**, *179*, 53–59.

Dui, W.; Lu, W.; Ma, J.; Jiao, R. A Systematic Phenotypic Screen of F-box Genes through a Tissue-specific RNAi-based Approach in Drosophila. *J. Genet. Genom.* **2012**, *39*, 397–413.

Eeken, J. C.; Sobels, F. H. The Effect of Two Chemical Mutagens ENU and MMS on MR-mediated Reversion of an Insertion-sequence Mutation in *Drosophila melanogaster*. *Mutation Res.* **1983**, *110*, 297–310.

FAOSTAT. *Agriculture Data*. 2012. Available at: www.fao.org.

Feng, Z.; Mao, Y.; Xu, N.; Zhang, B.; Wei, P.; Yang, D. L; Wang, Z.; Zhang, Z.; Zheng, R.; Yang, L.; Zeng, L.; Liu, X.; Zhu, J. K. Multi-generation Analysis Reveals the Inheritance, Specificity, and Patterns of CRISPR/Cas9-induced Gene Modifications in Arabidopsis. *Proc. Nat. Acad. Sci. U.S.A.* **2014**, *111*, 632–4637.

Flor, H. H. Inheritance of Pathogenicity in *Melampsora lini*. *Phytopathology* **1942**, *32*, 653–669.

Gao, D. Y.; Liu, A. M.; Zhou, Y. H.; Cheng, Y. J.; Xiang, Y. H.; Sun, L. H.; Zhai, W. X. Molecular Mapping of a Bacterial Blight Resistance Gene *Xa-25* in Rice. *Acta Genet. Sin.* **2005**, *32*, 183–188.

Gao, D. Y.; Xu, Z. G.; Chen, Z. Y.; Sun, L. H.; Sun, Q. M.; Lu, F.; Hu, B. S.; Liu, Y. F.; Tang, L. H. Identification of a New Gene for Resistance to Bacterial Blight in a Somaclonal Mutant HX-3 (indica). *Rice Genet. Newslett.* **2001**, *18*, 66.

Giovannoni, J. J.; Wing, R. A.; Ganal, M. W.; Tanksley, S. D. Isolation of Molecular Markers from Specific Chromosomal Intervals Using DNA Pools from Existing Mapping Populations. *Nucl. Acid Res.* **1991**, *19*, 6553–6558.

Gnanamanickam, S. S.; Brindha, P. V.; Narayanan, N. N.; Vasudevan, P.; *Kavitha, S.* An overview of bacterial blight disease of rice and strategies for its management. *Curr. Sci.* **1999**, *77*, 1435–1443.

Gonzalez, C.; Szurek, B.; Manceau, C.; Mathieu, T.; Sere, Y.; Verdier, V. Molecular and Pathotypic Characterization of New *Xanthomonas oryzae* Strains from West Africa. *Mol. Plant Microbe Int.* **2007**, *20*, 534–546.

Goto, T.; Matsumoto, T.; Furuya, N.; Tsuchiya, K.; Yoshimura, A. Mapping of Bacterial Blight Resistance Gene *Xa11* on Rice Chromosome 3. *Japan Agric. Res. Q.* **2009**, *43*, 221–225.

Gu, K.; Sangha, J. S.; Li, Y.; Yin, Z. High-resolution Genetic Mapping of Bacterial Blight Resistance Gene *Xa10. Theor. Appl. Genet.* **2008**, *116*, 155–163.

Gu, K.; Tian, D.; Yang, F.; Wu, L.; Sreekala, C.; Wang, D.; Wang, G. L.; Yin, Z. High-resolution Genetic Mapping of *Xa27*(t), a New Bacterial Blight Resistance Gene in Rice, *Oryza sativa* L. *Theor. Appl. Genet.* **2004**, *108*, 800–807.

Gu, K.; Yang, B.; Tian, D.; Wu, L.; Wang, D.; Sreekala, C.; Yang, F.; Chu, Z.; Wang, G. L.; White, F. F.; Yin, Z. R Gene Expression Induced by a Type-III Effector Triggers Disease Resistance in Rice. *Nature* **2005**, *435*, 1122–1125.

Guo, S.; Zhang, D.; Lin, X. Identification and Mapping of a Novel Bacterial Blight Resistance Gene *Xa35*(t) Originated from *Oryza minuta*. *Sci. Agric. Sin.* **2010**, *43*, 2611–2618.

Guvvala, L. D.; Koradi, P.; Shenoy, V.; Marella, L. S. Making an Indian Traditional Rice Variety Mahsuri, Bacterial Blight Resistant Using Marker-assisted Selection. *J. Crop Sci. Biotech.* **2013**, *6*, 111–121.

He, Q.; Li, D.; Zhu, Y.; Tan, M.; Zhang, D.; Lin, X. Fine Mapping of *Xa2*, a Bacterial Blight Resistance Gene in Rice. *Mol. Breed.* **2006**, *17*, 1–6.

Huang, N.; Angeles, E. R.; Domingo, J.; Magpantay, G.; Singh, S.; Zhang, G.; Kumaravadivel, N.; Bennett, J.; Khush, G. S. Pyramiding of Bacterial Blight Resistance Genes in Rice: Marker-assisted Selection Using RFLP and PCR. *Theor. Appl. Genet.* **1997**, *95*, 313–320.

Hur, Y. J.; Jeung, J. U.; Kim, Y. S.; Park, H. S.; Cho, J. H.; Lee, J. Y.; Sohn, Y. B.; Song, Y. C.; Park, D. S.; Lee, C. W.; Sohn, J. G.; Nam, M. H.; Le, J. H. Functional Markers for Bacterial Blight Resistance Gene *Xa3* in Rice. *Mol. Breed.* **2013**, *31*, 981–985.

IASSS. IASSS Brief 49. Executive Summary: Global Status of Commercialized Biotech/GM Crops. 2014. Available at: www.iasss.org.

Ishiyama, S. Studies on Bacterial Blight of Rice. *Rep. Agric. Exp. Stat., Tokyo* **1922**, *45*, 233–261.

Iyer, A. S.; McCouch, S. R. The Rice Bacterial Blight Resistance Gene *xa5* Encodes a Novel Form of Disease Resistance. *Mol. Plant Microb. Int.* **2004**, *17*, 1348–1354.

Jeung, J. U.; Heu, S. G.; Shin, M. S.; Vera; Cruz, C. M.; Jena, K. K. Dynamics of *Xanthomonas oryzae* pv. *oryzae* Populations in Korea and Their Relationship to Known Bacterial Blight Resistance Genes. *Phytopathology* **2006**, *96*, 867–875.

Jin, X.; Wang, C.; Yang, Q.; Jiang, Q.; Fan, Y.; Liu, G.; Zhao, K. Breeding of Near-isogenic Line CBB30 and Molecular Mapping of *Xa30(t)*, a New Resistance Gene to Bacterial Blight in Rice. *Sci. Agric. Sin.* **2007**, *40*, 1094–1100.

Jones, J. D.; Dangl, J. L. The Plant Immune System. *Nature* **2006**, *444*, 323–329.

Kachroo, A.; He, Z.; Patkar, R.; Zhu, Q.; Zhong, J.; Li, D.; Ronald, P.; Lamb, C.; Chattoo, B. B. Induction of H_2O_2 in Transgenic Rice Leads to Cell Death and Enhanced Resistance to both Bacterial and Fungal Pathogens. *Transgenic Res.* **2003**, *12*, 577–586.

Kamoun, S.; Kado, C. I. A Plant-inducible Gene of *Xanthomonas campestris* pv. *campestris* Encodes an Exocellular Component Required for Growth in the Host and Hypersensitivity on Nonhosts. *J. Bacteriol.* **1990**, *172*, 5165–5172.

Khush, G. S.; Angeles, E. R. A New Gene for Resistance to Race 6 of Bacterial Blight in Rice *Oryza sativa*. *Rice Genet. Newlett.* **1999**, *16*, 92–93.

Khush, G. S. What It Will Take to Feed 5.0 Billion Rice Consumers in 2030. *Plant Mol. Biol.* **2005**, *59*, 1–6.

Khush, G. S.; Bacalangco, E.; Ogawa, T. A New Gene for Resistance to Bacterial Blight from *O. longistaminata*. *Rice Genet. Newslett.* **1990**, *7*, 121–122.

Kim, S. M.; Suh, J. P.; Qin, Y.; Noh, T. H.; Reinke, R. F.; Jena, K. K. Identification and Fine-mapping of a New Resistance Gene, *Xa40*, Conferring Resistance to Bacterial Blight Races in Rice (*Oryza sativa* L.). *Theor. Appl. Genet.* **2015**, *128*, 1933–1943.

Kumar, P. N.; Sujatha, K.; Laha, G. S.; Rao, K. S.; Mishra, B.; Viraktamath, B. C.; Hari, Y.; Reddy, C. S.; Balachandran, S. M.; Ram, T.; Madhav, M. S.; Rani, N. S.; Neeraja, C. N.; Reddy, G. A.; Shaik, H.; Sundaram, R. M. Identification and Fine-mapping of *Xa33*, a Novel Gene for Resistance to *Xanthomonas oryzae* pv. *oryzae*. *Phytopathology* **2012**, *102*, 222–228.

Kumar, P.; Rosegrant, M. W.; Hazell, P. B. *Cereals Prospects in India to 2020.* International Food Policy Research Institute (IFPRI), 1995; p 23.

Leach, J. E.; White, F. F. Bacterial Avirulence Genes. *Ann. Review Phytopathol.* **1996**, *34*, 153–179.

Lee, K. S.; Rasabandith, S.; Angeles, E. R.; Khush, G. S. Inheritance of Resistance to Bacterial Blight in 21 Cultivars of Rice. *Phytopathology* **2003**, *93*, 147–152.

Lee, S. W.; Han, S. W.; Sririyanum, M.; Park, C. J.; Seo, Y. S.; Ronald, P. C. A Type I-secreted, Sulphated Peptide Triggers *XA21*-mediated Innate Immunity. *Science* **2009**, *326*, 850–853.

Li, W.; Xu, Y. P.; Zhang, Z. X.; Cao, W. Y.; Li, F.; Zhou, X.; Chen, G. Y.; Cai, X. Z. Identification of Genes Required for Non-host Resistance to *Xanthomonas oryzae pv. oryzae* Reveals Novel Signaling Components. *PLoS One* **2012**, 7, e42796.

Lima, J. M.; Dass, A.; Sahu, S. C.; Behera, L.; Chauhan, D. K. A RAPD Marker Identified a Susceptible Specific Locus for Gall Midge Resistance Gene in Rice Cultivar ARC5984. *Crop Prot.* **2007**, *26*, 1431–1435

Lin, X. H.; Zhang, D. P.; Xie, Y. F.; Gao, H. P.; Zhang, Q. Identifying and Mapping a New Gene for Bacterial Blight Resistance in Rice Based on RFLP Markers. *Phytopathology* **1996**, *86*, 1156–1159.

Liu, Q.; Yuan, M.; Zhou, Y.; Li, X.; Xiao, J.; Wang, S. A Paralog of the MtN3/saliva Family Recessively Confers Race-specific Resistance to *Xanthomonas oryzae* in Rice. *Plant Cell Environ.* **2011,** *34,* 1958–1969.

Marx, J. L. Gene transfer into the Drosophila germ line. *Science* **1982,** *218,* 364–365.

Mew, T. W.; Vera Cruz, C. M.; Medalla, E. S. Changes in Race Frequency of *Xanthomonas oryzae* pv. *oryzae* in Response to Rice Cultivars Planted in the Philippines. *Plant Dis.* **1992,** *76,* 1029–1032.

Miao, L.; Wang, C.; Zheng, C.; Che, J.; Gao, Y.; Wen, Y.; Li, G.; Zhao, K. Molecular Mapping of a New Gene for Resistance to Rice Bacterial Blight. *China Agric. Sci.* **2010,** *43,* 3051–3058.

Mir, G. N.; Khush, G. S. Genetics of Resistance to Bacterial Blight in Rice Cultivar DV86. *Crop Res.* **1990,** *3,* 194–198.

Nakayama, A.; Fukushima, S.; Goto, S.; Matsushita, A.; Shimono, M.; Sugano, S.; Jiang, C. J.; Akagi, A.; Yamazaki, M.; Inoue, H.; Takatsuji, H. Genome-wide Identification of WRKY45-regulated Genes that Mediate Benzothiadiazole-induced Defense Responses in Rice. *BMC Plant Biol.* **2013,** *13,* 150.

Noda, T.; Ohuchi, A. A New Pathogenic Race of *Xanthomonas campestris* pv. *oryzae* and Its Inheritance of Differential Rice Variety Tetep to It. *Ann. Phytopathol. Soc. Japan* **1989,** *55,* 201–207.

Noda, T.; Li, C.; Li, J.; Ochiai, H.; Ise, K.; Kaku, H. Pathogenic Diversity of *Xanthomonas oryzae* pv. *oryzae* Strains from Yunnan Province, China. *Japan Agric. Res. Q.* **2001,** *35,* 97–103.

Noda, T.; Yamamoto, T.; Kaku, H.; Horino, O. Geographical Distribution of Pathogenic Races of *Xanthomonas oryzae* pv. *oryzae* in Japan in 1991 and 1993. *Annu. Phytopathol. Soc. Japan* **1996,** *62,* 549–553.

Ogawa, T.; Kaku, H.; Yamamoto, T. Resistance Gene of Rice Cultivar, *Asaminori* to Bacterial Blight of Rice. *Japan. J. Breed.* **1989,** *39,* 196–197.

Ogawa, T.; Lin, L.; Tabien, R. E.; Khush, G. S. A New Recessive Gene for Resistance to Bacterial Blight of Rice. *Rice Genet. Newslett.* **1987,** 4, 98–100.

Oryzabase. Integrated Rice Science Database. 2011. Available at http://www.shigen.nig. ac.jp/rice/oryzabase.

Petpisit, V.; Khush, G. S.; Kauffman, H. E. Inheritance of Resistance to Bacterial Blight in Rice. *Crop Sci.* **1977,** *17,* 551–554.

Pinta, W.; Toojinda, T.; Thummabenjapone, P.; Sanitchon, J. Pyramiding of Blast and Bacterial Leaf Blight Resistance Genes into Rice Cultivar RD6 Using Marker Assisted Selection. *Afr. J. Biotech.* **2013,** 12, 4432–4438.

Porter, B. W.; Chittoor, J. M.; Yano, M.; Sasaki, T.; White, F. F. Development and Mapping of Markers Linked to the Rice Bacterial Blight Resistance Gene *Xa7*. *Crop Sci.* **2003,** *43,* 1484–1492.

Rajpurohit, D.; Kumar, R.; Kumar, M.; Paul, P.; Awasthi, A.; Basha, O. P.; Puri, A.; Jhang, T.; Singh, K.; Dhaliwal, H. S. Pyramiding of Two Bacterial Blight Resistance and a Semidwarfing Gene in Type 3 Basmati Using Marker-assisted Selection. *Euphytica* **2011,** *178,* 111–126.

Ray, S. K.; Rajeshwari, R.; Sonti, R. V. Mutants of *Xanthomonas oryzae* pv. *oryzae* Deficient in General Secretory Pathway are Virulence Deficient and Unable to Secrete Xylanase. *Mol. Plant Microb. Int.* **2000,** *13,* 394–401.

Rubin, G. M.; Spradling, A. C. Genetic Transformation of Drosophila with Transposable Element Vectors. *Science* **1982,** *218,* 348–353.

Ryals, J. A.; Neuenschwander, U. H.; Willits, M. G.; Molina, A.; Steiner, H. Y.; Hunt, M. D. Systemic Acquired Resistance. *Plant Cell* **1996**, *8*, 1809–1819.

Sakaguchi, S. *Linkage Studies on the Resistance to Bacterial Leaf Blight, Xanthomonas oryzae* (*Uyed et Ishiyama*) *Dowson, in Rice. Bull. Nat. Inst. Agricul. Sci. Ser. D, 1967, 16*, 1–18.

Salzberg, S. L.; Sommer, D. D.; Schatz, M. C.; Phillippy, A. M.; Rabinowicz, P. D.; Tsuge, S.; Furutani, A.; Ochiai, H.; Delcher, A. L.; Kelley, D.; Madupu, R.; Puiu, D.; Radune, D.; Shumway, M.; Trapnell, C.; Aparna, G.; Jha, G.; Pandey, A.; Patil, P. B.; Ishihara, H.; Meyer, D. F.; Szurek, B.; Verdier, V.; Koebnik, R.; Dow, J. M.; Ryan, R. P.; Hirata, H.; Tsuyumu, S.; Won; Lee, S.; Seo, Y. S.; Sriariyanum, M.; Ronald, P. C.; Sonti, R. V.; Van; Sluys, M. A.; Leach, J. E.; White, F. F.; Bogdanove, A. J. Genome Sequence and Rapid Evolution of the Rice Pathogen *Xanthomonas oryzae* pv. *oryzae* PXO99A. *BMC Genomics* **2008**, *9*, 204.

Sanchez, A. C.; Ilag, L. L.; Yang, D.; Brar, D. S.; Ausubel, F.; Khush, G. S.; Yano, M.; Sasaki, T.; Li, Z.; Huang, N. Genetic and Physical Mapping of *xa13*, a Recessive Bacterial Blight Resistance Gene in Rice. *Theor. Appl. Genet.* **1999**, *98*, 1022–1028.

Shan, Q.; Wang, Y.; Chen, K.; Liang, Z.; Li, J.; Zhang, Y.; Zhang, K.; Liu, J.; Voytas, D. F.; Zheng, X.; Zhang, Y.; Gao, C. Rapid and Efficient Gene Modification in Rice and Brachypodium Using TALENs. *Mol. Plant.* **2013**, *6*, 1365–1368.

Sharma, A.; Sharma, R.; Imamura, M.; Yamakawa, M.; Machii, H. Transgenic Expression of Cecropin B, an Antibacterial Peptide from *Bombyx mori*, Confers Enhanced Resistance to Bacterial Leaf Blight in Rice. *FEBS Lett.* **2000**, *484*, 7–11.

Shen, W.; Song, C.; Chen, J.; Fuy, A.; Wu, J.; Jiang, S. Transgenic Rice Plants Harboring Genomic DNA from *Zizania latifolia* Confer Bacterial Blight Resistance. *Rice Sci.* **2011**, *18*, 17–22.

Shen, Y. W.; Chern, M. S.; Silva, F. G.; Ronald, P. Isolation of a *Xanthamonas oryzae* pv. *oryzae* Flagellar Operon Region and Molecular Characterization of *flhF. Mol. Plant Microb. Int.* **2001**, *14*, 204–213.

Singh, A. K.; Gopalakrishnan, S.; Singh, V. P.; Prabhu, K. P.; Mohapatra, T.; Singh, N. K.; Sharma, T. R.; Nagarajan, M.; Vinod, K. K.; Singh, D.; Singh, U. D.; Chander, S.; Atwal, S. S.; Seth, R.; Singh, V. K.; Ellur, R. K.; Singh, A.; Anand, D.; Khanna, A.; Yadav, S.; Goel, N.; Singh, A.; Shikari, A. B.; Singh, A.; Marathi, B. Marker Assisted Selection: A Paradigm Shift in Basmati Breeding. *Ind. J. Genet. Plant Breed.* **2011**, *71*, 120–128.

Singh, R. K.; Chattopadhyay, A.; Pandey, S. K. Status and Perspectives of Biological Control of Rice Diseases. In: *Microbial Empowerment in Agriculture: A Key to Sustainability and Crop Productivity*; Sarma, B. K., Singh, A., Eds.; Biotech Books: New Delhi, 2016; pp 335–372.

Singh, S.; Sidhu, J. S.; Huang, N.; Vikal, Y.; Li, Z.; Brar, D. S.; Dhaliwal, H. S.; Khush, G. S. Pyramiding Three Bacterial Blight Resistance Genes (*xa5, xa13* and *Xa21*) Using Marker-assisted Selection into Indica Rice Cultivar PR106. *Theor. Appl. Genet.* **2001**, *102*, 1011–1015.

Solnica-Krezel, L.; Schier, A. F.; Driever, W. Efficient Recovery of ENU-induced Mutations from the Zebrafish Germline. *Genetics* **1994**, *136*, 1401–1420.

Song, W. Y.; Wang, G. L.; Chen, L. L.; Kim, H. S.; Pi, L. Y.; Holsten, T.; Gardner, J.; Wang, B.; Zhai, W. X.; Zhu, L. H.; Fauquet, C.; Ronald, P. A Receptor Kinase-like Protein Encoded by the Rice Disease Resistance Gene, *Xa21. Science* **1995**, *270*, 1804–1806.

Suh, J. P.; Noh, T. H.; Kim, K. Y.; Kim, J. J.; Kim, Y. G.; Jena, K. K. Expression Levels of Three Bacterial Blight Resistance Genes against K3a Race of Korea by Molecular and Phenotype Analysis in Japonica Rice (*O. sativa* L.). *J. Crop Sci. Biotechnol.* **2009**, *12*, 103–108.

Sun, X.; Cao, Y.; Yang, Z.; Xu, C.; Li, X.; Wang, S.; Zhang, Q. *Xa26*, a Gene Conferring Resistance to *Xanthomonas oryzae* pv. *oryzae* in Rice, Encodes an LRR Receptor Kinase-like Protein. *Plant J.* **2004**, *37*, 517–527.

Sun, X.; Yang, Z.; Wang, S.; Zhang, Q. Identification of a 47 kb DNA Fragment Containing *Xa4*, a Locus for Bacterial Blight Resistance in Rice. *Theor. Appl. Genet.* **2003**, *106*, 683–687.

Sundaram, R. M.; Chatterjee, S.; Oliva, R.; Laha, G. S.; Cruz, L. J. E.; Sonti, R. V. Update on Bacterial Blight of Rice: Fourth International Conference on Bacterial Blight. *Rice* **2014**, *7*, 12.

Sundaram, R. M.; Vishnupriya, M. R.; Biradar, S. K.; Laha, G. S.; Reddy, G. A.; Rani, N. S.; Sarma, N. P.; Sonti, R. V. Marker Assisted Introgression of Bacterial Blight Resistance in Samba Mahsuri, an Elite Indica Rice Variety. *Euphytica* **2008**, *160*, 411–422.

Tagami, Y.; Mizukami, T. Historical Review of the Researches on Bacterial Leaf Blight of Rice Caused by Xanthomonas. *Special Report on Plant Disease and Insect Pest Forecasting Service.* Ministry of Agriculture: Japan, 2008, Vol. 10, 1–112.

Takai, R.; Isogai, A.; Takayama, S.; Che, F. S. Analysis of Flagellin Perception Mediated by *flg22* Receptor *OsFLS2* in Rice. *Mol. Plant Microb. Int.* **2008**, *21*, 1635–1642.

Takakura, Y.; Che, F. S.; Ishida, Y.; Tsutsumi, F.; Kurotani, K.; Usami, S.; Isogai, A.; Imaseki, H. Expression of a Bacterial Flagellin Gene Triggers Plant Immune Responses and Confers Disease Resistance in Transgenic Rice Plants. *Mol. Plant Pathol.* **2008**, *9*, 525–529.

Takatsuji, H.; Jiang, C. J.; Sugano, S. Salicylic Acid Signaling Pathway in Rice and the Potential Applications of its Regulators. *Japan Agric. Res. Q.* **2010**, *44*, 217–223.

Tan, G. X.; Ren, X.; Weng, Q. M.; Shi, Z. Y.; Zhu, L. L.; He, G. C. Mapping of a New Resistance Gene to Bacterial Blight in Rice Line Introgressed from *Oryza officinalis*. *Yi Chuan Xue Bao* **2004**, *31*, 724–729.

Taura, S.; Ogawa, T.; Yoshimura, A.; Ikeda, R.; Iwata, N. Identification of a Recessive Resistance Gene to Rice Bacterial Blight of Mutant Line XM6, *Oryza sativa* L. *Jpn. J. Breed.* **1992a**, *42*, 7–13.

Taura, S.; Ogawa, T.; Yoshimura, A.; Ikeda, R.; Omura, T. Identification of a Recessive Resistance Gene in Induced Mutant Line XM5 of Rice to Rice Bacterial Blight. *Jpn. J. Breed.* **1991**, *41*, 427–432.

Taura, S.; Tabien, R. E.; Khush, G. S.; Yoshimura, A.; Omura, T. *Resistance Gene of Rice Cultivar Taichung Native 1 to Philippine Races of Bacterial Blight Pathogens. Jpn. J. Breed.* **1992b**, *42*, 195–201.

Thomas, K. R.; Capecchi, M. R. Site-directed Mutagenesis by Gene Targeting in Mouse Embryo-derived Stem Cells. *Cell* **1987**, *51*, 503–512.

Thomma, B. P.; Nurnberger, T.; Joosten, M. H. Of PAMPs and Effectors: The Blurred PTI-ETI Dichotomy. *Plant Cell* **2011**, *23*, 4–15.

Tian, D.; Wang, J.; Zeng, X.; Gu, K.; Qiu, C.; Yang, X.; Zhou, Z.; Goh, M.; Luo, Y.; Murata-Hori, M.; White, F. F.; Yin, Z. The Rice TAL Effector-dependent Resistance Protein XA10 Triggers Cell Death and Calcium Depletion in the Endoplasmic Reticulum. *Plant Cell* **2014**, *26*, 497–515.

Tiwari, K. K.; Pattnaik, S.; Singh, A.; Sandhu, M.; Mithra, S. A.; Abdin, M. Z.; Singh, A.; Mohapatra, T. Allelic Variation in the Microsatellite Marker Locus RM6100 Linked to Fertility Restoration of WA Based Male Sterility in Rice. *Indian J. Genet. Plant Breed.* **2015**, *74*, 409–413.

van den Ackerveken, G.; Marois, E.; Bonas, U. Recognition of the Bacterial Avirulence Protein *avrBs3* Occurs Inside the Host Plant Cell. *Cell* **1996**, *87*, 1307–1316.

Verhagen, B. W. M.; Glazebrook, J.; Zhu, T.; Chang, H. S.; Van Loon, L. C.; Pieterse, C. M. J. The Transcriptome of Rhizobacteria-induced Systemic Resistance in *Arabidopsis. Mol. Plant Microb. Int.* **2004**, *17*, 895–908.

Vikal, Y.; Chawla, H.; Rajiv, S.; Lore, J. S.; Singh, K. Mapping of Bacterial Blight Resistance Gene *xa8* in Rice (*Oryza sativa* L.). *Ind. J. Genet. Plant Breed.* **2014**, *74*, 589–595.

Vikal, Y.; Das, A.; Patra, B.; Goel, R. K.; Sidhu, J. S.; Singh, K. Identification of New Sources of Bacterial Blight (*Xanthomonas oryzae* pv. *oryzae*) Resistance in Wild *Oryza* Species and *O. glaberrima. Plant Genet. Res.* **2007**, *5*, 108–112.

Voytas, D. F. Plant Genome Engineering with Sequence-specific Nucleases. *Ann. Rev. Plant Biol.* **2013**, *64*, 327–350.

Wang, C.; Fan, Y.; Zheng, C.; Qin, T.; Zhang, X.; Zhao, K. High-resolution Genetic Mapping of Rice Bacterial Blight Resistance Gene *Xa23. Mol. Genet. Genomics* **2014**, *289*, 745–753.

Wang, C.; Tan, M.; Xu, X.; Wen, G.; Zhang, D.; Lin, X. Localizing the Bacterial Blight Resistance Gene, *Xa22*(t), to a 100-Kilobase Bacterial Artificial Chromosome. *Phytopathology* **2003**, *93*, 1258–1262.

Wang, C.; Wen, G.; Lin, X.; Liu, X.; Zhang, D. Identification and Fine Mapping of the New Bacterial Blight Resistance Gene, *Xa31*(t), in Rice. *Eur. J. Plant Pathol.* **2009**, *123*, 235–240.

Wang, D.; Amornsiripanitch, N.; Dong, X. A Genomic Approach to Identify Regulatory Nodes in the Transcriptional Network of Systemic Acquired Resistance in Plants. *PLoS Pathol.* **2006**, *2*, e123.

Wang, G. L.; Song, W. Y.; Wu, R. L.; Sideris, S.; Ronald, P. C. The Cloned Gene, *Xa27*, confers Resistance to Multiple *Xanthomonas oryzae* pv. *oryzae* Isolates in Transgenic Plants. *Mol. Plant Microb. Int.* **1996**, *9*, 850–855.

Wei, Y.; Yao, F.; Zhu, C.; Jiang, M.; Li, G.; Song, Y.; Wen, F. Breeding of Transgenic Rice Restorer Line for Multiple Resistance against Bacterial Blight, Striped Stem Borer and Herbicide. *Euphytica* **2008**, *163*, 177–184.

White, F. F.; Yang, B. Host and Pathogen Factors Controlling the Rice-*Xanthomonas oryzae* Interaction. *Plant Physiol.* **2009**, *150*, 1677–1686.

Wu, L.; Goh, M. L.; Sreekala, C.; Yin, Z. *XA27* Depends on an Amino-terminal Signal-anchor-like Sequence to Localize to the Apoplast for Resistance to *Xanthomonas oryzae* pv. *oryzae. Plant Physiol.* **2008a**, *148*, 1497–1509.

Wu, X.; Li, X.; Zu, C.; Wang, S. Fine Genetic Mapping of *xa24*, a Recessive Gene for Resistance against *Xanthomonas oryzae* pv. *oryzae* in Rice. *Theor. Appl. Genet.* **2008b**, *118*, 185–191.

Xiang, Y.; Cao, Y.; Xu, C.; Li, X.; Wang, S. *Xa3*, Conferring Resistance for Rice Bacterial Blight and Encoding a Receptor Kinase-like Protein, is the Same as *Xa26. Theor. Appl. Genet.* **2006**, *113*, 1347–1355.

Yang, B.; White, F. Diverse Members of the *AvrBs3/PthA* Family of Type III Effectors are Major Virulence Determinants in Bacterial Blight Disease of Rice. *Mol. Plant Microbes Int.* **2004**, *17*, 1192–200.

Yang, B.; White, F. Avoidance of Host Recognition by Alterations in the Repetitive and C-terminal Regions of *AvrXa7*, a Type III Effector of *Xanthomonas oryzae* pv. *oryzae. Mol. Plant Microbes Int.* **2005**, *18*, 142–149.

Yang, B.; Sugio, A.; White, F. F. *Os-8N3* is a Host Disease Susceptibility Gene for Bacterial Blight of Rice. *Proc. Nat. Acad. Sci. U.S.A.* **2006**, *103*, 10503–10508.

Yang, B.; Zhu, W.; Johnson, L. B.; White, F. F. The Virulence Factor *AvrXa7* of *Xanthomonas oryzae* pv. *oryzae* is a Type III Secretion Pathway-dependent Nuclear-localized double-stranded DNA-binding Protein. *Proc. Nat. Acad. Sci. U.S.A.* **2000**, *97*, 9807–9812.

Yang, Y.; Gabriel, D. W. Intragenic Recombination of a Single Plant Pathogen Gene Provides a Mechanism for the Evolution of New Host Specificities. *J. Bacteriol.* **1995,** *177,* 4963–4968.

Yang, Z.; Sun, X.; Wang, S.; Zhang, Q. Genetic and Physical Mapping of a New Gene for Bacterial Blight Resistance in Rice. *Theor. Appl. Genet.* **2003,** *106,* 1467–1472.

Yoshimura, A.; Mew, T. W.; Khush, G. S.; Omura, T. Genetics of Bacterial Blight Resistance in a Breeding Line of Rice. *Phytopathology* **1984,** *74,* 773–777.

Yoshimura, A.; Mew, T.; Khush, G.; Omura, T. Inheritance of Resistance to Bacterial Blight in Rice Cultivar Cas 209. *Phytopathology* **1983,** *73,* 1409–1412.

Yoshimura, S.; Yamanouchi, U.; Katayose, Y.; Toki, S.; Wang, Z. X.; Kono, I.; Kurata, N.; Yano, M.; Iwata, N.; Sasaki, T. Expression of *Xa1,* a Bacterial Blight-resistance Gene in Rice, is Induced by Bacterial Inoculation. *Proc. Nat. Acad. Sci. U.S.A.* **1998,** *95,* 1663–1668.

Yoshimura, S.; Yoshimura, A.; Iwata, N.; McCouch, S. R.; Abenes, M. L.; Baraoidan, M. R.; Mew, T. W.; Neson, R. J. Tagging and Combining Bacterial Blight Resistance Genes in Rice Using RAPD and RFLP Markers. *Mol. Breed.* **1995,** *1,* 375–378.

Yuan, B. S.; Pu, T. M.; Hua, L. X. Genetic Mapping of a Bacterial Blight Resistance Gene *Xa14* in Rice. *Acta Agron. Sin.* **2010,** *36,* 422–427.

Zhai, W.; Chen, C.; Zhu, X.; Chen, X.; Zhang, D.; Li, X.; Zhu, L. Analysis of T-DNA-*Xa21* Loci and Bacterial Blight Resistance Effects of the Transgene *Xa21* in Transgenic Rice. *Theor. Appl. Genet.* **2004,** *109,* 534–542.

Zhang, F.; Maeder, M. L.; Unger-Wallace, E.; Hoshaw, J. P.; Reyon, D.; Christian, M.; Li, X.; Pierick, C. J.; Dobbs, D.; Peterson, T.; Joung, J. K.; Voytas, D. F. High Frequency Targeted Mutagenesis in *Arabidopsis thaliana* Using Zinc Finger Nucleases. *Proc. Nat. Acad. Sci. U.S.A.* **2010,** *107,* 12028–12033.

Zhang, F.; Zhuo, D. L.; Zhang, F.; Huang, L. Y.; Wang, W. S.; Xu, J. L.; Cruz, C. V.; Li, Z. K.; Zhou, Y. L. *Xa39,* a Novel Dominant Gene Conferring Broad-spectrum Resistance to *Xanthomonas oryzae* pv. *oryzae* in Rice. *Plant Pathol.* **2015,** *64,* 568–575.

Zhang, H.; Wang, S. Rice versus *Xanthomonas oryzae* pv. *oryzae*: A Unique Pathosystem. *Curr. Opin. Plant Biol.* **2013,** *16,* 188–195.

Zhang, Q. Genetics of Quality Resistance and Identification of Major Resistance Genes to Rice Bacterial Blight. In: Genetics and Improvement of Resistance to Bacterial Blight in Rice; Zhang, Q., Ed.; Science Press: Beijing, 2007; pp 130–177.

Zhang, X.; Wang, C.; Zheng, C.; Che, J.; Li, Y.; Zhao, K. *HrcQ* is Necessary for *Xanthomonas oryzae* pv. *oryzae* HR-induction in Non-host Tobacco and Pathogenicity in Host Rice. *Crop J.* **2013,** *1,* 143–150.

Zheng, C. K.; Wang, C. L.; Yu, Y. J.; Zhao, K. J. Identification and Molecular Mapping of *Xa32(t),* A Novel Resistance Gene for Bacterial Blight (*Xanthomonas oryzae* pv. *oryzae*) in rice. *Acta Agron. Sin.* **2009,** *35,* 1173–1180.

Zhu, W.; Yang, B.; Chittoor, J. M.; Johnson, L. B.; White, F. F. *AvrXa10* Contains an Acidic Transcriptional Activation Domain in the Functionally Conserved C-terminus. *Mol. Plant Microbes Int.* **1998,** *11,* 824–832.

CHAPTER 3

MOLECULAR APPROACHES FOR CONTROLLING BLAST DISEASE IN RICE

DEEPTI SRIVASTAVA[*,†], PRAMILA PANDEY, N. A. KHAN, and K. N. SINGH

Department of Plant Molecular Biology and Genetic Engineering, N. D. University of Agriculture and Technology, Kumarganj, Faizabad 224229, Uttar Pradesh, India

[*]*Corresponding author. E-mail: deeptifzd@gmail.com*

[†]*Present address: CSIR-National Botanical Research Institute, 226 001 Lucknow, Uttar Pradesh, India.*

CONTENTS

ABSTRACT

Blast disease caused by the fungus pathogen *Magnaporthe oryzae* is one of the most severe diseases of rice. Many control procedures are used for the management of rice blast disease, but due to the instability of blast fungus, rice production remains threatened by blast disease. This chapter outlines the application of traditional and molecular approaches for the control of blast disease. Traditionally, chemical control method is most effective; however, the use of chemicals is not generally desired due to the serious environmental threat it poses in rice. Although biocontrol agents for blast have been successfully deployed to combat the disease in the laboratory, greenhouse, and fields, the feasibility of such strategies on a commercial scale still remains to be tested in the natural fields condition. For effective management of blast disease, breeding program should be focused on utilizing the broad spectrum of resistance genes and pyramiding of genes and quantitative trait loci. The availability of rice and *M. oryzae* genome sequence data is facilitating blast resistance management program to new paradigms which includes isolation and characterization of R and Avr genes. With the identification, isolation and characterization of blast resistance genes in rice, it is now possible to dissect the actual allelic variants of these genes within an array of rice cultivars via allele mining. In present chapter, role of molecular breeding, transgenics, and few new methods like miRNA in controlling rice blast disease is also discussed. All these update information will be helpful guidance for rice breeders to develop durable blast resistant rice variety through advanced molecular techniques.

3.1   INTRODUCTION

Rice (*Oryza sativa*) is one of the major food crops that constitute the staple diet all over the world. It is cultivated everywhere in the world except Antarctica and has tremendous economic importance. More than 23% of the calories consumed by the world population comes from rice. Of the total area under rice cultivation, 92% of the rice is grown in Asia, which is the home to more than half of the world population. Nowadays, yield of rice reduces significantly due several phytopathogens. Among several phytopathogens, rice blast is the most destructive disease broadly prevalent in rice fields, leading to significant grain yield and quality reduction. Losses caused by the blast disease are dependent on the growth stage of the plant at which infection occurs, level of resistance, and existing environmental conditions. It occurs more frequently in rain-fed areas in wet season due to favorable environmental conditions for disease development. In India, blast was first recorded in 1913 and the first devastating epidemic was reported in 1919 in the Tanjore delta of erstwhile Madras state. A 4% reduction in yield due to blast was estimated for the first time in India. During 1960–1961, the total loss due to blast was 265,000 t. Thirty percent of rice yields are lost due to blast disease caused by fungal pathogen *Magnaporthe oryzae* (Spence et al., 2014), so rice production will be required to increase by more than 30% to meet the staple food requirements by 2030 due to rapidly increasing world population, limitations to increase cultivated land and nonavailability of water for irrigation. Reducing the loss due to blast can prove to be a critical component toward mitigating the world food security. Although many efforts are given on the study of genetics of rice, it continues to be the most destructive disease of rice. Therefore, strategies for the reduction of yield losses in an environmentally sustainable and economical manner need to be implemented urgently. The present chapter describes blast pathogen and its management with various traditional and molecular approaches.

3.2   THE BLAST PATHOGEN

The fungus that causes rice blast is called anamorph form *M. oryzae* (formerly, *Magnaporthe grisea*). It is a haploid filamentous *Ascomycete* with a relatively small genome of ~40 Mb divided into seven chromosomes (Dean et al., 2005). It is an ascomycete because it produces sexual spores (ascospores) in structures called asci and is classified in the family Magnaporthaceae. The asci are found within specialized structures called perithecia. Pyriform macroconidia are produced on conidiophores which protrude from lesions on plants. The name *Pyricularia* refers to the pyriform shape of the conidia. The conidiogenous cells of *Pyricularia* are polyblastic; integrated on the conidiophores; and are sympodial, cylindrical, geniculate, and denticulate. These germinate and develop an appressorium at the tip of the germ tube, which attaches to the surface of plant tissues; an infection-peg from the appersorium penetrates into plant tissues. The wall of conidiophores and appersorium are pigmented by melanin.

3.3   HOST RANGE OF THE PATHOGEN

M. oryzae is capable of infecting many grass species but individual isolate exhibits a limited host range, infecting one, or at most a few, grass species

(Asuyama, 1965; Kato, 1978). Although the host range of the fungus is restricted (Kato and Yamaguchi, 1982), occasional reports of cross infection of rice by isolates from weed hosts have led to speculation that the pathogen population on weed host could be a source of inoculum for the rice blast. Although rice (*Oryza sativa*) is its main host, *M. oryzae* can survive on the following plants: *Agropyron repens*, *Agrostis palustris*, *A. tenuis*, *Alopecurus pratensis*, *Andropogon* sp., *Anthoxanthum odoratum*, *Arundo donax*, *Avena byzantina*, *A. sterilis*, *A. sativa*, *Brachiaria mutica*, *Bromus catharticus*, *B. inermis*, *B. sitchensis*, *Canna indica*, *Chikushichloa aquatica*, *Costus speciosus*, *Curcuma aromatica*, *Cynodon dactylon*, *Cyperus rotundus*, *C. compressus*, *Dactylis glomerata*, *Digitaria sanguinalis*, *Echinochloa crus-galli*, *Eleusine indica*, *Eragrostis* sp., *Eremochloa ophiuroides*, *Eriochloa villosa*, *Festuca altaica*, *F. arundinacea*, *F. elatior*, *F. rubra*, *Fluminea* sp., *Glyceria leptolepis*, *Hierochloe odorata*, *Holcus lanatus*, *Hordeum vulgare*, *Hystrix patula*, *Leersia hexandra*, *L. japonica*, *L. oryzoides*, *Lolium italicum*, *L. multiflorum*, *L. perenne*, *Muhlenbergia* sp., *Musa sapientum*, *Oplismenus undulatifolius*, *Panicum miliaceum*, *P. ramosum*, *P. repens*, *Pennisetum typhoides*, *Phalaris arundinacea*, *P. canariensis*, *Phleum pratense*, *Poa annua*, *P. trivialis*, *Saccharum officinarum*, *Secale cereale*, *Setaria italica*, *S. viridis*, *Sorghum vulgare*, *Stenotaphrum secundatum*, *Triticum aestivum*, *Zea mays*, *Zingiber mioga*, *Z. officinale*, and *Zizania latifolia*.

3.3.1 SYMPTOMS OF DISEASE

The fungus produces spots or lesions on leaves, nodes, panicles, and collar of the flag leaves. On the leaves, lesions are typically spindle-shaped, wide in the center, and pointed toward either end. Large lesions usually develop a diamond shape with a grayish center and brown margin. Under favorable conditions, lesions on the leaves of susceptible lines expand rapidly and tend to coalesce, leading to complete necrosis of infected leaves. Leaf lesions range from diamond to elongate shape with tapered, pointed ends (Fig. 3.1a). Under heavy dew, all aerial parts of the plant can be affected; leaf surfaces become speckled with oval lesions, plants are liable to lodging if stems are infected (Fig. 3.1b). If the panicle is infected, then a severe yield loss results (Ou, 1985). The conidia are solitary, dry, acropleurogenous, simple, variously shaped, and hyaline to pale brown (Ellis, 1971, 1976). The lesion morphology commonly observed with the rice infection is the typical eye shaped with grayish center and brown margin. On rice, the lesions which were formed were very long and thin. The aged spots did not show water-soaking symptoms. The highly variable-specific virulence of the fungus and its genetic plasticity make its control and management difficult. Thus, *M. oryzae* is one of the most devastating threats to food security worldwide.

1a   **1b**

FIGURE 3.1   (a) Blast-infected rice plant in field and (b) typical symptom of blast disease on leaves of rice.

3.3.2 DISEASE CYCLE

A disease cycle begins when a blast spore infects and produces a lesion on the rice plant and ends when the fungus sporulates repeatedly for about 20 days and disperses many new airborne spores. Under favorable moisture and temperature conditions (long periods of plant surface wetness, high humidity, little or no wind at night and night temperatures (between 12 and 32°C) the infection cycle can continue. In the canopy of rice plants, newly developed leaves act as receptors for the spores. The maximum number of spores produced was 20,000 on 1 lesion on leaves and 60,000 on 1 spikelet in 1 night. Lesions on leaves become an inoculums source for panicles (Royal Society of Chemistry).

3.4 RICE BLAST MANAGEMENT

Many of the control practices useful in reducing plant diseases are of limited use to control rice blast. Since blast is present in most rice growing areas, and it has such a wide host range, eradication and crop rotation are of little value. Lots of work on developing effective rice blast management strategies has been done over a century. The control measures found effective and utilized in the fields are described below (Fig. 3.2). The following procedures broadly used for the control of blast disease as follows.

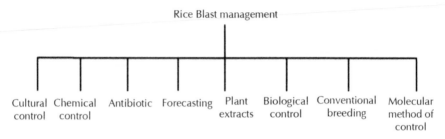

FIGURE 3.2   Various methods for controlling rice blast disease.

3.4.1   CULTURAL CONTROL

When there were no methods of disease management in the past, cultivation practices were the only means to control the diseases. These include nutrient management, water management, time of planting, spacing, etc.

3.4.1.1   NUTRIENT MANAGEMENT

In case of rice blast, two nutrients, namely, nitrogen and silicon have been found to affect the disease occurrence and development significantly. Since a long time back, studies have shown that high N supply always induces heavy incidence of rice blast (Hori, 1898). Delayed or large top dressings are often responsible for severe disease (Ikeda, 1933; Murata et al., 1933). A limit of 15 kg N/ha is recommended for upland rice in Brazil, specifically to reduce vulnerability to blast (Prabhu and Morais, 1986). Plant receiving large amount of N are found to have fewer silicated epidermal cells and thus have lower resistance (Miyake and Ikeda, 1932). The correlation between silica content and disease incidence was also studied on different cultivars of rice, and it was observed that plants with high silica content or large number of silicated epidermal cells had slight damage from blast disease (Onodera, 1917). So it is suggested that resistance of rice to blast can be increased by applying silica slag in the field (Kawashima, 1927). Studies conducted at

University of Florida USA, showed that reduction in the rice blast with the application of silica (calcium silicate slag) was comparable to that of fungicide (Benomyl) and now silicon fertilization has become a routine practice in Florida rice production (Datnoff et al., 1997). Singh and Singh (1980) reported that application of water hyacinth compost to soil reduces the rice blast disease. Positive effects of Si application in controlling blast disease have also been reported in many studies (Prabhu et al., 2003; Seebold, 1988; Seebold et al., 2000). In Malaysia, Ashtiani et al. (2012) investigated the effects of different rates and sources of Si on rice variety MR219 toward controlling the blast disease. Silicon was applied to the soil prior to planting using two different Si sources which were silica gel (0, 60, 120, 180 g/5 kg soil) and liquid sodium silicate (0, 1, 2, 3 mL/L), respectively. Results revealed that there was a significant decline on disease severity and incidence for all rice plants that were treated with Si compared to the nontreated rice plants. The Si depositions were more intensive on the dumbbell-shaped cells in the Si-treated plants compared to the nontreated plants. Highest reduction (75%) of disease severity was observed for plants receiving silica gel application at the rate of 120 g. This result confirmed that Si displays a significant role in controlling rice blast fungus.

3.4.1.2   WATER MANAGEMENT

The availability of water also affects the susceptibility of host plant to *P. oryzae*. Rice grown under upland conditions is more susceptible than rice grown in flooded soil (Kahn and Libby, 1958). Under upland conditions, susceptibility is increased further with increasing drought stress. Hence, flooding the field in upland rice can reduce the severity of blast.

3.4.1.3   TIME OF PLANTING

Planting time also has a noticeable effect on the development of blast within a rice crop. For rice blast control, early planting is recommended. In tropical upland rice, crops sown early during the rainy season generally have a higher probability of escaping blast infection than late-sown crops, which are often blasted severely. In upland areas of Brazil, farmers are advised to sow early to escape inoculums produced on neighboring farms (Prabhu and Morais, 1986).

3.4.2   CHEMICAL CONTROL

Chemicals, mainly fungicides are the most frequently and widely used method of plant disease management worldwide. For rice blast, most aggressive and successful chemical control program in world has been shown by Japan. The copper fungicides were first effectively used in Japan shortly after the turn of the century and continued to be used until the Second World War (Thurston, 1998), but as they are highly phytotoxic, a more attractive alternative was sought. Subsequently, copper fungicides were used in mixture with phenylemercuric acetate (PMA) which was more effective than copper alone in rice blast control and were less toxic to the rice plant. Later, discovery was made by Ogawa (1953) that a mixture of PMA and slaked lime provides much more effective control of rice blast and was less toxic and cheap, hence used extensively. However, these fungicides are toxic to mammals and are severe environmental pollutants, so banned by Japanese Government in mid-1968 (Ou, 1985). Then the organophosphorus fungicides were introduced to control blast in Japan. Additional studies revealed that resistance to one organophosphorus fungicide did not necessarily confer resistance to other specific fungicides. So it was suggested that rotating the use of fungicides or mixing them, rather than continuously relying on single compound, greatly reduces the risk of developing highly resistant populations (Uesugi, 1978). At the same time, development and implication of new systemic fungicides was also on progress. The phosphorothiolate fungicides, including iprobenfos and edifenphos, were introduced in Japan as rice blast fungicides in 1963. Iprobenfos and isoprothiolane have systemic action and are used mainly as granules for application on the surface of paddy water (soil application). Copper fungicides were found effective for rice blast control in India as well, but it was seen that high-yielding varieties (HYVs) were copper-shy; hence, the emphasis was shifted to another group of fungicides, namely, dithiocarbamate and edifenphose, but they were having shorter residual activity. So in 1974–1975, the first generation systemic fungicides benomyl, carbendazim, and others were evaluated and found effective. Following these, many systemic fungicides with different mode of action, like antimitotic compounds, melanin inhibitors, ergosterol biosynthesis inhibitor, and other organic compounds were discovered for rice blast control (Siddiq, 1996). In a chemical scheduling trial, Bavistin 1 g/L spray at tillering + Hinosan 1 g/L at heading and after flowering provided the best yield increase. Tricyclazole and pyroquilon fungicides as seed dressers have been found effective to provide protection to seed up to 8 weeks after sowing. Some of the recently developed chemicals for blast control are

    i.   Carpropamid (1999, melanin biosynthesis inhibitor),

    ii.   Fenoxanil (2002, melanin biosynthesis inhibitor), and

    iii. Tiadinil (2004, plant activator).

In the most recent field, evaluation of commercial fungicidal formulations rabicide (tetrachlorophthalide), nativo (tebuconazole + trifloxystobin), and score (difenoconazole) are found most effective (Usman et al., 2009). The site-specific fungicides are recommended to be used in mixture or in rotation due to the development of resistance in the pathogen. The non-fungicidal agents are supposedly specific to the target organism and are less likely to lead to resistance problems (Yamaguchi, 2004).

3.4.3 ANTIBIOTICS

The first antibiotic which was found to inhibit the growth of rice blast fungus on rice leaves was "cephalothecin," produced by a species of *Cephalothecium* (Yoshii, 1949). Following this, "antiblastin" (Suzuki, 1954), "antimycin-A" (Harada, 1955), "blastmycin" (Watanabe et al., 1957) and "blasticidin-A" (Fukunaga et al., 1968) were found and tested but due to their chemical instability and toxicity to fish, none of them was put to practical use. Then in 1955, a new systemic antibiotic, Blasticidin S was developed by Fukunaga which is produced by *Streptomyces griseochromogenes*. It was found to be superior for blast control and effective mainly in postinfectional control. But it was an inferior protectant and highly toxic to plants and mammals (Ou, 1985). Shortly after the discovery of blasticidin S, a new antibiotic Kasugamycin, produced by *Streptomyces kasugaensis*, was discovered. It gave excellent control of rice blast and had very less toxicity to mammals and rice plant (Okamoto, 1972). In around 1970, in the areas where the antibiotics have been used extensively and exclusively for blast control, population of *P. oryzae* began to show resistance to antibiotic compounds (Uesugi, 1978). However, after halting the use of antibiotics in the areas with resistant populations of *P. oryzae*, the population of resistant types reduced to nearly zero and later the use of antibiotics in some areas successfully resumed (Uesugi, 1978). Katagiri and Uesugi (1978) reported the frequency of emergence of resistant mutants of *P. oryzae* against different chemicals. It was highest in kasugamycin, followed by edifenphose and isoprothiolane and was lowest in benomyl.

3.4.4   DISEASE FORECASTING

For the economic and most effective use of fungicides, it is best to follow the forecast. In the early works, many studies on methods of forecasting have been made, based upon the information on fungus, host plant, and environment (Ou et al., 1971). Using 13-year data, Padmanabhan (1963) concluded that whenever the minimum temperature of 24°C or below was associated with RH of 90% or above, the conditions were favorable to blast infection. Attempts to correlate spore content and blast incidence was also done in India. Later on, EI-Refaei (1971) found that number of lesions was more closely correlated with the dew point than the number of air born spores. Number of computer simulation-based forecast models are available such as

   i.   Leafblast (Choi et al., 1988),
   ii. Epiblast (Kim and Kim, 1993), and
   iii. Epibla (Manibhushanrao and Krishnan, 1991).

The work on forecasting, through machine learning technique based on support vector machines (SVMs) method have been found better than existing machine learning technique and conventional multiple regression approaches in forecasting plant diseases. An online SVM-based web-server for rice blast prediction was the first of its kind worldwide and helping the plant science community and farmers in their decision-making process (Kaundal et al., 2006).

3.4.5   PLANT EXTRACTS/BOTANICALS

In the recent years, some botanicals were evaluated for their antifungal activity against *P. oryzae*, and few of them were found very effective. The leaf extract of *Atalantia monophylla* was found to control disease up to 82.22% followed by *Plumbago rosea*, 70.57%. The biochemical studies showed that *A. monophylla* have higher content of phenols (4.8 mg/g) and flavinoids (24.5 mg/g) compared to others (Parimelazhagan, 2001). In India, in the same year, that is in 2001, experiments conducted at UAS Dharwad, to find out best bioagents, fungicides, and neem-based formulations showed the neem-based formulations such as Nimbicidine and Neem gold were most effective among them tested ones. At CRRI Cuttak (Central Rice Research Institute, India), *Ocimum sanctum* (Tulasi), and *Aegle marmelos* (Bael) were found very effective in blast control. When tried on field the plots treated with leaf extracts of bael and tulsi had only 2% disease intensity as compared to Henosan treated (25% DI) and control (85% DI) (CRRI 2007–2008). Hubert et al. (2015) studied the efficacy of selected plant extracts against *Pyricularia oryzae*. Studies were conducted to determine the effect of aqueous extracts of *Aloe vera*, *Allium sativum*, *Annona muricata*, *Azadirachta indica*, *Bidens pilosa*, *Camellia sinensis*, *Chrysanthemum coccineum*, processed *Coffee arabica*, *Datura stramonium*, *Nicotiana tabacum*, and *Zingiber officinalis* for control of rice blast disease (*P. oryzae*) in vitro and in vivo. The results indicate that processed *C. arabica* at 10% and 25% (v/v) had the highest (81.12%) and (89.40%) inhibitory effect, respectively, against *P. oryzae* aqueous extract from *N. tabacum* at 10% concentration ranked third (80.35%) in inhibiting *P. oryzae*. These were followed by extracts from 25% *A. vera* (79.45%) and 25% *C. coccineum* flower (78.83%). The results also indicate that, extracts from *A. indica*, *A. vera*, *A. sativum*, *C. arabica*, *D. stramonium*, *C. sinensis*, *Z. officinalis*, and *N. tabacum* did not have any phytotoxic effect on seed germination, shoot height, root length, dry weight, seedling growth, and seedling vigor index. These plant extracts can thus be used for rice seed treatment to manage rice blast disease.

3.4.6   BIOLOGICAL CONTROL

Chemical pesticides offer marginal protection from the disease, yet pose environmental risks and may put nonpathogenic organisms, including humans, at risk (Aktar et al., 2009). Thus, the control strategies currently employed are limited in effectiveness and may lead to further problems. An alternative means of crop protection would be through the use of biological control agents. The search for the biological agents which can control the rice blast started in end of 1980s. The first report of a biological agent found effective in control of *P. oryzae* was of *Chaetomium cochliodes*. When the rice seeds were coated with spore suspension of *C. cochlioides* the early infection by blast was controlled and seedlings were healthy and taller than the control. It has been found that rice blast incidence can be reduced by mass vaccination method with avirulent isolates of *P. oryzae*. In India, the studies on bacterial agents for the control of rice blast were conducted at "Center for Advanced Studies in Botany, University of Madras," and it was found that among the 400 bacterial isolates collected from rice fields of IRRI, 3 strains of *Pseudomonas fluorescens*, 5 of *Bacillus* spp., and one of *Enterobacter* spp., were inhibitory under in-vitro conditions. Microbes have also been engineered to control rice blast. An epiphytic bacterium *Erwinia ananas* transformed with the chitinolytic enzyme gene (*Chi A*) from an antagonistic bacterium *Serratia marcescens* strain B2, a tomato epiphytic bacterium, was found inhibitory against *P. oryzae* (Someya et al., 2004). Recent studies on biocontrol of rice blast showed that *Bacillus subtillis* strain B-332 (Mu et al., 2007), 1Pe2, 2R37, and 1Re14 (Yang et al., 2008) and *Streptomyces sindenius* isolate 263 have good antagonistic activity against *P. oryzae*. Exploitation of host resistance is the most cost-effective. Naturally occurring rhizospheric microbe also suppress rice blast infection (Spence et al., 2014). Naturally occurring root-associated rhizospheric bacteria were isolated from California field grown rice plants (M-104), 11 of which were taxonomically identified by16S rRNA gene sequencing and fatty acid methyl ester analysis. Out of 11 bacteria isolated from rice soil, pseudomonad EA105 most effectively inhibited the growth and appressoria formation of *M. oryzae* through a mechanism that is independent of cyanide production.

In developing countries, poor farmers cannot afford application of chemicals and pesticides to control blast disease. Chemical control of plant pathogens is most effective and but the use of chemicals is not generally preferred due to the serious environmental threat it poses. Besides, their continuous use leads to the reappearance of resistant races of the pathogen under selection pressure. Although biocontrol agents for blast have been successfully deployed to combat the disease in the laboratory, greenhouse, and field tests, the feasibility of such strategies on a commercial scale still remains to be tested. Use of resistant cultivars is the best alternative to overcome yield losses. The variability of the pathogen and the history of resistance breakdown have led to the development of a number of different plant breeding approaches to achieve durable blast resistance. In the present chapter, following approaches are used in the management of blast resistance in rice.

3.5 CONVENTIONAL BREEDING APPROACH FOR BLAST RESISTANCE

Conventional approaches are important for producing novel genetic variants, conserving wild germplasm, sexual hybridization between contrasting parental lines and mutation. Over the last 30 years, conventional breeding has given IRRI's elite cultivars with vast range of genes for resistance to diseases (Bonman et al., 1992; Khush et al., 1989). In conventional breeding program following methods are used:

(1) Pedigree method
(2) Backcrossing
(3) Recurrent selection
(4) Mutation breeding

3.5.1 PEDIGREE METHOD

Pedigree method is suitable when resistance is governed by major genes. Pedigree is the most widely used in rice improvement and highly suitable to develop rice with resistance to insects and diseases. With the help of pedigree method, it is possible to combine genes for resistance to six or seven major diseases and insects in a short period (Khush, 1978). The major drawback of pedigree breeding is that requires much time to evaluate lines periodically throughout the growing season and to keep records on which selection is based at maturity. Besides this, it also requires pedigree method requires the greatest familiarity with the material and with the relative effects of geno-type and environment on character expression as well as this breeding is not the most effective approach when traits are governed by polygenes. For example, resistance to stem borers and sheath blight appears to be under polygenic control. For these two traits, di-allele selective mating system is suitable (Khush 1978; Jensen, 1970).

3.5.2 BACKCROSSING

Backcrossing is a widely most common technique in rice breeding for introgression or substitution of a target gene from donor parent to recip-ient. It provides a defined way to improve varieties (Allard, 1960, 1999). The main purpose of backcrossing is to decline the donor genome content into the progenies (Xi et al., 2008). Backcross breeding has been adopted in the South and Southeast Asia (Joseph et al., 2004; Toojinda et al., 2005) as breeding strategy to improve elite varieties such as KDML105, Basmati and Manawthukha for their resistances to blast (Sreewongchai et al., 2010).

3.5.3 RECURRENT SELECTION

Recurrent selection is another traditional breeding method used in rice for male sterility (Fujimaki, 1979). It allows defined and shorter breeding cycles, more precise follow-up of genetic gains, and provides opportunity to develop wide-range genetic diversity breeding lines. Using this method, upland cultivar CG-91 was developed with resistance to rice blast (Guimaraes & Correa-Victoria, 2000). In order to assess the efficacy of this method in rice, an evaluation study observed 6.65% genetic gain considering two cycles of recurrent selection in the irrigated rice population CNA-IRAT 4 (Rangel et al., 2005). On the other hand, 6.2% gains was observed after selecting for rice blast resistance comparing cycles 1 and 2 of the upland rice population CNA-7 (de Badan et al., 2005). In almost all self-pollinated crops including rice, breeders chose to use pedigree selection which is alternative to recurrent selection.

3.5.4   MUTATION BREEDING

Mutation breeding in rice is used to match conventional breeding, since this technique is very effective for improving major traits, such as agronomic traits, resistance to pests and diseases, and grain physical parameters and eating quality. In classical mutation breeding, induced mutations are used for developing a new variety, whereby it is difficult to trace the mutated genes in subsequent breeding. It is now possible to tag mutated genes, pyramid them into a single elite breeding line, and follow up them in subsequent breeding programs (Shu, 2009). The advantages of mutation breeding include creation of new gene alleles that do not exist in germplasm pools, and the induction of new gene alleles into the new varieties that can be used directly as a commercial variety. The disadvantage of mutation breeding is limited power in generating the dominant alleles that might be desired; it is also less effective than cross breeding for a trait needs for a combination of multiple alleles. Many attempts have been made to improve disease resistance in rice through mutation breeding. Positive results, particularly for resistance to blast (*P. oryzae*) have been reported from several countries. The Department of Agriculture, Thailand, officially released the variety RD6, a radiation induced glutinous mutant of the popular non-glutinous variety Khao Dawk Mali 105 (KDML 105). Selections in the M2 populations were made for glutinous and blast-resistant mutants (Khambanonda, 1978). An attempt was made to induce blast resistance in the HYV Ratna (IR8/TKm 6) through chemomutagenesis with EMS 0.1 and 0.2% concentrations. The mutagen treatment induced great variability; in different generators M2–M5, higher resistance and higher susceptibility were found than the parent (Kaur et al., 1971). Blast resistance mutant R917 was derived from the F_1 progeny radiated by 10 krad 60Co c-ray (Zhang et al., 2003). The Mtu 17 mutants possessing desirable agronomic characters through chemomutagenesis with dES; some of the mutants had blast resistance, whereas the parent Mtu 17 was susceptible (Gangadharan and Mathur, 1976). Through mutation breeding, several mutant lines, such as Mahsuri Mutant SPM 129, SPM 130, and SPM 142 (Azlan et al., 2004) have produced successfully for blast resistance in Malaysia (Mohamad et al., 2006). In China, the mutant rice variety "Zhefu 802" deriving from var. "Simei No. 2," induced by Gamma rays, has a high resistance to rice blast (Ahloowalia et al., 2004; Shu et al., 1997). Today, the technique became part of the tools kit breeders have to enhance specific rice characteristics in well-adapted varieties. Through conventional breeding programs major genes *Pib*, *Pita*, *Pia*, *Pi1*, *Pikh*, *Pi2*, and *Pi4* have been introduced into rice varieties for blast resistance (Kiyosawa, 1982). Identifying key genomic regions associated with blast resistance against a broad spectrum of isolates in backcross introgression lines have been developed through conventional breeding program (Korinsaka et al., 2011).

Some components of breeding strategies suggest prolong durability of resistance which generally can be adopted for stabilization and control of blast disease in rice are discussed below:

3.5.4.1 BACKCROSSING FOR CONCENTRATION OF SLOW-BLASTING (MINOR GENES) COMPONENTS

Breakdown of varietal resistance to rice blast disease attributed to the failure of varieties to capture the entire complement of genetic factors for disease resistance from the respective parent sources in their parentage (Nottegham, 1993). Existence of slow-blasting characters, originally presented as horizontal resistance (van der Plank, 1975), mainly found only in tall, upland rice varieties (Bidaux, 1976; Villareal, 1980) and identified several slow blasting components in several varieties (Villareal, 1980). Identification of slow blasting sergeants in segregating populations is difficult particularly in bulk breeding systems. It might be somewhat easier in a pedigree system of breeding where discrete progeny rows can be evaluated for identification of lines with slow blasting components.

3.5.4.2   COMBINATION OF MAJOR GENES WITH SLOW-BLASTING (MINOR GENES) COMPONENTS

The combination of major genes (vertical resistance) with slow-blasting components (minor genes) is believed to provide increased stability of the resistance mechanism to blast, because the genes for vertical and horizontal resistance in combination increase the effectiveness of each other. Centre International de Agricultural Tropical (CIAT) rice breeding program (CIAT, 1982) attempted adopting this strategy, but found difficult to detect the combinations of the two types of resistance in a given pedigree line, as the lower level of horizontal resistance is masked by the presence of vertical genes. Under such circumstances, it is proposed to select for vertical resistance and hope for the best. Therefore, the practical outcome of using this strategy in a breeding program is not predictable.

3.5.4.3   MIXTURES OF VARIETY

Varietal mixtures are the way of reducing the development of blast races consisting of 80–90% resistant plants and 10–20% susceptible plants of similar varietal background. This strategy is easier to introduce but need to ensure their agronomical uniformity. In Yunan province of Southeast China, highly susceptible glutinous plants were mixed and planted with non-glutinous hybrid indica rice reduced the development of blast in glutinous rice (Leung et al., 2003; Zhu et al., 2005). But measuring panicle blast resistance is difficult because the panicle infection is influenced by weather, and even small differences in maturity period between lines can result inaccurate assessment of their level of resistance. Several researchers mentioned that minor genes that play an important role in maintaining an acceptable level of disease under field conditions. Such genes (minor) are difficult to identify and characterize in the presence of major genes due to epistatic interactions among themselves. Their presence could also affect the accuracy of classification of lines for complete resistance to blast (Wang et al., 1994).

3.5.4.4   MULTIPLE LINES

The durability resistance of multiline varieties depends upon the rate of blast races develop, the number of lines component in a mixture, and the extent of planted area. Development of multiline varieties using blast resistant isogenic lines had been attempted for "Nipponbare" (Higashi et al., 1981; Horisue et al., 1984) "Toyonishiki" (Nakajima, 1994) and "Sasanishiki" (Matsunaga, 1996). Blast control effects by the use of these multiple line varieties have been confirmed (Ise, 1990; Koizumi and Fuji, 1994; Koizumi et al., 1996; Nakajima et al., 1996). New isogenic lines have been bred and elaborated variation analysis in the distribution of the races of the blast pathogen, which is essential for stable utilization (Ashizawa et al., 2001; Tsuji et al., 1999). A cross combination of Koshihikari blast-resistant isogenic lines (BLs) was made by (Ishizaki et al., 2005). The BLs were bred by crossing with Sasani-shiki (*Pia*), Todorokiwase (*Pii*), *Pi4 (Pita-2)*, Niigatawase *(Piz)*, Koshimi-nori (*Pik*), Tsuyuake (*Pik-m*), Toride 1 (*Piz-t*), and BL1 (*Pib*) as the donor parent, respectively, and then repeated backcrossings with "Koshihikari" as the pollen parent were performed. Individuals used for backcrossing after the first filial generation of the first backcross generation (BC_1F_1) were sprayinoculated with race 001 of the blast pathogen for the nursery test. About 30 heterozygotes showing resistance were selected for cultivation in the field. Individuals that resembled "Koshihikari" were selected prior to the backcrossing procedure. Selected individuals were transplanted to 1/5000 a Wagner pots and about 50 seeds were obtained per stump. Backcrossing was performed six times for BL No. 4 and five times for the other lines, following examples of BLs for Toyonishiki, Nipponbare and Sasanishiki, which have been bred so far. Seedlings of B_5F_2 or B_6F_2 after selfed genera-tion were spray-inoculated with race 001 for selecting homozygotes with true resistance as well as for fixation. All the Koshihikari BLs was found to be identical with the original "Koshihikari" in terms of practical agronomic traits and clearly superior in suppressing blast.

3.5.4.5 DEPLOYMENT OF GENE

This strategy involves the use of the distinct type of blast resistance mecha-nism in each variety and use of varieties in a predetermined pattern (temporal or spatial). Based on rice cultivation practices, seasonal and regional prefer-ences for different location specific varieties are used. As a result, distinct varieties could be developed using diverse sources of blast resistance. Even among varieties used for a particular season, variety with different maturity period should consist of distinct sources of blast resistance. This situation will slow down the development of new virulent races, and improve the durability of blast resistance in present varieties. Among many strategies, distinct gene deployment in different maturity groups may help to improve the durability of blast resistance in newly developing rice varieties. Nevertheless, the conventional resistance breeding has apparent weakness, such as long breeding cycle, low selection efficiency, and difficulty in distant crossing, leading to the lag between the development of new resistant cultivars and the emergence of virulent pathotypes of the causal pathogen.

3.6 BIOTECHNOLOGICAL AND MOLECULAR APPROACHES FOR BLAST RESISTANCE

3.6.1 BLAST-RESISTANT GENE

Conventional genetic analysis identified donors with resistance, availability of pure isolates of the blast pathogen and use of advanced molecular analysis techniques, These have been designated as *Pi1, Pi2, Pi3, Pi4, Pi5, Pi6, Pi7, Pi9, Pi10*, and *Pi11* (Causse et al., 1994; Wang et al., 1994), *Pia, Pib, Pik, Pit, Pita, Pita 2, Pi12, Pi17, Pi18, Pi19, Pi20, Pi23, Pi57, Pi62* (Nagato and Yoshimura, 1998), *Pii* and *Pi15* (Pan et al., 2003), *pi21* (Fukuoka and Okuno, 2001), *Pi25* (Yang et al., 2001), *Pi27* (Zhou et al., 2004), *Pi24, Pi25, Pi26, Pi27, Pi28, Pi29, Pi30, Pi31* and *Pi32* (Sallaud et al., 2003), *Pi33* (Berruyer et al., 2003), *Pish* (Fukuta et al., 2004), *Pi35* (Nguyen et al., 2006), *Pi36* (Liu et al., 2005), *Pi37* (Chen et al., 2006), *Pi38* (Gowda et al., 2006), *Pi39* (Lin et al., 2007), and *Pi40* (Jeung et al., 2007). Out of 100, 5 blast resistance gene, that is, *Pi-kh* or *Pi54* (Sharma et al., 2005), *Pitp(t)* (Barman et al., 2004), *Pi38* (Gowda et al., 2006), *Pi42(t)* (Kumar et al., 2010) and, *Pi-Wn(t)* (Rathore et al., 2012) have been mapped in India. Nine blast resistance genes—*Pi-b* (Wu et al., 2004), *Pi-ta* (Bryan et al., 2000) and *Pi-kh* (Sharma et al., 2005), *Pi37* (Lin et al., 2007), *Piz-5* and *Piz-t* (Zhou et al., 2005), *Pi9* (Qu et al., 2006), *Pid2* (Chen et al., 2006), and *Pi36* (Liu et al., 2007) have been cloned. The wide genetic variation available in blast fungus may be the main reason why many resistance genes in rice have evolved. Majority of the quantitative trait loci (QTLs) are associated with qualitative genes. Quantitative resistance is generally considered as partial resistance in a particular cultivar (Parlevliet, 1979) which is controlled by multiple loci, known as QTLs. Approximately, 350 leaf blast resistances QTLs have been mapped. These QTLs were identified in 15 different populations, most of which are derived from indica japonica crosses (Chen et al., 2003; Lopez-Gerena, 2006; Sallaud et al., 2003; Sato et al., 2006; Wu et al., 2004; Talukder et al., 2005). Partial resistance is characterized by compatibility between the pathogen and the plant with reduced development of disease compared to plants with no partial resistance (Bonman and Ahn, 1990; Parlevliet, 1998). Four partial-resistance genes have been identified and described as specific, *Pif* (Yunoki et al., 1970), *Pi21* (Fukuoka and Okuno, 1997), *Pb1* (Fujii et al., 1995), and *Pi34* (Zenbayashi-Sawata et al., 2005).

3.6.2   HOST RESISTANCE

Exploitation of host resistance is the most cost-effective and reliable method of disease management. In some instances, resistant varieties have provided effective and durable disease control. But in the case of rice blast, success is short-lived or not easily achieved. It is because of the presence of lineages (that may consist of different physiologic races) overcoming host resistance (IRRI, 2010). Early studies on host resistance were more concentrated on nature of resistance. Miyake and Ikeda (1932) reported that the cultivar Bozu, resistant to rice blast contains a large amount of silicon than the suscep-tible cultivar. Further studies showed that degree of resistance increases in proportion to the amount of silica applied and also to the amount of silicon accumulated in the plant. Ito and Sakamoto (1939) found that resistance to mechanical puncture of the leaf epidermis was positively related with resis-tance to blast. They found that puncture resistance was reduced by applica-tion of nitrogen fertilizer and by low soil moisture, but was increased as the plant become older. Hori et al. (1960) reported that distribution of starch in the leaf sheath is related to resistance that is, longer accumulation indi-cates more resistance. It is known that resistance to penetration of fungus is obviously less important than resistance to its spread within the host plant after penetration. A hypersensitive reaction is common in resistant cultivars. Kawamura and Ono (1948) were able to isolate *P. oryzae* from hypersensi-tive lesion 2 days after inoculation but not after 4 days. Low toxins pyricu-larin and α-picolinic acid produced by *P. oryzae* are toxic to rice plant and cause stunting of seedlings, leaf spotting and other injurious effects. Earlier, Tamari and Kaji (1955) found that when combined with chlorogenic acid or ferulic acid, both present in the rice plant, they (pyricularin and α-picolinic acid) become nontoxic to rice plant. So they believed that ability of rice plant to biosynthesize chlorogenic acid is related to resistance. All these findings not only generated the knowledge of host pathogen interaction but also contributed in searching resistance sources and setting the strategies of breeding for blast resistance. First, most important step in resistance breeding is the evaluation of germplasm for disease resistance sources. In 1969, Link and Ou (1969) proposed a system of standardization of race numbers of

P. oryzae. IRRI also stepped forward and planted uniform blast nurseries in 50 testing stations in 22 different countries for pathogen race evaluation and till 1975 more than 260 physiologic races of *P. oryzae* were reported from the different parts of world. Resistance to *P. oryzae* in rice is usually dominant and controlled by one or few pairs of genes (Thurston, 1998). At IRRI, in 1979, almost 100,000 lines and accessions were tested and not a single one was found to be completely resistant to all races. Host plant resistance can be broadly categorized as follows.

3.6.2.1 *VERTICAL RESISTANCE*

Vertical Resistance (also known as Complete resistance, specific resistance or true resistance), in which the pathogen fails to produce sporulating lesions, can be manipulated easily by breeders. But it also has been known to break down, sometimes with serious economic consequences. In Korea, the resistance of the Tongil varieties was effective for 5 years before a virulent race appeared in 1976 (Lee et al., 1976). The variety Reiho had complete resistance to Japanese races upon its release in Japan in 1969. Its area of cultivation increased until 1973, when it was damaged severely by blast (Matsumoto, 1974). In Japan, the longevity of complete resistance seems to be about 3 years. Similarly, with var. Reiho was later released in Egypt as a blast resistant variety in 1984, it occupied about 25% of the rice crop area within a year but resistance was overcome in the first year, resulting in a blast epidemic of some consequence (Bonman and Rush, 1985). In Colombia, a series of resistant varieties was released from 1969 to 1986, but their resistance lasted only a year or two before being overcome by previously unidentified virulent races (Ahn and Mukelar, 1986).

3.6.2.2 *HORIZONTAL RESISTANCE*

Assuming the gene-for-gene relationship (Flor, 1956) and given the variability of the pathogen, it is not difficult to understand why the effectiveness of complete blast resistance is short-lived. It has been observed that when complete resistance was overcome by the pathogen, usually some level of residual resistance remains. This residual resistance has been referred to variously as, horizontal resistance (HR), general resistance, field resistance, slow-blasting, and partial resistance, among others. The general, HR to *P. oryzae* was reported in 1971 (Ou et al., 1971). Efforts to identify, characterize, and exploit this type of resistance which is effective against all races of pathogen were undertaken by IRRI. But the 1978 epidemic of *P. oryzae* in Korea altered the attitude of IRRI breeders and pathologists toward HR. The improved indica–japonica hybrid rice cultivars grown in Korea were possessing vertical (monogenic) as well as horizontal (polygenic) resistance suddenly became susceptible to *P. oryzae* in 1978 (Crill et al., 1982). Korean pathologists had defined the HR as varieties with disease ratings 4–5. Since HR studies in Korea (Crill et al., 1982) may have been defined qualitatively and not quantitatively, the Korean experience should not be used as a reason to discontinue the search for rice cultivars with HR (Thurston, 1998). There are many examples of partial resistance that appear to be effective and durable under field conditions. The varieties IR36 and IR50 are susceptible to the same races of *P. oryzae* (Bonman et al., 1986), but when inoculated with the same isolates, IR36 produces fewer and smaller lesions than does IR50 (Yeh and Bonman, 1986). Frequently, the main strategy of breeder and pathologist, given the choice, is to save only the most resistant-appearing lines in a screening nursery, and usually these are lines with complete resistance to the races present in the nursery. Because complete resistance masks partial resistance, there is no way to evaluate such lines without either challenging them with isolates of *P. oryzae* that are virulent (i.e., the complete resistance is overcome), or by progeny-testing of a cross with a highly susceptible variety. Using the IRRI 1975 blast rating scale, lines with ratings of 3–6 probably represent those with usable levels of partial resistance that are not masked by complete resistance. So by introducing such lines into a breeding program and avoiding lines with little partial resistance, a strong pool of genes that contribute to race-nonspecific partial resistance could be gradually accumulated in the breeding population. Effective resistance can be achieved by combining into the same cultivar, different race-specific genes and genes conferring quantitative resistance. Another method is by deploying resistance genes in mixed plant populations. Recent studies indicated that use of cultivar mixture is an effective tool in blast management. IRRI scientists introduced the practice of inter-planting glutinous rice varieties with blast-resistant hybrid varieties in Yunnan province, China. Blast caused great yield loss on traditional glutinous rice varieties in China and farmers were spraying fungicides for up to seven times. Interplanting has prevented the fungus from continuous build-up of inoculum that had previously occurred in the monoculture fields of the glutinous varieties.

3.6.3 MARKER-ASSISTED SELECTION

Conventional breeding for disease resistance is laborious, time consuming, and highly dependent on environmental conditions in comparison to molecular breeding particularly Marker assisted selection (MAS), which is simpler, highly efficient, and precise (Koizumi, 2007). In addition, conventional breeding generally depend upon the phenotype of artificial identification, and the performance of field resistance is too long to catch up with the frequent emergences of new virulent races of the pathogen and the release of improved varieties cannot be guaranteed (Werner et al., 2005; Zhang et al., 2006). MAS is extremely powerful in blast resistance breeding because resistance phenotypes are often simple or encoded by single or few genes (Young, 1994). MAS have the advantage for the blast control by governing definite interaction of a particular resistance (R) gene with a particular avirulence gene in the pathogen (Silue et al., 1992). Molecular markers have greater opportunity to improve the efficiency of conventional breeding by carrying out selection not directly on the trait of interest but also on linked molecular markers of that particular trait. Some SSR markers (RM168, RM8225, RM1233, RM6836, RM5961, and RM413) have been found by Ashkani et al. (2011) that could be used in MAS programs (Table 3.1). Availability of molecular markers along with MAS strategies are essential to develop durable blast-resistant variety against different races of *M. oryzae* (Ashkani et al., 2012). The high cost of MAS will be a major obstacle for it adoption in the developing countries in the nearest future. The initial cost of using markers is more expensive compared to conventional breeding; though time savings could lead to expedite variety release which could lead to greater profit. The low reliability of markers to determine phenotype is another important factor obstructing the successful application of markers for line development.

3.6.4 MARKER-ASSISTED BACKCROSS BREEDING

Marker-assisted backcross breeding (MABC) as another technique recently has been given attention in rice breeding for the introgression of blast-resistance genes (one or a few genes) into the susceptible or in an adapted or elite varieties. MABC is the process of using markers to select for target loci, minimize the length of the donor segment containing a target locus, and accelerate the recovery of the recurrent parent (RP) genome during backcrossing (Charcosset, 1994; Hasan et al., 2015; Hospital, 2001). The

TABLE 3.1 List of Available Blast-resistance Genes and Tightly Linked Markers in Rice.

Chromosome	Gene	Tightly linked marker		Map position (cM)	Donor rice	Resistance type
		Marker type	Marker name			
1	Pit	SNP	t311, t256, t8042	12.2	Tjahaja	Complete
	Pi27(t)	Microsatellite	RM151.RN259	28.4–38.8	Q14	Complete
	Pi24(t)	—	—	64.4	Azucena	—
	Pitp(t)	Microsatellite	RM246	114.1	Tetep	—
	Pi35(t)	Microsatellite	RM1216, RM 1003	132.0–136.6	Hokkai 188	Partial
	Pi37	Microsatellite	RM302, RM212, FPSM1, FPSM2, FPSM4	136.1	St. No. 1	Japonica
		STS	S15628, FSTS1, FSTS2, FSTS3, FSTS4			
2	Pi(t)	—	—	—	—	—
	Pid1(t)	Microsatellite	RM262	87.5–89.9	Digu	—
	Pig(t)	Microsatellite	RM166, RM208	142.0–154.1	Guangchangzhan	—
	Pitq5	—	—	150.5–157.9	Teqing	Complete
	Piy1(t)	Microsatellite	RM3248, RM20	153.2–154.1	Yanxian No. 1	—
	Piy2(t)	Microsatellite	RM3248, RM20	153.2–154.1	Yanxian No. 1	—
	Pib	SNP	b213, b28, b2, b3989, Pibdom	154.1	Tohoku IL9	Complete
		Micosatellite	RM138, RM166, RM208, RM266, RM138, RM166, RM 208, RM 266			
	Pi25(t)	—	—	157.9	IR64	—
	Pi14(t)	—	—	—	Maowang	Complete
	Pi16(t)	—	—	—	AUS373	Complete
4	Pi21	STS	P702D03–79	58.6	Owarihatamochi	Partial
	Pikur1	—	—	86	Kuroka	—

TABLE 3.1 *(Continued)*

Chromo- some	Gene	Marker type	Tightly linked marker Marker name	Map position (cM)	Donor rice	Resistance type
	Pi39(t)	Microsatellite	RM3843, RM5473	107.4–108.2	Chubu 111 (Haonaihuan)	–
	Pi(t)	–	–	–	–	–
5	Pi26(t)	–	–	22.5–24.7	Azucena	–
	Pi23(t)	–	–	59.3–99.5	Sweon 3655	–
	Pi10	InDel	OPF62700	88.5–102.8	Tongil	Complete
6	Pi22(t)	–	–	38.7–41.9	Sweon 365	–
	Pi26(t)	–	–	51.0–63.2	Gumei 2	–
	Pi27(t)	–	–	51.9	IR64	–
	Pi40(t)	Microsatellite	RM3330, RM527	54.1–61.6	IR65482-4-136-2-2	–
		CAPS	S2539			
	Piz-5	–	–	58.7	Tadukan	Complete
	Piz	InDel	z4794	58.7	Zenith	Complete
		SNP	Z60510, z5765, z56592, z565962			
	Piz-t	InDel	Z4794	58.7	Tordel	Complete
	Pi9	–	–	58.7	75-1-127 (101141)	Complete
	Pi25(t)	–	–	63.2–64.6	Gumei2	–
	Pid2	–	–	65.8	Digu	complete
	Pigm(t)	CAPS	C26348	65.8	Gumei4	–
	Pitq1	–	–	103.0–124.4	Teqing	Complete
	Pi8	–	–	–	Kasalath	Complete
	Pi13(t)	–	–	–	Mawong	Complete

TABLE 3.1 (Continued)

Chromo-some	Gene	Marker type	Tightly linked marker Marker name	Map position (cM)	Donor rice	Resistance type
	Pi13	–	–	–	Kasalath	–
7	Pi17(t)	–	–	94.0–104.0	DJ123	Complete
8	Pi36	Microsatellite	RM5647	21.6–25.2	Q61	–
		CAPS	CRG2, CRG3, CRG4			
	Pi33	Microsatellite	RM 72, RM44	45.4	IR64, Bala	Complete
	Pizh-			53.2–84.8	Zhai-Ye-Quing	Complete
	Pi29(t)			69	IR64	–
	PiGD–1(t)				Sanhuangzan 2	
	Pii2(t)				Ishikari shiroke	
9	Pi5(t)	CAPS	94A20r, 76B14f, 40 N23r	31.3–33.0	RIL125, RIL249, RIL260 (Moroberekan)	Complete
		SNP	JJ817			
	Pi3(t) Pa			31.3–33.0	Kan-Tao	Complete
	Pi15			31.3–34.9	GA25	Complete
	Pii				Ishikari shiroke	Complete
10	Pi28(t)			114.7	Azucena	–
	PiGD–2(t)				Sanhuangzhan2	–
11	Pia	CAPS	Yca72	36	Aichi Asahi	Complete
	PiCO39(t)	CAPS	RGA8, RZ141, RGACO39	49.1	CO39	Complete
	Pilm2			56.2–117.9	Lemont	Complete
	Pi30(t)			59.4–60.4	IR64	–

TABLE 3.1   (Continued)

Chromosome	Gene	Marker type	Tightly linked marker Marker name	Map position (cM)	Donor rice	Resistance type
	Pi7(t)	—	—	71.4–84.3	RIL29 (Moroberekan)	Complete
	Pi34	—	—	79.1–91.4	Chubu 32	Partial
	Pi38	Microsatellite	RM206, RM21	79.1–88.7	Tadukan	—
	PBR	—	—	80.5–120.3	St No.1	—
	Pb1	—	—	85.7–91.4	Modan	Partial
	Pi44(t)	—	—	91.4–117.9	RIL29 (Moroberekan)	Complete
	Pikh	Microsatellite	RM206, TRS26, TRS33, RM144, RM224, RM1233	101.9	Tetep	Complete
	Pi54	—	RM206, TRS26, TRS33, RM144, RM224, RM1233		Taipei 309 (TP)	—
	Pi1	—	—	112.1–117.9	C101LAC (Lac23)	
	Pik-m	InDel	K6861, k2167	115.1–117	Tsuyake	Complete
		SNP	K641, k6441, k473, k7237			
	Pi18(t)	—	—	117.9	Sweon365	Complete
	Pik	InDel	K6816, k2167	119.9–120.3	Kusabue	Complete
	Pik-p	SNP	K641, k39575, k403, k3957	119.9–120.3	HR22	Complete
	Pik-s	Microsatellite	RM144, RM224, RM1233	115.1–117.3	Shin2	Complete
	Pik-g	—	—	—	GA20	Complete
	Pise1	—	—	—	Sensho	—
	Pif	—	—	—	Chugoku 31-1 (St. No. 1)	Partial
	Mpiz	—	—	—	Zenith	—

TABLE 3.1 *(Continued)*

Chromosome	Gene	Tightly linked marker		Map position (cM)	Donor rice	Resistance type
		Marker type	Marker name			
	Pikur2	–	–	–	Kuroka	–
	Piisi	–	–	–	Imochi shirazau	–
12	Pi24(t)	SNP	–	10.3	Zhong 156	Complete
	Pi62(t)	–	–	12.2–26.0	Yashiromochi	–
	Pitq6	–	–	29.2–47.5	Teqing	Complete
	Pi6(t)	–	–	32.6–63.2	Apura	Complete
	Pi12(t)	–	–	42.8–53	RIL10 (Moroberekan)	Complete
	Pi31(t)	–	–	44.3	IR64	–
	Pi32(t)	–	–	47.5	Ir64	–
	Pi12(t)	–	–	47.6–48.2	K80 (Hong-jiaozhan)	–
	Ipi(t)	–	–	47.6–58.3		–
	IPi3(t)	–	–	47.6–58.3		–
	Pi157	–	–	49.5–62.2	Moroberecan	–
	Pita	SNP	Ta642, ta801, ta3, ta577, ta5, Pita 440, pi-ta 1042, Pi-ta403	50.4	Taducan	Complete
	Pita2	SNP	Ta642, ta801, ta3, ta577	50.4	Shimokita	Complete
	Pi19(t)	–	–		Aichi Asahi	Complete
	Pi39(t)	CAPS	39M6, 39M7	50.4	Q15	Complete
	Pi20(t)	Microsatllite	RM1337, RM5364, RM7102	51.5–51.8	IR24	Complete
	PiGD-3(t)	–	–	55.8	Sanhuangzhan2	–

Adapted from Koide et al. (2009).

main purpose of MABC is to transfer the desired character/or targeted gene along with recovering the RP characters/or genes. MABC is now playing an important role for the development of blast-resistant cultivars (Sundaram et al., 2009) and is superior to conventional backcrossing in precision and efficiency and time saving. Molecular markers which are tightly linked with important traits are used in MABC. Therefore, molecular markers are the tools that can be used to detect the presence of desire character in backcrossing and greatly increases the efficiency of selection (Table 3.2). The methods and potential application of MAS and MABC for the improvement of rice have been reviewed (Collard and Mackill, 2008; Hasan et al., 2015).

3.6.5 GENE PYRAMIDING

Pyramiding is the accumulation of genes into a single line or cultivar. In a gene pyramiding, strategy is to cumulate genes identified in multiple parents into a single genotype. The end product of a gene-pyramiding program is a genotype with all of the target genes. Pyramiding multiple-resistance genes provides durable stress-resistance expression in crops. Gene pyramiding technique generally used for combining multiple disease or pest-resistance genes for specific races of a pathogen or insect to develop long-lasting resistance. It helps in crop improvement program by decreasing the breeding duration. Different R-genes often have resistance to different isolates, races, or biotypes. So, by combining these different R genes will expand the resistance spectrum against the different number of races or isolates in a variety at the same time.

Molecular markers help in the identification and selection of desired traits in early developmental stages of the plant. Such molecular markers greatly enhances the gene pyramiding efficiency by reducing the time of selection of plants and savings of resources such as greenhouse or field space, water, and fertilizer Therefore, MAS-based gene pyramiding will help in the pyramiding of different genes into a single genetic background (Joshi and Nayak, 2010; Koide et al., 2009; Shinoda et al., 1971). Hittalmani et al. (2000) have successfully pyramided three genes, *Pi1*, *Piz5*, and *Pita* in a susceptible rice variety, Co39. Chen et al. (2004) successfully adopted MAS and pyramided three blast-resistance genes—*Pi-d(t)1*, *Pi-b*, and *Pi-ta2*—in rice G46B from Digu, BL-1, and Pi-4, respectively. Similarly Singh et al. (2011) also pyramided blast-resistance genes *Piz 5* and *Pi54* into "PRR78" basmati restorer line of rice hybrid Pusa RH10 by Marker-assisted simultaneous

TABLE 3.2 Examples for Application of Marker-assisted Selection (MAS) and Marker-assisted Backcrossing (MABB) in Rice.

Trait	Gene(s)/QTLs	Marker(s) used	Technique used	Application	Reference
Blast resistance	*Pi1, Piz-5, Pita*	RFLP	MAS	Pyramiding of three near isogenic lines (C101LAC, C101A51 and C101PKT) for blast resistance in into a single cultivar Co-39, each carrying the major genes, *Pi1, Piz-5,* and *Pita,* respectively	Hittalmani et al. (2000)
Blast resistance	*Pi1*	SSR and ISSR	MAS	Applied for backcross breeding of variety	Liu et al. (2002)
Bacterial blight resistance + blast resistance	*Xa21, Piz*	SSR	MAS	MAS functional for pyramiding of target traits	Narayanan et al. (2002)
Blast resistance	*Pid1, Pib, Pita,* and *Pi2*	SSR	MAS	*Pid1, Pib,* and *Pita* genes were introduced into G46B cultivar, while *Pi2* Zhenshan97B cultivars of rice	Chen et al. (2004)
Blast resistance	*Pi-z*	SSR	MAS	Closely linked with *Pi-z* locus has been successfully used for selection of blast resistance in a wide array of rice germplasm	Fjellstrom et al. (2006)
Blast resistance + bacterial blight resistance + sheath blight resistance	*Xa13, Xa21, Pi54,* qSBR11	SSR [for blast resistance (*Xa13* and *Xa21*), for bacterial blight resistance (*Pi54*), and sheath blight resistance (qSBR11)]	MAS	MAS-assisted transfer of genes conferring the resistance toward three different diseases in rice	Singh et al. (2012a)
Blast resistance + Bacterial blight resistance	*Pi-genes, Xa5*	SSR	MAS	SSR MAS Near-isogeniclines (NILs) derived from two blast resistant crosses (RD6 × P0489 and RD6 × JaoHomNin) were pyramided with IR62266 (*xa5*), to transfer bacterial leaf blight resistance to RD6 lines	Pinta et al. (2013)

TABLE 3.2 *(Continued)*

Trait	Gene(s)/QTLs	Marker(s) used	Technique used	Application	Reference
Blast resistance	Pita	Gene specific marker	MAS	Existence of the *Pi-ta* gene in 141 rice germplasm has been successfully determined, but the results were more articulated when *Pi-ta* gene was introduced through advanced breeding lines	Wang et al. (2007)
Submergence tolerance + brown plant hopper resistance + blast resistance + bacterial blight resistance	chr9 QTL, *Xa21*, *Bph* and QTLs blast, and quality loci	SSR and STS	MABB	MABB confirmed the transfer of gene and QTL for into cultivar KDML105	Toojinda et al. (2005)
Blast resistance	*Pi1*, *Pi2*, *Pi33*	SSR	MABB	Introgressed into Jin23B cultivar through MABB	Chen et al. (2008)
Blast resistance + bacterial blight	*Pi1*, *Pi2*, *Xa23*	SSR for blast resistance (*Pi1*, *Pi2*), for bacterial blight resistance (*Xa23*)	MABB	Successfully applied for breeding the variety (Rongfeng B)	Fu et al. (2012)
Blast resistance	*Piz-5*, *Pi54*	SSR	MABB	Combination of blast resistance gene from donor lines (C101A51 and Tetep) into cultivar PRR78 to develop Pusa1602 (PRR78 + *Piz5*) and Pusa1603 (PRR78 + *Pi54*), respectively	Singh et al. (2012b)
Blast resistance	*Pi-9(t)*	pB8	MABB	MABB applied to introgress the cultivar Luhui17	Wen and Gao (2011)
Blast resistance	*Pi-1*, *Pi-z*	SSR	MABB	Pyramiding of *Pi-1* and *Piz-5* genes into introduced PRR78 cultivars	Gouda et al. (2013)

but stepwise backcross breeding (MASS-BB). *Piz5* and *Pi54*, taken from non-Basmati donors, C101A51 and Tetep, respectively, and transferred into PRR78. Marker-assisted foreground selection coupled with stringent pheno-typic selection and background analysis was carried out for recovery of RP phenome and genome (RPG) in two separate backcross series to produce BC_2F_1 plants with individual blast-resistance gene. Background analysis revealed that the RPG recovery was up to 91.6% (Table 3.3).

Pyramiding is the strategy for the accumulation of different genes into a single line or cultivar. The developed end product of a gene-pyramiding program is a genotype with all of the target accumulated genes. Gene pyra-miding of multiple resistance genes provides durable resistance in crops for the multiple diseases/stress. Presently, gene pyramiding technique broadly employed for combining multiple disease, pest resistance of different abiotic resistance genes for specific races of a pathogen or insect or stresses to develop durable resistance. Gene pyramiding helps in the crop improvement program rapidly and reduces breeding duration. Different R-genes confer resistance to different isolates, races, or biotypes in the plants, so combining these resistance gene broadens the number of races or isolates that a more than one character in a variety at the same time.

For the development of new elite breeding lines and varieties, plant breeders will combine desirable traits from the multiple parental lines. Nowadays, gene pyramiding can be accelerated by using molecular markers for the identification and selection of plants that contain the desired allele combination in very early stage. The above advantage resulting in obvious savings of resources including greenhouse or field space, water, and fertil-izer. Molecular marker can help in the existing plant breeding programs which allows plant breeders to admittance, transfer, and combine different genes with precision manners. MAS-based gene pyramiding makes possible in pyramiding of different genes effectively into a single genetic background of a crop (Joshi and Nayak, 2010). There are several factors that is, number of genes to be transferred, distance between the target genes and flanking markers, number of genotype selected in each breeding generation, and nature of germplasm used is decisive for successful gene pyramiding program. Presently, gene pyramiding programs are consid-ered as one of the most valuable strategies for attaining durable resistance against blast disease in rice (Hittalmani et al., 2000; Koide et al., 2009; Shinoda et al., 1971). There are different successful stories of the accumu-lation of different blast resistance genes in elite rice cultivars by different researchers (Table 3.3).

TABLE 3.3 Examples for Application of Gene Pyramiding in Controlling Blast Disease in Rice.

Traits	Parental lines	Pyramided genes	DNA (markers) used	Reference
Blast resistance	C101LAC., C101A51	*Pi1, Pi2,* and *Pi33*	SSR	Chen et al. (2008)
Blast resistance	IR5, IR8, IR20, IR22, IR24, IR26, IR28, IR29, IR30, IR32, IR34, IR36, IR38, IR40, IR42, IR43, IR44, IR45, IR46, IR48, IR50 IR52, IR54, IR56, IR58, IR60, IR62, IR64, IR65, IR66, IR68, IR70, IR72, IR74	*Pib* and *Pita*	SSR	Fujita et al. (2009)
Blast resistance	CO39	*Pish* and *Pib*	SSR	Koide et al. (2009)
Blast resistant	IR64, JHN	Multiple resistance QTLs	Multiple resistance QTLs	Sreewongchai et al. (2010)
Blast resistant	Rongfeng B	*Pi1, Pi2 Xa23*		Fu et al. (2012)
Blast resistant	C101LAC, C101A51	*Pi-1* and *Pi-2*	RG64 and C481	Mahdian and Shah-savari (2013)
Blast and bacterial leaf blight resistance	RD6 × P0489; RD6 × JHN	Four QTLs for blast resistance and one gene for bacterial leaf blight (xa5)	SSR	Pinta et al. (2013)
Blast resistant	C101A51, Tetep	*Piz5* and *Pi54*	SSR	Singh et al. (2011)
Blast resistance	Carnaroli, Baldo, Arborio	*Piz* and *Pi5*	SSR	Urso et al. (2013)
Leaf blast resistance	Koshihikari	*Pi21, Pi34,* and *Pi35*	SSR	Yasuda et al. (2014)
Blast resistance	GZ63-4S	*Pi2* and *Xa23*	SSR (*M-Xa23*)	Jiang et al. (2012)

3.6.6 QTLS

Blast resistance was broadly classified into complete and partial resistance (Wang et al., 1994). The complete resistance is denoted as qualitative character and in race-specific manners controlled by a major resistance gene (R genes). However, partial resistance is a quantitative character and non-race specific, which is controlled by several genes known as quantitative resistance loci (Young, 1996). If the resistance is highly partial, then it can also be controlled by a major gene and is race specific. Qualitative and quantitative blast resistances have been reported in different rice germplasm (Ou, 1985). Several qualitative resistance major genes (~100 genes) for blast resistance have been identified and mapped in the rice genome by several researchers (Ashkani et al., 2014; Sharma et al., 2012). There are up to 22 R-genes have been successfully cloned, mapped, and molecularly characterized. In the practice of resistance breeding program, a single R gene which has a broad resistance spectrum is more effective against the pathogen. There are several information on introgression of *Pi* genes related to blast disease into commercial and elite rice varieties. There are some reports, like *Pi-9* gene that exists in the indica rice line 75-1-127 (Liu et al., 2002), was introgressed from the wild species *O. minuta* (Amante-Bordeos et al., 1992). The *Pi-ta* allele was identified in wild rice *O. rufipogon* and *O. nivara*, or in their hybrids with *O. sativa* (Jena and Khush, 2000). These resistance genes function in a manner of gene-for-gene trend, so the pathogen can acclimatize by mutating or deleting the corresponding avirulence gene, respectively. Consequently varieties, carrying R genes confer high levels of resistance usually lose their resistance after a few years in regular manner (Chen et al., 2003). Quantitative resistances governed by the QTLs are elongated enduring disease resistance in opposition to a wide-range of pathogens, promising for sustainable rice production (Song and Goodman, 2001). QTL mapping is a modern type of lessons to establish genes controlling a quantitative trait. After the first report of a QTL analysis of rice resistant to blast (Wang et al., 1994), several QTLs connected to blast resistance have been identified by using different type of molecular markers. Detection of QTLs has been employed to map different major or minor genes involved in the resistance against the disease (Ashkani et al., 2013a,b; Chen et al., 2003; Fukuoka and Okuno, 2001; Miyamoto et al., 2001; Sallaud et al., 2003; Tabien et al., 2002; Talukder et al., 2004; Wang et al., 1994; Wu et al., 2005; Zenbayashi et al., 2002). Detection of different QTLs connected with blast resistance has been distributed the efficient genetics confirmations for the molecular marker-assisted breeding and cloning of the major resistance genes (Table 3.4). QTL mapping is also

TABLE 3.4 Examples for Application of QTLs in Controlling Blast Disease in Rice.

Mapping population	Parents used in crossing	Total no. of QTLs detected	Used markers	References
Recombinant in-bred lines (RILs)	CT9993-5-10-1-m × KDML105(F8); Zhenshan 97 × Minghui63 (RILs); Moroberekan × Co39 (F7); Lemont × Teqing (F8); Lemont × Teqing (F14); Bala × Azucena (F6); Zhong156 × Gumei2 (F8); OryzicaLlanos5 × Fanny (F5 and F6); SHZ-2 × Lijiangxin-tuan-heigu (LTH) (RILs); KDML 105 × JHN (F6); Suweon365 × Chucheong (RILs)	186	RFLPs, SSR, RAPD, Isozymes, AFLPs, DR gene markers	Sirithunya et al. (2002), Chen et al. (2003), Wang et al. (1994), Tabien et al. (2002), Liu et al. (2004), Talukder et al. (2005), Wu et al. (2005), Lopez-Gerena (2006), Noenplab et al. (2006), Cho et al. (2008)
Double haploid (DH)	IR64 × Azucena; IR64 × Azucena; ZYQ8 × JX17	146	RFLP, RAPD, isozymes	Xu et al. (2004), Sallaud et al. (2003), Bagali et al. (2000)
Single-segment substitution lines (SSSLs) SSR back cross population	Developed by the use of HXJ74 as recipient and 24 accessions as donors 11	11	SSR	Zhang et al. (2012)
Back cross population	Way Rarem × Oryzica Llanos5 (IRGC117017); Oryza sativa cv. MR219 × O. rufipogon IRGC 105491; SHZ-2 × TXZ-13; Oryza rufipogon × cultivated rice IR64	45	SSR, SNP	Lestari et al. (2011), Rahim et al. (2012), Utani et al. (2008), Liu et al. (2011)
F2, F3, and F4	Nipponbare × Owarihatamochi (F4 lines); Kahei × Koshihikari (F2:3); Tainung69 × Koshihikari (F2); URN12 × Koshihikari (F2); Norin29 × Chu-bu32 (F3); PongsuSeribu2 × Mahsuri (F2:3); TAM × KHZ (F2:3); Junambyeo × O. minuta introgres-sion line IR71033-121-15 (F2:3); Danghang-Shali × Hokkai 188 (F2:3)	60	RFLPs, SSR STS	Fukuoka and Okuno (2001), Sato et al. (2006), Zenbayashi et al. (2002), Ashkani et al. (2013a,b), Nguyen et al. (2006), Rahman et al. (2011)

useful for the identification of multiple loci controlling complete resistance in a lofty resistant cultivar as well as in estimating the number, location and effect of genomic region involved in partial blast resistance against the pathogens (Sallaud et al., 2003). Different rice improvement programs now aspire to include quantitative or polygenic resistance into elite rice varieties. Several previous studies have confirmed that genetic linkage maps constructed with different molecular markers are very helpful for the analysis and recognition for the qualitative trait loci (Bao et al., 2000; Price et al., 2000).

3.6.7 ASSOCIATION MAPPING (GENETICS)

Association mapping also known as "linkage disequilibrium mapping", is a method of mapping QTLs that takes advantage of historic linkage disequilibrium to link phenotypes (observable characteristics) to genotypes (the genetic constitution of organisms). Association mapping is based on the idea that traits that have entered a population only recently will still be linked to the surrounding genetic sequence of the original evolutionary ancestor, or in other words, will more often be found within a given haplotype, than outside of it. It is most often performed by scanning the entire genome for significant associations between a panel of SNPs (which, in many cases are spotted onto glass slides to create "SNP chips") and a particular phenotype. These associations must then be independently verified in order to show that they either (a) contribute to the trait of interest directly, or (b) are linked to/in linkage disequilibrium with a QTL that contributes to the trait of interest (Gibson & Muse, 2009). Association mapping also requires extensive knowledge of SNPs within the genome of the organism of interest, and, therefore, difficult to perform in species that have not been well studied or do not have well-annotated genomes (Yu et al., 2008). Association mapping for blast resistance was performed on 226 *japonica* rice cultivars with 118 pairs of SSR markers. The blast resistance was evaluated by inoculating with two isolates, DB22 and DB77, at the tillering stage in 2013 and 2014, separately. A total of 31 associations with 17 different SSRs were significantly ($P < 0.05$) associated with blast resistance based on the mixed linear model, of which nine markers could be detected in both 2013 and 2014, including two markers that were simultaneously associated with the two isolates. Five of the nine stable markers were consistent with the genome regions identified by linkage mapping in previous reports. Phenotypic effects of each allele of the nine stable markers were compared, and 18 favorable alleles were identified. Five elite parental combinations were designed for improving blast resistance in rice. Results demonstrate that association mapping can complement and enhance previous QTL information for MAS and breeding by design (Guo et al., 2015).

3.7 OTHER BIOTECHNOLOGICAL APPROACHES

Breeding work utilizing both phenotypic and genotypic markers are more reliable and fast. Conventional breeding are based on gene expression due to which many limitations, for example, epistatic effect exist. Conventional breeding methods may create a resistance variety which is time consuming and intensive task. Biotechnological approaches are important in introducing genes which provide resistance against *M. grisea*. Rice breeders are looking at basic bioscience and biotechnology for solving some important problems that conventional breeding methods have not satisfactorily solved. Therefore, future breeding strategies should focus at broadening the genetic and cytoplasmic background of new varieties that are being developed not only for blast resistance but also for other important pests and diseases as well. These information greatly advance our understanding of molecular mechanisms that govern race specificity. The application of advanced molecular technologies could speed up crop improvement. There are some biotechnological approaches that can be used for the development of blast resistance rice.

3.7.1 TISSUE CULTURE

The various tissue-culture methods and gene-transfer techniques now available could significantly shorten the breeding process and overcome some of the substantial agronomic and environmental problems that have not been solved using conventional methods (Zapata et al., 1995). Tissue culture is one of the fundamental tools of crop science research. Cell culture is one of the alternative methods of inducing resistance to diseases in susceptible cultivars which are well adapted to local soil and climatic conditions. The genetic variation produced in tissue culture, termed somaclonal variation, has been reported in many crop species (Larkin and Scowcroft, 1981). Fortunately during the last 30 years, extensive work has been done on selecting the disease resistant plants against different pathogens. The cell-free culture filtrate (CF) or the pure toxins of the pathogens and direct infection by the pathogen or all of them together could be used for the selection of novel disease-resistant plants (El-Kazzaz, 2001; El-Kazzaz and Ashour, 2004). However, the information on somaclonal variation for blast resistance is very little. According to Pachon (1989), there was no variation for blast resistance in somaclones. On the other hand, Bouharmon et al. (1991) obtained R_2 lines resistant to blast from the calli derived from mature embryos. In Brazil, a high degree of partial resistance has been reported in progenies of regenerated plants derived from immature panicles of a susceptible upland rice cultivar IAC47 (Araujo et al., 1997). These discrepancies may be partly attributed to the test conditions and disease pressure under which the somaclones were assessed, and the nature of resistance. Also various other factors may affect somaclonal variation such as genotype, explants source, composition of the culture medium, and time of cultivation (Evans et al., 1984). The alterations in the genome have been attributed to expression of mutant cells in explants, meiotic crossing over, and cytological changes (Evans et al., 1984). The studies on somaclonal variation may permit accomplishment of breeding objectives in relation to rapid development of blast-resistant lines from existing commercial suscep-tible rice cultivars. An experiment was conducted by (Araujo et al., 2000) on Basmati-370 rice which was susceptible to some pathotypes of *P. oryzae* in Brazil, to assess the degree of blast resistance and some agronomic char-acteristics in the advanced generations of its somaclones. These evaluations were carried out in R5–R9 generations, in field trials, in rice blast nursery, and greenhouse. Significant variations in grain quality and other agronomic characteristics were not observed. However, some of the somaclones showed higher degree of blast resistance. Two somaclones, SCBAS04 and SBAS16 exhibited higher degree of partial resistance to blast. The degree of blast resistance of upland rice (*Oryza sativa* L.) cultivar Araguaia has decreased significantly due to yield loss obtained from resistant somaclones, adaptation to greenhouse and field selection procedures (Araujo et al., 2002). Green-house selection with two specific physiologic races yielded 44 somaclones with slow-blasting resistance, similar plant type and yield potential as that of Aragaia. A step-by-step protocol followed for resistant calli selection via a tissue culture technique under stress of *P. oryzae* CFs (El-Kazzaz et al., 2009). The results reveal that the resistance in regenerated rice plantlets to *P. oryzae* pathogen segregated as 1: resistant, 2: moderate resistant, 1: suscep-tible to *P. oryzae* may be controlled by one pair of genes.

3.7.2   GENE TRANSFORMATION

Transgenic plants can acquire a single desired trait without any alteration of the original genetic make-up and can overcome the limitation of traditional breeding. Transgenic technologies allow multiple genes insertion simultaneously into genome to obtain broad-spectrum-resistant lines. Genetic engineering are important approaches in the management of fungi diseases by introducing and overexpressing of genes that encode proteins involved in the synthesis of compounds toxic to fungi and with direct inhibitory effect on the growth of fungi (Cornelissen and Melchers, 1993). Some transgenic strategies for improving blast resistance based on the host–pathogen gene-for-gene interaction system and antifungal protein genes have been developed (Campbell et al., 2002; Tan et al., 2004). To date, there have been reports on increasing rice blast resistance through transformation of chitinase gene (Nishizawa et al., 1999), plant antitoxin gene (Stark-Lorenzen et al., 1997), chitinase–glucanase gene (Feng et al., 1999), trichosanthin gene (Ming et al., 2000), wasabi phytoalexin gene (Kanzaki et al., 2002), and rice blast-resistance genes *Pi-ta*, *Pi-9*, *Pi-2*, etc. (Chen et al., 2006; Qu et al., 2006; Wang et al., 1999; Zhou et al., 2006). Transgenic approach has been used as an attractive alternative to conventional techniques for the genetic improvement of Basmati rice (Jain and Jain, 2000). During the last 10 years, a rapid progress has been made toward the development of transformation methods in rice. Several transformation methods including Agrobacterium, biolistic, and DNA uptake by protoplasts have been employed to produce transgenic rice. Stark-Lorenzen et al. (1997) reported active transcription of grapevine stilbene synthase gene in transgenic rice after incubation with the fungus of the rice blast *P. oryzae*. Results indicated enhanced resistance of transgenic rice to *P. oryzae*. Stilbene synthase in some plants species synthesizes a phytoalexin trans-reserveartaol that seems to have a role in the early protection of plants against fungal pathogen (Coutos-Thevenot et al., 2001).

Five types of toxin have been purified from the crude toxin of rice blast fungi, namely, picularin, picolinic acid, tenuazonic acid, piriculol, and coumarin. The crude toxin has strong inhibition on seed germination and extension of plumule and radicle of rice, and can serve as a selection pressure for disease-resistant mutant selection at the callus level (Xu et al., 2003). The transgenic rice blast-resistant gene improves the resistance to rice blast crude toxin, the transgenic *Pi-d2* stable rice line TP-Zh01-62-5 was used under different selection pressures of rice blast crude toxin. The results showed that as the concentration of rice blast crude toxin increased, the embryonic callus induction rate decreased obviously. When the concentration of crude toxin reached 50%, the embryonic callus induction rate of the control decreased to zero, and the growth of embryos was completely suppressed, whereas that of the transgenic plants was still at 30.1%, suggesting that the resistance of transgenic plants to the rice blast crude toxin was increased compared with the control. This provided another evidence for the feasibility of rice blast crude toxin as measures for resistance to rice blast. Recently, chloroplast-expressed MSI-99 in tobacco improves disease resistance and displays inhibitory effect against rice blast fungus The antimicrobial peptide MSI-99 has been suggested as an antimicrobial peptide conferring resistance to bacterial and fungal diseases. A vector harboring the MSI-99 gene was constructed and introduced into the tobacco chloroplast genome via particle bombardment. Transformed plants were obtained and verified to be homoplastomic by PCR and southern hybridization. In plants, assays demonstrated that the transgenic tobacco plants displayed an enhanced resistance to the fungal disease. The 23-amino-acid long antimicrobial peptide MSI-99 is a derivative of Magainin II, with a His7 to Lys substitution that prevents the gene from recognition and digestion by plant endonucleases, while maintaining its normal functions. The evaluation of the antimicrobial activity revealed that the crude protein extracts from the transgenic plants manifested an antimicrobial activity against *E. coli*, even after incubation at 120°C for 20 min, indicating significant heat stability of MSI-99. The MSI-99-containing protein extracts were first proved in vitro and in vivo to display significant suppressive effects on two rice-blast isolates. These findings provide a strong basis for the development of new biopesticides to combat rice blast (Wang et al., 2015).

3.8   ALLELE MINING

Allele mining is used to identify novel alleles or allelic variants of a gene/ or candidate genes of interest, based on the available information about the genes, from a wide range of germplasm. Genetic material used for allele mining should be diverse and gene sequence information of a particular crop species should be known. For allele mining generally wild relatives. Wild relatives and local landraces are reservoirs of various beneficial alleles that are hidden in their phenotype (Tanksley et al., 1996). Therefore, wild relatives and local landraces are generally used for allele mining. Nowadays, rice genome is completely sequenced that helps in the mining of allelic diversity throughout the rice germplasm with the help of several bioinformatics tools. Allele mining can be divided into approaches that is EcoTilling and sequence-based allele mining where sequence-based allele mining is simple as well as cost-effective (Ashkani et al., 2014; Ramkumar et al., 2012).

To date, many works on identification of novel blast resistance by allele mining are done by many researchers. Allele mining of genes from wild

and cultivated rice species aims to detect superior alleles for blast resistance (Kumari et al., 2013). Through allele mining techniques, functional
marker to discriminate the resistance and susceptible alleles of *Pi54* has
been developed (Ramkumar et al., 2012). Costanzo and Jia (2010) analyzed
the sequence level similarity for *Pikm* alleles, derived from 15 different rice
cultivars. *M. oryzae* has also been differentiated from *M. grisea* by using
allele mining (Couch and Kohn, 2002). Vasudevan et al. (2015) identified
novel alleles of the rice blast-resistance gene *Pi54*. They explored the allelic
diversity of the *Pi54* gene among 885 Indian rice genotypes that were found
resistant in screening against field mixture of naturally existing *M. oryzae*
strains as well as against five unique strains. Sequence-based allele mining
was used to amplify and clone the *Pi54* allelic variants. Nine new alleles of
Pi54 were identified based on the nucleotide sequence comparison to the
Pi54 reference sequence as well as to already known *Pi54* alleles. DNA
sequence analysis of the newly identified *Pi54* alleles revealed several single
polymorphic sites, three double deletions, and an eight base pair deletion. A
SNP-rich region was found between a tyrosine kinase phosphorylation site
and the nucleotide binding site domain. Newly identified *Pi54* alleles will
expand the allelic series and are candidates for rice blast-resistance breeding
programs (Table 3.5).

3.9   miRNA IN BLAST DISEASE MANAGEMENT

miRNAs are small regulatory RNAs of 20–22 nt that are encoded by endogenous MIR genes. Their primary transcripts form precursor RNAs, which
have a partially double-stranded stem-loop structure and which are processed
by DCL proteins to release mature miRNAs (Voinnet et al., 2009). In the
miRNA biogenesis pathway, primary miRNAs (pri-miRNAs) are transcribed
from nuclear-encoded MIR genes by RNA polymerase II (Pol II) (Lee et al.,
2004) leading to precursor transcripts with a characteristic hairpin structure
(Fig. 3.3). In plants, the processing of these pri-miRNAs into pre-miRNAs
is catalyzed by DCL1 and assisted by HYPONASTIC LEAVES 1 (HYL1)
and SERRATE (SE) proteins (Voinnet et al., 2009). The pre-miRNA hairpin
precursor is finally converted into 20–22-ntmiRNA/miRNA duplexes by
DCL1, HYL1, and SE. The duplex is then methylated at the 3′ terminus by
HUA ENHANCER 1 (HEN1) and exported into the cytoplasm by HASTY
(HST1), an exportin protein (Kim et al., 2008). In the cytoplasm, one strand
of the duplex (the miRNA) is incorporated into an AGO protein, the catalytic
component of RISC, and guides RISC to bind to cognate target transcripts by

TABLE 3.5 Example of Allele Mining for Blast Resistance in Rice.

R-Genes/Locus	Chromosome	Rice germplasm	Reference
Pi-ta	12	From wild rice species [O. rufipogon (Griff) and from O. rufipogon (ETOR)]	Yang et al. (2007), Geng et al. (2008)
Pi-ta	12	From O. rufipogon	Huang et al. (2008)
Pi-ta	12	From cultivated (AA) and wild species and invasive weedy rice	Lee et al. (2009, 2011)
Pi-ta	12	In 26 accessions, consisting of wild rice (O. rufipogon), cultivated rice (O. sativa) and related wild rice species (O. meridionalis and O. officinalis) collected from 10 different countries of the world	Yoshida and Miyashita (2009)
Pi-ta	12	From land races and wild Oryza species	Ramkumar et al. (2012)
Pi-ta	12	In Indian land races of rice	Sharma et al. (2010)
Pi-ta	12	From Indian landraces of rice collected from different ecogeographical regions including the northwestern Himalayan region of India	Thakur et al. (2013a)
Pi-kh (Pi54)	11	From wild and cultivated species of rice	Rai et al. (2011)
Pi-kh (Pi54)	11	From the blast-resistant wild species of rice, O. rhizomatis	Das et al. (2012)
Pi-kh (Pi54)	11	From six cultivated rice lines and eight wild rice species	Kumari et al. (2013)
Pi-kh (Pi54)	11	In Indian land races of rice	Sharma et al. (2010)
Pi-z(t)	06	In Indian landraces of rice	Sharma et al. (2010)
Pi-z(t)	06	In 529 land races of rice collected at three different geographical locations of India	Thakur et al. (2013b)
Pid3	06	From 36 accessions of wild rice O. rufipogon	Shang et al. (2009), Xu et al. (2014)
Pid3-A4	06	From wild rice A4 (O. rufipogon)	Lv et al. (2013)
Pi9		Indifferent rice species, five AA genome Oryza species including two cultivated rice species (O. sativa and O. glaberrima) and three wild rice species (O. nivara, O. rufipogon, and O. barthii)	Liu et al. (2011)
AC134922	11	Rice lines from various sources	Wang et al. (2014)

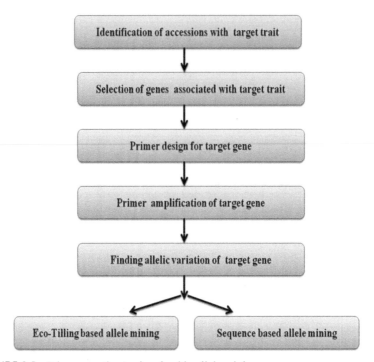

FIGURE 3.3   Diagrammatic step involved in allele mining.

sequence complementarity (Fig. 3.4). In addition to the control of targets at the posttranscriptional level, miRNAs regulate gene expression by causing epigenetic changes such as DNA and histone methylation (Khraiwesh et al., 2010; Bao et al., 2004).

A number of miRNAs have been linked to biotic stress responses in plants. Role of miRNAs in plants after infection by pathogenic bacteria, viruses, nematodes, and fungi has been reported (Katiyar-Agarwal & Jin, 2010; Ruiz-Ferrer & Voinnet, 2009). Sequenced small RNA libraries from rice tissues (leaves, roots) that had been treated with elicitors obtained from the rice blast fungus *M. oryzae*, the causal agent of the rice blast disease (Talbot, 2003) has been studied. Functional characterization of a novel miRNA from rice, osa-miR7695, which negatively regulates an alternatively spliced transcript of the OsNramp6 (Natural resistance-associated macrophage protein 6) metal transporter gene. Overexpression of the newly identified miRNA results in enhanced resistance to pathogen infection in rice plants.

MicroRNAs (miRNAs) are indispensable regulators for development and defense in eukaryotes. However, the miRNA species have not been explored for rice (*Oryza sativa*) immunity against the blast fungus *M. oryzae*. Deep sequencing small RNA libraries from susceptible and resistant lines in normal conditions and upon *M. oryzae* infection, Li et al. (2014) identified a group of known rice miRNAs that were differentially expressed upon *M. oryzae* infection. They were further classified into three classes based on their expression patterns in the susceptible japonica line Lijiangxin Tuan Hegu and in the resistant line International Rice Blast Line Pyricularia-Kanto51-m-Tsuyuake that contains a single-resistance gene locus, *Pyricularia-Kanto 51-m (Pikm)*, within the Lijiangxin Tuan Hegu background. RNA-blot assay of nine of them confirmed sequencing results. Real-time reverse transcription-polymerase chain reaction assay showed that the expression of some target genes was negatively correlated with the expression of miRNAs. Moreover, transgenic rice plants over expressing miR160a and miR398b displayed enhanced resistance to *M. oryzae*, as demonstrated by decreased fungal growth, increased hydrogen peroxide accumulation at the infection site, and upregulated expression of defense-related genes. Taken together, their data indicate that miRNAs are involved in rice immunity against *M. oryzae* and that overexpression of miR160a or miR398b can enhance rice resistance to the disease.

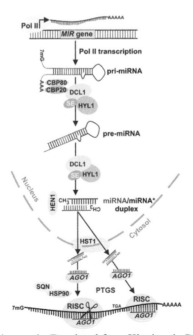

FIGURE 3.4   miRNA biogenesis (Reprinted from Khraiwesh, B.; Zhu, J. K.; Zhu, J. Role of miRNAs and siRNAs in Biotic and Abiotic Stress Responses of Plants. *Biochim. Biophys. Acta* **2012**, *1819*, 137–148. © 2012 with permission from Elsevier.)

3.10   FUTURE PROSPECTS

Molecular biology and biotechnological tools have totally changed the research on rice blast management. The availability of genome sequences of both, the host rice (Dean et al., 2005) and the pathogen has opened many doors for further research. Introduction of new sciences like nanotechnology in agricultural research and management could be proved very beneficial in future as a nanotech-based company *viz.,* NANO GREEN has reported the control of rice blast using nanomolecules. Cloning of R and Avr genes and study of their gene products will add to the knowledge of host pathogen interaction. Enhancing the host plant resistance is being considered as the best approach to handle the rice blast disease. Rice cultivars containing only a single *R* gene to a specific pathogen race often become susceptible over time due to the emergence of new virulent races. Understanding of genetic identity of contemporary *M. oryzae* is important for accurate deployment of rice cultivars with different *R* genes. Stacking *R* genes with overlapped resistance spectra can lead to long-lasting resistance. For combinations of different blast-resistance genes in host plant in the rice blast breeding programs, superior alleles of the targeted genes should be considered. During the evolution and artificial selection processes, a significant portion of beneficial alleles have been left behind in the land races and wild species (McCouch et al., 2007), which can be used for the development of better rice varieties. Effective blast management also requires international cooperation. The knowledge gained by collaborative effort ought to lead to more effective methods to reduce crop loss due to blast disease worldwide. Although considerable progress has been made toward understanding the nature of disease-resistance genes, defense responses, and the signal transduction leading to activation of defense responses in rice, the whole story is still far from clear. The development of genetically engineered bioagents will supplement the environment friendly ways of management of rice blast. There is still a need for the further development of noble fungicides and fungistats with longer residual effect which can be better assisted by biotechnology in future. For resistance management, strategies like gene rotation, gene pyramiding, spatial and temporal gene deployment, and use of varietal mixtures will be the best mean to reduce the epidemics. The development and use of transgenic rice would be the best way of rice blast management in the future. Allele mining possesses good potential to be applied in molecular plant breeding. With the ever-increasing sequence data in GenBank and ever-expanding crop gene banks, it is highly essential to develop novel and efficient mining strategies to screen GenBank collections more efficiently for DNA sequence variation and the management of genome resources. The success of allele mining mainly depends on the type of genetic materials used for screening and should be as diverse as possible. To this end, wild relatives and local landraces are used because they are reservoirs of useful alleles hidden in their phenotype (Tanksley et al., 1996). The challenge to build the association between sequence polymorphisms and putative function is of importance for the application of allele mining for R (resistant) genes. GWAS (genome-wide association study) analysis and high-throughput functional validation, like RNAi approaches, could facilitate the identification of new R alleles. Overall, to improve food security, to promote fundamental research and practical breeding new technologies must be developed and advanced methods for identification of novel alleles for functional genomics and crop improvement may become an effective tool for helping in feeding of the twenty-first-century world.

3.11 CONCLUSION

Due to the increase in population there is urgent need to increase rice production globally. To increase rice production, development of high-yielding rice varieties with tolerance to biotic and abiotic stresses is required. Among the biotic stresses blast disease is the most harmful threat to high productivity of rice. Yet lots of management tools and practices are available for the disease but the effectiveness is dependent on their integrated use. Sanitation measures, fertilizer practices, and other aspects of rice culture, as they relate to the onset and development of blast needs further research. Durable resistance is influenced by environmental factors, so other means of disease management must be applied to assist host–plant resistance. The use of resistant rice cultivars is a powerful tool to reduce the use of environmentally destructive pesticides. Using classical plant breeding techniques, plant breeders have developed a number of blast-resistant cultivars adapted to different rice growing region worldwide. Recent advances in rice genomics provide additional tools for plant breeders to develop rice production systems that could be sustainable and environmentally favorable. The breeding strategies such as pyramiding of genes, gene rotation, and multiline varieties have been found effective in resistance management. Biotechnological tools and techniques, assisting the development of control measures have very bright future. Due to the highly variable nature of pathogen, need for the continuous research on development of durably resistant cultivars will always be there. Present chapter outlines the different traditional methods and as well as molecular approaches for the control of rice blast disease and by which rice yield will meet up the demand in the coming years and decades globally. Finally, the knowledge gained through research must be communicated and demonstrated to the farmers so that they can use it.

KEYWORDS

- lesions
- resistance gene
- alleles
- quantitative trait
- transgenic

REFERENCES

Ahloowalia, B. S.; Maluszynski, M.; Nichterlein, K. Global Impact of Mutation-derived Varieties. *Euphytica* **2004**, *135*, 187–204.

Ahn, S.; Mukelar, A. *Rice Blast Management Under Upland Conditions. Progress in Upland Rice Research*; Int. Rice Res. Inst.: Manila, 1986; pp 363–374.

Aktar, M. W.; Sengupta, D.; Chowdhury, A. Impact of Pesticides Use in Agriculture: Their Benefits and Hazards. *Interdiscip. Toxicol.* **2009**, *2*, 1–12.

Allard, R. W. Principles of Plant Breeding, 2nd ed. Wiley: New York, 1999.

Allard, R. W. *Principles of Plant Breeding*. Wiley: New York, 1960.

Amante-Bordeos, A.; Sitch, L.; Nelson, R.; Dalmacio, R.; Oliva, N.; Aswidinnoor, H.; Leung, H. Transfer of Bacterial Blight and Blast Resistance from the Tetraploid Wild Rice *Oryza minuta* to Cultivated Rice, *Oryza sativa. Theor. Appl. Genet.* **1992**, *84*, 345–354.

Araujo, G. L.; Prabhu, A. S.; Freire, A. B. Development of Blast Resistant Somaclones of the Upland Rice Cultivar Araguaia. *Pesq. Agro. Bras.* **2002**, *35*, 357–367.

Araujo, L. G.; Prabhu, A. S.; Freire, A. B. Variacao somaclonal na cultivar de arroz IAC-47 para resistência á brusone. *Fitop. Bras. Bras.* **1997**, *22*, 125–130.

Araujo, L. G.; Prabhu, A. S.; Freire, A. D. Development of Blast Resistant Somaclones of the Upland Rice Cultivar Araguaia. *Pesq. Agropec. Bras.* **2000**, *35,* 357–367.

Ashizawa, T.; Zenbayashi, K.; Koizumi, S. Development of a Simulation Model for Forecasting Rice Blast Epidemics in Multiline. *Jpn. J. Phytopathol.* **2001**, *67*, 194.

Ashkani, S.; Rafii, M. Y.; Rahim H. A.; Latif, M. A. Mapping of the Quantitative Trait Locus (QTL) Conferring Partial Resistance to Rice Leaf Blast Disease. *Biotechnol. Lett.* **2013b**, *35*, 799–810.

Ashkani, S.; Rafii, M. Y.; Rahim, H. A.; Latif, M. A. Genetic Dissection of Rice Blast Resistance by QTL Mapping Approach Using F3 Population. *Mol. Biol. Rep.* **2013a**, *40*, 2503–2515.

Ashkani, S.; Rafii, M. Y.; Rusli, I.; Sariah, M.; Abdullah, S. N. A.; Harun, A. R.; Latif, M. A. SSRs for Marker-assisted Selection for Blast Resistance in Rice (Oryza sativa L.). *Plant Mol. Biol. Rep.* **2012,** *30,* 79–86.

Ashkani, S.; Rafii, M. Y.; Sariah, M.; Abdullah, S. N. A.; Rusli, I.; Harun, A. R.; Latif, M. A. Analysis of Simple Sequence Repeat Markers Linked with Blast Disease Resistance Genes in a Segregating Population of Rice (*Oryza sativa*). *Genet. Mol. Res.* **2011,** *10*(3), 1345–1355.

Ashkani, S.; Rafii, M.; Shabanimofrad, M.; Ghasemzadeh, A; Ravanfar, S.; Latif, M. Molecular Progress on the Mapping and Cloning of Functional Genes for Blast Disease in Rice (*Oryza sativa* L.): Current Status and Future Considerations. *Crit. Rev. Biotechnol.* **2014,** *4,* 1–15.

Ashtiani, F. A.; Kadir, J. B.; Nasehi, A.; Rahaghi, S. R. H.; Sajili, H. Effect of Silicon on Rice Blast Disease. *Pert. J. Tropic. Agric. Sci.* **2012,** *35,* 1–12.

Asuyama, H. Morphology, Taxonomy, Host Range, and Life Cycle of *Pyricularia oryzae*, In: *Rice Blast Disease*; Chandler, R. F., Ed.; Johns Hopkins Press: Baltimore, MD, 1965; pp 9–22.

Azlan, S.; Alias, I.; Saad, A.; Habibuddin, H. Performance of Potential Mutant Lines of MR 180. In: Modern Rice Farming; Sivaprasagam et al., Eds.; *Proceed. Int. Rice Conf. Serdang*, MARDI: Malaysia, 2004; pp 293–296.

Bagali, P. G.; Hittalmani, S.; Shashidhar, S. Y.; Shashidhar; H. Identification of DNA Markers Linked to Partial Resistance for Blast Disease in Rice Across Four Locations, In: *Adv. Rice Blast Res.,* 2000.

Bao, J.; Zheng, X.; Xia, Y.; He, P.; Shu, Q.; Lu, X.; Chen, Y.; Zhu, L. H. QTL Mapping for the Paste Viscosity Characteristics in Rice (*Oryza sativa* L.). *Theor. Appl. Genet.* **2000,** *100,* 280–284.

Bao, N.; Lye, K. W.; Barton, M. K. MicroRNA Binding Sites in Arabidopsis Class III HD-ZIP mRNAs are Required for Methylation of the Template Chromosome. *Dev. Cell* **2004,** *7,* 653–662.

Barman, S. R.; Gowda, M.; Venu, R. C.; Chattoo, B. B. Identification of a Major Blast Resistance Gene in the Rice Cultivar Tetep. *Plant Breed.* **2004,** *123,* 300–302.

Berruyer, R.; Adreit, H.; Milazzo, J.; Gaillard, S.; Berger, A.; Dioh, W.; Lebrun, M. H.; Tharreau, D. Identification and Fine Mapping of *Pi33*, the Rice Resistance Gene Corresponding to the *Magnaporthe grisea* Avirulence Gene ACE1. *Theor. Appl. Genet.* **2003,** *107,* 1139–1147.

Bidaux, J. M. IRAT. Rice Pathology. *Annual Report BP 636*, Bouake: Ivory Coast, 1976.

Bonman, J. D.; Dios, T. V.; Khin, M. Physiologic Specialization of *Pyricularia oryzae* in the Philippines. *Plant Dis.* **1986,** *70,* 767–769.

Bonman, J. M.; Ahn, S. W. *Proceedings of the International Rice Research Conference*. IRRI: Philippines, 1990.

Bonman, J.; Khush, G.; Nelson, R. Breeding Rice for Resistance to Pests. *Ann. Rev. Phytopathol.* **1992,** *30,* 507–528.

Bonman, J.; Rush, M. *Report on Rice Blast in Egypt*. Nat. Rice Inst.: Egypt, 1985.

Bouharmon, J.; Dekeyser, A.; Van Sint Jan, V.; Dogbe, Y. S. Application of Somaclonal Variation and In Vitro Selection to Rice Improvement. In: *Rice Genetics II, Proceedings of the Second International Rice Genetics Symposium*, 14–18 May 1990, International Rice Research Institute: Philippines, 1991; pp 271–277.

Bryan, G. T.; Wu, K. S.; Farrall, L.; Hershey, H. P.; McAdams, S. A.; Faulk, K. N.; Donaldson, G. K.; Tarchini, R.; Valent, B. A Single Amino Acid Difference Distinguishes Resistant and Susceptible Alleles of the Rice Blast Resistance Gene *Pi-ta*. *Plant Cell* **2000**, *12*(11), 2033–2046.

Campbell, M. A.; Heather, A. F.; Pamela, C. Engineering Pathogen Resistance in Crop Plants. *Transgen. Res.* **2002**, *11*(6), 599–613.

Causse, M. A.; Fulton, T. M.; Cho, Y. G.; Ahn, S. N.; Chunwongse, J.; Wu, K. S.; Xiao, J. H.; Yu, Z. H.; Ronald, P. C.; Harrington, S. E.; Second, G.; McCouch, S. R.; Tanksley, S. D. Saturated Molecular Map of the Rice Genome Based on an Interspecific Backcross Population. *Genetics* **1994**, *138*, 1251–1274.

Charcosset, A. Marker-assisted Introgression of Quantitative Trait Loci. *Genetics* **1994**, *147*, 1469–1485.

Chen, H.; Chen, Z.; Ni, S.; Zuo, S.; Pan, X.; Zhu, X. Pyramiding Three Genes with Resistance to Blast by Marker Assisted Selection to Improve Rice Blast Resistance of Jin23B. *Chin. J. Rice Sci.* **2008**, *22*, 23–27.

Chen, H.; Wang, S.; Xing, Y.; Xu, C.; Hayes, P. M.; Zhang, Q. Comparative Analyses of Genomic Locations and Race Specificities of Loci for Quantitative Resistance to *Pyricularia grisea* in Rice and Barley. *Proc. Natl. Acad. Sci. U.S.A.* **2003**, *100*, 2544–2549.

Chen, J.; Henny, R.; Devanand, P; Chao, C. AFLP Analysis of Nephthytis (*Syngonium podophyllum* Schott) Selected from Somaclonal Variants. *Plant Cell Rep.* **2006**, *24*(12), 743–749.

Chen, X. W.; Li, S. G.; Ma, Y. Q.; Li, H. Y.; Zhou, K. D.; Zhu, L. H. Marker-assisted Selection and Pyramiding for Three Blast Resistance Genes, *Pi-d (t) 1, Pi-b, Pi-ta2*, in rice. *Chin. J. Biotechnol.* **2004**, *20*, 708–714.

Cho, Y. C.; Kwon, S. W.; Suh, J. P.; Kim, J. J.; Lee, J. H.; Roh, J. H.; Oh, M. K.; Kim, M. K.; Ahn, S. N.; Koh, H. J. QTLs Identification and Confirmation of Field Resistance to Leaf Blast in Temperate Japonica Rice (*Oryza sativa* L.). *J. Crop Sci. Biotechnol.* **2008**, *11*, 269–276.

Choi, W.; Park, E.; Lee, E. Leaf Blast: A Computer Simulation Model for Leaf Blast Development on Rice. *Kor. J. Plant Pathol. (Korea R.)* **1988**, *4*, 25–32.

CIAT (Centre International de Agricultural Tropical). *Annual Report, P128*. CIAT: Cali, Colombia, 1982.

Collard, B. C.; Mackill, D. J. Marker-assisted Selection: An Approach for Precision Plant Breeding in the Twenty-first Century. *Philos. Trans. R. Soc. B* **2008**, *363*, 557–572.

Cornelissen, B. J. C.; Melchers, L. S. Strategies for Control of Fungal Diseases with Transgenic Plants. *Plant Physiol.* **2004**, *101*, 709–712.

Costanzo, S.; Jia, Y. Sequence Variation at the Rice Blast Resistance Gene *Pi-km* Locus: Implications for the Development of Allele Specific Markers. *Plant Sci.* **2010**, *178*, 523–530.

Couch, B. C.; Kohn, L. M. A Multilocus Gene Genealogy Concordant with Host Preference Indicates Segregation of a New Species, *Magnaporthe oryzae*, from *M. grisea*. *Mycologia* **2002**, *94*, 683–693.

Coutos-Thevenot, P.; Poinssot, B.; Bonomelli, A.; Yean, H.; Breda, C.; Buffard, D.; Esnault, R.; Hain, R.; Boulay, M. In Vitro Tolerance to *Botrytis cinerea* of Grapevine 41B Rootstock in Transgenic Plants Expressing the Stilbene Synthase Vst1 Gene under the Control of a Pathogen-inducible PR 10 Promoter. *J. Exp. Bot.* **2001**, *52*, 901–110.

Crill, P.; Ham, Y.; Beachell, H. The Rice Blast Disease in Korea and its Control with Race Prediction and Gene Rotation. In: *Evolution of the Gene Rotation Concept for Rice Blast Control*. International Rice Research Institute: Los Banos, Philippines, 1982; pp 123–130.

Das, A.; Soubam, D.; Singh, P.; Thakur, S.; Singh, N.; Sharma, T. A Novel Blast Resistance Gene, *Pi54rh* Cloned from Wild Species of Rice, *Oryza rhizomatis* Confers Broad Spectrum Resistance to *Magnaporthe oryzae. Funct. Int. Gen.* **2012**, *12*, 215–228.

Datnoff, L.; Deren, C.; Snyder, G. Silicon Fertilization for Disease Management of Rice in Florida. *Crop Prot.* **1997**, *16*(6), 525–531.

de Badan, A. C.; Guimaraes, E. P.; Ramis, C. Genetic Gain for Resistance to Blast in a Rice Population. In: *Population Improvement, a Way of Exploiting Rice Genetic Resources in Latin America*; Guimaraes, E. P., Ed.; Food and Agriculture Organization of the United Nations (FAO): Rome, **2005**; pp 299–329.

Dean, R. A.; Talbot, N. J.; Ebbole, D. J.; Farman, M. L.; Mitchell, T. K.; Orbach, M. J.; Pan, H. The Genome Sequence of the Rice Blast Fungus *Magnaporthe grisea. Nature* **2005**, *434*, 980–986.

El-Refaei, M. I. Epidemiology of Rice Blast Disease in the Tropics with Special Reference to the Leaf Wetness in Relation to Disease Development. In: Ph. D. thesis, Indian Agricultural research Institute: New Delhi, India, 1971.

El-Kazzaz, A. A. Inheritance of Resistance to *Fusarium oxysporum* f. sp. *Lycopersici* in F2 Tomato Plants via Tissue Culture. *Egypt J. Genet. Cytol.* **2001**, *30*, 51–59.

El-Kazzaz, A. A.; El-Mougy, N. S. Inheritance of Disease Resistance in Cucumber Plants to Root Rot Caused by *Fusarium solani* Using Tissue Culture Techniques. *Egypt. J. Phytopathol.* **2001**, *29*, 57–68.

El-Kazzaz, A. A.; Hanafy, M. S.; Abdel-Kader, M. M. In Vitro Selection of Resistant Rice Plants against Rice Blast Caused by *Pyricularia oryzae* via Tissue Culture Technique. *Arch. Phytopathol. Plant Prot.* **2009**, *42*(9), 847–856.

Ellis, M. B. *Dematiaceous Hyphomycetes*; Commonwealth Mycological Institute: Kew, England, 1971; p 680.

Ellis, M. B. *More Dematiaceous Hyphomycetes*; Commonwealth Mycological Institute: Kew, England, 1976; p 507.

Evans, D. A.; Sharp, W. R.; Medina-Filho, H. P. Somaclonal and Gametoclonal Variation. *Ann. J. Bot.* **1984**, *71*, 759–774.

Feng, D. R.; Wei, J. W.; Xu, X. P.; Xu, Y.; Li, B. J. Introduction of Multiple Antifungal Protein Genes into Rice and Preliminary Study on Resistance to *Pyricularia oryzae* of Transgenic Rices. *Acta Sci. Nat. Univ. Sunyatsen* **1999**, *38*(4), 62–66.

Fjellstrom, R.; Mc-clung, A. M.; Shank, A. R. SSR Markers Closely Linked to the Pi-z Locus are Useful for Selection of Blast Resistance in Abroad Array of Rice Germplasm. *Mol. Breed.* **2006**, *17*, 149–157.

Flor, H. The Complementary Genic Systems in Flax and Flax rust. *Adv. Genet.* **1956**, *8*, 29–54.

Fu, C.; Wu, T.; Liu, W.; Wang, F.; Li, J.; Zhu, X.; Huang, H.; Liu, Z. R.; Liao, Y.; Zhu, M. Genetic Improvement of Resistance to Blast and Bacterial Blight of the Elite Maintainer Line Rongfeng B in Hybrid Rice (*Oryza sativa* L.) by Using Marker-assisted Selection. *Afr. J. Biotechnol.* **2012**, *11*, 13104–13114.

Fujii, K.; Hayano-Saito, Y.; Shumiya, A.; Inoue, M. Genetical Mapping based on the RFLP Analysis for the Panicle Blast Resistance Derived from a Rice Parental Line St. No. 1. *Breed Sci.* **1995**, *45*, 209 (In Japanese).

Fujimaki, H. Recurrent Selection by Using Male Sterility for Rice Improvement. *Japan Agric. Res.* **1979**, *13*(3), 153–156.

Fujita, D.; Santos, R. E.; Ebron, L. A.; Telebanco-Yanoria, M. J.; Kato, H.; Kobayashi, S.; Uga, Y.; Araki, E.; Takai, T.; Tsunematsu, H.; Imbe, T.; Khush, G. S.; Brar, D. S.; Fukuta,

Y.; Kobayashi N. Development of Introgression Lines of an Indica-type Rice Variety, IR64, for Unique Agronomic Traits and Detection of the Responsible Chromosomal Regions. *Field Crops Res*. **2009**, *114*, 244–254.

Fukunaga, K.; Misato, T.; Ishii, I; Asakawa, M.; Katagiri, M. Research and Development of Antibiotics for Rice Blast Control. *Bull. Nat. Inst. Agric. Sci. Tokyo* **1968**, *22*, 1–94.

Fukuoka, S.; Okuno, K. QTL Analysis and Mapping of *pi21*, a Recessive Gene for Field Resistance to Rice Blast in Japanese Upland Rice. *Theor. Appl. Gent.* **2001**, *103*, 185–190.

Fukuoka, S.; Okuno, K. QTL Analysis and Mapping of *pi21*, a Recessive Gene for Field Resistance to Rice Blast in Japanese Upland Rice. *Theor. Appl. Genet.* **2001**, *103*, 185–190.

Fukuoka, S.; Okuno, K. QTL Analysis for Field Resistance to Rice Blast Using RFLP Markers. *Rice Genet. Newsl.* **1997**, *14*, 99.

Fukuta, Y.; Araki, E.; Yanoria, M. J. T.; Imbe, T.; Tsunematsu, H.; Kato, H.; Ebron, L. A.; Mercado-Escueta, D.; Khush, G. S. Development of Varieties for Blast Resistance in IRRI-Japan Collaborative Research Project. In: *Rice Blast Interaction with Rice and Control*; Kawasaki, S., Ed.; Kluwer: Dordrecht, 2004; pp 229–233.

Gangadharan, C.; Mathur, S. C. Di-ethyl Sulphate Induced Blast Resistant Mutants in Rice Variety Mtu. *Sci. Cult.* **1976**, *42*(4), 226–228.

Geng, X. S.; Yang, M. Z.; Huang, X. Q.; Cheng, Z. Q.; Fu, J.; Sun, T. Cloning and Analyzing of Rice Blast Resistance Gene *Pi-ta+* Allele from Jinghong Erect Type of Common Wild Rice (*Oryza rufipogon Griff*) in Yunnan. *Yi Chuan* **2008**, *30*, 109–114.

Gibson, G.; Muse, S. V. *A Primer of Genome Science*. Sinauer Associates: Sunderland, MA, **2009**.

Gouda, P. K.; Saikumar, S.; Varma, C. M.; Nagesh, K.; Thippeswamy, S.; Shenoy, V. Marker-assisted Breeding of *Pi-1*and *Piz-5* Genes Imparting Resistance to Rice Blast in PRR78 Restorer Line of PusaRH-10 Basmati Rice Hybrid. *Plant Breed.* **2013**, *132*, 61–69.

Gowda, M.; Barmanm, S. R.; Chattoo, B. B. Molecular Mapping of a Novel Blast Resistance Gene *Pi38* in Rice Using SSLP and AFLP Markers. *Plant Breed.* **2006**, *125*, 596–599.

Guimaraes, E. P.; Correa-Victoria, F. Use of Recurrent Selection for Develop Resistance *Pyricularria grisea* Sacc. On rice. In: *Advances in Rice Population Improvement*; Guima-raes, E. P., Ed.; Embrapa Rice and Beans: Santo Antonio de Goias, 2000; pp 165–175.

Guo, L.; Zhao, H.; Wang, J.; Liu, H.; Zheng, H.; Sun, J.; Yang, L.; Sha, H.; Zou, D. Dissection of QTL Alleles for Blast Resistance Based on Linkage and Linkage Disequilibrium Mapping in *Japonica* rice Seedlings. *Aust. Plant Pathol.* **2015**, *45*, 209–218.

Harada, Y. Studies on a New Antibiotic for Rice Blast Control. In: Lecture Given at the Annual Meeting of the Agricultural Chemical Society of Japan, **1955**.

Hasan, M.; Rafi, M. Y.; Ismail, M. R.; Mahmood, M.; Rahim, H. A.; Alam, M. Marker Assisted Backcrossing: A Useful Method for Rice Improvement. *Biotechnol. Biotechnol.. Equip.* **2015**, *29*, 237–254.

Higashi, T.; Sato, H.; Horisue, N.; Fujimaki, H. Breeding of Isogenic Lines for Blast Resistance in Rice. 1. Comparison of Characters Between B4F$_2$ Lines and their Recurrent, "Nipponbare." *Breed. Sci.* **1981**, *31*, 46 (in Japanese).

Hittalmani, S.; Parco, A.; Mew, T.; Zeigler, R.; Huang, N. Fine Mapping and DNA Marker-assisted Pyramiding of the Three Major Genes for Blast Resistance in Rice. *Theor. Appl. Genet.* **2000**, *100*, 1121–1128.

Hori, M.; Arata, T.; Inoue, Y. Studies on the Forecasting Method of Blast Disease. VI. Forecasting by the Degree of Accumulated Starch in the Sheath of Rice Plant. *Ann. Phytopathol. Soc. Jpn.* **1960**, *25*(1), 2.

Hori, S. *Blast Disease of Rice Plants. Special Report*; Imperial Agricultural Experimental Station: Tokyo, 1898, Vol. 1; pp 1–36.

Horisue, N.; Higashi, T.; Sato, H.; Koizumi, S. Breeding of Isogenic Lines for Blast Resistance in Rice. 2. Agronomical Characteristics of Kanto-IL1-14. *Breed Sci.* **1984**, *34*, 316 (in Japanese).

Hospital, F. Size of Donor Chromosome Segments around Introgressed Loci and Reduction of Linkaged Ragin Marker-assisted Backcross Programs. *Genetics* **2001**, *158*, 1363–1379.

Huang, C. L.; Hwang, S. Y.; Chiang, Y. C.; Lin, T. P. Molecular Evolution of the *Pi-ta* Gene Resistant to Rice Blast in Wild Rice (*Oryza rufipogon*). *Genetics* **2008**, *179*, 1527–1538.

Hubert, J.; Mabagala, R.; Mamiro, D. Efficacy of Selected Plant Extracts against *Pyricularia grisea*, Causal Agent of Rice Blast Disease. *Am. J. Plant Sci.* **2015**, *6*, 602–611.

Ikeda, M. Influence of Application of Various Nitrogenous Fertilizers on Silica Content of Rice Plants. *Tottor. Soc. Agric. Sci.* **1933**, *4*, 23–33.

International Rice Research Institute (IRRI). International Rice Research Institute, 2010. http://irri.org/component/itpgooglesearch/search?q=annual+reports [last accessed: 10th Nov 2013].

Ise, K. Effect of Mixing Planting of Near Isogenic Lines of 'Nippobare' Rice to Reduce Blast Disease. *Breed. Sci.* **1990**, *40*(Supp. 1), 288 (in Japanese).

Ishizaki, K.; Hoshi, T.; Abe, S.; Sasaki, Y.; Kobayashi, K.; Kasaneyama, H.; Matsui, T.; Azuma, S. Breeding of Blast Resistant Isogenic Lines in Rice Variety "Koshihikara" and Evaluation of their Characters. *Breed. Sci.* **2005**, *55*(3), 371–377.

Ito, S.; Sakamoto, M. Studies on Rice Blast. *Res. Hokkaido Univ. Bot. Lab. Fac. Agric. Rep.*, 1939.

Jain, R. K.; Jain, S. Transgenic Strategies for Genetic Improvement of Basmati Rice. *Ind. J. Exp. Biol.* **2000**, *38*(1), 6–17.

Jena, K. K.; Khush, G. S. Exploitation of Species in Rice Improvement Opportunities, Achievements and Future Challenges. In: *Rice Breeding and Genetic Research Priorities and Challenges*; Nanda, J. S., Ed.; Science Publication: Enfield, 2000; pp 69–284.

Jensen, N. F. A Diallel Selective Mating System for Cereal Breeding. *Crop Sci.* **1970**, *10*(6), 629–635.

Jeung, J. U.; Kim, B. R.; Cho, Y. C. A Novel Gene, *Pi40(t)*, Linked to the DNA Markers Derived from NBS-LRR Motifs Confers Broad Spectrum of Blast Resistance in Rice. *Theor. Appl. Genet.* **2007**, *115*, 1163–1177.

Jiang, H.; Feng, Y.; Bao, L.; Li, X.; Gao, G.; Zhang, Q. Improving Blast Resistance of Jin 23B and its Hybrid Rice by Marker-assisted Gene Pyramiding. *Mol. Breed.* **2012**, *30*, 1679–1688.

Joseph, M.; Gopalakrishnan, S.; Sharma, R. K.; Singh, V. P.; Singh, A. K.; Singh, N. K.; Mohapatra, T. Combining Bacterial Blight Resistance and Basmati Quality Characteristics by Phenotypic and Molecular Marker-assisted Selection in Rice. *Mol. Breed.* **2004**, *13*(4), 377–387.

Joshi, R.; Nayak, S. Gene Pyramiding—A Broad Spectrum Technique for Developing Durable Stress Resistance in Crops. *Biotechnol. Mol. Biol. Rev.* **2010**, *5*, 51–60.

Kahn, R. P.; Libby, J. L. The Effect of Environmental Factors and Plant Age on the Infection of Rice by the Blast Fungus, *Pyricularia oryzae*. *Phytopathology* **1958**, *48*, 25–30.

Kanzaki, H.; Nirasasawa, S.; Saitoh, H.; Ito, M. Over Expression of the Wasabi Defensin Gene Confers Enhanced Resistance to Blast Fungus (*Magnaporthe grisea*) in Transgenic Rice. *Theor. Appl. Genet.* **2002**, *105*, 809–814.

Katagiri, M.; Uesugi, Y. *In Vitro* Selection of Mutants of *Pyricularia oryzae* Resistant to Fungicides Rice. *Ann. Phytopathol. Soc. Jpn.* **1978**, *44*, 218–219.

Katiyar-Agarwal, S.; Jin, H. Role of Small RNAs in Host–Microbe Interactions. *Annu. Rev. Phytopathol.* **2010**, *48,* 225–246.

Kato, H. Biological and Genetic Aspects in the Perfect Stage of Blast Fungus *Pyricularia oryzae* and Its Allies. In: Mutation Breeding for Disease Resistance: Gamma Field Symposia, 1978, No. 17.

Kato, H.; Yamaguchi, T. The Perfect State of *Pyricularia oryzae* CAV in Culture. *Ann. Phytol. Soc. Jpn.* **1982**, *142*, 507–510.

Kaundal, R.; Kapoor, A. S.; Raghava, G. P. Machine Learning Techniques in Disease Forecasting: A Case Study on Rice Blast Prediction. *BMC Bioinf.* **2006**, *7*(1), 485.

Kaur, S.; Padmanabhan, S. Y.; Rao, M. Induction of Resistance to Blast Disease (*Pyricularia oryzae*) in the High Yielding Variety, Ratna (IRE 9 TKM 6). In: Proceedings of the IAEA Research Coordination Geoling, Ames, IA, 1971; pp 141–145.

Kawamura, E.; Ono, K. Study on the Relation between the Pre-infection Behavior of Rice Blast Fungus, *Pyricularia oryzae*, and Water Droplets on Rice Plant Leaves. *Bull. Nat. Agric. Exp. Stat.* **1948**, *4*, 1–12.

Kawashima, R. Influence of Silica on Rice Blast Disease. *Japanese J. Soil Sci. Plant Nutr.* **1927**, *1*, 86–91.

Khambanonda, P. Mutation Breeding in Rice for High Yield and Better Blast Resistance. *Thai Agric. Sci.* **1978**, *11*(4), 263–271.

Khraiwesh, B.; Arif, M. A.; Seumel, G. I.; Ossowski, S.; Weigel, D.; Reski, R.; Frank, W. Transcriptional Control of Gene Expression by MicroRNAs. *Cell* **2010**, *140*, 111–122.

Khraiwesh, B.; Zhu, J. K.; Zhu, J. Role of miRNAs and siRNAs in Biotic and Abiotic Stress Responses of Plants. *Biochim. Biophys. Acta* **2012**, *1819*, 137–148.

Khush, G. S. Breeding Methods and Procedures Employed at IRRI for Developing Rice Germplasm with Multiple Resistance to Diseases and Insects. In: Symposium on Methods of Crop Breeding. *Trop. Agric. Res. Ser.* **1978**, *11*, 69–76.

Khush, G. S.; Mackill, D. J.; Sidhu, G. S. Breeding Rice for Resistance to Bacterial Blight. In: Pages 207–217, In: Proc. International Workshop on Bacterial Blight of Rice, International Rice Research Institute: Los Banos, Philippines, 1989.

Kim, C. K.; Kim, C. H. *The Rice Leaf Blast Simulation Model EPIBLAST Systems Approaches for Agricultural Development*; 1993; pp 309–321.

Kim, S.; Yang, J. Y.; Xu, J.; Jang, I. C.; Prigge, M. J.; Chua, N. H.; Two Cap-binding Proteins CBP20 and CBP80 are Involved in Processing Primary MicroRNAs. *Plant Cell Physiol.* **2008**, *49*, 1634–1644.

Kiyosawa, S. Gene Analysis for Blast Resistance. *Oryza* **1982**, *18*, 196–203.

Koide, Y.; Kobayashi, N.; Xu, D.; Fukuta, Y. Resistance Genes and Selection DNA Markers for Blast Disease in Rice (*Oryza sativa* L.). *Jpn. Agric. Res. Q.* **2009**, *43*(4), 255–280.

Koizumi, S. Durability of Resistance to Rice Blast Disease. *JIRCAS Working Rep.* **2007**, *53*, 1–10.

Koizumi, S.; Fuji, S. Variation of Field Resistance to Leaf Blast in a Rice Strain, Chubu 32, Due to Isolates of the Pathogen. *Res. Bull. Aichi. Agric. Res. Cent.* **1994**, *27*, 85–93.

Koizumi, S.; Tani, T.; Fuji, S. Control of rice blast by multilines. *J. Agric. Sci.* **1996**, *51*, 89–93 (in Japanese).

Korinsaka, S.; Sirithunya, P.; Meakwatanakarnd, P.; Sarkarunge, S.; Vanavichitc, A.; Toojinda, T. Changing Allele Frequencies Associated with Specific Resistance Genes to Leaf Blast in Backcross Introgression Lines of Khao Dawk Mali 105 Developed from a Conventional Selection Program. *Field Crop. Res.* **2011**, *122*, 32–39.

Kumar, P.; Pathania, S.; Katoch, P.; Sharma, T. R.; Plaha, P.; Rathore, R. Genetics and Physical Mapping of Blast Resistance Gene *Pi42(t)* on the Short Arm of Rice Chromosome 12. *Mol. Breed.* **2010**, *25*, 217–228.

Kumari, A.; Das, A.; Devanna, B.; Thakur, S.; Singh, P.; Singh, N.; Sharma, T. Mining of Rice Blast Resistance Gene *Pi54* Shows Effect of Single Nucleotide Polymorphisms on Phenotypic Expression of the Alleles. *Eur. J. Plant Pathol.* **2013**, *137*, 55–65.

Larkin, P. J.; Scowcroft, W. R. Somaclonal Variation—A Novel Source of Variability from Cell Cultures for Plant Improvement. *Theor. Appl. Genet.* **1981**, *60*(4), 197–214.

Lee, E. J.; Kim, H. K.; Ryn, J. D. Studies on the Resistance of Rice Varieties to the Blast Fungus, *Pyricularia oryzae* Cav. [in Korean]. Research Report for 1976. Institute of Agricultural Sciences, Office of Rural Development: Korea, 1976.

Lee, S.; Costanzo, S.; Jia, Y.; Olsen, K. M.; Caicedo, A. L. Evolutionary Dynamics of the Genomic Region around the Blast Resistance Gene *Pi-ta* in AA Genome *Oryza* Species. *Genetics* **2009**, *183*, 1315–1325.

Lee, S.; Jia, Y.; Jia, M.; Gealy, D. R.; Olsen, K. M.; Caicedo A. L. Molecular Evolution of the Rice Blast Resistance Gene *Pi-ta* in Invasive Weedy Rice in the USA. *PLoS One* **2011**, *6*, e26260.

Lee, Y.; Kim, M.; Han, J.; Yeom, K. H.; Lee, S.; Baek, S. H.; Kim, V. N. MicroRNA Genes are Transcribed by RNA Polymerase II. *EMBO J.* **2004**, *23*, 4051–4060.

Lestari, P.; Trijatmiko, K. R.; Warsun, A.; Ona, I.; Cruz, C. V.; Bustamam, M. Mapping Quantitative Trait Loci Conferring Blast Resistance in Upland indica Rice (*Oryza sativa* L.). *J. Crop Sci. Biotechnol.* **2011**, *14*, 57–63.

Leung, H.; Zhu, Y.; Resulla-Molina, I.; Fan, J. X.; Chen, H.; Pangga, I.; Vera, C. C.; Mew, T. W. Using Genetic Diversity to Achieve Sustainable Rice Disease Management. *Plant Dis.* **2003**, *87*(10), 1156–1169.

Li, Y.; Lu, Y. G.; Shi, Y.; Wu, L.; Xu, Y. J.; Huang, F.; Guo, X. Y.; Zhang, Y.; Fan, J.; Zhao, J. Q.; Zhang, H. Y.; Xu, P. Z.; Zhou, J. M.; Wu, X. J.; Wang, P. R.; Wang, W. M. Multiple Rice MicroRNAs Are Involved in Immunity against the Blast Fungus *Magnaporthe oryzae*. *Plant Physiol.* **2014**, *164*, 1077–1092.

Lin, F.; Liu, Y.; Wang, L.; Liu, X.; Pan, Q. A High-resolution Map of the Rice Blast Resistance Gene Pi15 Constructed by Sequence Ready Markers. *Plant Breed.* **2007**, *126*(3), 287–290.

Link, K. C.; Ou, S. H. Standardization of the International Race Numbers of *Pyricularia oryzae*. *Phytopathology* **1969**, *59*, 339–342.

Liu, P.; Zhu, Z.; Lu, Y. Marker-Assisted Selection in Segregating Generations of Self-fertilizing Crops. *Theor. Appl. Genet.* **2004**, *109*, 370–376.

Liu, S.; Li, X.; Wang, C.; Li, X.; He, Y. Improvement of Resistance to Rice Blast in Zhenshan 97 by Molecular Marker-aided Selection. *Acta Bot. Sin.* **2002**, *45*, 1346–1350.

Liu, X. Q.; Wang, L.; Chen, S.; Lin, F.; Pan, Q. H. Genetic and Physical Mapping of *Pi36(t),* a Novel Rice Blast Resistance Gene Located on Rice Chromosome 8. *Mol. Genet. Genom.* **2005**, *274*(4), 394–401.

Liu, X.; Lin, F.; Wang, L.; Pan, Q. The *In Silico* Map-based Cloning of *Pi36*, a Rice Coiled-Nucleotide-Binding Site Leucine-rich Repeat Gene that Confers Race Specific Resistance to the Blast Fungus. *Genetics* **2007**, *176*, 2541–2549.

Liu, Y.; Zhu, X. Y.; Zhang, S.; Bernardo, M.; Edwards, J.; Galbraith, D. W.; Leach, J; Zhang, G; Liu, B; Leung, H. Dissecting Quantitative Resistance against Blast Disease Using Heterogeneous Inbred Family Lines in Rice. *Theor. Appl. Genet.* **2011**, *122*, 341–353.

Lopez-Gerena, J. Mapping QTL Controlling Durable Resistance to Rice Blast in the Cultivar Oryzica Llanos 5. Ph. D. Thesis, Universidad del Valle, Plant Pathology College of Agriculture, Cali, Colombia and Kansas State University: Manhatten, KS, 2006.

Lv, Q.; Xu, X.; Shang, J.; Jiang, G.; Pang, Z.; Zhou, Z.; Wang, J.; Liu, Y.; T Li, T.; Li, X.; Xu, J.; Cheng, Z.; Zhao, X.; Li, X.; Zhu, L. Functional Analysis of Pid3-A4, an Ortholog of Rice Blast Resistance Gene Pid3 Revealed by Allele Mining in Common Wild Rice. *Phytopathology* **2013**, *103*, 594–599.

Mahdian, S.; Shahsavari, A. Pyramiding of Blast Resistance Genes *Pi-1* and *Pi-2* in Tarom Mahalli Rice Cultivar. *Seed Plant Improv. J.* **2013**, *29*, 391–395.

Manibhushanrao, K.; Krishnan, P. Epidemiology of Blast (EPIBLA): A Simulation Model and Forecasting System for Tropical Rice in India. In: Rice Blast Modeling and Forecasting; IRRI: Manila, Philippines, 1991; pp 31–38.

Matsumoto, S. Pathogenic Race Occurred on a Blast Resistant Variety Reiho [in Japanese, English Summary]. *Proc. Assoc. Plant Prot.* **1974**, *20*, 72–74.

Matsunaga, K. Breeding of a Multiline Rice Cultivar "Sasanishiki BL" and its Use for Control of Blast Disease in Miyagi Prefecture. *J. Agric. Sci.* **1996**, *51*, 173–176 (in Japanese).

McCouch, S. R.; Sweeney, M.; Li J.; Jiang H.; Thomson M.; Septiningsih E.; Edwards J.; Moncada, P.; Xiao, J.; Garris, A.; Tai, T.; Martinez, C.; Tohme, J.; Sugiono, M.; McClung, A.; Yuan L. P.; Ahn S. N. Through the Genetic Bottleneck: *O. rufipogon* as a Source of Trait-enhancing Alleles for *O. sativa. Euphytica* **2007**, *154*, 317–339.

Ming, X. T.; Wang, L. J.; An, C. C.; Yuan, H. Y.; Zheng, H. H.; Chen, Z. L. Introducing Trichosanthin Gene into Rice Mediated by *Agrobacterium tumefacien* and Testing the Activity of Resistance to Blast. *Chin. Sci. Bull.* **2000**, *45*, 1080–1084.

Miyake, K.; Ikeda, M. Influence of Silica Application on Rice Blast. *Jpn. J. Soil Sci. Plant Nutr.* **1932**, *6*, 53–76.

Miyamoto, M..; Yano, M.; Hirasawa, H. Mapping of Quantitative Trait Loci Conferring Blast Field Resistance in the Japanese Upland Rice Variety Kahei. *Breed. Sci.* **2001**, *51*, 257–261.

Mohamad, O.; Nazir, B. M.; Alias, I.; Azlan, S.; Abdul, Rahim, H.; Abdullah, M. Z.; Othman, O.; Hadzim, K.; Saad, A.; Habibuddin, H.; Golam, F. Development of Improved Rice Varieties through the Use of Induced Mutations in Malaysia. *Plant Mut. Rep.* **2006**, *1*(1), 27–33.

Mu, C.; Liu, X.; Lu, Q.; Jiang, X.; Zhu, C. Biological Control of Rice Blast by *Bacillus subtilis* B-332 strain. *Acta Phytophyl. Sin.* **2007**, *34*(2), 123–128.

Murata, J.; Kuribayashi, K.; Kawai, I. Studies on the Control of Rice Blast III. Influence of Homemade Manure on Blast Disease. *Nogi Kairy. Shiry.* **1933**, *64*, 1–138.

Nagato, Y.; Yoshimura, A. Report of the Committee on Gene Symbolization, Nomenclature and Linkage Groups. *Rice Genet. Newslett.* **1998**, *15*, 13–74.

Nakajima, T. Mechanism of Rice Blast Disease Control by Multilines. *J. Agric. Sci.* **1994**, *49*, 390–395.

Nakajima, T.; Sonoda, R.; Yaegashi, H. Effect of a Multiline of Rice Cultivar Sasanishiki and its Isogenic Lines on Suppressing Rice Blast Disease. *Ann. Phytopathol. Soc. Jpn.* **1996**, *62*, 227–233.

Narayanan, N.; Baisakh, N.; Cruz, V.; Gnanamanickam, S.; Datta, K.; Datta, S. Molecular Breeding for the Development of Blast and Bacterial Blight Resistance in Rice cv. *IR*50. *Crop Sci.* **2002**, *42*, 2072–2079.

Nguyen, T. T. T.; Koizumi, S.; La, T. N.; Zenbayashi, K. S.; Ashizawa, T.; Yasuda, N.; Imazaki, I.; Miyasaka, A. *Pi35*(t), a New Gene Conferring Partial Resistance to Leaf Blast in the Rice Cultivar Hokkai 188. *Theor. Appl. Genet.* **2006**, *113*, 697–704.

Nishizawa, Y.; Nishio, Z.; Nakazono, K.; Soma, M.; Nakajima, E.; Ugaki, M.; Hibi, T. Enhanced Resistance to Blast (*Magnaporthe grisea*) in Transgenic Japonica Rice by Constitutive Expression of Rice Chitinase. *Theor. Appl. Genet.* **1999**, *99*, 383–390.

Noenplab, A.; Vanavichit, A.; Toojinda, T.; Sirithunya, P.; Tragoonrung, S.; Sriprakhon, S. QTL Mapping for Leaf and Neck Blast Resistance in Khao Dawk Mali105 and Jao Hom Nin Recombinant Inbred Lines. *Sci. Asia* **2006**, *32*, 133–142.

Nottegham, J. L. Durable Resistance to Blast Disease. In: *Durability of Disease Resistance*; Jacobs, Th., Parlievliet, J. E.; Kluwer: London, 1993; pp 125–134.

Ogawa, M. Studies on Blast Control of Ceresan Lime. *Ohug. Shiko. Agric. Res.* **1953**, *3*, 1–5.

Okamoto, M. On the Characteristics of Kasumin, Antibiotic Fungicide. *Jpn. Pest. Inform.* **1972**, *10*, 66–69.

Onodera, I. Chemical Studies on Rice Blast (*Dactylaria parasitance Cavara*). *J. Sci. Agric. Soc.* **1917**, *180*, 606–617.

Ou, S. H. *Rice Diseases*. International Rice Research Institute: Manila, Philippines, 1985.

Ou, S.; Nuque, F.; Ebron, T.; Awoderu, V. A Type of Stable Resistance to Blast Disease of Rice. *Phytopathology* **1971**, *61*(6), 703–706.

Pachon, J. G. *Evaluation of the Potential Use of Somaclonal Variation in the Improvement of Some Characters of Economic Importance in Rice (Oryza sativa L.)*; Pontificia Universidad Javeriana: Bogota, 1989; p 94.

Padmanabhan, S. The Role of Therapeutic Treatments in Plant Disease Control with Special Reference to Rice Diseases. *Ind. Phytol. Soc. Bull.* **1963**, *1*, 79–84.

Pan, Q. H.; Hu, Z. D.; Takatoshi, T.; Wang, L. Fine Mapping of the Blast Resistance Gene *Pi15*, Linked to *Pii*, on Rice Chromosome 9. *Acta Bot. Sin.* **2003**, *45*, 871–877.

Parimelazhagan, T. Botanical Fungicide for the Control of Rice Blast Disease. *Bioved* **2001**, *12*, 11–15.

Parlevliet, J. E. Components of Resistance that Reduce the Rate of Epidemic Development. *Ann. Rev. Phytopathol.* **1979**, *17*, 203–222.

Parlevliet, J. E. Identification and Evaluation of Quantitative Resistance. In: *Plant Disease Epidemiology: Genetics, Resistance, and Management*; Leonard, K. J.; Fry, W. E., Eds.; Vol. 2; McGraw-Hill: New York, 1988; pp 215–247.

Pinta, W.; Toojinda, T.; Thummabenjapone, P.; Sanitchon, J. Pyramiding of Blast and Bacterial Leaf Blight Resistance Genes in to Rice Cultivar RD6 Using Marker Assisted Selection. *Afr. J. Biotechnol.* **2013**, *12*, 4432–4438.

Prabhu, A. S.; Filippi, M. C.; Zimmermann, F. J. P. Cultivar Response to Fungicide Application in Relation to Rice Blast Control, Productivity and Sustainability. *Pesq. Agropec. Bras.* **2003**, *38*, 11–17.

Prabhu, A.; Morais, O. Blast Disease Management in Upland rice in Brazil. In: Proceedings of Symposium on Progress in Upland Rice Research, International Rice Research Institute: Manila, Philippines, 1986.

Qu, S. H;, Liu, G. F.; Zhou, B.; Bellizzi, M.; Zeng, L. R.; Dai, L. Y.; Han, B.; Wang, G. L. The Broad-spectrum Blast Resistance Gene Pi9 Encodes an NBS-LRR Protein and is a Member of a Multigene Family in rice. *Genetics* **2006**, *172*(3), 1901–1914.

Rahim, H. A.; Zarifth, S. K.; Bhuiyan, M. A. R.; Narimah, M. K.; Wickneswari, R.; Abdullah, M. Z.; et al. Evaluation and Characterization of Advanced Rice Mutant Line of Rice (*Oryza sativa*), MR219-4 and MR219-9 under Drought Condition. In: *Proceedings of the Research and Development Seminar*, 26–28 September 2012, Vol. 44 Bangi; pp 1–15.

Rahman, L.; Khanam, S; Jaehwan, R; Mapping of QTLs Involved in Resistance to Rice Blast (*Magnaporthe grisea*) Using *Oryza minuta* Introgression Lines. *Czech. J. Genet. Plant Breed.* **2011**, *47*, 85–94.

Rai, A. K.; Kumar, S. P.; Gupta, S. K.; Gautam, N.; Singh, N. K.; Sharma, T. R. Functional Complementation of Rice Blast Resistance Gene *Pi-kh* (*Pi54*) Conferring Resistance to Diverse Strains of *Magnaporthe oryzae. J. Plant Biochem. Biotechnol.* **2011**, *20*, 55–65.

Ramkumar, G.; Biswal, A.; Mohan, K. M.; Sakthivel, K.; Sivaranjani, A.; Neeraja, C.; Ram, T.; Balachandran, S.; Sundaram, R.; Prasad, M. Identifying Novel Alleles of Rice Blast Resistance Genes *Pikh* and *Pita* through Allele Mining. *Int. Rice Res. Notes* **2012**, *35*, 1–6.

Rangel, P. H. N.; Cordeiro, A. C. C.; Lopes, S. I. G.; de Morais, O. P.; Brondani, C.; Brondani, R. P. V.; Yokoyama, S.; Schiocchet, M.; Bacha, R.; Ishy, T. Advances in Population Improvement of Irrigated Rice in Brazil. In: *Population Improvement, a Way of Exploiting Rice Genetic Resources in Latin America*; Guimaraes, E. P., Eds.; Food and Agriculture Organization of the United Nations (FAO): Rome, 2005; pp 145–186.

Rathore, R.; Awasthi, A.; Katoch, A.; Swaranjli; Sharma, T. R. Identification and Mapping of a New Blast Resistance Gene "*Pi*-W(T)" from Wild Rice (*Oryza rufipogon*). In: International Symposium on Plant Biotechnology for Food Security, Feb. 21–24, 2012, NASC Complex: New Delhi, India, 2012.

Ruiz-Ferrer, V.; Voinnet, O. Roles of Plant Small RNAs in Biotic Stress Responses. *Ann. Rev. Plant Biol.* **2009**, *60*, 485–510.

Sallaud, C.; Lorieux, M.; Roumen, E.; Tharreau, D.; Berruyer, R.; Svestasrani, P.; Garsmeur, O.; Ghesquiere, A.; Notteghem, J. L. Identification of Five New Blast Resistance Genes in the Highly Blast-resistant Rice Variety IR64 Using a QTL Mapping Strategy. *Theor. Appl. Genet.* **2003**, *106*, 794–803.

Sato, H.; Takeuchi, Y.; Hirabayashi, H.; Nemoto, H.; Hirayama, M.; Kato, H.; Imbe, T. A. Mapping QTLs for Field Resistance to Rice Blast in the Japanese Upland Rice Variety Norin12. *Breeding Sci.* **2006**, *56*, 415–418.

Seebold, K. W. The Influence of Silicon Fertilization on the Development and Control of Blast, Caused by *Magnaporthe Grisea* (Hebert) Barr, in Upland Rice. Ph. D. Thesis, University of Florida: Gainesville, 1988.

Seebold, K. W.; Datnoff, L. E.; Correa-Victoria, F. J.; Kucharek, T. A.; Snyder, G. H. Effect of Silicon Rate and Host Resistance on Blast, Scald, and Yield of Upland Rice. *Plant Dis.* **2000**, *84*, 871–876.

Shang, J. J.; Tao, Y.; Chen, X. W.; Zou, Y.; Lei, C. L.; Wang, J.; Li, X. B.; Zhao, X. F.; Zhang, M. J.; Lu, Z. K.; Xu, J. C.; Cheng, Z. K.; Wan, J. M.; Zhu, L. H. Identification of a New Rice Blast Resistance Gene, *Pid3*, by Genomewide Comparison of Paired Nucleotide-binding Site-leucine-rich Repeat Genes and their Pseudogene Alleles between the Two Sequenced Rice Genomes. *Genetics* **2009**, *182*, 1303–1311.

Sharma, T. R.; Madhav, M. S.; Singh, B. K.; Shanker, P.; Jana, T. K.; Dalal, V.; Pandit, A.; Singh, A.; Gaikwad, K.; Upreti, H. C.; Singh, N. K. High-resolution Mapping, Cloning and Molecular Characterization of the *Pi-kh* Gene of Rice, which Confers Resistance to *M. grisea. Mol. Genet. Gen.* **2005**, *274*, 569–578.

Sharma, T. R.; Rai, A. K.; Gupta, S. K.; Singh, N. K. Broad-spectrum Blast Resistance Gene *Pi-k*[h] Cloned from Rice Line Tetep Designated *Pi54. J. Plant Biochem. Biotechnol.* **2010**, *191*, 87–89.

Sharma, T. R.; Rai, A. K.; Gupta, S. K.; Vijayan, J.; Devanna, B. N.; Ray, S. Rice Blast Management through Host–Plant Resistance: Retrospect and Prospects. *Agric. Res.* **2012**, *1*, 37–52.

Shinoda, H.; Toriyama, K.; Yunoki, T.; Ezuka, A.; Sakurai, Y. Studies on the Varietal Resistance of Rice to Blast. Linkage Relationship of Blast Resistance Genes. *Bull. Chugoku Agric. Exp. Stn. Set. A* **1971**, *20*, 1–25.

Shu, Q. Y. Induced Plant Mutations in the Genomics Era. Food and Agriculture Organization of the United Nations: Rome, 2009; pp 425–427.

Shu, Q.; Wu, D.; Xia, Y. The Most Widely Cultivated Rice Variety 'Zhefu 802' in China and its Geneology. *MBNL* **1997**, *43*, 3–5.

Siddiq, E. A. Rice. In: *50 Years of Crop Science Research in India*; In: Paroda, R. S.; Chadha, K. L., Eds.; ICAR (Indian Council of Agricultural Research): India, 1996.

Silue, D.; Tharreau, D.; Notteghem, J. L. Identification of *Magnaprothe grisea* Avirulence Genes to Seven Rice Cultivars. *Phytopathology* **1992**, *82*, 1462–1467.

Singh, A.; Singh, V. K.; Singh, S.; Pandian, R.; Ellur, R. K.; Singh, D.; Bhowmick, P. K.; Gopala Krishnan, S.; Nagarajan, M.; Vinod, K. K.; Singh, U. D.; Prabhu, K. V.; Sharma, T. R.; Mohapatra, T.; Singh, A. K. Molecular Breeding for the Development of Multiple Disease Resistance in Basmati Rice. *AoB Plants* **2012a**, pls029. DOI:10.1093/aobpla/pls029.

Singh, N.; Singh, R. S. Lysis of *Fusarium oxysporium* f. *udum* Caused by Soil Amended with Organic Matter. *J. Mycol. Plant Pathol.* **1980**, *10*, 146–150.

Singh, V. K.; Singh, A.; Singh, S. P.; Ellur, R. K.; Singh, D.; Gopalakrishnan, S.; Nagarajan, M.; Vinod, K. K.; Singh, U. D.; Rathore, R.; Prasanthi, S. K.; Agrawal, P. K.; Bhatt, J. C.; Mohapatra, T.; Prabhu, K. V.; Singh, A. K. Incorporation of Blast Resistance Gene in Elite Basmati Rice Restorer Line PRR78, Using Marker Assisted Selection. *Field Crop Res.* **2012b**, *128*, 8–16.

Singh, V. K.; Singh, A.; Singh, S.; Ellur, R. K.; Singh, D.; Gopala, K. S. Marker-assisted Simultaneous but Step Wise Back Cross Breeding for Pyramiding Blast Resistance Genes *Piz5* and *Pi54* into an Elite Basmati Rice Restorer Line 'PRR78'. *Plant Breed.* **2011**, *132*, 486–495.

Sirithunya, P.; Tragoonrung, S.; Vanavichit, A.; Pa-in, N.; Vongsaprom, C.; Toojinda, T. Quantitative Trait Loci Associated with Leaf and Neck Blast Resistance in Recombinant Inbred Line Population of Rice (*Oryza sativa*). *DNA Res.* **2002**, *9*, 79–88.

Someya, N.; Numata, S.; Nakajima, M.; Hasebe, A; Akutsu, K. Influence of Rice-isolated Bacteria on Chitinase Production by the Biocontrol Bacterium *Serratia marcescens* strain B2 and the Genetically Modified Rice Epiphytic Bacterium. *J. Gen. Plant Pathol.* **2004**, *70*(6), 371–375.

Song, F.; Goodman, R. M. Molecular Biology of Disease Resistance in Rice. *Physiol. Mol. Plant Pathol.* **2001**, *59*, 1–11.

Spence, C.; Alff, E.; Johnson, C.; Ramos, C.; Donofrio, N.; Sundaresan, V.; Bais, H. Natural Rice Rhizospheric Microbes Suppress Rice Blast Infections. *BMC Plant Biol.* **2014**, *14*, 130.

Sreewongchai, T.; Toojinda, T.; Thanintorn, N.; Kosawang, C.; Vanavichit, A.; Tharreau, D.; Sirithunya, P. Development of Elite Indica Rice Lines with Wide Spectrum of Resistance to Thai blast Isolates by Pyramiding Multiple Resistance QTLs. *Plant Breed.* **2010**, *129*, 176–180.

Stark-Lorenzen, P.; Nelke, B.; Hanler, G.; Mühlbach, H. P.; Thomzik, J. E. Transfer of a Grapevine Stilbene Synthase Gene to Rice (*Oryza sativa* L.). *Plant Cell Rep.* **1997**, *16*, 668–673.

Stark-Lorenzen, P.; Nelke, B.; Hanler, G.; Muhlbach, H. P.; Thomzik, J. E. Transfer of a Stilbene Synthase Gene to rice (*Oryza sativa* L.). *Plant Cell Rep.* **1997**, *16*, 668–673.

Sundaram, R. M.; Vishnupriya, M.; Laha, G. S.; Rani, N. S., Rao, P. S., Balachandran, S. M., Reddy, G. A.; Sarma, N. P.; Shonti, R. V. Introduction of Bacterial Blight Resistance into Triguna, a High Yielding, Mid-early Duration Rice Variety. *Biotechnol. J.* **2009**, *4*, 400–407.

Suzuki, H. Studies on Antiblastin (I–IV). *Ann. Phytopathol. Soc. Jpn.* **1954**, *18*, 138.

Tabien, R. E.; Li, Z.; Paterson, A. H.; Marchetti, M. A.; Stansel, J. W.; Pinson, S. R. M. Mapping QTLs for Field Resistance to the Rice Blast Pathogen and Evaluating their Individual and Combined Utility in Improved Varieties. *Theor. Appl. Genet.* **2002**, *105*, 313–324.

Talbot, N. J. On the Trail of a Cereal Killer: Exploring the Biology of *Magnaporthe grisea*. *Annu. Rev. Microbiol.* **2003**, *57*, 177–202.

Talukder, Z. I.; McDonald, A. J. S.; Price, A. H. Loci Controlling Partial Resistance to Rice Blast Do Not Show Marked QTL × Environment Interaction When Plant Nitrogen Status Alters Disease Severity. *New Phytol.* **2005**, *168*, 455–464.

Talukder, Z. I.; Tharreau, D.; Price, A. H. Quantitative Trait Loci Analysis Suggests that Partial Resistance to Rice Blast is Mostly Determined by Race-specific Interactions. *New Phytol.* **2004**, *162*, 197–209.

Tamari, K.; Kaji, J. Biochemical Studies of the Blast Fungus (*Pyricularia oryzae* Cavara). Part 2. Studies on the Physiological Action of Pyricularin, a Toxin Produced by the Blast Fungus on Rice Plants. *J. Agricul. Chem. Soc. Jpn.* **1955**, *29*, 185–190.

Tan, Y. N.; Yi, Z. L.; Jiang, J. X.; Qin, J. P.; Xiao, L. Strategies and Advances in Improving Resistance to Rice Blast by Transgenic Approaches. *Mol. Plant Breed.* **2004**, *2*(6), 847–852.

Tanksley, S.; Grandillo, S.; Fulton, T.; Zamir, D.; Eshed, Y.; Petiard, V.; Lopez, J.; Beck-Bunn, T. Advanced Backcross QTL Analysis in a Cross between an Elite Processing Line of Tomato and its Wild Relative *L. pimpinellifolium. Theor. Appl. Genet.* **1996**, *92*, 213–224.

Thakur, S. Gupta, Y.; Singh, P.; Rathour, R.; Variar, M.; Prashanthi, S. Molecular Diversity in rice Blast Resistance Gene *Pi-ta* Makes It Highly Effective Against Dynamic Population of *Magnaporthe oryzae. Funct. Integr. Genom.* **2013b**, *13*, 309–322.

Thakur, S.; Singh, P.; Rathour, R.; Variar, M.; Prashanthi, S.; Singh, A. Positive Selection Pressure on Rice Blast Resistance Allele Piz-t Makes it Divergent in Indian Land Races. *J. Plant Int.* **2013a**, *8*, 34–44.

Thurston, H. D. *Tropical Plant Diseases.* Amer. Phytopathol. Soc. (APS Press), 1988.

Toojinda, T.; Tragoonrung, S.; Vanavichit, A.; Siangliw, J. L.; Pa-In, N.; Jantaboon, J.; Siangliw, M.; Fukai, S. Molecular Breeding for Rainfed Lowland Rice in the Mekong Region. *Plant Prod. Sci.* **2005**, *8*(3), 330–333.

Tsuji, H.; Sasahara, M.; Kanno, H.; Ohba, A; Kanagawa, M. Change of Pathogenic Races of Rice Blast Fungus on Multiline Cultivars 'Sasanishiki BL' in Recent Years. *Ann. Rep. Plant Protect. North Jpn.* **1999**, *50*, 16–20.

Uesugi, Y. *Resistance of Phytopathogenic Fungi to Fungicides.* Japan Pesticide Information: Japan, 1978.

Urso, S.; Orasen, G; Perrini, R.; Tacconi, G.; Delfanti, S.; Biselli, C. Pyramiding of *Pi* Resistance Genes to Increase Blast Resistance in Italian Rice Varieties Using Marker-assisted Selection Approaches. In: Proceedings of the 57th Italian Society of Agricultural Genetics Annual Congress 16th–19th, September, Foggia, 2013.

Usman, G. M.; Wakil, W.; Sahi, S. T.; Saleemil, Y. Influence of Various Fungicides on the Management of Rice Blast Disease. *Mycopathology* **2009**, *7*(1), 29–34.

Utani, D. W.; Moeljopawiro, S.; Aswidinnoor, H.; Setiawan, A.; Hanarida, I. Blast Resistance Genes in Wild rice *Oryza rufipogon* and Rice Cultivar IR64 Indonesian. *J. Agric.* **2008**, *1*, 71–76.

van der Plank, J. E. Horizontal Resistance: Six Suggested Projects in Relation to Blast Disease in Rice. In: *Horizontal Resistant to Blast Disease in Rice, CIAT Series*; CE-9, Colombia: Cali, 1975; pp 21–26.

Vasudevan, K.; Gruissem, W.; Bhullar, N. K. Identification of Novel Alleles of the Rice Blast Resistance Gene *Pi54*. *Sci. Rep.* **2015,** *5*, 15678.

Villareal, R. L. Slow Leaf Blast Infection in Rice (*Oryza sativa* L.). Ph. D. Thesis, Pennsylvania University, USA, 1980; p 123.

Voinnet, O. Origin, Biogenesis, and Activity of Plant MicroRNAs. *Cell* **2009,** *136*, 669–687.

Wang, D.; Guo, C.; Huang, J.; Yang, S.; Tian, D.; Zhang, X. Allele Mining of Rice Blast Resistance Genes at AC134922 Locus. *Biochem. Biophys. Res. Commun.* **2014,** *446*, 1085–1090.

Wang, G. L.; Mackill, D. J.; Bonman, J. M.; McCouch, S. R.; Champoux, M. C.; Nelson, R. J. RFLP Mapping of Genes Conferring Complete and Partial Resistance to Blast in a Durably Resistant Rice Cultivar. *Genetics* **1994,** *136*, 1421–1434.

Wang, Y. P.; Wei, Z. Y.; Zhang, Y. Y.; Lin, C. J.; Zhong, X. F.; Wang, Y. L.; Ma, J. Y.; Ma, J.; Xing, S. C. Chloroplast-expressed MSI-99 in Tobacco Improves Disease Resistance and Displays Inhibitory Effect against Rice Blast Fungus. *Int. J. Mol. Sci.* **2015,** *216*(3), 4628–4641.

Wang, Z. X.; Yano, M.; Yamanouchi, U.; Iwamoto, M.; Monna, L.; Hayasaka, H.; Katayose, Y.; Sasaki, T. The Pib Gene for Rice Blast Resistance Belongs to the Nucleotide Binding and Leucinerice Repeat Class of Plant Disease Resistance Genes. *Plant J.* **1999,** *19*, 55–64.

Wang, Z.; Jia, Y.; Rutger, J.; Xia, Y. Rapid Survey for Presence of a Blast Resistance Gene Pi-t a in Rice Cultivars Using the Dominant DNA Markers Derived from Portions of the Pi-ta Gene. *Plant Breed.* **2007,** *126*, 36–42.

Watanabe, K.; Tanaka, T.; Fukuhara, K.; Miyairi, N.; Yonehara, H.; Umezawa, H. Blastmycin, a New Antibiotic from *Streptomyces* sp. *J. Antibiot., Ser. A* **1957,** *10*(1), 39–45.

Wen, S.; Gao, B. Introgressing Blast Resistant Gene Pi-9(t) in to Elite Rice Restorer Luhui17 by Marker-assisted Selection. *Rice Genom. Genet.* **2011,** *2*, 31–36.

Werner, K.; Friedt, W.; Ordon, F. Strategies for Pyramiding Resistance Genes against the Barley Yellow Mosaic Virus Complex (BaMMV, BaYMV, BaYMV-2). *Mol. Breed.* **2005,** *16*, 45–55.

Wu, J. L.; Fan, Y. Y.; Li, D. B.; Zheng, K. L.; Leung, H.; Zhuang, J. Y. Genetic Control of Rice Blast Resistance in the Durably Resistant Cultivar Gumei 2 against Multiple Isolates. *Theor. Appl. Genet.* **2005,** *111*, 50–56.

Wu, J. L.; Sinha, P. K.; Varivar, M.; Zheng, K. L.; Leach, J. E.; Courtois, B.; Leung, H. Association between Molecular Markers and Blast Resistance in an Advanced Backcross Population of Rice. *Theor. Appl. Genet.* **2004,** *108*(6), 1024–1032.

Xi, Z. Y.; He, F. H.; Zeng, R. Z.; Zhang, Z. M.; Ding, X. H.; Li, W. T.; Zhang, G. Q. Development of a Wide Population of Chromosome Single-segment Substitution Lines in the Genetic Background of an Elite Cultivar of Rice (*Oryza sativa* L.). *Genome* **2008,** *49*(5), 476–484.

Xu, M. H.; Li, C. Y.; Li, J. B.; Tan, X. L.; Tian, W. Z.; Tang, Z. S. Analysis of Resistant Spectrum to Rice Blast in Transgenic Rice Lines Introduced Lysozyme Gene from T4 Phage. *Agric. Sci. China* **2003,** *2*, 273–279.

Xu Y. H.; Wang, J. W.; Wang, S.; Wang, J. Y.; Chen, X. Y. Characterization of GaWRKY1, a Cotton Transcription Factor That Regulates the Sesquiterpene Synthase Gene (+)-delta-cadinene Synthase-A. *Plant Physiol.* **2004,** *135*, 507–515.

Xu, X.; Lv, Q.; Shang, J.; Pang, Z.; Zhou, Z.; Wang, J. Excavation of Pid3 Orthologs with Differential Resistance Spectra to *Magnaporthe oryzae* in Rice Resource. *PLoS One* **2014,** *9*, e93275.

Yamaguchi, I. Overview on the Chemical Control of Rice Blast Disease. In: *Rice Blast: Interaction with rice and control: Proceedings of the 3rd International Rice Blast Conference*, Tsukuba Science City: Ibaraki, Japan, 2004.

Yang, J. H.; Liu, H. X.; Zhu, G. M.; Pan, Y. L.; Xu, L. P.; Guo, J. H. Diversity Analysis of Antagonists from Rice-associated Bacteria and their Application in Biocontrol of Rice Diseases. *J. Appl. Microb.* **2008**, *104*(1), 91–104.

Yang, M. Z.; Cheng, Z. Q.; Chen, S. N.; Qian, J.; Xu, L. L.; Huang, X. Q. A Rice Blast-resistance Genetic Resource from Wild Rice in Yunnan, China. *J. Plant Physiol. Mol. Biol.* **2007**, *33*, 589–595.

Yang, Q. Z.; Saito, K.; Yang, P. W.; Wang, Q.; Sunohara, Y.; Zheng, F. P. Ye, C. R.; Li, J. R.; Kato, A. Molecular Mapping of a New Blast Resistance Gene *Pi25*(t) Possessed in a *Japonica* Rice Cultivar, *Oryza sativa* L. cv Yunxi 2. In: Proceedings of the 1st Rice Blast Congress in China, Kunming, 2001; pp 49–55.

Yasuda, N.; Mitsunaga, T.; Hayashi, K.; Koizumi, S.; Fujita, Y. Effects of Pyramiding Quantitative Resistance Genes *pi21*, *Pi34*, and *Pi35* on Rice Leaf Blast Disease. *Plant Dis.* **2014**, *99*, 904–909.

Yeh, W.; Bonman, J. Assessment of Partial Resistance to *Pyricularia oryzae* in Six Rice Cultivars. *Plant Pathol.* **1986**, *315*(3), 319–323.

Yoshida, K.; Miyashita, N. T. DNA Polymorphism in the Blast Disease Resistance Gene *Pita* of the Wild Rice *Oryza rufipogon* and Its Related Species. *Genes Genet. Syst.* **2009**, *814*, 121–136.

Yoshii, K. Studies on Cephalothecium as a Means of Artificial Immunization of Agricultural Crops. *Ann. Phytopathol. Soc. Jpn.* **1949**, *13*, 37–40.

Young, N. D. Constructing a Plant Genetic Linkage Map with DNA Markers. In: *DNA-Based Markers in Plants*; Ronald, I. K. V., Phillips, L.; Kluwer: Dordrecht, 1994; pp 39–57.

Young, N. D. QTL Mapping and Quantitative Disease Resistance in Plants. *Ann. Rev. Phytopathol.* **1996**, *34*, 479–501.

Yu, J.; Holland, J. B.; McMullen, M. D.; Buckler, E. S. Genetic Design and Statistical Power of Nested Association Mapping in Maize. Ge*netics* **2008**, *178*(1), 539–551.

Yunoki, T.; Ezuka, A.; Sakurai, Y.; Shinoda, H.; Toriyama, K. Studies on the Varietal Resistance to Rice Blast. 3. Testing Methods for Field Resistance on Young Seedling Grown in Greenhouse. *Bull. Chugoku Natl. Agric. Exp. Stn.* **1970**, *E6*, 1–19.

Zapata-Arias, F. J.; Torrizo, L. B.; Ando, A. Current Developments in Plant Biotechnology for Genetic Improvement: The Case of Rice (*Oryza sativa* L.). *World J. Microbiol. Biotechnol.* **1995**, *11*(4), 393–399.

Zenbayashi, K.; Ashizawa, T.; Tani, T.; Koizumi, S. Mapping of the QTL (Quantitative Trait Locus) Conferring Partial Resistance to Leaf Blast in Rice Cultivar Chubu 32. *Theor. Appl. Genet.* **2002**, *104*, 547–552.

Zenbayashi-Sawata, K. *Pi34-AVRPi34:* A New Genefor–Gene Interaction for Partial Resistance in Rice to Blast Caused by *Magnaporthe grisea. J. Genet. Plant Pathol.* **2005**, *71*, 395–401.

Zhang, M. X.; Xu, J. L.; Luo, R. T.; Shi, D.; Li, Z. K. Genetic Analysis and Breeding Use of Blast Resistance in a Japonica Rice Mutant R917. *Euphytica* **2003**, *130*(1), 71–76.

Zhang, Y. S.; Luo, L. J.; Xu, C. G.; Zhang, Q. F.; Xing, Y. Z. Quantitative Trait Loci for Panicle Size, Heading Date and Plant Height Co-segregating in Trait-performance Derived Near-isogenic Lines of Rice (*Oryza sativa*). *Theor. Appl. Genet.* **2006**, *113*, 361–368.

Zhang, Y.; Yang, J; Shan. Z.; Chen, S.; Qiao, W.; Zhu, X. Substitution Mapping of QTLs for Blast Resistance with SSSLs in Rice (*Oryza sativa* L.). *Euphytica* **2012**, *184*, 141–150.

Zhou, B.; Qu, S.; Liu, G.; Dolan, M.; Sakai, H.; Lu, G.; Bellizzi, M.; Wang, G. L. The Eight Amino-acid Differences within Three Leucine-rich Repeats between Pi2 and Piz-t Resistance Proteins Determine the Resistance Specificity to *Magnaporthe grisea*. *Mol. Plant–Microbe Interact.* **2006,** *19*, 1216–1228.

Zhou, E.; Jia, Y.; Lee, F. N.; Lin, M.; Jia, M.; Correll, J. C.; Cartwright, R. D. Evidence of the Instability of a Telomeric *Magnaporthe grisea* Avirulence Gene AVR-Pita in the US. *Phytopathology* **2005,** *95*(6), 118.

Zhou, J. H.; Wang, J. L.; Xu, J. C.; Lei, C. L.; Ling, Z. Z. Identification and Mapping of a Rice Blast Resistance Gene *Pi-g(t)* in the Cultivar Guangchangzhan. *Plant Pathol.* **2004,** *53*, 191–196.

Zhu, Y. Y.; Fang, H.; Wang, Y. Y.; Fan, J. X.; Yang, S. S.; Mew, T. W.; Mundt, C. C. Panicle Blast and Canopy Moisture in Rice Cultivar Mixtures. *Phytopathology* **2005,** *95*, 433–438.

CHAPTER 4

MOLECULAR TOOLS FOR CONTROLLING OF SHEATH BLIGHT DISEASE OF RICE AND ITS MANAGEMENT

MD. SHAMIM[1†], DIVAKAR SHARMA[2], DEEPA BISHT[2], RAJA HUSSAIN[1], N. A. KHAN[1], PRAMILA PANDEY[1], RAVI KESARI[2], and K. N. SINGH[1*]

[1]*Department of Plant Molecular Biology and Genetic Engineering, N. D. University of Agriculture and Technology, Kumarganj, Faizabad 224229, Uttar Pradesh, India*

[2]*National JALMA Institute for Leprosy and Other Mycobacterial Diseases, Tajganj, Agra 282001, Uttar Pradesh, India*

*Corresponding author. E-mail: kapildeos@hotmail.com

†Present address: Department of Molecular Biology and Genetic Engineering, Bihar Agricultural University, Sabour, Bhagalpur 813210, Bihar, India

CONTENTS

ABSTRACT

Sheath blight (ShB) of rice, caused by *Rhizoctonia solani* Kuhn (teleomorph *Thanatephorus cucumeris* [Frank] Donk), is now a serious disease in rice-growing countries. We review the occurrence and spread of this disease on the different host; however, a brief discussion was made only on rice. The taxonomy of *R. solani*, classification of the pathogen, and strategies for disease management are briefly described in their host. Chemical control is nowadays is commonly used to control this disease. Till date that there is no high resistance in cultivated rice was observed; however, three rice genotypes namely Jasmine 85, Tatep, and Tequing were reported as moderate resistance. However, some wild-rice accessions have reported that they have promising level of resistance against this disease.

4.1   INTRODUCTION

Rice (*Oryza sativa* L.), a member of the family Graminae, is widely grown in tropical and subtropical region. Approximately 50% of the world's population consumes rice as their main staple food (www.irri.org/). Disease damage is one of the most serious limiting factors for rice production. Just about 90% of the world's rice is grown in the Asian continent and constitutes a staple food for 2.7 billion people worldwide. It is however unfortunate that such an important crop is attacked by many kinds of diseases, of which sheath blight (ShB) caused by *R. solani* Kuhn is one of the most destructive diseases throughout the world (Mew & Rosales, 1984).

Incidence of ShB in rice fields is dependent on the method of planting and plant population density. Investigations at farmers' fields and experimental fields revealed that square method of transplantation resulted in optimum high-yield density, higher leaf area index and dry matter production. This method of transplantation also contributed to increased ShB resistance and higher grain yields (Yang et al., 2008). Sparse planting resulted in lower ShB occurrence and greater lodging resistance in rice. The other important effects of sparse planting included fewer number of stems/m², more stems/hill, delay in date of maximum tillering stage, heading time, ripening time, greater number of pods per head, and more pods on secondary rachis-branches (Sugiyama et al., 2007). Planting of rice seedlings far from the bund resulted in reduced ShB incidence since bunds have weed hosts of *R. solani*. Root diseases of maize also caused and occurred more frequently in Southern Germany where maize and sugar beet are cultivated in narrow crop rotations. The disease may cause severe losses in localized areas and may reduce grain yield up to 30%. *Rhizoctonia* root rot in wheat and barley, caused by the soil borne fungal pathogen *R. solani* AG-8, was first diagnosed as a problem in direct-seeded wheat and barley in the PNW in themid-1980s. Wheat is generally less severely affected than barley, and spring-seeded crops are more vulnerable to infection than fall-sown crops. Yield losses associated with *Rhizoctonia* root rot are especially pronounced in direct-seed systems. SB, caused by the fungal pathogen *R. solani* Kuhn, causes significant yield loss and reduction in grain quality for rice (*O. sativa* L.) in the southern United States and other regions of the world (Lee & Rush, 1983; Rush & Lindberg, 1996). All current US rice cultivars are susceptible to *R. solani* with costly fungicide applications as the primary means of control. Various studies have shown that response of different rice lines to infection by *R. solani* is expressed as partial resistance (Liu et al., 2009), also referred to as incomplete, quantitative, field, or horizontal resistance (Wang et al., 2010).

4.2   SHEATH BLIGHT DISEASE

Rice ShB pathogen also produces toxin that induce characteristic symptoms on rice leaves, wilting of seedlings, and inhibited rice radicle growth. A positive correlation was noted between crude toxin production and the virulence of the pathogen. The radicles and seedlings of resistant rice cultivars were more tolerant to the crude toxin compared to susceptible cultivars, indicating the scope of resistance screening through treatment of rice radicles with the crude toxin (Xu et al., 2004).

4.3   SHEATH BLIGHT PATHOGEN

R. solani Kuhn (teleomorph *Thanatephorus cucumeris* [Frank] Donk) infects at least 200 plant species and is one of the most common soil-borne pathogens in crop plants (Gvozdeva, 2006). *Rhizoctonia* belongs to the Basidiomycetes, with *R. solani* being multinucleate. *R. solani* species [teleomorph *T. cucumeris* (Frank)] represent a collective species (Sneh et al., 1996), which has been divided into 13 anastomosis groups (AGs) (AG-1–AG-13) and AG-BI (the bridging isolate AG) (Carling, 1996; Carling et al., 2002). The anastomosis group AG-1 can be further subdivided into three intraspecific groups based on disease symptoms, cultural characteristics, rDNA similarity, and isozymes (Linde et al., 2005; Liu & Sinclair, 1993; Mohammadi et al., 2003). The intraspecific groups are AG-1 IA (ShB on rice), AG-1 IB (web blight), and AG-1 IC (damping off) (Sneh et al., 1991). Rice ShB is a particularly important component of the rice disease complex, occurring in most rice-producing areas, including India. From hyphal anastomosis reactions, isolates are divided into AGs. Strains of *R. solani* belong to at least 14 different, genetically defined populations of AGs determined by anastomosis between hyphae of strains belonging to the same AG. The AGs themselves do not necessarily give information on the genetic variation and taxonomic relationships within and between AGs. Among the different types of symptoms, ShB is the most prominent and common one. Because of its semi-saprophytic nature, *R. solani* has a wide range and uncharacterized pathogenecity mechanisms. Even though *R. solani* is causing a wide range of economically important diseases in different plan species, there is a lack of knowledge concerning genes and their function in relation to pathogenicity (Lubeck, 2004). Extending the current focus on genomics to include *R. solani* would be of great utility for building up knowledge on genes and gene expression from this important plant pathogen. Wu et al. (2014) reported that high nitrogen (N) rate and dense planting were conducive to ShB development. Application of silicon fertilizer under high N rate failed to suppress the disease epidemic, especially when silicon concentration of the soil is high or there is enough plant-available silicon.

4.3.1 HOST RANGE

R. solani has a wide host range in cereal crops. Kozaka (1965) in Japan recorded 188 species in 32 families and Tsai (1970) in Taiwan listed 20 sp. belonging to 11 families as potential host upon inoculation. According to Meena and Muthusamy (1998), *R. solani* have broad host range of many cereal plant (Table 4.1).

TABLE 4.1 Host Range of the *R. solani*.

Host	Mode of resistance
Cajanus cajan, Capsicum annuum, Curcuma longa, Dolichos biflorus, Lycopersicom esculentum, Panicum miliaceum, Paspa/ um scrobicu/atum, Setaria italica, Sorghum vulgare, Zea mays	Moderately susceptible
Brachiaria mutica, Cynodon dactylon, Cyperus rotundus, Echinochloa colona, Eleusine corocana, Phaseolus aureus	Susceptible
Dolichos lablab var. typicus, Vigna sinensis	Highly susceptible

Adapted from Sneh et al. (1991).

4.4   MODE OF INFECTION AND TRANSMISSION OF PATHOGEN

4.4.1   TRANSMISSION OF PATHOGEN

The fungus survives either as sclerotia or mycelia in plant debris, forms infection cushions surface of floodwater, germinates, and forms infection cushions and/or lobate appressoria on the plant surface for infection (Fig. 4.1). After the initial infection, the pathogen moves up the plant by surface hyphae and develops new infection structures over the entire plant, causing significant necrotic damage (Ou, 1985). The infected rice seeds may produce 4–6.6% seedling infection in India (Mathur, 1983; Ou, 1985). But on transplantation, the infected seedlings were unable to develop disease (Naidu, 1992). Disease cycle takes place predominantly through sclerotia in the humid tropics. Sclerotia, the dormant are shed before/or during the harvest operation and remain in soil and survive for a long time. When the buoyant sclerotia tend to accumulate in undisturbed standing water at the plant–water interface, the aerobic fungus creeps up several centimeters in 24 h and the primary infections are caused in wetland rice. Rain water runoff and flood irrigation permit good dispersal of floating sclerotia (Lee, 1979) and consequently provide the primary foci of infection through the stretches of rice fields. Further, with the increasing size of sclerotia on their fragments, number, and size of lesions also increased (Gangopadhyay, 1983). The pathogen induced lesions on leaf blades and leaf sheaths of infected plants.

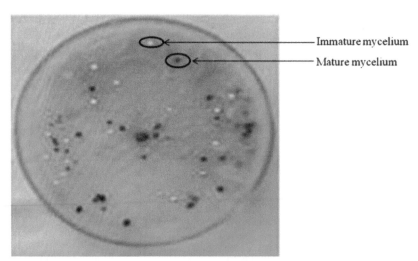

FIGURE 4.1   Morphology of the *R. solani* (D-14 belonging to AG1-IA) obtained from G. B Pant University of Agriculture and Technology, Pant Nagar India (photograph: Md. Shamim).

It produces sclerotia on both abaxial and adaxial leaf sheath surfaces but not in the tissue. The pathogen forms infection cushions and lobate appresoria on leaf sheath and directly penetrates the cuticle or through stomata (Kim & Ishii, 1992). Once infection occurs, secondary spread takes place through direct contact (role of basidiospores uncertain). Sclerotia may move from one field to another through irrigation water, and during movement, they may produce mycelia and secondary or tertiary sclerotia (IRRI, 1973).

4.4.2 DETECTION OF SHB

Rice sheath diseases caused by *Rhizoctonia* species are relatively difficult to diagnose by visual observation alone due to the similarity of the symptoms with those caused by other disorders. Moreover, various *Rhizoctonia* species have been isolated from rice sheaths showing similar symptoms. *R. oryzae*, the causal agent of "bordered sheath spot" and *R. oryzae-sativae*, the causal agent of "aggregate sheath spot" have been reported on rice from Eastern and Southeastern Asia (Inagaki et al., 2004; Inagaki, 1996). These pathogens produce very similar symptoms in the field. In India, scientific information on rice ShB and related diseases is scanty and the population diversity of the causal agents has not yet been surveyed. However, knowledge of the populations of pathogenic *Rhizoctonia* species is essential for integrated control strategies; along with the understanding of the influence of other characteristics, including pathogenicity, host range, and adaptability to environmental conditions. Sayler and Yang (2007) studied a real-time, quantitative polymerase chain reaction (QPCR) assay to detect and quantify *R. solani* AG-1 IA DNA from infected rice plants. A specific primer pair was designed based on the internal transcribed spacer region of the fungal ribosomal DNA. The specific detection of *R. solani* DNA was successful with quantities as low as 1 pg. The QPCR assay could be used for detecting the rice ShB pathogen, quantifying fungal aggressiveness, and evaluating the resistance level of rice cultivars. Grosch et al. (2007) reported that RAPD-PCR was used for identifying a specific fragment from which SCAR primers were developed and used for PCR detection of the subgroup AG 1-IB. The designed SCAR primer N18-rev/N18-for allowed the unequivocal detection of the specific DNA fragment of 324 bp from field-grown lettuce plants with bottom rot symptoms or artificially inoculated plant species and from different types of inoculated field soils. A specific diagnosis PCR assay for *R. solani* subgroup AG 1-IB was established, which can be used as a highly specific, reproducible, and applicable test system in plant disease diagnosis.

The designed primer pair may have applications in a multiplex detection tool for *R. solani* or soil-borne pathogens. Lees et al. (2002) used a conventional primer set (Rs1F2 and Rs2R1) was designed from the nuclear ribosomal internal transcribed spacer (ITS1 and ITS2) regions of *R. solani*. Following PCR amplification, a 0.5-kb product was amplified from DNA of all isolates of AG-3 using primers Rs1F2 and Rs2R1. No product was amplified when DNA from isolates belonging to a range of other *R. solani* AGs or from a selection of other potato pathogens was tested, confirming the specificity of the primers for AG-3 only. *R. solani* AG-3 was also detected in potato tissue with varying black scurf severity, and in soil inoculated with sclerotia of *R. solani* to a minimum detection level of 5×10^{-4} g sclerotia/g soil. In addition, specific primers RsTqF1 (based on the Rs1F2 sequence) and RsTqR1, and a TaqMan™ fluorogenic probe RQP1, were designed to perform real-time quantitative (TaqMan) PCR. The conventional PCR and real-time PCR assays were compared and combined with direct DNA extraction from soil and a seed-baiting method to determine the most reliable method for the detection and quantification of AG-3 in both artificially inoculated field soil and naturally infested soils. It was shown that direct DNA extractions from soil could be problematic, although AG-3 was detectable using this method combined with the real-time PCR assay. Okubara et al. (2008) developed SYBR Green I-based real-time QPCR assays specific to internal transcribed spacers ITS1 and ITS2 of the nuclear ribosomal DNA of *R. solani* and *R. oryzae*. The assays were diagnostic for *R. solani* AG-2-1, AG-8, and AG-10, three genotypes of *R. oryzae*, and an AG-I-like binucleate *Rhizoctonia* species. Quantification was reproducible at or below a cycle threshold (C_t) of 33, or 2–10 fg of mycelial DNA from cultured fungi, 200–500 fg of pathogen DNA from root extracts, and 20–50 fg of pathogen DNA from soil extracts. However, pathogen DNA could be specifically detected in all types of extracts at about 100-fold below the quantification levels.

4.5 DISEASE MANAGEMENT

4.5.1 CHEMICAL CONTROL

Inorganic nutrient management is a major factor determining rice ShB disease. Tang et al. (2007) reported that plant variety and nitrogen fertilizers are the major factors influencing ShB disease and concomitant yield losses in rice, both during wet and dry seasons. Varieties with taller stature, fewer tillers, and lower leaf N concentration, such as IR68284H, generally had lower ShB lesion height (LH), ShB index, and consequently lower yield loss from the disease. Greenhouse and field studies with the fungicide Lustre (37.5SE) (flusilazole + carbendazim) against ShB revealed that application of the triazole mix could reduce disease severity and increase yields. Further, it was proved that the test fungicide was a safe combination fungicide without any phytotoxic symptoms. Its prophylactic application gave better results than as a curative application (Reddy & Muralidharan, 2007). Certain new fungicidal formulations were also found effective against rice ShB. Among them, Amistar 25 SC@1.0 mL L^{-1} (30.6%) and RIL-010/FI 25 SC at 0.75 mL L^{-1} (30.1%) showed a high degree of efficacy in reducing the disease severity and were superior over the standard fungicides (validamycin at 2.5 mL L^{-1}). Highest grain yields were also reported in these fungicide treatments (Ranjan et al., 2005).

4.5.2 BIOLOGICAL CONTROL

Leaf extracts of certain plant species were also used for effective management of rice ShB. Among them, the leaf extract of *Pithecellobium dulce* was highly effective in inhibiting mycelia growth of test pathogen (2.5 cm over 8.9 cm in control). Both the leaf extracts of *P. dulce* and *Prosopis juliflora* were equally effective in inhibiting sclerotial number, dry weight, and germination of the pathogen and also in controlling ShB with a disease incidence of 32.3% and 33.3%, respectively, over 76.2% in control (Meena et al., 2002).

4.5.3 HOST RESISTANCE

Disease resistance in plants can be classified into two major categories. Various terms have been used to describe the two categories of resistance, such as vertical versus horizontal resistance (Van der Plank, 1968), qualitative versus quantitative resistance (Ou et al., 1975), and complete versus partial resistance (Jia et al., 2007; Parlevliet, 1979; Shamim et al., 2014).

There are different rice lines reported resistance like Brimful, Jasmine 85, LSBR-5, LSBR-33, Marsi, Minghui 63, Saza, Tetep, Tadukan, Teqing, Tauli, and ZYQ8, in which a high degree of quantitative resistance is found against ShB pathogen under field conditions (Groth & Nowick, 1992; Khush, 1977; Li et al., 1995; Pan et al., 1999; Sato et al., 2004; Wasano, 1988). However, in most cases, qualitative resistance is modulated by direct or indirect interaction between the products of a major disease resistance (R) gene and an avirulence gene; this type of resistance is specific to pathogen race and is lifetime limited in a particular cultivar due to the strong selection pressure against and the rapid evolution of the pathogen (McDonald & Linde, 2002). In contrast, quantitative resistance is conferred by quantitative trait loci (QTLs) and is presumably race nonspecific and durable (Roumen, 1994).

Several groups have attempted to identify sources of ShB resistance by screening local accessions, cultivars, landraces, and/or advanced breeding lines (Table 4.2). Sources of ShB resistance have been sought for in different rice-growing regions by many different research groups. These studies resulted in the identification of genotypes with moderate-to-high levels of resistance. In rice, only partial resistance to rice ShB has been identified, as evidenced by a survey of 6000 rice cultivars from 40 countries from which no cultivar exhibiting a major gene for rice ShB resistance was identified (Jia et al., 2002). More recently, QTLs analysis identified six QTLs associated with ShB resistance on 6 of the 12 rice chromosomes, but only 1 QTL appeared to be independent of plant height, a morphological trait associated with ShB resistance (Li et al., 1995). Additional research suggests it is feasible to identify major genes conferring high levels of partial resistance (Pan et al., 1999), pyramid these genes, and achieve nearly complete ShB resistance.

TABLE 4.2   Resistance Rice Cultivars for Sheath Blight Resistance.

Sl. No.	Rice cultivar for sheath blight	References
1.	NC 678, Dudsor, Bhasamanik	Das (1970)
2.	Chin-kou-tsan, Zenith, CO.17, Dinominga, Puang Nahk 16, Baok, Toma-112, R.T.S.31, Kele Kala	Wu (1971)
3.	Lalsatkara	Roy (1977)
4.	ARC15762, ARC 18119, ARC 18275, ARC 18545	Bhaktavatsalam et al. (1978)
5.	IR24, IR26, IR29, Jaya, Jaganath, Mashoori, Pankaj, Rajeshwari, Supriya, Sabari, TKM6	Rajan and Nair (1979)
6.	Nizersail, Rajasail, Tabend, Ta-poo-cho-z, Kattachambha, DA 29, ARC 5925, ARC 5943, ARC 14529, ARC 10572, ARC 10618, ARC 10836	Manian and Rao (1979)
7.	Tapoochoz, Bahagia, Laka	Crill et al. (1982)

TABLE 4.2 *(Continued)*

Sl. No.	Rice cultivar for sheath blight	References
8.	Taraboli 1, Dholamula, Supkheru, Chidon	Borthakur and Addy (1988)
9.	Bharati, Rohini	Gokulapulan and Nair (1983)
10.	Bog II, Aduthurni, Chinese galendopuram, Arkavati, Saket-4, Neela, MTU-3, MTU-7, MTU-13, MTU-3642, BPT-6	Ansari et al. (1989)
11.	Tetep, Tapoo-cho-z, Guyanal	Sha and Zhu (1990)
12.	LSBR-5, LSBR-33	Xie et al. (1992)
13.	RU8703196, B82-761	Marchetti et al. (1995, 1996)
14.	KK2, Dodan, IR40 and Camor	Singh and Dodan (1995)
15.	Chingdar, As 93-1, Mairan, N–22, Panjasali, Up-52, Upland-2	Singha and Borah (2000)
16.	TIL:455, TIL:514, TIL:642	Pinson et al. (2008)
17.	MCR10277	Nelson et al. (2012)
18.	WSS3, Jarjan, Nepal 555 and Nepal 8	Taguchi-Shiobara et al. (2013)
19.	298 induced mutated (by gamma radiation) Pusa Basmati lines	Meena et al. (2013)
20.	Moderately resistant rice cultivars, Teqing, Jasmine85, Tetep, Pecos, Azucena and Taducan	Hossain et al. (2014)
Wild rice		
1.	*O. latifolia* (DRW 37004), *O. punctata* (DRW 32002), and *O. rufipogon* accession DRW 22017-5	Ram et al. (2008)
2.	Seven Oryza spp. accessions moderately resistant, three were *O. nivara* accessions (IRGC104705, IRGC100898, and IRGC104443), *O. barthii* (IRGC100223), *O. meridionalis* (IRGC105306), O. nivara/O. sativa (IRGC100943), and *O. officinalis* (IRGC105979)	Prasad and Eizenega (2008)
3.	*O. australiensis* and *O. grandiglumis*	Shamim et al. (2014)

Adapted and modified from Ram et al. (2008), Srinivasachary et al. (2010), and Shamim et al. (2014).

Screening has also been conducted at the Centro International de Agricultura Tropical (CIAT); 63% of the genotypes tested were found to be promising candidates as sources of ShB resistance. ShB resistance sources were also sought by Raj et al. (1987) and Hein (1990). Lee et al. (1999) tested 282 recently introduced accessions into the USA and reported that 25 showed high levels of resistance. Similarly, Biswas (2005) and Jia et al. (2009) identified eight and two resistant genotypes of rice cultivars.

4.5.4 RESISTANCE IN WILD RELATIVES

Both wild species and landraces of the *Oryza* genus possess under-exploited alleles that may have a strong potential for the improvement of Asian rice (*O. sativa* L.) and African rice (*Oryza glaberrima* Steud.). Wild rice accessions have been used to successfully develop resistance against many rice diseases (Brar & Khush, 1997). Over the years, a very large number of accessions from different species of *Oryza* have been tested at IRRI to identify sources for ShB resistance (Table 4.2). From a total of 233 accessions tested, 76 were found to contain a high level of resistance to ShB and 29 showed moderately resistance. The latter accessions belonged to the African rice, *O. glaberrima* (2n = 24 AA), a close relative of *O. sativa* (2n = 24 AA). The relatively high resistant accessions belonged to mixed genetic groups (IRRI, 1992). In addition to the studies mentioned above, Amante et al. (1990) and Lakshmanan and Velusamy (1991) evaluated wild accessions or their derivatives for ShB resistance. In the USA, Prasad and Eizenega (2008) evaluated 73 *Oryza* spp. accessions with three different screening methods and identified seven accessions (three *O. nivara*) Sharma and Shastry and one each of *O. barathi* A. Chev, *O. meridionalis* Ng, *O. nivara*, *O. sativa* L., and *O. officinalis* Wall ex Watt) that showed moderate resistance. Similar efforts were made by Ram et al. (2008), who screened 22 accessions belonging to 11 different species of *Oryza*, identifying the accessions of *O. latifolia* (Desv.), *O. grandiglumis* (Doell) Prod, *O. nivara*, and *O. rufipogon* as having a higher level of resistance, and Shamim et al. (2014) also reported two wild rice accessions *O. australiensis* and *O. grandiglumis* (Fig. 4.2).

FIGURE 4.2 (a) Disease reaction of *R. solani* on wild rice accession *O. australiansis* and (b) disease reaction of *R. solani* on wild rice accession *O. grandiglumis*, obtained from IRRI, Philippines (photograph: Md. Shamim).

4.5.5 INDIAN CONTINENTS

Ram et al. (2008) screened 32 accessions belonging to 11 different species of *Oryza*, namely, 11 accessions of *O. rufipogon*, 8 of *O. nivara*, 3 of *O. eichingeri*, 2 each of *O. officinalis* and *O. latifolia*, and 1 each of *O. longistaminata*, *O. minuta*, *O. aha*, *O. meridionalis*, *O. punctate*, and *O. grandiglumis* against ShB along with susceptible variety of *O. saliva* (cv. Ajaya). Crosses of susceptible *O. sativa* (HM 36-6-4-F) and resistant *O. latifolia* (DRW 37004) and *O. sativa* (HM 36-6-4-F) × *O. punctata* (DRW 32002) and F$_1$s were produced using embryo rescue technique were screened. One BC$_1$F$_1$ plant of each *O. sativa* × *O. latifolia* and *O. sativa* × *O. punctata* showed high level of resistance with score 3 and one each from *O. sativa* × *O. latifolia* and *O. sativa* × *O. punctata* showed moderate level of resistance with score 5. Finally, they reported that the ShB resistance is heritable and there is scope for introgression of genes from distantly related species. The *O. rufipogon* accession DRW 22017-5 provides an important source for ShB resistance, which can be exploited to improve the modern high-yielding cultivars and pyramiding it with the genes for moderate resistance in cultivated germplasm would certainly increase the level of resistance. Shamim et al. (2014) also reported that wild rice accessions, *O. australiensis* and *O. grandiglumis* have resistance source against *R. solani*, belonging to AG 1 IA anastomosis group.

4.6   BIOCHEMICAL RESPONSES IN RICE AFTER INFECTION BY *R. SOLANI*

Sareena et al. (2006) studied defense response in transgenic Pusa Basmati 1 (PB1) rice lines engineered with rice chitinase gene (*chi11*) against the *R. solani*. After inoculation, with *R. solani* enhanced production of phenylalanine ammonia lyase, peroxidase, and polyphenoloxidase enzyme activities in resulted followed by reduced symptom development in transgenic rice lines in comparison to non-transgenic control plants. After infection with *R. solani*, loss of chlorophyll was resulted in non-transgenic line in comparison to transgenic rice line. Sayari et al. (2014) studied the role of NH-1, several PR genes, phenylalanine ammonia-lyase, and lipoxygenase in the defense responses of rice against *R. solani*, the causal agent of rice ShB disease. The induction of PR-5, PR-9, PR-10, PR-12, PR-13, and NH-1 was observed in the resistant and susceptible Iranian cultivars of rice Tarom and Khazar rice cultivars after infection by *R. solani*. Even though plant–pathogens, their hosts and the interactions between them have been studied using classical biochemical, genetic, molecular biological, and plant pathology approaches, systems biology approaches such as genomics and proteomics are essential to provide global information on the various cellular genomic and proteomic networks. Defense responsive proteins, including different enzymes that can directly act on pathogen components have been linked to basal resistance and these resistance governed by different quantitative traits both of which are associated with broad-spectrum resistance. Rice proteomics research has made considerable progress recently in providing functional information of proteins expressed in the various developmental stages, tissues, cells, and abiotic and biotic stress environments.

Chitinases are an example of defense response enzymes that have been linked to basal resistance. Lee et al. (2006) identified a specific 3-β HSD proteins in resistant rice varieties LSBR-5 associated with response to infection by *R. solani* after 2-dimensional gel electrophoresis and electrospray ionization quadrupole-time of flight mass spectrometry (ESI Q-TOF MS). Sixteen additional proteins identified with the above studied have been previously reported to be involved in antifungal activity, signal transduction, energy metabolism, photosynthesis, protein folding and degradation, signal transduction, and antioxidation, further localization of the expressed protein was located on their respective chromosome (Fig. 4.3).

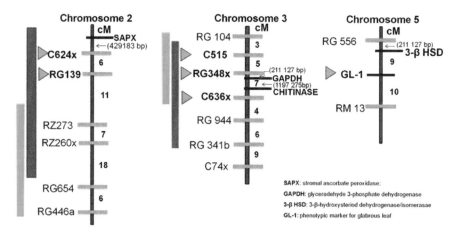

FIGURE 4.3 Composite genetic and physical maps for QTL regions and DNA markers on chromosomes 2, 3, and 5 associated with sheath blight disease resistance and proteins (modified from Pinson et al. 2005; Lee et al., 2006). Left side of chromosome shows QTL regions and genetic distance between markers in Kosambi map units (cM). Right side of chromosome shows proteins detected in this study and their physical distance to DNA markers in base pairs obtained from the Gramene website (HYPERLINK "http://www. gramene.org/" http://www.gramene.orgHYPERLINK "http://www.gramene.org/" /). ▷ Markers identified by Pinson, Oard, and Capdevielle (unpublished data) and unpublished results using discriminant analysis and step-wise multiple regression.

4.7 RESISTANCE QTLS AGAINST SHEATH BLIGHT

Over the past decades, studies on resistance to ShB have been conducted by many researchers who have had diverse objectives, including screening the germplasm of cultivated rice/rice wild relatives, assessment of genetically engineered plants with genes for resistance, and phenotyping for QTL mapping or validation (Fig. 4.4 and Table 4.3). To this end, a broad spectrum of methods has been employed, which can be described by four main components: the biological hierarchy level addressed (from organs to field plots), the inoculation method used, the incubation conditions, and the disease assessment methodology. The choice of these components is critical to the outcomes of the studies, since it is underpinned by the (presupposed) biological processes involved in disease resistance. Methodological choices also have major consequences on the accuracy, precision, repeatability, and ultimately usefulness of the results. Quantitative resistance, in contrast with qualitative resistance, is generally considered as partial resistance in a particular cultivar (Parlevliet, 1979). This type of disease resistance is controlled by multiple loci, referred to as QTLs and does not comply with simple Mendelian inheritance. Thus, selecting for these QTLs is difficult. However, several studies have indicated that pyramiding resistance QTLs can achieve the same level or even a higher level of resistance than that conferred by an R gene (Castro et al., 2003; Gangopadhyay, 1983; Groth & Nowick, 1992; Gvozdeva, 2006; IRRI, 1973, 1992; Jia et al., 2012; Khush, 1977; Lee & Rush, 1983; Meena & Muthusamy, 1998; Richardson et al., 2006; Rush & Lindberg, 1992; Silva et al., 2012; Wang et al., 2010; Wasano, 1988; Wen et al., 2015; Xie et al., 2008; Zeng et al., 2011; Zuo et al., 2008).

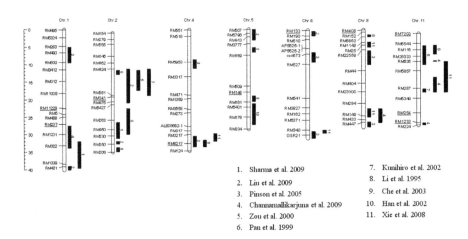

1. Sharma et al. 2009
2. Liu et al. 2009
3. Pinson et al. 2005
4. Channamallikarjuna et al. 2009
5. Zou et al. 2000
6. Pan et al. 1999

7. Kunihiro et al. 2002
8. Li et al. 1995
9. Che et al. 2003
10. Han et al. 2002
11. Xie et al. 2008

FIGURE 4.4 Physical position of marker loci significantly associated with sheath blight in till study (with underlines) in comparison with previous studies. The black bars show the estimated location according to their flanking markers. (Reprinted from Jia, L.; Yan, W.; Zhu, C.; Agrama, H. A.; Jackson, A.; Yeater, K.; Li, X.; Huang, B.; Hu, B.; McClung, A.; Wu, D. Allelic Analysis of Sheath Blight Resistance with Association Mapping in Rice. *PLoS ONE* **2012,** 7, e32703. http://journals.plos.org/plosone/article?id=10.1371/journal.pone.0032703. Public domain.)

Li et al. (1995) identified six QTLs for ShB resistance in an F_4 population of Teqing/Lemont, but one allele on chromosome 8 for the resistance

TABLE 4.3 Resistance Details on rice ShB-QTL Identified in Different Mapping Populations and their Association with Other Traits.

QTLs name	Position of QTLs on chromosome	Associated with	Resistance sources	Recipient	Genetic material/mapping population(s)	References
qSB-2, qSB-3, qSB-4, qSB-8, qSB-9, and qSB-12	2–4, 8, 9, and 12	PH, HD, PH	Teqing (indica)	Lemont (tropical japonica)	255 F_4	Li et al. (1995)
qSB-2, qSB-3, and qSB-7	2, 3, and 7	HD	Jasmine 85 (indica)	Lemont (tropical japonica)	F_2 clonal families	Pan et al. (1999)
qSB-2 (2 years), qSB-3 (1 year), qSB-7 (1 year), qSB-9.1 (1 year), qSB-9.2 (1 year), and qSB-11 (2 years)	2, 3, 7, 9, and 11	NA	Jasmine 85 (indica)	Lemont (tropical japonica)	128 F_2 clonal families	Zou et al. (2000)
qSBR-2, qSBR-3, qSBR-7, and qSBR-11	2, 3, 7, and 11	CL and ND	Zhai Ye Qing 8 (ZYQ8) (indica)	Jing Xi 17 (JX17) (japonica)	DH, 127 HD	Kunihiro et al. (2002)
qSB-5 and qSB-9	5 and 9	ND	Minghui 63(indica)	Zhenshan 97 (indica)	RILs 240 lines	Han et al. (2002)
Rsb 1	5	ND	Xiangzaoxian19 4011a (indica) indica)		1032 F2	Che et al. (2003)
qSB-3 and qSB-12	3 and 12	CL	WSS2 (Tetep) (indica)	Hinohikari (japonica)	60 BC1F1	Sato et al. (2004)
qSB-9 and qSB-11			Teqing (indica)	(tropical japonica)	115 F2 clonal population	Tan et al. (2005)
qSB-1, qSB-2, qSB-3.1, qSB-3.2, qSB-4.1, qSB-4.2, qSB-5, qSB-6.1, qSB-6.2, qSB-7, qSB-8.1, qSB-8.2, qSB-9, qSB-10, and qSB-12	1–10 and 12	HD, PH	Teqing (indica)	(tropical japonica)	F10 and F11	Pinson et al. (2005)
qSB-1, qSB-2, qSB-3, and qSB-9	1–3 and 9	PH and HD	Pecos (tropical japonica)	Rosemont (tropical japonica)	279 F2:3	Sharma et al. (2009)

TABLE 4.3 *(Continued)*

QTLs name	Position of QTLs on chromosome	Associated with	Resistance sources	Recipient	Genetic material/ mapping population(s)	References
qSB-1 (both), qSB-2-1 (mist), qSB-2-2 (mist), SB-3-1 (both), qSB-3-3 (mist), qSB-5 (microch), qSB-6 (microch), qSB-9-1 (microch), and qSB-9-2 (both)	1–6 and 9	ND	Jasmine 85 (indica)	Lemont (tropical japonica)	250 F5 RILs	Liu et al. (2009)
qSBR1-1, qSBR3-1, qSBR7-1, qSBR8-1, qSBR9-1, qSBR3-11-1, qSBR3-11-2, and qSBR3-11-3	1, 3, 7–9, and 11	NA	Tetep (indica)	HP2216 (indica)	127 RIL (F2:10), 96 varieties 192 F2 population Derived from Pusa Basmathi I/ Tetep	Channamallikarjuna et al. (2010)
qSB-11LE	11		Backcross population	Lemont	Near-isogenic lines (NILs) six BC1F1 populations and one BC2F1	Zuo et al. (2011)
qSB92 and qSB121	9 and 11		Teqing	Lemont	backcross introgression lines (TILs)	Wang et al. (2012)
qShB9-2	9		Tetep	Pusa6B	backcross inbred lines	Pandian et al. (2012)
			Jasmine 85	Lemont	216 LIRILs	Liu et al. (2013)
qDR-1a, qDR-1b, qDR-4, qDR-5, qDR-6, qDR-12, qLL-1a, qLL-1b, qLL-3, qLL-9, qLH-1a, qLH-1b, qLH-1c, qLH-1d, qRLL-1a, qRLL-1b, qRLL-1c, qRLL-3, qRLL-4, qRLL-6a, qRLL-6b, qRLL-9, qRLH-1a, qRLH-1b, qRLH-1c, qRLH-2, qRLH-4, qRLH-6a, qRLH-6b, qPH-1a, qPH-1b, qPH-1c, qPH-3, qPH-7a, qHD-6, and qHD-7	1–9 and 12		HH1B	RSB02	RIL $F_{8:11}$ population consisting of 155 lines	Liu et al. (2014)

TABLE 4.3 (Continued)

QTLs name	Position of QTLs on chromosome	Associated with	Resistance sources	Recipient	Genetic material/ mapping population(s)	References
qSB-7 and qSB-9	7 and 9	HD and PH	Teqing	WLJ1	Advanced backcrossed lines	Chen et al. (2014)
TAC1TQ and *qSB9TQ*	**1 and 9**	TA and GY	IL55	IR24	NILs	Zuo et al. (2014)
qSB9	9	NA	Jarjan	Koshihikari	BILs	Taguchi-Shiobara et al. (2013)
qshb7.3 and qshb9.2	7 and 9		ARC10531	BPT-5204	mapping populations (F_2 and BC_1F_2)	Yadav et al. (2015)
qshb 1, qshb-3, qshb-5, qshb-6, *qHZaLH8*, and *qHZaDR8*	1, 3, 5, 6, and 8	LH and DR	CJ06	TN1	doubled haploid (DH) population	Zeng et al. (2015)

PH (plant height) and HD (heading date) are given in bold when the peak LOD value is exactly at the same position on the chromosome. QTL, quantitative trait locus (loci); PVE, phenotypic variation explained; ND, not determined; TA, tiller angle; NA, not associated with ShB-QTL; CL, culm length; LL, lesion length; ALL, actual lesion length; NILs, near-isogenic lines; BILs, backcrossed inbred lines; DH, doubled haploid. Adapted and modified from Srinivasachary et al. (2010).

contributed by Lemont could not be identified in our clonal population, of which one parental variety was also Lemont. On the other hand, a major resistance QTL, *qSB*-11, on chromosome 11, which explained 31.2% of the total phenotypic variation, was identified in our study, and Li et al. also indicated that there might have been a putative resistant QTL in the same interval of qSB-11, though they did not give any further information about the effect of this locus possibly due to its low LOD score. In addition, they identified three QTLs for heading date and four QTLs for plant height in the resistance loci interval, and thus thought that the QTLs for ShB resistance were closely associated with the QTLs for heading date or plant height.

qShB9-2, a QTL for ShB, was mapped to a region at the bottom of chromosome 9 consisting of ≈1.2 Mbp flanked by SSR markers RM215 and RM245 (Liu et al., 2009). The majority of variants in *qShB9-2* were classified as sSNPs (73%), a substantially smaller percentage as nsSNPs (26%), and the smallest fractions identified were insertions (1.0%) or deletions (0%). When the CV selection procedure was carried out to identify candidate nsSNPs for SB resistance within *qShB9-2*, relatively few selected nsSNPs (10) were found that mapped throughout most (≈1.1 Mbp) of the QTL. The nsSNPs were detected in a total of 10 genes that were placed into seven groups based on gene ontology/gene function. The physical location of selected nsSNPs within *qShB9-2* along with corresponding genes and QTL *qShB9-2* explained ≈25% of the observed variation for SB resistance when Jasmine 85 was used as the resistant parent (Liu et al., 2009).

Li et al. (1995a,b) identified six QTLs for disease score in the F4 population of Lemont/Teqing, but one allele on chromosome 8 for resistance. The main QTL (QSbr4c) controlling LH and actual lesion length (ALL) associated with RM280, on chromosome 4, was located near the chromosomal region of resistance QTL and QSbr2a controlling LH and ALL associated with RM341, on chromosome 2, was approximately mapped on the same chromosomal region of qSB2 identified by Zou et al. (2000).

Loan et al. (2004) constructed 266 Near Isogenic Introgression Lines with randomly introgressed Lemont segment of a cross between Lemont × Teqing. Further, 15 M-QTLs detected for LH and ALL over assessment times were mapped on seven chromosomes (1, 2, 3, 4, 5, 9, and 12), explaining 35.8%–93.8% of the phenotypic variation. The QTLs with high additive effects for most resistance traits were found at the markers RM341 (on chromosome 2), RM156 (on chromosome 3), and RM280 (on chromosome 4). The four QTLs, namely, QSbr1a, QSbr2a, QSbr4c, and QSbr9b that were found not associated with plant morphology or heading date are potentially useful in breeding programs for ShB resistance.

Zuo et al. (2007) reported introgression of the QTL, qSB-11LE and observed reduced grain loss by 10.71% in Lemont background under severe disease infestation in field trials. Pinson et al. (2005) predicted that qSB-9TQ and qSB-3TQ could possibly reduce the crop loss due to ShB by 15% when introduced into Lemont. Xie et al. (2008) studied the resistance QTLs by crossing the Lemont and Teqing (LT-ILs and TQ-ILs). Lemont further a total of 10 main-effect QTLs (M-QTLs) and 13 epistatic QTLs (E-QTLs) conferring ShB resistance (SBR) were mapped using data obtained from different years and genetic backgrounds. Among them, 6 M-QTLs detected in 2006 were verified in 2007, suggesting that these M-QTLs had reliable performance across years. *QRlh4* was the only M-QTL expressed under the reciprocal backgrounds. On chromosome 10, *QRlh0a* between RM216 and RM 311 was detected in TQ-ILs and *QRlh0b* between RM222 and RM 216 was detected in LT-ILs and regarded as different gene because their directions of additive affect were opposite. Most QTLs identified in TQ-ILs were not expressed in LT-ILs, indicating the presence of a significant effect of genetic background. By comparative mapping, 8 M-QTLs detected in this study were located in the same or near regions that were associated with SBR identified in the previous studies. These M-QTLs have great potential to be applied in rice breeding for SBR by marker-assisted selection (MAS), and M-QTLs expressed stably in different backgrounds are favorable for gene pyramiding in SBR improvement in rice.

Yin et al. (2008) and Wang et al. (2012) have found that pyramiding of diverse ShB R-QTLs could help achieve higher levels of resistance to ShB. The $F_{2:3}$ progenies of the cross ARC10531 and BPT-5204 (210 progenies) were phenotyped for ShB during wet season 2012. The frequency distribution curve of $F_{2:3}$ progenies for disease were continuous and near to normal distribution. The range of relative LH in percentage was 21–75%. In the $F_{2:3}$ progenies, more individuals were distributed toward 40–50% of relative LH and population appears to be skewed more toward susceptible side. The mean value recorded for relative LH was 38.51%. The results revealed that variability for the morphological traits, *namely*, plant height, heading date, number of tillers, panicle length, and disease score (RLH %) ranges from 70 to 127, 69 to 33, 5 to 31, 5 to 25, 14 to 25, and 21 to 75, respectively. By using the BSA method, two bulks having distinct and often contrasting phenotypes for the trait of interest are generated from a segregating population from a single cross. Seventy polymorphic markers were used for screening of parents ARC10531, BPT-5204, resistant bulk (RB), and susceptible bulk (SB) along with individuals of $F_{2:3}$ populations used in respective bulks. Two markers RM 205 (chromosome 9) and RM 336 (chromosome 7) clearly distinguished SBs from RBs.

Yin et al. (2008) reported that the three resistance QTLs could significantly improve the resistance to rice ShB separately or jointly. A rice ShB resistance QTL qSB7 (superscript Tq) on rice chromosome 7 of Teqing was confirmed by using the backcross between Teqing and Lemont. The effects and pyramiding effects of qSB7 (superscript Tq), qSB9 (superscript Tq) (a rice ShB resistance QTL mapped on chromosome 9 of Teqing) and qSB11 (superscript Le) (a rice ShB resistance QTL mapped on chromosome 11 of Lemont) were studied by using a set of near-isogenic lines (NILs) under the background of Lemont. Sharma et al. (2009) constructed a population of 279 $F_{2:3}$ progeny derived from a cross between two tropical *japonica* U.S. rice cultivars, Rosemont (semi-dwarf, SB susceptible) and Pecos (tall, SB resistant), was used to map SB resistance.

Liu et al. (2009) detected SB resistance QTLs in a cross derived from two rice cultivars, that is, susceptible cv. Lemont and the resistant cv. Jasmine 85. Further, disease occurrence screenings of 250 F_5 recombinant inbred lines (RILs) were investigated under greenhouse conditions. Further, 10 ShB-QTLs were identified on six different chromosomes namely chromosome 1, 2, 3, 5, 6, and 9. Screening under micro-chamber identified four of five new ShB-QTLs, one on each of chromosomes 1, 3, 5, and 6. Both micro-chamber and mist-chamber methods identified two ShB-QTLs, qShB1 and qShB9-2. Four of the ShB-QTLs or ShB-QTL regions identified on chromosomes 2, 3, and 9 were previously reported in the literature. The major ShB-QTL qShB9-2, which co-segregated with simple sequence repeat marker RM245 on chromosome 9, contributed to 24.3% and 27.2% of total phenotypic variation. ShB QTLs qShB9-2, a plant-stage-independent QTL, was further confirmed in nine haplotypes of 10 resistant Lemont/Jasmine 85 RILs using haplotype analysis.

Channamallikarjuna et al. (2010) examined 127 recombinant inbred lines in seven environmental conditions at three locations across 4 years, but the QTL with the largest effect was detected. One QTL, qSBR11-1, was detected commonly in three conditions ($r^2 = 12–14\%$), but no QTL was detected in more than three. Zuo et al. (2011) studied ShB resistance and its potential in breeding programs by using NILs and found the three different genotypes at the qSB-11LE locus and seven backcross populations sowed positive effect. Observation from the field disease evaluation data under artificial inoculation revealed that the inheritance of resistance of qSB-11LE to ShB is controlled by additive gene action and corresponding genes have a dosage effect on ShB resistance. Further in greenhouse evaluations, the resistance effect of *qSB-11LE* was expressed at 11 and 14 days after inoculation at the tillering stage. Finally, analysis of field resistance of six BC_1F_1 populations and one BC_2F_1 population, developed by backcrosses between Lemont as the donor parent and six commercial *O. indica* rice cultivars as recurrent parents, significantly indicated that *qSB-11LE* could be effectively used to enhance these cultivars against ShB resistance.

Wang et al. (2012) selected three Teqing into Lemont backcross introgression lines (TILs) with more resistant than their susceptible parent (Lemont). Further these QTLs were molecularly verified to contain Teqing alleles at *qSB92* and/or *qSB121*. By comparing the ShB resistance in microchamber evaluations and inoculated field plots, the phenotypic values of the QTL were measured. Under both study conditions, disease resistance ranked *qSB92* + *qSB121* > *qSB92* > *qSB121* > no QTL, with both *qSB92* and *qSB121* acting as dominant resistance genes. In micro-chamber studies, *qSB92 TQ* reduced disease with an average of 1.0 disease index units and *qSB121TQ* by 0.7 using a scale of 0–9. Field effects of *qSB92 TQ* and *qB121TQ* were less pronounced, with average phenotypic gains of 0.5 and 0.2 units, respectively. TIL:642 proved to contain *qSB92 TQ* in an introgression so small that it was tagged by just RM205 on the tip of chromosome 9. These studies verify that the *indica* introgression of *qSB92 TQ* or *qSB121 TQ* can measurably improve resistance to ShB disease in a highly susceptible tropical *japonica* cultivar, and fine mapped the *qSB92* locus.

Pandian et al. (2012) identified that Tetep carry 12 QTLs governing ShB resistance. Further, parental screening for the ShB resistance by using a highly virulent isolate Kapurtala was done. Evaluation of 186 advanced backcross inbred lines for ShB resistance revealed that 9 Pusa6B-derived inbred lines were resistant, 11 Pusa1460-derived inbred lines, and 12 PRR78-derived lines were moderately resistant to ShB. The varying quantum of resistance depicted by the field screening implies that, varying number of QTLs present in the residual donor segments of the ABLs.

Liu et al. (2013) confirmed the major ShB-QTL *qShB9-2* based on the field data and also identified one new ShB-QTL between markers RM221 and RM112 on chromosome 2 in the RIL population derived from the cross of Lemont × Jasmine 85 (LJRIL). Based on the field verification of ShB evaluations, the micro-chamber and mist-chamber assays were simple, effective, and reliable methods to identify major ShB-QTLs like *qShB9-2* in the greenhouse at early vegetative stages. The markers RM215 and RM245 were found to be closely linked to *qShB9-2* in greenhouse and field assays, indicating that they will be useful for improving ShB resistance in rice breeding programs using MAS.

Taguchi-Shiobara et al. (2013) collected three landraces from the Himalayas, Jarjan, Nepal 555 and Nepal 8, with resistance to ShB. Further, they developed backcrossed inbred lines derived from a cross between Jarjan and the leading Japanese cultivar Koshihikari and further were used in QTL analyses, since later-heading lines showed fewer lesions. Eight QTLs were further identified, and only one QTL on chromosome 9 (between markers Nag08KK18184 and Nag08KK18871) was detected. Chromosome segment substitution lines (CSSLs) carrying it showed resistance in field tests. Thirty F2 lines derived from a cross between Koshihikari and one CSSL supported the QTL.

Zuo et al. (2014) identified two important QTLs namely *qSB9TQ* and *TAC1TQ*, which control SB resistance and tiller angle (TA). These two QTLs, *TAC1TQ* and *qSB9TQ*, have high breeding potential, and pyramiding SB resistance QTL and morphological trait QTL is a potential approach in improving rice SB resistance. Hossain et al. (2014a,b) selected nine rice cultivars and screened at greenhouse conditions. Results showed that Tetep and Teqing had the lowest disease ratings. UKMRC2, a new high yielding cultivar, was as recipient parent. Crosses between UKMRC2 and Teqing, and UKMRC2 and Tetep were made and confirmed. Subsequently four-way crosses between the two F_{1s} were performed to develop pyramidal lines.

Zuo et al. (2014) reported that qSB9 is a major quantitative trait locus that confers significant resistance to rice ShB. However, the precise location has not yet been determined. They reported the fine mapped location of *qSB9 TQ*, the resistant allele(s) underlying qSB9 derived from *indica* rice variety Teqing (TQ). A population containing 235 CSSLs that integrated TQ donor segments specific to the qSB9 region in the Lemont genetic background were developed and studied. These CSSLs contained identical genetic backgrounds, as monitored with 111 molecular markers and showed similar morphologies except for TA. They also identified a gene controlling TA, *TAC1 TQ*, in the qSB9 region by comparing the TA phenotype and the genotype of each CSSL. Although *TAC1 TQ* only showed a very mild effect on SB resistance, it affected the accurate evaluation of the contribution of *qSB9 TQ*. The development of new molecular markers in this region and accurate determination of the SB resistance phenotypes of these 10 CSSLs by conducting both field and greenhouse tests allowed us to finemap *qSB9 TQ* to a 146-kb region defined by markers CY85 and Y86.

Wen et al. (2015) reported eight different QTLs for disease rating (four in E1, four in E2, and three in E3), six QTLs for LH (one in E1, three in E2, and two in E3), and seven QTLs for percentage of LH (one in E1, four in E2, and two in E3). Sixteen of the ShB-QTLs co-localized as six clusters on chromosomes 3, 7, 11, and 12. Four of the six clusters contained ShB-QTLs that were detected in two environments, while the other two clusters with ShB-QTLs were detected in one environment. Three ShB-QTLs (*qSBD-3-2*, *qSBL-3-1*, and *qSBPL-3-1*) were delimited to a 581-kb region flanked by markers D333B and D334 on chromosome 3.

Yadav et al. (2015) studied 40 different rice germplasm including 8 wild, 4 landraces, 26 cultivated, and 2 advanced breeding lines for ShB resistance. Except two rice varieties, Tetep and ARC10531 expressed moderate level of resistance against ShB. Further two mapping populations (F_2 and BC_1F_2) were developed from the cross BPT-5204/ARC10531 for QTL mapping. With the utilization of composite interval mapping analysis, 9 QTLs were mapped to 5 different chromosomes with phenotypic variance ranging from 8.40% to 21.76%. Two SSR markers RM336 and RM205 were noted to be closely related with the major QTLs qshb7.3 and qshb9.2. A hypothetical β 1–3 glucanase with other 31 candidate genes were identified in-silico study by utilizing rice database RAP-DB.

Zeng et al. (2015) developed a doubled haploid population that was constructed from a cross between a *japonica* variety CJ06 and an *indica* variety TN1 and analyzed the QTLs for SB resistance under three different environments. They identified QTLs for LH on chromosomes 1, 3, 4, 5, 6, and 8 and explained 4.35–17.53% of the phenotypic variation against ShB. The ShB resistance allele of *qHNLH4* from TN1 decreased LH by 3.08 cm and contributed to 17.53% of the variation at environment 1. The QTL for LH (*qHZaLH8*) detected on chromosome 8 in environment 2 explained 16.71% of the variation, and the resistance allele from CJ06 reduced LH by 4.4 cm. Eight QTLs for DR were identified on chromosomes 1, 5, 6, 8, 9, 11, and 12 under three conditions with the explained variation from 2.0% to 11.27%. The QTL for disease rating (*qHZaDR8*), which explained variation of 11.27%, was located in the same interval as that of *qHZaLH8*; both QTLs were detected.

4.8   TRANSGENIC RICE AGAINST SHB

There is several transgenic rice lines with different defense related gene with increased resistance to ShB (Table 4.4) have been reported (Datta et al., 2000; Lin et al., 1995; Mao et al., 2014; Xinping et al., 2001). Pathogenesis-related (PR) proteins are produced in response to an attack by a pathogen and are known to play key roles in the plant defense mechanisms (Datta et al., 1999a,b). Over-expression of PR proteins, including chitinase (PR-3),

TABLE 4.4 Overview of Historical Efforts to Develop Transgenic Resistance in Rice against ShB.

Sl. No.	Cultivars transformed	Transformed gene(s)	Origin of concerned gene(s) and other additional features	References
1.	Yamahoushi, Nipponbare	Bar	Herbicide tolerance gene, reduced ShB infection when plants spayed with bialaphos or phosphinothricin	Uchimiay et al. (1993)
2.	Chinsurah Boro II	Chi 11	Chitinase-containing rice genomic DNA (1.1 kb)	Lin et al. (1995)
3.	Chinsurah Boro II, IR72, IR1500	TLP-D34	Rice thamatin-like protein, a member of PR-5 group	Datta et al. (1999b)
4.	IR64, IR72, IR688998, MH63 Chinsurah Boro II	RC 7	Rice chitinase	Datta et al. (2000, 2001)
5.	M202	pinA, pinB	Structural protein from *Triticum aestivum*	Krishnamurthy et al. (2001)
6.	Swarna	Chi 11	Rice chitinase	Baisakh et al. (2001)
7.	IR72	Chi, Xa21, Bt	Chitinase, receptor-like kinase, and *Bt* toxin	Datta et al. (2002)
8.	Kenfong	MODI, RCH0	Modified maize ribosome-inactivating protein gene and basic chitinase	Kim et al. (2003)
9.	Pusa Basmati 1	Chi 11	Rice chitinase	Sridevi et al. (2003)
10.	ADT38, ASD16, IR50 Pusa Basmati 1	Chi 11, tip	Enhanced resistance to both ShB and ShR	Kalpana et al. (2006)
11.	Pusa Basmati 1	Ace-AMP1	A non-lipid transfer protein with antimicrobial property isolated from *Allium cepa* showed enhanced resistance against ShB, Blast and BLB	Patkar and Chattoo (2006)
12	Pusa Basmati 1, White Ponni ADT38, Co43	RC 7	Rice chitinase	Nandakumar et al. (2007)
13.	ASD16, ADT38, IR72, IR64, White Ponni	Chi 11, tlp, Xa21	Rice chitinase, thaumatin-like protein and serine-threonine kinase enhanced resistance to both ShB and BLB	Maruthasalam et al. (2007)
14.	Pusa Basmati 1	Chi 11, b-1,3-glucanase	Rice chitinase and tobacco b-1,3-glucanase	Sridevi et al. (2008)

TABLE 4.4 *(Continued)*

Sl. No.	Cultivars transformed	Transformed gene(s)	Origin of concerned gene(s)and other additional features	References
15.	Pusa Basmati 1	Chi 11	Rice chitinase	Sripriya et al. (2008)
16.	Pusa Basmati 1	Rs-AFP2	A defensin gene from *Raphanus sativus*	Jha and Chattoo (2009a)
17.	Pusa Basmati 2	Dm-AMP1	A defense gene from *Dalia merkii*	Jha et al. (2009)
18.	Pusa Basmati 1	Dm-AMP1, Rs-AFP2	Defensin genes from *D. merkii* and *R. sativus*, respectively	Jha and Chattoo (2009b)
19.	JinHui35	McCHIT	A class I chitinase gene of bitter melon	Li et al. (2009)
20.	Pusa Basmati 1	Cht 42	A chitinase gene from *Trichoderma* spp.	Shah et al. (2009)
21.	White Ponni	tlpD34, PR5, *chi*11, PR3	Thaumatinlike protein gene (*tlp*D34, PR5) combination with the chitinase gene (*chi*11, PR3)	Shah et al (2013)
22.	Kitaake	ACS2 (1-aminocyclopropane-1-carboxylic acid synthase	Rice ACS2 (1-aminocyclopropane-1-carboxylic acid synthase, a key enzyme of ET biosynthesis)	Helliwell et al. (2013)
23.	Taipei 309	chitinase gene (*RCHI0*) β1,3-glucanase gene (*AGLU1*)	Rice basic chitinase gene (*RCHI0*) and the alfalfa β1,3-glucanase gene (*AGLU1*)	Mao et al. (2014)
24.	Zhonghua 11	OsPGIP1	Over expressed OsPGIP1	Wang et al. (2015)

Adapted and modified from Srinivasachary et al. (2010).

β,3-glucanases (PR-2), thaumatin-like proteins (PR-5), and other plant- or microbe-derived antifungal proteins have been used to develop transgenic plants against fungal infection. Chitinases that hydrolyze the b-1,4 linkages of nacetyl glucosamine (chitin) have been well characterized. Overexpression of different chitinases in rice cultivars has been found to result in enhanced resistance against ShB (Datta et al., 2001). The expression of pinA and/or pinB (Krishnamurthy et al., 2001), Ace-AMP1 (Patkar & Chattoo, 2006), and Dm-AMP1 (Jha et al., 2009) resulted in not only enhanced resistance against ShB but also against other rice diseases. There have also been efforts to combine resistance genes to generate plants with increased resistance to ShB. These researchers suggested that Dm-AMP1 and Rs-AFP2 may be the best genes used to date in transgenic approaches. To date, more than 12 rice cultivars, including IR72, IR64, Chinsurah Boro II, Basmati 122, Swarna, and IR58, have been transformed with genes for ShB resistance. Lin et al. (1995) reported the transgenic plant with a chitinase gene under the control of the CaMV 35S promoter showed resistance to the ShB pathogen, *R. solani.* Yuan et al. (2004) constructed transgenic rice, Zhongda 2, by rice chitinase gene (*RC24*), showed high resistance to rice ShB (*R. solani*) in laboratory and a 2-year field experiment. The *R. solani* could invade sheath of Zhongda 2 and induce symptoms of the disease. No difference was noted in time of penetration or incubation period between Zhongda 2 and non-transgenic rice control, Zhuxian B, but the hyphae lysate could be observed earlier than control.

Kalpana et al. (2006) engineered the different lines of elite indica rice cultivars, ADT38, ASD16, IR50, and PB1 by constitutively overexpressing rice tlp encoding a thaumatin-like protein. The putative transformants and their progenies expressing tlp showed enhanced resistance against the ShB pathogen, *R. solani*, when compared to the non-transformed plants. The use of rice chi11, encoding a chitinase, as a cotransgene along with tlp produced a tlp–chi11 co-transformant that showed enhanced resistance against *R. solani* than the ones that express tlp or chi11 transgene alone. Maruthasalam et al. (2007) transformed indica rice cultivars were with two genes rice chitinase (*chi11*) and a thaumatin-like protein (*tlp*) coexpression of chitinase and thaumatin-like protein in the progenies of a transgenic PB1 line revealed an enhanced resistance to the ShB pathogen, *R. solani*, as compared to that in the lines expressing the individual genes. The transgenic PB1 line pyramided with the genes *chi11*, *tlp*, and *Xa21* showed enhanced resistance against ShB and bacterial blight. Datta et al. (2001) generated transgenic elite indica rice cultivars with a PR-3 rice chitinase gene (RC7) showed higher resistance to rice ShB disease caused by *R. solani.* Xinping et al. (2001) generated transgenic rice by introducing a basic chitinase gene (RC24) into the elite indica variety Zhuxian B and stably integrated in the genome of transgenic rice from R_0 generation to R_6 generation and expressed. Two transgenic strains, Zhuzhuan 68 and Zhuzhuan 70, and 43 zy transgenic lines were obtained, showing significantly higher resistance against rice blast and ShB.

Sripriya et al. (2008) generated transgenic *O. sativa* L. var. PB1 by using *Agrobacterium tumefaciens*. The TDNA of the cointegrate vector pGV2260::pSSJ1 carried the hygromycin phosphotransferase (hph) and betaglucuronidase genes. The binary vector pCamchi11, without a plant selectable marker gene, harbored the rice chitinase (chi11) gene under maize ubiquitin promoter. Co-transformation of the gene of interest (chi11) with the selectable marker gene (hph) occurred in 4 out of 20 $T_{(0)}$ rice plants (20%). Segregation of hph from chi11 was accomplished in two (CoT6 and CoT23) of the four co-transformed rice plants in the $T_{(1)}$ generation. The selectable marker free lines C_0T_6 and C_0T_{23} contained single copies of chi11. The lines C_0T_6 and C_0T_{23} exhibited 38% and 40% reduction in ShB disease.

Shah et al. (2009) constructed a transgenic rice line with 42 kDa endo-chitinase (*cht*42) gene from the mycoparasitic fungus, *Trichoderma virens*. Eight different transgenic plants containing single copies of complete TDNA were identified by Southern blot analysis. Homozygous transgenic plants were further identified for five lines in the T_1 generation. Homozygous T_2 plants constitutively accumulated high levels of the *cht*42 transcript, showed 2.4–4.6-fold higher chitinase activity after infection with *R. solani*. Infection assays with *R. solani* showed up to 62% ShB disease index reduction.

Shah et al. (2013) developed transgenic rice (cv. White Ponni) with thaumatin like protein gene (*tlpD*34, PR5) combination with the chitinase gene (*chi*11, PR3). The homozygous T_2 plants harboring *tlpD*34 + *chi*11 genes showed 2.8–4.2-fold higher chitinase activity. Upon infection with *R. solani*, the disease index reduced from 100% in control plants to 65% in a T_3 homozygous transgenic line T_4 expressing the *tlpD*34 gene alone. Disease index reduced up to 39% in the T2 homozygous transgenic line CT22 co-expressing *tlpD*34 and *chi*11 genes.

Helliwell et al. (2013) produced transgenic lines with inducible production of ET by expressing the rice ACS2 (1-aminocyclopropane-1-carboxylic acid synthase, a key enzyme of ET biosynthesis) transgene under control of a strong pathogen-inducible promoter. The OsACS2-overexpression lines showed significantly increased levels of the OsACS2 transcripts, endogenous ET and defense gene expression in comparison to wild rice, especially in response to pathogen infection. The transgenic lines further exhibited increased resistance to a field isolate of *R. solani*, as well as different races of *M. oryzae*.

Mao et al. (2014) transformed rice basic chitinase gene (*RCH10*) and the alfalfa β-1,3-glucanase gene (*AGLU1*) were tandemly inserted into transformation vector pBI101 under the control of 35S promoter with its enhancer sequence to generate a double-defense gene expression cassette pZ100. The pZ100 cassette was transformed into rice (cv. Taipei 309) by *Agrobacterium*-mediated transformation. More than 160 independent transformants were obtained and confirmed by PCR. Northern analysis of inheritable progenies revealed similar levels of both *RCH10* and *AGLU1* transcripts in the same individuals. Disease resistance to both ShB and blast was challenged in open field inoculation. Immunogold detection revealed that RCH10 and AGLU1 proteins were initially located mainly in the chloroplasts and were delivered to the vacuole and cell wall upon infection, suggesting that these subcellular compartments act as the gathering and execution site for these antifungal proteins.

Wang et al. (2015) used OsPGIP1 against the PGase from *R. solani* for the transformation purpose. In addition, the location of OsPGIP1 was also determined by subcellular localization and subsequently, over expressed OsPGIP1 in a rice cultivar Zhonghua 11 (*O. sativa* L. ssp. japonica). Field testing of *R. solani* inoculation showed that the ShB resistance of the transgenic rice was significantly improved. Furthermore, the levels of ShB resistance were in accordance with the expression levels of OsPGIP1 in the transgenic lines. The results revealed the functions of OsPGIP1 and its resistance mechanism against rice ShB.

Richa et al. (2016) have analyzed molecular and functional analysis of the resistance genes with the major *R. solani*-resistance QTL qSBR11-1in indica rice genotypes Tetep. Sequencing and further study revealed the presence of a set of 11 tandem repeats containing genes with a high degree of homology to class III chitinase defense-response genes. Comparison between the resistant Tetep and the susceptible HP2216 lines shows that the induction of the chitinase genes is much higher in the Tetep line. Recombinant protein produced in vitro for 6 of the 11 genes showed chitinolytic activity in gel assays, but we did not detect any xylanase inhibitory activity. All the six *in vitro* expressed proteins show antifungal activity with a clear inhibitory effect on the growth of the *R. solani* mycelium. The characterized chitinase genes can provide an important resource for the genetic improvement of *R. solani* susceptible rice lines for ShB resistance breeding.

4.8.1   RNAI FOR SHEATH BLIGHT RESISTANCE

Manosalva et al. (2011) studied the role of GF14e in rice disease resistance by suppressing its expression using an RNA interference (RNAi)-silencing approach. GF14e-silenced transgenic plants showed spontaneous HR-like lesions and enhanced resistance to a virulent strain of *Xanthomonas oryzae* pv. oryzae. The enhanced resistance correlates with high expression of a rice peroxidase gene and accumulation of ROS. Silencing GF14e also enhanced resistance to the necrotrophic ShB pathogen *R. solani*.

Xia (2016) studied a gene encoding, a nucleoporin, named as cloned from rice Nipponbare (*O. sativa* L. spp. *japonica*, var. *nippobare*). Further, the expression of *OsSeh1* gene was induced by salicylic acid or ShB agent *R. solani*. The highest expression of *OsSeh1* was observed at 24 h as 3.5 times more expression, treated with *R. solani*. RNAi rice lines of *OsSeh1* gene were more susceptible to *R. solani*. In the transgenic line of T1 generation, relative expression quantity of *OsSeh1* was found 6–11 with significant resistance, which was higher than 2.47 in wild type rice Nipponbare. In the RNAi rice plants, relative expression quantity of *OsSeh1* is 0.2–0.6 in with obvious susceptibility. The results showed that rice resistance to *R. solani* was positively correlated with *OsSeh1* expression levels.

4.9   FUTURE DIRECTION

Silva et al. (2012) conducted a research to exploit whole genome sequences of 13 rice (*O. sativa* L.) inbred lines to identify non-synonymous SNPs (nsSNPs) and candidate genes for resistance to ShB. Two filtering strategies were developed to identify nsSNPs between two groups of known resistant and susceptible lines. A total of 333 nsSNPs detected in the resistant lines were absent in the susceptible group. Selected variants associated with resistance were found in 11 of 12 chromosomes. More than 200 genes with selected nsSNPs were assigned to 42 categories based on gene family/gene ontology. Three new regions with novel candidates were also identified. A subset of 24 nsSNPs detected in 23 genes was selected for further study. Individual alleles of the 24 nsSNPs were evaluated by PCR whose presence or absence corresponded to known resistant or susceptible phenotypes of nine additional lines. Sanger sequencing conformed presence of 12 selected nsSNPs in two lines. "Resistant" nsSNP alleles were detected in two accessions of *O. nivara* that suggests sources for resistance occur in additional *Oryza* sp.

Jia et al. (2012) identified ShB QTLs via association mapping in rice using 217 subcore entries from the USDA rice core collection, which were phenotyped with a micro-chamber screening method and genotyped with 155 genome-wide markers. Structure analysis divided the mapping panel into five groups, and model comparison revealed that PCA5 with genomic control was the best model for association mapping of ShB. Ten marker loci on seven chromosomes were significantly associated with response to the ShB pathogen. Among multiple alleles in each identified loci, the allele contributing the greatest effect to ShB resistance was named the putative resistant allele. Among 217 entries, entry GSOR 310389 contained the most putative resistant alleles, 8 out of 10. The number of putative resistant alleles presented in an entry was highly and significantly correlated with the decrease of ShB rating ($r = 20.535$) or the increase of ShB resistance. Majority of the resistant entries that contained a large number of the putative resistant alleles belonged to indica, which is consistent with a general observation that most ShB resistant accessions are of indica origin. These findings demonstrate the potential to improve breeding efficiency by using MAS to pyramid putative resistant alleles from various loci in a cultivar for enhanced ShB resistance in rice.

KEYWORDS

- **sheath blight**
- **wild rice**
- **soil-borne pathogen**
- **resistance gene**
- **quantitative trait loci**

REFERENCES

Amante, A. D.; de la Pena, R.; Stich, L. A.; Leung, H.; Mew, T. W. Sheath Blight (ShB) Resistance in Wild Rice. *Int. Rice Res. Newsl.* **1990**, *15*, 5.

Ansari, M. M.; Sharma, A.; Thangal, M. H. Evaluation of Rice Cultures against Sheath Blight. *J. Andaman Sci. Assoc.* **1989**, *5*, 89–90.

Baisakh, N.; Datta, K.; Oliva, N.; Ona, I.; Rao, G. J. N.; Mew, T. W.; Datta, S. K. Rapid Development of Homozygous Transgenic Rice using Anther Culture Harboring Rice Chitinase Gene for Enhanced Sheath Blight Resistance. *Plant Biotechnol.* **2001**, *18*, 101–108.

Bhaktavatsalam, G.; Satyanarayana, K.; Reddy, A. P. K.; John, V. T. Evaluation of Sheath Blight Resistance in Rice. *Int. Rice Res. Newsl.* **1978**, *3*, 9–10.

Biswas, A. Screening of Rice Varieties for Sheath Blight (ShB) Disease Tolerance in West Bengal, India. *Oryza.* **2005**, *42*, 83–84.

Borthakur, B. K.; Addy, S. K. Screening of Rice (*Oryza sativa*) Germplasm for Resistance to Sheath Blight (*Rhizoctonia solani*). *Indian J. Agric. Sci.* **1988**, *58*, 537–538.

Brar, D.; Khush, G. S. Alien Introgression in Rice. *Plant Mol. Biol.* **1997**, *35*, 35–47.

Carling, D. E. Grouping in *Rhizoctonia solani* by Hyphal Anastomosis Interactions. In: *Rhizoctonia Species: Taxonomy, Molecular Biology, Ecology, Pathology and Disease Control*; Sneh, B., Jabaji-Hare, S., Dijst, G., Eds.; Kluwer Academic Publishers: Dordrecht, The Netherlands, 1996; pp 35–48.

Carling, D. E.; Kuninaga, S.; Brainard, K. A. Hyphal Anastomosis Reactions rDNA-internal Transcribed Spacer Sequences, and Virulence Levels among Subsets of *Rhizoctonia solani* Anastomosis Group-2 (AG-2) and AG-BI. *Phytopathology* **2002**, *92*, 43–50.

Castro, A. J.; Capettini, F.; Corey, A. E.; Filichkina, T.; Hayes, P. M.; Kleinhofs, A.; Kudrna, D.; Richardson, K.; Sandoval-Islas, S.; Rossi, C.; Vivar, H. Mapping and Pyramiding of Qualitative and Quantitative Resistance to Stripe Rust in Barley. *Theor. Appl. Genet.* **2003**, *107*, 922–930.

Channamallikarjuna, V.; Sonah, H.; Prasad, M.; Rao, G. J. N.; Chand, S.; Upreti, H. C.; Singh, N. K.; Sharma, T. R. Identification of Major Quantitative Trait loci *qSBR11*-1 for Sheath Blight Resistance in Rice. *Mol. Breed.* **2010**, *25*, 155–166.

Che, K. P.; Zhan, Q. C.; Xing, Q. H.; Wang, Z. P.; Jin, D. M.; He, D. J.; Wang, B. Tagging and Mapping of Rice Sheath Blight Resistant Gene. *Theor. Appl. Genet.* **2003**, *106*, 293–297.

Chen, Z. X.; Zhang, Y. F.; Feng, F.; Feng, M. H.; Jiang, W.; Mac, Y. Y.; Pand, C. H.; Hu, H. L.; Li, G. S.; Pan, X. B.; Zuo, S. M. Improvement of *japonica* Rice Resistance to Sheath Blight by Pyramiding *qSB-9*^TQ and *qSB-7*^TQ. *Field Crop. Res.* **2014**, *161*, 118–127.

Crill, P.; Nuque, F. L.; Estrada, B. A.; Bandong, J. M. The Role of Varietal Resistance in Disease Management. In *Evolution of Gene Rotation Concept for Rice Blast Control*; IRRI, Ed.; International Rice Research Institute: Los Banos, 1982; pp 103–121.

Das, N. P. Resistance of Some Improved Varieties of Rice (*Oryza sativa* L.) to Sheath Blight Caused by *Rhizoctonia solani* Kuhn. *Indian J. Agric. Sci.* **1970**, *40*, 566–568.

Datta, K.; Koukolıkova-Nicola, Z.; Baisakh, N.; Oliva, N.; Datta, S. K. *Agrobacterium* Mediated Engineering for Sheath Blight Resistance of Indica Rice Cultivars from Different Ecosystems, *Theor. Appl. Genet.* **2000**, *100*, 832–839.

Datta, K.; Muthukrishnan, S.; Datta, S. K. Expression and Function of PR-protein Genes in Transgenic Plants. In Datta, S. K.; Muthukrishnan, S., Eds., Pathogenesis Related Proteins in Plants. CRS: Boca Raton, FL, 1999a; pp 261–277.

Datta, K.; Baisakh, N.; Maung, Thet, K.; Tu, J.; Datta, S. K. Pyramiding Transgenes for Multiple Resistance in Rice Against Bacterial Blight, Yellow Stem Borer and Sheath Blight. *Theor. Appl. Genet.* **2002**, *106*, 1–8.

Datta, K.; Tu, J.; Oliva, N.; Ona, I.; Velazhahana, R.; Mew, T. W.; Muthukrishnan, S.; Datta, S. K. Enhanced Resistance to Sheath Blight by Constitutive Expression of Infection Related Rice Chitinase in Transgenic Elite Indica Rice Cultivars. *Plant Sci.* **2001**, *160*, 405–414.

Datta, K.; Velazhahan, R.; Oliva, N.; Ona, I.; Mew, T.; Khush, G. S.; Muthukrishnan, S.; Datta, S. K. Over-expression of the Cloned Rice Thaumatin-like Protein (PR-5) Gene in Transgenic Rice Plants Enhances Environmental Friendly Resistance to *Rhizoctonia solani* causing Sheath Blight Disease. *Theor. Appl. Genet.* **1999b**, *98*, 1138–1145.

Fu, J.; Huang, Z.; Wang, Z.; Yang, J.; Zang, J. Pre-anthesis Non-structural Carbohydrate Reserve in the Stem Enhances the Sink Strength of Inferior Spikelets During Grain Filling of Rice. *Field Crop. Res.* **2011,** *123,* 170–182.

Gokulapulan, C.; Nair, M. C. Field Screening of Sheath Blight and Rice Root Nematode. *Int. Rice Res. Newsl.* **1983,** *8,* 4.

Grosch, R.; Schneider, J. H. M.; Peth, A.; Waschke, A.; Franken, P.; Kofoet, A.; Jabaji-Hare, S. H. Development of a Specific PCR Assay for the Detection of *Rhizoctonia solani* AG 1-IB Using SCAR Primers. *J. Appl. Microbiol.* **2007,** *102,* 806–819.

Han, P. Y.; Xing, Z. Y.; Chen, X. Z.; Gu, L. S.; Pan, B. X.; Chen, L. X.; Zhang, F. Q. Mapping QTLs for Horizontal Resistance to Sheath Blight in an Elite Rice Restorer Line, Minghui 63. *Acta Genet. Sin.* **2002,** *29,* 622–626.

Hein, Reaction of Germplasm to Sheath Blight of Rice. *Myanmar J. Agric. Sci.* **1990,** *2,* 1–12.

Helliwell, E. E.; Wang, Q.; Yang, Y. Transgenic Rice with Inducible Ethylene Production Exhibits Broad-spectrum Disease Resistance to the Fungal Pathogens *Magnaporthe oryzae* and *Rhizoctonia solani. Plant Biotechnol. J.* **2013,** *11,* 33–42.

Hossain, M. K.; Jena, K.; Bhuiyan, M. A. R.; Ratnam, W. Development of Pyramidal Lines with Two Major QTLs Conferring Resistance to Sheath Blight in Rice (*Oryza sativa* L.) AIP Conf. Proc. **2014a,** *1614,* 765.

Hossain, M. K.; Tze, O. S.; Nadarajah, K.; Jena, K.; Bhuiyan, M. A. R.; Ratnam, W. Identification and Validation of Sheath Blight Resistance in Rice (*Oryza sativa* L.) Cultivars Against *Rhizoctonia solani.* Can. *J. Plant Pathol.* **2014b,** *36,* 482–490.

Inagaki, K. Distribution of Strains of Rice Bordered Sheath Spot Fungus, *Rhizoctonia oryzae*, in Paddy Fields and their Pathogenicity to Rice Plants. *Ann. Phytopathol. Soc. Jpn.* **1996,** *62,* 386–392.

Inagaki, K.; Qingyuan, G.; Masao, A. Overwintering of Rice Sclerotial Disease Fungi, *Rhizoctonia* and *Sclerotium* spp. in Paddy Fields in Japan. *Plant Path. J.* **2004,** *3,* 81–87.

IRRI. *Standard Evaluation System for Rice*; INGER Genetic Resources Centre, International Rice Research Institute: Manila, 1996; 52 pp.

Jha, S.; Chattoo, B. B. Expression of a Plant Defensin in Rice Confers Resistance to Fungal Phytopathogens. *Transgen. Res.* **2009a,** *19,* 373–384.

Jha, S.; Chattoo, B. B. Transgene Stacking and Coordinated Expression of Plant Defensins Confer Fungal Resistance in Rice. *Rice* **2009b,** *2,* 143–154.

Jha, S.; Tank, H. G.; Prasad, B. D.; Chattoo, B. B. Expression of Dm-AMP1 in Rice Confers Resistance to *Magnaporthe oryzae* and *Rhizoctonia solani. Transgen. Res.* **2009,** *18,* 59–69.

Jia, L.; Yan, W.; Zhu, C.; Agrama, H. A.; Jackson, A.; Yeater, K.; Li, X.; Huang, B.; Hu, B.; McClung, A.; Wu, D. Allelic Analysis of Sheath Blight Resistance with Association Mapping in Rice. *PLoS ONE* **2012,** *7,* e32703.

Jia, L. M.; Agrama, H.; Yeater, K.; McClung, A.; Wu, D. Evaluation of the USDA Rice Core Collection for Sheath Blight Disease using Micro-chamber. In: International Annual Meeting of Footprints in the Landscape: Sustainability through Plant and Soil Sciences, Pittsburgh, 2009 Available at: http://a-c-s.confex.com/crops/2009am/webprogram/Paper52830.html.

Jia, Y.; Correa-Victoria, F.; McClung, A.; Zhu, L.; Liu, G.; Wamishe, Y.; Xie, J.; Marchetti, M. A.; Pinson, S. R. M.; Rutger, J. N.; Correll, J. C.; Rapid Determination of Rice Cultivar Response to the Sheath Blight Pathogen *Rhizoctonia solani* Using a Micro-chamber Screening Method. *Plant Dis.* **2007,** *91,* 485–489.

Kalpana, K.; Maruthasalam, S.; Rajesh, T.; Poovannan, K.; Kumar, K. K.; Kokiladevi, E.; Raja, J. A. J.; Sudhakar, D.; Velazhahan, R.; Samiyappan, R.; Balasubramanian, P. Engineering Sheath Blight Resistance in Elite Indica Rice Cultivars Using Genes Encoding Defense Proteins. *Plant Sci.* **2006**, *170*, 203–215.

Kim, J. K.; Jang, I. C.; Wu, R.; Zuo, W. N.; Boston, R. S.; Lee, Y. H.; Ahn, I. P.; Nahm, B. H. Co-expression of a Modified Maize Ribosome-inactivating Protein and a Rice Basic Chitinase Gene in Transgenic Rice Plants Confers Enhanced Resistance to Sheath Blight. *Transgen. Res.* **2003**, *12*, 475–484.

Krishnamurthy, K.; Balconi, C.; Sherwood, J. E.; Giroux, M. J. Wheat Puroindolines Enhance Fungal Disease Resistance in Transgenic Rice. *Mol. Plant Microb. Int.* **2001**, *14*, 1255–1260.

Kunihiro, Y.; Qian, Q.; Sato, H.; Teng, S.; Zeng, D. L.; Fujimoto, K.; Zhu, L. H. QTL Analysis of Sheath Blight Resistance in Rice (*Oryza sativa* L.). *Acta Genet. Sin.* **2002**, *29*, 5.

Lakshmanan, P.; Velusamy, R. Resistance to Sheath Blight (ShB) and Brown Spot (BS) in Lines Derived from *Oryza officinalis*. *Int. Rice Res. Newslett.* **1991**, *16*, 8.

Lee, F. N.; Dilday, R. H.; Moldenhauer, K. A. K.; Rutger, J. N.; Yan, W. Sheath Blight and Rice Blast Resistance in Recently Introduced Rice Germplasm. *Res. Ser. Ark. Agric. Exp. Stn.* **1999**, *468*, 195–210.

Lee, J.; Bricker, T. M.; Lefevre, M.; Pinson, S. R. M.; Oard, H. J. Proteomic and Genetic Approaches to Identifying Defence-related Proteins in Rice Challenged with the Fungal Pathogen *Rhizoctonia solani*. *Mol. Plant Pathol.* **2006**, *7*, 405–416.

Lees, A. K.; Cullen, D. W.; Sullivan, L.; Nicolson, M. J. Development of Conventional and Quantitative Real-time PCR Assays for the Detection and Identification of *Rhizoctonia solani* AG-3 in Potato and Soil. *Plant Pathol.* **2002**, *51*, 293–302.

Li, Z. K.; Pinson, S. R. M.; Marshetti, M. A.; Stansel, J. W.; Park, W. D. Characterization of Quantitative Trait Loci (QTLs) in Cultivated Rice Contributing to Field Resistance to Sheath Blight (*Rhizoctonia solani*). *Theor. Appl. Genet.* **1995**, *91*, 374–381.

Li, P.; Pei, Y.; Sang, X. C.; Ling, Y. H.; Yang, Z. L.; He, G. H. Transgenic Indica Rice Expressing a Bitter Melon (*Momordica charantia*) Class I chitinase Gene (McCHIT1) Confers Enhanced Resistance to *Magnaporthe grisea* and *Rhizoctonia solani*. *Eur. J. Plant Pathol.* **2009**, *125*, 533–543.

Li, Z. K.; Pinson, S. R. M.; Marchetti, M. A.; Stansel, J. W.; Park, W. D. Characterization of Quantitative Trait loci (QTLs) in Cultivated Rice Contributing to Field-resistance to Sheath Blight (*Rhizoctonia solani*). *Theor. Appl. Genet.* **1995**, *91*, 382–388.

Lin, W.; Anuratha, C. S.; Datta, K.; Potrykus, I.; Muthukrishnan, S.; Datta, S. K. Genetic-engineering of Rice for Resistance to Sheath Blight. *Biotechnology* **1995**, *13*, 686–691.

Linde, C. C.; Zala, M.; Paulraj, R. S. D.; McDonald, B. A.; Gnanamanickam, S. S. Population Structure of the Rice Sheath Blight Pathogen *Rhizoctonia solani* AG-1 IA from India. *Eur. J. Plant Path.* **2005**, *112*, 113–121.

Liu, Z. L.; Sinclair, J. B. Differentiation of Intraspecific Groups Within Anastomosis Group-1 of *Rhizoctonia solani* Using Ribosomal DNA Internal Transcribed Spacer and Isozyme Comparisons. *Can. J. Plant Pathol.* **1993**, *15*, 272–280.

Liu, G.; Jia, Y.; Correa-Victoria, F. J.; Prado, G. A.; Yeater, K. M.; McClung, A.; Correll, J. C. Mapping Quantitative Trait loci Responsible for Resistance to Sheath Blight in Rice. *Phytopathology* **2009**, *99*, 1078–1084.

Liu, G.; Jia, Y.; McClung, A.; Oard, J. H.; Lee, F. N.; Correll, J. C. Confirming QTLs and Finding Additional loci Responsible for Resistance to Rice Sheath Blight Disease. *Plant Dis.* **2013**, *97*, 113–117.

Liu, Y.; Chen, L.; Fu, D.; Lou, Q.; Mei, H.; Xiong, L.; Li, M.; Xu, X.; Mei, X.; Luo, L. Dissection of Additive, Epistatic Effect and QTL × Environment Interaction of Quantitative Trait Loci for Sheath Blight Resistance in Rice. *Hereditas* **2014**, *151*, 28–37.

Lore, J. S.; Hunjan, M. S.; Singh, P.; Willocquet, L.; Sri, S.; Savary, S. Phenotyping of Partial Physiological Resistance to Rice Sheath Blight. *J. Phytopathol.* **2013**, 161, 224–229.

Lubeck, M. Molecular Characterization of *Rhizoctonia solani*. *Appl. Mycol. Biotechnol.* **2004**, *4*, 205–224.

Manian, S.; Rao, K. M. Resistance to Sheath Blight Disease in India. *Int. Rice Res. Newsl.* **1979**, *4*, 5–6.

Manosalva, P. M.; Bruce, M.; Leach, J. E. Rice 14-3-3 Protein (GF14e) Negatively Affects Cell Death and Disease Resistance. *Plant J.* **2011**, *68*, 777–787.

Mao, B.; Liu, X.; Hu, D.; Li, D. Coexpression of *RCH10* and *AGLU1* Confers Rice Resistance to Fungal Sheath Blight *Rhizoctonia solani* and Blast *Magnorpathe oryzae* and Reveals Impact on Seed Germination. *World J. Microbiol. Biotechnol.* **2014**, *30*, 1229–1238.

Marchetti, M. A.; Bollich, C. N. Quantification of the Relationship between Sheath Blight Severity and Yield Loss in Rice. *Plant Dis.* **1991**, *75*, 773–775.

Marchetti, M. A.; Bollich, C. N.; McClung, A. M.; Scott, J. E.; Webb, B. D. Registration of RU8703196 Disease-resistant Rice Germplasm. *Crop Sci.* **1995**, *35*, 601.

Marchetti, M. A.; McClung, A. M.; Webb, B. D.; Bollich, C. N. Registration of B82-761 Long-grain Rice Germplasm Resistant to Blast and Sheath Blight. *Crop Sci.* **1996**, *36*, 815.

Maruthasalam, S.; Kalpana, K.; Kumar, K. K.; Loganathan, M.; Poovannan, K.; Raja, J. A. J.; Kokiladevi, E.; Samiyappan, R.; Sudhakar, D.; Balasubramanian, P. Pyramiding Transgenic Resistance in Elite Indica Rice Cultivars against the Sheath Blight and Bacterial Blight. *Plant Cell Rep.* **2007**, *26*, 791–804.

McDonald, B. A.; Linde, C. Pathogen Population Genetic, Evolutionary Potential, and Durable Resistance. *Ann. Rev. Phytopathol.* **2002**, *401*, 349–379.

Meena, B.; Ramamoorthy, V.; Muthusamy, M. Effect of Some Plant Extracts on Sheath Blight of Rice. *Curr. Res.* **2002**, *31*, 49–50.

Meena, S. C.; Singh, V.; Adhipathi, P.; Chand, R. Screening for Sheath Blight Resistant Genotypes Among Mutated Population of Rice cv. Pusa Basmati-1. *Bioscan* **2013**, *8*, 919–924.

Mew, T. W.; Rosales, A. M. Relationship of Soil Microorganisms to Rice Sheath Blight Development in Irrigated and Dryland Rice Cultures; Technical Bulletin ASPAC Food and Fertilizer Technology Center: Taipei City, Taiwan, 1984; 79; p 11.

Mohammadi, M.; Banihashemi, M.; Hedjaroude, G. A.; Rahimian, H. Genetic Diversity among Iranian Isolates of *Rhizoctonia solani* Kuhn Anastomosis Group 1 Subgroups Based on Isozyme Analysis and Total Soluble Protein Pattern. *J. Phytol. Phytopathol. Zeit.* **2003**, *151*, 162–170.

Nandakumar, R.; Babu, S.; Kalpana, K.; Raguchander, T.; Balasubramanian, P.; Samiyappan, R. Agrobacterium-mediated Transformation of Indica Rice with Chitinase Gene for Enhanced Sheath Blight Resistance. *Biol. Plant.* **2007**, *51*, 142–148.

Nelson, J. C.; Oard, J. H.; Groth, D.; Utomo, H. S.; Jia, Y.; Liu, G.; Moldenhauer, K. A. K.; Correa-Victoria, F. J.; Fjellstrom, R. G.; Scheffler, B.; Prado, G. A. Sheath blight resistance QTLS in japonica rice germplasm. *Euphytica* **2012**, *184*, 23–34.

Okubara, P. A.; Schroeder, K. L.; Paulitz, T. C. Identification and Quantification of *Rhizoctonia solani* and *R. oryzae* using Real-time Polymerase Chain Reaction. *Phytopathology* **2008**, *98*, 837–847.

Ou, S. H. *Rice Disease*, 2nd ed. Commonwealth Mycological Institute Publication: Kew, Surrey, 1985.

Ou, S. H.; Nuque, F. L.; Bandong, J. M. Relationship between Qualitative and Quantitative Resistance in Rice Blast. *Phytopathology* **1975**, *65*, 1315–1316.

Pan, X. B.; Zou, J. H.; Chen, Z. X.; Lu, J. F.; Yu, H. X.; Li, H. T.; Wang, Z. B.; Pan, X. Y.; Rush, M. C.; Zhu, L. H. Tagging Major Quantitative Trait Loci for Sheath Blight Resistance in a Rice Variety, Jasmine 85. *Chin. Sci. Bull.* **1999**, *44*, 1783–1789.

Pandian, R. T. P.; Sharma, P.; Singh, V. K.; Singh, A.; Ellur, R. K.; Singh, A. K.; Singh, U. D. Validation of Sheath Blight Resistance in Tetep Derived Basmati and Parental Lines of Rice Hybrid. *Indian Phytopathol.* **2012**, *65*, 233–237.

Parlevliet, J. E. Components of Resistance that Reduce the Rate of Epidemic Development. *Ann. Rev. Phytopathol.* **1979**, *17*, 203–222.

Patkar, R. N.; Chattoo, B. B. Transgenic Indica Rice Expressing ns-LTP-like Protein Shows Enhanced Resistance to Both Fungal and Bacterial Pathogens. *Mol. Breed.* **2006**, *17*, 159–171.

Pinson, S. R. M.; Capdevielle, F. M.; Oard, J. H. Confirming QTLs and Finding Additional Loci Conditioning Sheath Blight Resistance in Rice Using Recombinant Inbred Lines. *Crop Sci.* **2005**, *45*, 503–510.

Pinson, S. R. M.; Oard, J. H.; Groth, D.; Miller, R.; Marchetti, M. A.; Shank, A. R.; Jia, M. H.; Jia, Y.; Fjellstrom, R. G.; Li, Z. Registration of TIL:455, TIL:514, and TIL:642, Three Rice Germplasm Lines Containing Introgressed Sheath Blight Resistance Alleles. *J. Plant Regist.* **2008**, *2*, 251–254.

Prasad, P.; Eizenega, G. C. Rice Sheath Blight Disease Resistance Identified in *Oryza* spp. Accessions. *Plant Dis.* **2008**, *92*, 1503–1509.

Raj, R. B.; Wahab. T.; Rao, G. V.; Rao, A. S.; Reddy, T. C. V. Evaluation of Rice Cultures against Bacterial Blight and Sheath Blight Diseases. *Ind. Phytopathol.* **1987**, *40*, 397–399.

Rajan, K. M.; Nair, P. V. Reaction of Certain Rice Varieties to Sheath Blight and Sheath Rot Diseases. *Agric. Res. J. Kerala* **1979**, *17*, 259–260.

Ram, T.; Majumder, N. D.; Laha, G. S.; Ansari, M. M.; Kar, C. S.; Mishra, B. Identification of Donors for Sheath Blight Resistance in Wild Species of Rice. *Indian J. Genet. Plant Breed.* **2008**, *68*, 317–319.

Ranjan, N.; Laha, S. K.; Bhattacharya, P. M.; Dutta, S. Evaluation of New Fungicidal Formulation for Controlling the Rice Sheath Blight Disease. *J. Mycopathol. Res.* **2005**, *43*, 113–115.

Reddy, C. S.; Muralidharan, K. Lustre 37.5 SE—An Effective Combination Product of Flusilazole and Carbendazim against Sheath Blight of Rice. *Indian J. Plant Prot.* **2007**, *35*, 287–290.

Richa, K.; Tiwari, I. M.; Kumari, M.; Devanna, B. N.; Sonah, H.; Kumari, A.; Nagar, R.; Sharma, V.; Botella, J. R.; Sharma, T. R. Functional Characterization of Novel Chitinase Genes Present in the Sheath Blight Resistance QTL: qSBR11-1 in Rice Line Tetep. *Front. Plant Sci.* **2016**, *7*, 244.

Richardson, K. L.; Vales, M. I.; Kling, J. G.; Mundt, C. C.; Hayes, P. M. Pyramiding and Dissecting Disease Resistance QTL to Barley Stripe Rust. *Theor. Appl. Genet.* **2006**, *113*, 485–495.

Roumen, E. C. In a Strategy for Accumulating Genes for Partial Resistance to Blast Disease in Rice within a Conventional Breeding Program. In: *Rice Blast Disease*; Zeigler, R. S., Leong, S. A., Teng, P. S., Eds.; CAB International: Cambridge, 1994, pp 245–265.

Roy, A. K. Screening of Rice Cultures against Sheath Blight. *Indian J. Agric. Sci.* **1977**, *47*, 259–260.

Sareena, S.; Poovannan, K.; Kumar, K. K.; Raja, J. A. J.; Samiyappan, R.; Sudhakar, D.; Bala-subramania, P. Biochemical Responses in Transgenic Rice Plants Expressing a Defence Gene Deployed against the Sheath Blight Pathogen, *Rhizoctonia solani. Curr. Sci.* **2006,** *91*, 1529–1532.

Sato, H.; Ideta, O.; Ando, I.; Kunihiro, Y.; Hirabayashi, H.; Iwano, M.; Miyasaka, A.; Nemoto, H.; Imbe, T. Mapping QTLs for Sheath Blight Resistance in the Rice Line WSS2. *Breed. Sci.* **2004,** *54*, 265–271.

Sayari, M.; Babaeizad, V.; Ghanbari, M. A. T.; Rahimian, H. Expression of the Pathogenesis Related Proteins, NH-1, PAL, and Lipoxygenase in the Iranian Tarom and Khazar Rice Cultivars, in Reaction to *Rhizoctonia solani*—The Causal agent of Rice Sheath Blight. *J. Plant Prot. Res.* **2014,** *54*, 36–43.

Sayler, R. J.; Yang, Y. Detection and Quantification of *Rhizoctonia solani* AG-1 IA, the Rice Sheath Blight Pathogen, in Rice using Real-time PCR. *Plant Dis.* **2007,** *91*, 1663–1668.

Sha, X. Y.; Zhu, L. H. Resistance of Some Rice Varieties to Sheath Blight (ShB). *Int. Rice Res. Newsl.* **1990,** *15*, 7–8.

Shah, J. M.; Raghupathy, V.; Veluthambi, K. Enhanced Sheath Blight Resistance in Trans-genic Rice Expressing an Endochitinase Gene from *Trichoderma virens. Biotechnol. Lett.* **2009,** *31*, 239–244.

Shah, J. M.; Singh, R.; Veluthambi, K. Transgenic Rice Lines Constitutively Coexpressing tlpD34 and Chi11 Display Enhancement of Sheath Blight Resistance. *Biol. Plant.* **2013,** *57*, 351–358.

Shamim, M.; Kumar, D.; Srivastava, D.; Pandey, P.; Singh, K. N. Evaluation of Major Cereal Crops for Resistance against *Rhizoctonia solani* under Green House and Field Conditions. *Indian Phytol.* **2014,** *67*, 42–48.

Sharma, A.; McClung, A. M.; Pinson, S. R. M.; Kepiro, J. L.; Shank, A. R.; Tabien, R. E.; Fjellstrom, R. Genetic Mapping of Sheath Blight Resistance QTL within Tropical Japonica Rice Cultivars. *Crop Sci.* **2009,** *49*, 256–264.

Singh, R.; Dodan, D. S. Reactions of Rice Genotypes to Bacterial Leaf Blight, Stem Rot, and Sheath Blight in Haryana. *Indian J. Mycol. Plant Pathol.* **1995,** *25*, 224–227.

Singha, K. D.; Borah, P. Screening of Local Upland Cultivars of Assam against Sheath Blight. *Ann. Biol.* **2000,** *16*, 161–162.

Sneh. B.; Burpee, L.; Ogoshi, A. *Identification of Rhizoctonia Species.* American Phytopatho-logical Press: St. Paul, MN, 1991.

Sneh, B.; Jabaji-Hare, S.; Neate, S.; Dijst, G. *Rhizoctonia Species: Taxonomy Molec-ular Biology, Ecology, Pathology and Disease Control.* Kluwer Academic Publishers: Dordrecht, The Netherlands, 1996.

Sridevi, G.; Sabapathi, N.; Meena, P.; Nandakumar, R.; Samiyappan, R.; Muthukrishnan, S.; Veluthambi, K. Transgenic Indica Rice Variety Pusa Basmati 1 Constitutively Expressing a Rice Chitinase Gene Exhibits Enhanced Resistance to *Rhizoctonia solani. J. Plant Biochem. Biotechnol.* **2003,** *12*, 93–101.

Sridevi, G.; Parameswari, C.; Sabapathi, N.; Raghupathy, V.; Veluthambi, K. Combined Expression of Chitinase and β-1, 3-glucanase Genes in Indica Rice (*Oryza sativa* L.) Enhances Resistance against *Rhizoctonia solani. Plant Sci.* **2008,** *175*, 283–290.

Srinivasachary; Willocquet, L.; Savary, S. Resistance to Sheath Blight (*Rhizoctonia solani* Kuhn) [(Teleomorph *Thanatephorus cucumeris* (A. B. Frank) Donk.] Disease: Current Status and Perspectives. *Euphytica* **2010,** *178*, 1–22.

Sripriya, R.; Raghupathy, V.; Veluthambi, K. Generation of Selectable Marker-free Sheath Blight Resistant Transgenic Rice Plants by Efficient Co-transformation of a Cointegrate

Vector T-DNA and a Binary Vector T-DNA in One *Agrobacterium tumefaciens* Strain. *Plant Cell Rep.* **2008**, *27*, 1635–1644.

Sugiyama, T.; Doi, M.; Nishio, K. Sparse Planting of Rice Cultivar 'Hinohikari' in Nara. *Bull. Nara Prefectur. Agric. Exp. Stat. Jpn.* **2007**, *37*, 41–46.

Taguchi-Shiobara, F.; Ozaki, H.; Sato, H.; Maeda, H.; Kojima, Y.; Ebitani, T.; Yano, M. Mapping and Validation of QTLs for Rice Sheath Blight Resistance. *Breed. Sci.* **2013**, *63*, 301–308.

Tan, C. X.; Ji, X. M.; Yang, Y.; Pan, X. Y.; Zuo, S. M.; Zhang, Y. F.; Zou, J. H.; Chen, Z. X.; Zhu, L. H.; Pan, X. B. Identification and Marker-assisted Selection of Two Major Quantitative Genes Controlling Rice Sheath Blight Resistance in Backcross Generations. *Acta Genet. Sin.* **2005**, *32*, 6.

Tang, Q. Y.; ShaoBing, P.; Buresh, R. J.; Zou, Y.; Castilla, N. P.; Mew, T. W.; Zhong, X. Rice Varietal Difference in Sheath Blight Development and Its Association with Yield Loss at Different Levels of N Fertilization. *Field Crop Res.* **2007**, *102*, 219–227.

Uchimiay, H.; Iwata, M.; Nojiri, C.; Samarajeewa, P. K.; Takasatsu, S.; Ooba, S.; Anzai, H.; Christensen, A. H.; Quail, P. H.; Toki, S. Bialaphos Treatment of Transgenic Rice Plants Expressing a Bar Gene Prevents Infection by the Sheath Blight Pathogen (*Rhizoctonia solani*). *Nat. Biotechnol.* **1993**, *11*, 835–836.

Van der Plank, J. E. *Disease Resistance in Plants.* Academic Press: London, 1968.

Wang, R.; Lu, L.; Pan, X.; Hu, Z.; Ling, F.; Yan, Y.; Liu, Y.; Lin, Y. Functional Analysis of OsPGIP1 in Rice Sheath Blight Resistance. *Plant Mol. Biol.* **2015**, *87*, 181–191.

Wang, Y.; Pinson, S. R. M.; Fjellstrom, R. G.; Tabien, R. E. Phenotypic Gain from Introgression of Two QTL, qSB9-2 and qSB12-1, for Rice Sheath Blight Resistance. *Mol. Breed.* **2012**, *30*, 293–303.

Wu, W.; Shah, F.; Shah, F.; Huang, J. Rice Sheath Blight Evaluation as Affected by Fertilization Rate and Planting Density. *Aust. Plant Path.* **2014**, *44*, 183–189.

Wu, Y. L. Varietal Differences in Sheath Blight Resistance of Rice Obtained in Southern Taiwan. *SABRAO Newsl.* **1971**, *3*, 5.

Xia, G. Cloning and Identification of *OsSeh1* Gene with Function in Rice Resistance to Sheath Blight. *J. Nucl. Agric. Sci.* **2016**, *30*, 231–239.

Xie, Q. J.; Linscombe, S. D.; Rush, M. C.; Jodarikarimi, F. Registration of LSBR-33 and LSBR-5 Sheath Blight Resistant Germplasm Lines of Rice. *Crop Sci.* **1992**, *32*, 507.

Xinping, X.; Jinting, C.; Jianzhong, Z.; Qiyun, Y.; Baojian, L. Novel Transgenic Rice Strains Resistant to Blast and Sheath Blight. *Acta Sci. Nat. Univ. Sun.* **2001**, *40*, 131–132.

Xu, J. Y.; Zhang, H.; Zhang, H.; Tong, Y.; Xu, Y.; Chen, X.; Ji, Z. Toxin Produced by *Rhizoctonia solani* and its Relationship with the Pathogenicity of the Fungus. *J. Yangzhou Univ.* **2004**, *25*, 61–64.

Yadav, S.; Anuradha, G.; Kumar, K. K.; Vemireddy, L. R.; Sudhakar, R.; Donempudi, K.; Venkata, D.; Jabeen, F.; Narasimhan, Y. K.; Marathi, B.; Siddiq, E. A. Identification of QTLs and Possible Candidate Genes Conferring Sheath Blight Resistance in Rice (*Oryza sativa* L.). *Spr. Plus* 2015, *4*, 175.

Yin, Y. J.; Zuo, S. M.; Wang, H.; Chen, Z. X.; Ma, Y. Y.; Zhang, Y. F.; Gu, S. L.; Pan, X. B. Pyramiding Effects of Three Quantitative Trait Loci for Resistance to Sheath Blight Using Near Isogenic Lines of Rice. *Chin. J. Rice Sci.* **2008**, *22*, 340–346.

Yuan, H. X.; Xu, X. P.; Zhang, J. Z.; Guo, J. F.; Li, B. J. Characteristics of Resistance to Rice Sheath Blight of Zhongda 2, a Transgenic Rice Line as Modified by Gene "RC24". *Rice Sci.* **2004**, *11*, 177–180.

Zeng, Y.; Ji, Z.; Ma, L.; Li, X.; Yang, C. Advances in Mapping Loci Conferring Resistance to Sheath Blight and Mining *Rhizoctonia solani* Resistance Resources. *Rice Sci.* **2011**, *18*, 56–66.

Zeng, Y. X.; Xia, L. Z.; Wen, Z. H.; Ji, Z. J.; Zeng, D. L.; Qian, Q.; Yang, C. D. Mapping Resistant QTLs for Rice Sheath Blight Disease with a Doubled Haploid Population. *J. Int. Agric.* **2015**, *14*, 801–810.

Zou, J. H.; Pan, X. B.; Chen, Z. X.; Xu, J. Y.; Lu, J. F.; Zhai, W. X.; Zhu, L. H. Mapping Quantitative Trait loci Controlling Sheath Blight Resistance in Two Rice Cultivars (*Oryza sativa* L.). *Theor. Appl. Genet.* **2000**, *101*, 569–573.

Zuo, S.; Yin, Y.; Zhang, L.; Zhang, Y.; Chen, Z.; Gu, S.; Zhu, L.; Pan, X. Effect and Breeding Potential of qSB-11[LE], a Sheath Blight Resistance QTL from a Susceptible Rice Cultivar. *Can. J. Plant Sci.* **2011**, *91*, 191–198.

Zuo, S.; Zhang, Y.; Chen, Z.; Jiang, W.; Feng, M.; Pan, X. Improvement of Rice Resistance to Sheath Blight by Pyramiding QTLs Conditioning Disease Resistance and Tiller Angle. *Rice Sci.* **2014**, *21*, 318–326.

CHAPTER 5

MOLECULAR APPROACHES FOR CONTROLLING OF FALSE SMUT DISEASE OF RICE: AN OVERVIEW

PRAMILA PANDEY[1*], NARENDRA SHANKAR PANDEY[2], and D. K. DWIVEDI[1]

[1]*Department of Plant Molecular Biology and Genetic Engineering, N. D. University of Agriculture and Technology, Kumarganj, Faizabad 224229, Uttar Pradesh, India*

[2]*Department of Environmental Sciences, Babu Banarsi Das University, Lucknow 226028, Uttar Pradesh, India*

Corresponding author. E-mail: pramila28@gmail.com

CONTENTS

ABSTRACT

Rice false smut (FS) caused by *Ustilaginoidea virens* is now one of the most devastating rice diseases worldwide. In present time, FS spreads in the several rice-growing regions like American, Italian, and Southern Asian rice-growing regions. The disease incidence rate was estimated up to more than the 16% in across northern India. There is heavy infestation of FS pathogen and the smut balls formed up to 100 grains per panicle in some rice fields with high disease severity. FS should be controlled in an integrated manner, keeping in mind epidemiological factors, sources and survival of inoculum in soil or on collateral hosts, the role of nutrients, and the stage of crop development. Control methods include the use of tolerant or resistant varieties, moderate application of nitrogenous fertilizers, and the use of effective fungicides at the appropriate time (before heading). Breeding and deployment of resistant varieties is considered as the most effective strategy to control this disease. However, little is known about the resistance gene(s) and quantitative trait loci for this important disease as well as molecular mechanisms for resistance against *U. virens*.

5.1 INTRODUCTION

Rice is the most economically important staple food crop in India, China, East-Asia, South-East Asia, Africa, and Latin America catering to nutritional needs of 70% of the population in these countries (FAO, 1995). In several developed countries such as North America and European Union (EU), rice consumption has increased due to food diversification and immigration (Faure & Mazaud, 1996). Worldwide, rice is grown on 161 million hectares, with an annual production of about 678.7 million tons of paddy (FAO, 2009). About 90% of the world's rice is grown and produced (143 million hectares of area with a production of 612 million tons of paddy) in Asia (FAO, 2009). Rice provides 30–75% of the total calories to more than 3 billion Asians (Braun & Bos, 2004; Khush, 2004). To meet the global rice demand, it is estimated that about 114 million tons of additional milled rice needs to be produced by 2035, which is equivalent to an overall increase of 26% in the next 25 years. The possibility of expanding the area under rice in the near future is limited. Therefore, this extra rice production has needed to come from a productivity gain (Kumar & Ladha, 2011). Maximum yields per unit area of land can be achieved and sustained only if along with high-yielding crop varieties there is also a provision for protection of the crop against its enemies (Srivastava et al., 2010).

Amongst the various biotic factors affecting rice production and productivity, rice diseases are one of the most important ones. The annual losses due to rice diseases are estimated to be 10–15% on an average basis worldwide. Therefore, judicious management of rice diseases can result in improved productivity and additional grain harvested. Rice diseases are caused by a wide variety of pathogen including fungus, bacteria, virus, and nematodes (Ling, 1980). In the pre-war period, diseases of rice were practically unimportant in Tropical Asia where ancient varieties were traditionally grown on soils of low fertility (Areygunawardena, 1968). However, with the increasing demands of world rice supplies and advent of green revolution resulting in use of improved varieties, high fertilization, irrigation, and intensive cultural practices have resulted a great increase in the occurrence and severity of diseases infesting rice in several countries (Teng, 1990). The rice diseases that often cause great economic losses are rice blast (*Magnaporthe grisea*), sheath blight (*Rhizoctonia solani*), bacterial blight (*Xanthomonas oryzae*), tungro virus disease, and false smut (FS) (*Ustilaginoidea virens*) especially in South and South-East Asia (Ling, 1980).

Among the diseases of rice, FS (green smut) is one of the most destructive diseases of rice now. FS caused by the ascomycete fungus *U. virens* (Cooke) Takah (Anamorph). *U. virens* was recently placed in Clavicipitaceae as the new name *Villosiclava virens* (Teleomorph) (Kepler et al., 2012; Tanaka et al., 2008), which can reproduce both sexually and asexually with multiple propagules (Fu et al., 2012; Singh & Dubey, 1984). It was also found that *U. virens* could invade rice coleoptiles and roots at the young seedling stage and the stamen filaments at the earlier booting stage (Ashizawa et al., 2012; Schroud & TeBeest 2005; Tang et al., 2013). Presently, the use of nitrogen fertilizers and large-scale planting of hybrid cultivars have been regarded as responsible for the increased disease severity of FS in rice-growing area (Deng, 1989).

5.2 DISEASE-CAUSING AGENT

The rice FS caused by *U. virens* (Cooke) Takahashi, an ascomycete in the class Sordariomycetes; order Hypocreales; monophyletic tribe *Ustilaginoideae*, closely related but distinct from the *Claviceptaceae* and Hypocreaceae clades (Bischoff et al., 2004). Tanaka et al. (2008) proposed the sexual reproductive stage, typically a fruiting bodies, that is, teleomorph *Villosiclava virens* because of taxonomic similarities with other *Villosiclava* spp. Due to the sexual reproductive stage of this fungus, the teleomorph name, *Claviceps oryzae-sativae*, has thus recently been amended because phylogenetic studies identified a close association with *Villosiclava* spp.

The fungus *U. virens* (Cooke) Takahashi is an ascomycetous fungus because it produces sexual spores (ascospores) in structures called asci. This fungus is also known as *V. virens* that causes rice FS, a devastating emerging disease worldwide. The genome size of *U. virens* is 39.4 Mb. This *pathogen* produces both sexual ascospores and asexual chlamydospores in its life cycle (Biswas, 2001; Wang, 1992). The asexual reproductive stage, that is, anamorph form of *U. virens* (Cooke) Takahashi, is widely used for description of the causing agent of FS in rice and maize (Abbas et al., 2002; Rush et al, 2000). The sexual reproduction, that is, teleomorph form had been named *Claviceps virens* Sakurai ex Nakata and *C. oryzae-sativae* Hashioka because the teleomorphic characteristics of *U. virens* are similar to those of *Claviceps*. Evolutionarily, this fungus is close to the entomopathogenic *Metarhizium* spp., suggesting potential host jumping across kingdoms. *U. virens* possesses reduced gene inventories for polysaccharide degradation, nutrient uptake, and secondary metabolism, which may result from adaptations to the specific floret infection and biotrophic lifestyles. However, phylogenetic analyses using sequences of the large subunit of the ribosomal RNA gene suggested that members of *Ustilaginoideae* are distinct from teleomorphic genera of *Clavicipitaceae* and should be recognized as a monophyletic group within Hypocreales (Bischoff et al., 2004).

Molecular phylogenetic studies have also revealed that *Ustilaginoidea* species formed a paraphyletic group, and not congeneric with Claviceps based on the ALDH1-1 gene, which encodes a member of the aldehyde dehydrogenase family (Tanaka & Tanaka, 2008). The pathogen of FS disease of rice *U. virens* produces conidia, secondary conidia, chlamydospores, sclerotia, and ascospores, all of which are capable of infection. This fungus overwinters by producing fungal structures called sclerotia, which contain chlamydospores (resting spores) and compact masses of mycelia.

5.2.1 FALSE SMUT BALL AND CHLAMYDOSPORE

FS balls on rice panicles are the typical symptom of the disease. The interiors of the FS balls are intertwined with hyphae at early stage and then chlamydospores are formed. The chlamydospores are almost smooth when young and become warty when mature (Kim & Park, 2007). Under bright-field light microscopy, the conidia are found to be round to elliptical and warty on the surface with diameters approximately ranging from 3 to 5 mm. Under scanning electron microscopy, the globose to irregularly rounded conidia are ornamented with prominent spines. The spines are pointed at the apex or irregularly curved, and approximately 200–550-nm long (Fu et al., 2012). Under transmission electron microscopy, both the spined conidia and hyphae displayed lipid globules and vacuoles in the cytoplasm enclosed by an electron-transparent cell wall, whereas on the surface of conidia, the electron-dense spines displayed obclavate means reversed club or irregularly protruding shapes with varying heights along the conidial cell wall. Hyphae had concentric bodies that showed an electron-transparent core surrounded by an electron-dense layer (Kim & Park, 2007). In addition, the chlamydospores produced in culture medium are found prone to germinate in distilled water and produces secondary spores. The conidia are holoblastically and sympodially produced at the apex of each conidiophore cell (Fu et al., 2012).

5.2.2 SCLEROTIUM

The fungus has several kinds of propagules. In addition to the chlamydospores in smut balls, sclerotia are often formed on the colony surfaces, especially in later autumn, with relatively lower temperatures and high-temperature differences between day and night (Ikegami, 1963; Miao, 1992). The sclerotia existing on or under the soil surface can germinate and form fruit-body, then produce ascospores that contribute to the primary infection sources (Singh & Dubey, 1984). The sclerotia produced on the mature FS balls are black horseshoe-shaped and irregular oblong or flat, ranged from 2 to 20 mm. The outer sclerotium wall appears rough at a low resolution under scanning electron microscope, whereas the surface is full of projections at a higher magnification. The interior of each sclerotium is intertwined with compact hyphae (Fu et al., 2012).

5.3 HOST RANGE

The Clavicipitaceae family consists of 43 genera such as *Claviceps*, *Cordyceps*, *Epichiloe*, *Metarhizium*, and *Ustilaginoidea*, which include pathogens across animal, plant, and fungal kingdoms. Such a broad host range makes this family unique for exploring evolution and host adaptation (Kepler et al., 2012). FS also infects the tassels of maize and possibly other cereal crops.

In addition to rice, maize and several weeds common to rice fields are found to be the alternative hosts of *U. virens*. It infects *Digitaria marginata*, a common rice weed, which occurs in 85% of the rice fields (Shetty & Shetty, 1985), and *Panicum trypheron*, a common grass around paddy field, *Echinochloa crusgalli* and *Imperata cylindrica*, two common weeds on irrigation canals in Egypt (Atia). Cross inoculation studies revealed that chlamydospores from *P. trypheron*-infected rice and vice versa. Thus *P. trypheron* is an important source of inoculums between seasons (Shetty & Shetty, 1987).

5.3.1　PATTERN OF INFECTION IN RICE

Knowing fungal infection process is critical for exploring the relationship between *U. virens* and rice host, it is also important to control the occurrence of the disease in rice production. However, Li et al. (2013) demonstrated the infection of *U. virens* in rice is a flower-infecting fungus that appeared FS balls in rice panicle. In vitro cocultivation of *U. virens* strain with young rice panicles proved that *U. virens* enters inside of spikelets from the apex and then grows downward to infect floral organs of rice. In response to *U. virens* infection process, rice host exhibits elevated reactive oxygen species accumulation and enhanced callose deposition on the infection site. The secreted compounds of *U. virens* finally suppress rice pollen germination. Examination of sectioning slides of freshly collected smut balls demonstrated that both pistil and stamens of rice flower are infected by *U. virens*, hyphae disgraced the contents of the pollen cells, and also invaded the filaments. Further, *U. virens* entered rice ovary through the thin-walled papillary cells of the stigma, then decomposed the integuments and infected the ovary of rice flower. The invaded pathogen could not penetrate the epidermis and other layers of the ovary of flower. Transverse section of the pedicel just below the smut balls showed that there were no fungal hyphae scrutinized in the vascular bundles of the pedicel, implicating that *U. virens* is not a systemic flower-infecting fungus.

5.4　HISTORY OF FALSE SMUT DISEASE IN RICE

FS disease has been reported from all the major rice-growing regions in Asia, Latin America, South Africa, Italy, and Europe (Qu, 1972). It was first reported in US rice-growing areas over a century ago in 1906. Since the late 1990s and early 2000s, the disease has increased in severity and has been severe in India, China, and the United States. In India, this disease was first reported from Tirunelveli district of Tamil Nadu in the 1870s (Cooke, 1878; Tanaka et al., 2008). This disease has been reported from Andhra Pradesh, Assam, Bihar, Haryana, Maharashtra, Karnataka, Orissa, Tamil Nadu, Uttar Pradesh. Rice FS disease was once categorized as a minor disease with sporadic occurrence in rice-growing areas because it occurs infrequently and irregularly. But yet, it has recently become one of the most devastating grain diseases in the majority of rice-growing areas of the world due to extensive planting of high-yielding cultivars and hybrids, over use of chemical fertilizers with increase application of nitrogen and also an apparent change in global and regional climates (Guo et al., 2012; Rush et al., 2000). This disease not only affect the loss of rice yield like in Ratna but also affect grain quality reduction and also causes the poisoning of livestock and humans who consume the rice grains contaminated by fungal mycotoxins (Koiso et al., 1994, 1998; Li et al., 1995; Nakamura et al., 1994; Singh & Dube, 1978). This pathogen produces two types of mycotoxins, ustiloxins and ustilaginoidins. Ustiloxins are antimitotic cyclic depsipeptides with a 13-membered core structure (Shan, 2012). Ustiloxins inhibit cell division, especially by inhibiting microtubule assembly and cell skeleton formation (Koiso et al., 1994). *Ustilaginoidins* are bi(naphtho-g-pyrone) derivatives, which exhibit weak antitumor cytotoxicity to human epidermoid carcinoma (Koyama & Natori, 1988; Koyama et al., 1998). It is very hard to control rice FS disease because when the disease symptoms appear, it is already at the late stage of infection and damage to rice plants has occurred. For controlling the FS disease of rice various methods are using it includes, use of resistant varieties, biological, chemical control and molecular approaches. So this chapter will cover nature of pathogen, how it causes infection in rice cultivars, methods of controlling FS disease, and will include chemical method, biological methods, molecular methods use of transgenic variety.

5.4.1 MODE OF INFECTION AND TRANSMISSION OF PATHOGEN

Rice FS has an exciting disease cycle. *U. virens* initiates infection through rice floral organs producing white hyphae and then develops powdery dark green chlamydospore balls in spikelets during the late phase of infection (Fig. 5.1). As diurnal temperatures fluctuate more in late autumn, sclerotia appear on the surface of spore balls (Ikegami, 1963). *U. virens* produces both sexual (ascospores) and asexual (chlamydospores) stages in its life cycle and thick-walled chlamydospores can survive up to 4 months in the field with sclerotia presumed to survive much longer. Sclerotia on or under soil surface can germinate and form fruiting bodies, and ascospores that contribute to primary infections are then produced after mating and meiosis (Singh & Dubey, 1998). *U. virens* hyphae enter floral organs primarily at upper parts of the stamen filaments between the ovary and lodicules (Tang et al., 2013). It occasionally infects the stigma and lodicules, but is not known to infect the ovaries. Primarily, the pathogen hyphae extend into the central vascular tissues without penetrating rice cell walls directly (Tang et al., 2013). Therefore, the pathogen is considered to be a biotrophic parasite (Tang et al., 2013). This disease infects the spikelets alone. The parasite may be seen only in the ears, where it develops it fructification in the ovary of individual grains. The grains are being transformed into large, velvety, green masses which may be more than twice the diameter of the normal grain. Embedded in this mass and visible on the two sides are the lemma and palea.

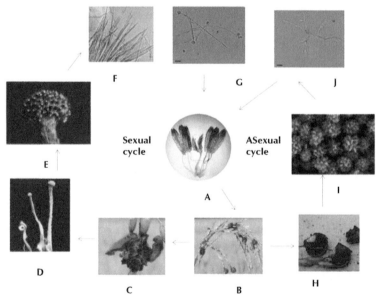

FIGURE 5.1 Major stages in the infection cycle of *U. virens*. (a) Stamen filaments in the florets infected by *U. virens*. Hyphae were stained with trypan blue to show primary infection sites. (b) FS balls formed in the rice spikelets. (c) Sclerotia formed on the surface of spore balls. (d) Stroma produced by a germinating sclerotium. (e) Ascocarp formed on the stroma. (f) Asci. (g) Ascospore germination, Scale bar, 10 mm. (h) Spore balls. (i) Chlamydospores under scanning electron microscopy. Scale bar, 3 mm. (j) Chlamydospore germination. Scale bar, 10 mm. (Reprinted by permission from Macmillan Publishers Ltd: Zhang, Y.; et al. Specific Adaptation of Ustilaginoidea virens in Occupying Host Florets Revealed by Comparative and Functional Genomics. *Nature Commun.* **2014,** *5,* 3849. © 2014.)

In the initial stages of development, it is orange-yellow in color, but soon turns dark green or almost black. The glumes are unaltered and are found closely applied to the center of the green mass, which burst out above and laterally from between them. Only a few grains in each panicle are usually affected, and they may occupy any part of the ear. The young ovary is infected by the parasite at an early stage of its development, and it transformed into a hard mass of closely united, fine, colorless hyphae. This sclerotium like body grows and burst out literally between the closely pressed glumes (Fan et al., 2010; Guo et al., 2012). Here, its outer layers develop spores of a brownish green color when ripe but orange in the immature mass. The center of the sclerotium consists of pseudoparenchymas which have entirely replaced the tissues of the grain. Toward the surface the hyphae are arranged radially, branched, and septate, whereas the exterior consists of a mass of loosely adhering spores. The spores are formed laterally and rarely terminally on the radial hyphae. The youngest spores are almost colorless and are found on the hyphae bounding the white center of sclerotium (Ikegami 1962; Schroud & TeBeest, 2005). The mature spores are greenish-brown and the outermost layer consists of these with a few fragments of the soporiferous hyphae. The mature spores have a rough, olive green, granular coating, soluble in alkali, alcohol, and mineral acids. The spores are developed on the minute projections from the sides of the radial hyphae. The spores are usually globose and measure 4–6 μm in diameter, occasionally elongated and 8×4 μg. The spores germinate in water and nutrient solutions. On germination they form slender, septate hyphae which later on form clusters of minute, pear-shaped, colorless conidia at and near their tips. In water, the germ tube may bear only a single conidium at its tip, but in nutrient solutions small heads of conidia are formed. The germ tube may branch also. The conidia also germinate in nutrient solutions with a short germ tube on which still smaller conidia (Fan et al., 2014).

Ashizawa et al. (2010) have reported that 6 days after infection, pathogen can reach the floral organs. Wang reported that the period from the late boot stage to the rupturing stage (that is, 1–4 days before the rupturing stage) is the crucial time for pathogen infection. At the time of 7–8 days after pathogen infection, normally, rice plants have completed anthesis, and the ovaries of the uninfected spikelets have been fertilized. After fertilization, the ovary needs large amounts of water, minerals, and photosynthesis products to develop. The channel for the ovary to obtain nutrients is the base of the pedicel connected to the rachis branches. For the infected spikelets, at this time fungal hyphae probably have used up the nutrients from degraded integuments and then may mimic the way used by the normally fertilized ovary to obtain nutrients from rice plants. Since large amounts of nutrients are taken by the infected pathogen, the adjacent spikelets may not be able to obtain enough nutrients for their developmental requirement, thus, leading to grain weight reduction and chaffiness. This is not surprising because the integuments are absorbed after ovary fertilization during the developmental process of rice grain. Pathogen infection caused the developmental arrestment and content decomposition of the pollen mother cells. In consequence, no mature pollen grains were produced. Furthermore, the supernatant of pathogen liquid culture inhibited pollen-grain germination and suppressed pollen-tube elongation. These inhibitory effects are probably due to the functions of the secreted compounds. Previous research reported that usti-loxins are able to not only inhibit tubulin polymerization in a concentration-dependent manner but also induce the depolymerization of preformed microtubules (Li et al., 1995). During the process of infection, *U. virens* secretes secondary compounds, which may permeate into cells and destroy cellular cytoskeleton, resulting in destruction of the infected cells and nutrient leakage for pathogen development since there are no special organs such as haustoria found for *U. virens*. Perhaps, only special types of cells at the specific developmental stage are sensitive to the secreted compounds, which may explain the limitation on the infection time of *U. virens*.

5.4.2 MOLECULAR DIAGNOSTICS OF FALSE SMUT IN RICE

The first and most important step in managing a plant disease is to correctly identify it. Although some diseases can be diagnosed quickly by visual examination, others require laboratory testing for diagnosis. These laboratory procedures may take days or even weeks to complete and are, in some cases, relatively insensitive. Delays are frustrating when a quick diagnosis is needed so that appropriate disease control measures may be taken to prevent plant injury, especially when high value cash crops are at stake. Fortunately, as the result of advances in biotechnology, new products and techniques are becoming available that will complement or replace time consuming laboratory procedures (Goud & Termorshuizen, 2003).

5.4.2.1 WITH HELP OF ENZYME-LINKED IMMUNOSORBENT ASSAY

A number of disease detection kits have been developed for use at the site where a disease is suspected. These kits, which in most cases do not require laboratory equipment, are especially useful to growers. Some tests only take 5 min to perform (Landgraf et al., 1991). The diagnostic kits are based on a method that uses proteins called antibodies to detect disease-causing organisms of plants (plant pathogens). The technique used is called ELISA (enzyme-linked immunosorbent assay). This assay is based on the ability of an antibody to recognize and bind to a specific antigen, a substance associated with a plant pathogen. The antibodies used in the diagnostic kits are highly purified proteins produced by injecting a warm-blooded animal (like a rabbit) with an antigen associated with one particular plant disease. The animal reacts to the antigen and produces antibodies. The antibodies produced recognize and react only with the proteins associated with the causal agent of that plant disease. Color changes on the unit's surface indicate a positive (disease present) reaction. Examples of diseases ELISA kits can detect include bacterial canker of tomato and soybean root rot (Bailey et al., 2002).

5.4.2.2   WITH THE HELP OF POLYMERASE CHAIN REACTION

A new technology, polymerase chain reaction (PCR) has great potential for raising the sensitivity of various assays that use nucleic acid probes. PCR is used to produce enormous numbers of copies of a specified nucleic acid sequence. This technique can allow the detection of very small amounts of a pathogen in a sample by amplifying the pathogen sequences to a detectable level. PCR is especially useful in plant quarantine point of view owing to its fastness. Recently, PCR primer pairs specific to *U. virens* based on the conserved internal transcribed spacer ribosomal DNA (rDNA) region have been used to determine the presence of *U. virens* in rice leaf, sheath and stem or panicle tissues before and after heading. The PCR-based techniques that selectively amplified *U. virens* rDNA in rice DNA extracts were initially developed by Zhou et al. (2003). In that study, panicles in the leaf sheaths of rice plants were artificially inoculated with secondary conidia at the booting stage and incubated as described by Fujita et al. (1989). Then, DNA was extracted from individual spikelets, leaf, sheath, and stem tissues before amplification with the US1-5/US3-3 (primary PCR) and US2-5/US4-3 (nested-PCR) primers. Further, Zhou et al. (2003) conducted histological examinations after dissecting plants under a stereomicroscope and staining with aniline blue. They observed conidia and hyphae on the inner surface of the leaf sheaths around the injection sites and on spikelets, some of which resulted in mycelial invasion of the spikelets. The results showed histological examinations were consistent with nested-PCR detection and that PCR "indicated that the specific primer pairs were useful in directly detecting *U. virens* from infected host tissues prior to the onset of visual symptoms" (Zhou et al., 2003).

5.4.2.3 MOLECULAR MECHANISM OF HOST–PATHOGEN INTERACTION

Infection of *U. virens* starts with the attachment of hyphae to the surface of floral organs. The early response of plants to pathogen invasion is initiated by recognition of the fungal PAMPs (Pattern-associated molecular patterns) such as chitin and pattern recognition receptors (PRRs) on mononuclear phagocytes. Fungal PAMPs are restricted to complex carbohydrates in the cell wall, including nanoproteins, phospholipomannan, beta-glucan, and possibly chitin. These PAMPs bind specifically to two classes of PRR in phagocyte membranes, Toll-like receptors and C-lectin-like receptors, through which they initiate signaling responses that terminate in release of pro- and anti-inflammatory cytokines like the innate immune response and in intiate phagocytosis and intracellular killing, and this type of defense response is critical since it may affect the final outcomes of pathogen invasion. Several studies have reported that chitinase genes and WRKY transcription factors are upregulated, that is, increase of cellular components in the early response of rice plants to infection by the filamentous fungus *Magnaporthe oryzae* (Bagnaresi et al., 2012; Mentlak et al., 2012). Recently, Chao et al. (2014) identified that some stage-specific biological processes of host plants, such as phosphorylation, programmed cell death, and cell-wall biogenesis, are largely influenced by *U. virens* infection, which suggests that the FS pathogen can manipulate the host cell's functions to benefit its development. Still there is no information reported on differential gene regulation between resistant and susceptible rice varieties during *U. virens* infection. In response to pathogen invasion, plants develop sophisticated mechanisms to modulate their growth and development in order to cope with pathogen infection. Plant hormones play important roles in regulating plant developmental processes and signaling networks involved in responses to plant–pathogen interactions. Salicylic acid, jasmonate, and ethylene (ET) are well-known hormones playing crucial roles in plant disease and pest resistance, while other growth-controlling hormones, such as auxin, gibberellic acid, and brassinosteroid, have recently been found to play active roles in fine-tuning a balance between immunity and growth/development in plants (Bari & Jones, 2009; Grant & Jones, 2009). Han et al. (2015) have explored the genetic basis of rice resistance to *U. virens*, differential expression profiles in resistant "IR28" and suscep-tible "LYP9" cultivars during early stages of *U. virens* infection by using RNA-seq data. This analysis showed that 748 genes were upregulated only in the resistant variety and 438 genes showed opposite expression patterns between the both genotypes. The genes encoding receptor-like kinases and cytoplasmic kinases were highly enriched in this pool of oppositely expressed genes. Many pathogenesis-related (PR) and diterpene phyto-alexin biosynthetic genes were specifically induced in the resistant variety. Interestingly, the RY repeat motif was significantly more abundant in the 5′-regulatory regions of these differentially regulated PR genes. Several WRKY transcription factors were also differentially regulated in the two genotypes, which is consistent that the cis-regulatory W-boxes were abun-dant in the promoter regions of upregulated genes in IR28. Furthermore, *U. virens* genes that are relevant to fungal reproduction and pathogenicity were found to be suppressed in the resistant cultivar. The results indicated that rice resistance to FS may be attributable to plant perception of PAMPs, acti-vation of resistance signaling pathways, induced production of PR proteins and diterpene phytoalexins, and suppression of pathogenicity genes in *U. virens* as well (Mei et al., 2006).

5.5 METHODS USED FOR CONTROLLING FALSE SMUT IN RICE

FS should be controlled in an integrated manner, keeping in mind epide-miological factors, sources, and survival of inoculum in soil or on collat-eral hosts, the role of nutrients, and the stage of crop development. Control methods include the use of tolerant or resistant varieties, moderate appli-cation of nitrogenous fertilizers, and the use of effective fungicides at the appropriate time (before heading).

5.5.1 CHEMICAL CONTROL

The aureofungin, captan, fentin hydroxide, furcarbanil (now superseded), mancozeb, and thiocyanomethylthiobenzothiozole as effective at inhib-iting conidial germination. Seed treatment with fungicides did not check the disease, but spraying the rice crop with carbendazim and copper fungicides at the time of tillering and pre-flowering effectively controlled the disease and yields increased. Copper oxychloride was most effective in decreasing disease incidence by 95.5% and 96.1% on the basis of infected tillers and grains, respectively, with a corresponding increase of 7.2% in grain yield (Dodan et al., 1996). In the USA, propiconazole or azoxystrobin applied during the boot stage of rice reduced the number of FS balls in harvested rice grain by 50–75%, but yield was not affected. Copper hydroxide fungicides reduced FS balls in harvested rice by 80%, but yield was also often reduced significantly. The use of fungicides to control the disease in the southern USA is not considered economical at present (Cartwright et al., 2000).

5.5.2 CULTURAL CONTROL

Cultural practices that have been investigated to reduce FS include no-till practices, groove irrigation, and reduced nitrogen applications (Brooks et al., 2009). Cultural practices like bunds and fields cleaning reduce the incidence as the disease has been reported on some of the weeds. Wenuning, a suspension of *Bacillus subtilis* in solution of validamycin, has been widely used in China for the control of FS disease (Liu et al., 2007).

5.5.3 BIOLOGICAL CONTROL

The increased reflection on environmental concern over pesticide use has been active in a large increase of biological disease control. Few antagonists are usually able to overcome the fungistatic effect of soil that results from the presence of metabolites produced by other species, including plants, and to survive under very extreme competitive conditions. Among the various antagonists used for the management of plant diseases, *Trichoderma* and *Pseudomonas* play a vital role. Among the various isolates of *Trichoderma*, *T. viride*, *T. harzianum*, *T. virens*, and *T. hamatum* are used against the management of various diseases of crop plants especially with dreaded soil-borne pathogens. It grows rapidly when inoculated in the soil, because they are naturally resistant to many toxic compounds, including herbicides, fungicides, and insecticides such as DDT and phenolic compound. The strain recovers very rapidly after the addition of sublethal doses of some of these compounds. Resistance to toxic compounds may be associated with the presence in Trichoderma strains of ABC transport systems. It has many advantages as a biocontrol agent owing to its high rhizosphere competence, ability to synthesize polysaccharide-degrading enzymes, amenability for mass multiplication, broad spectrum of action against various pathogens, and above all its environmental friendliness.

Fluorescent pseudomonads suppress the pathogens either directly through the production of various secondary metabolites or indirectly by inducing plant-mediated defense reactions. The crucial factor in the success of biological control by fluorescent pseudomonads is their ability to colonize the rhizosphere and their persistence throughout the growing season. Fluorescent pseudomonads are root colonizers because they occur in the natural habitat of rhizosphere and thus when they are rein-troduced to roots through seed or seed-piece inoculation, they colonize root-surface profusely. Fluorescent pseudomonads suppress the patho-gens by antibiosis through the production of various antibiotic substances such as 2,4-diacetyl phloroglucinol, phenazine-1-carboxylic acid, oomycin A, oxychlororaphine, pyoluteorin, pyrrolnitrin, and pyocyanine. Siderophores are extracellular, low molecular weight substances which selectively complex iron with high affinity. Fluorescent pseudomonads produce siderophores such as pseudobactin and pyoverdine which chelate the iron available in the soil and make it unavailable to pathogen, thus the pathogen dies for want of iron. In rice, seed treatment followed by root dipping and foliar spray with *Pseudomonas fluorescens* showed a higher induction of ISR against FS.

5.5.4 SOME RESISTANCE SOURCES AGAINST FALSE SMUT IN RICE

There are different regions in India where several resistant or tolerant rice cultivars have been identified: B3719C-TB-8-1-4, China 988, HPU 2202, HPU 5101, Nag 1-38, VL 501, and VRS 1 in Himachal Pradesh; Bogabordhan in Assam; and PR103 and PR108 in Punjab. Singh et al. (1987) identified 14 entries in Pantnagar (India): Homthong, IR22, IR26, IR28, IR1514A-E597, RP633-1, RP633-6, RP352-64-1-1-1, Sigadis, Surya, Tainan 3 mutant 587-4, TKM 6, Type 3, and Vijaya. In Arkansas, USA, common rice cultivars LaGrue, Drew, Wells, Cocodrie, Priscilla, and Cypress were susceptible; Bengal, Lemont, and Madison were moderately susceptible; and Jefferson and Kaybonnet were moderately resistant (Sutton et al., 2000). Some other important resistant varieties by using these can control FS disease (Table 5.1).

TABLE 5.1　Some Important Resistant/Moderately Rice Varieties for False Smut.

Score of infected germplasms	Rice germplasms resistant to false smut
Highly resistant (1%)	NOR 4031, OR 1206-21, CR 758-113*, NOR 30030*, NOR 30039*, CRK 7-11, OR 1356-RGA-OR 7, RP 8io. 150, Salivahana, CRK 7-17, CRK2-26, RAU 491-85-10, CN 1054, CN 847-1, CN 845-80-7-2, Pusa 1176-1*, Pusa 1301-95-1, OR 1559-11-1*, NOR 4004, NOR 4007, RAU 1306-4-3-2-2, CR 609-7-6, R 944-33, RP2731-105-36-19, RP 3152-1836, RP 1579-1, MRST 27-2027, CRK 13-11
Resistant (1–5%)	CRK 2-2, CRK 7-9, RAU 79-22-1, OR 1352-RGA-230, ORI 1358-RGA-4, ORI 534-RGA-129-14, CR 333-74,NOR 96001, NOR 96005, RP 3930 P12-111-1-2-1, CN 847-2, IR 4633-PM 1-32-2-1, CN 1082-A,CN 847-10-4-4, TOX 3098-2-2-1, IR 60251, NOR 40011-3-1, KMJ 3-144, OR 1575-1, NOR 4172,RAU 650-102-8, CR 835-9-3, CR 836-1-69, CR 683-1, IR 63429-23, ORR 47-4, CRK 13-11, CRK 4-1, RP 3148-1954, RP 2932-2350, CN 846-30-3-1, AN 31-39-7-1, CN 1035-59, CN 1035-60, CR 664-132-600, CR 780-1937, CRK 7-2-8, TRC 2229-F-4-1, OR 1335-7, OR 1360-RGA-OR 9, CN 847-5-5-N-7, CN 847-27-9-5, CN 1113-23-7-1, RP 8io. 146, Swarna

*Genotypes recording 0% infection (Singh & Pophaly, 2005).

5.6　BIOTECHNOLOGICAL APPROACHES FOR FALSE SMUT DISEASE MANAGEMENT

Nowadays, it has become a routine to transfer genes from one organism to another. It is possible to introduce genes conferring disease resistance into crop plants. Such gene transfers could be accomplished by two methods, direct methods and vector-mediated methods. Gene gun or biolistic method and Agrobacterium-mediated method are the best examples of the direct method and vector-mediated method, respectively (Tanaka et al., 2011).

5.6.1　RESISTANCE QTLS AND MOLECULAR BREEDING FOR FALSE SMUT

Li et al. (2008) developed a mapping population of 157 recombinant inbred lines (RILs) derived from the cross between resistant cultivar IR28 (*Oryza sativa* subsp. *indica*) and susceptible landrace Daguandao (*O. sativa* subsp. *japonica*). By using the RIL population, a model of mixed two major genes and polygenes (model E-1–3) is proposed to explain the inheritance of the resistance against rice FS disease. The two major genes have equivalent additive effect of 11.41, and their heritability is about 76.67%, while the heritability of polygenes is about 22.86%. Yu-Sheng et al. (2011) studied quantitative trait loci (QTLs) for resistance to rice FS. A RILs population with 157 lines derived from an inter-subspecies cross of Daguandao/ IR28 by the single seed descent method was used to detect QTLs. The RIL population further confirmed resistance to rice FS (*U. virens* (Cooke) Takahashi] in 2007 and 2009. QTL Cartographer scored the disease rate index of the two parents and 157 RILs caused by rice FS were scored and the QTLs for rice FS resistance were detected. Seven QTLs controlling FS resistance were detected on chromosomes 1, 2, 4, 8, 10, 11, and 12, respectively, with the phenotypic variance of 9.8–22.5%. There were five and four QTLs detected in 2007 and 2009, respectively, and only two QTLs were found in both years, the phenotypic variation was explained by individual QTL ranged from 18.0% to 19.3% for these two QTLs, and the additive effects of these two QTLs contributed to the 8.0–16.3% decrease of disease index and therefore the disease resistance increased. The direction of the additive effects at six loci qFsr1, qFsr2, qFsr8, qFsr10, qFsr11, and qFsr12 coincided with that predicted by phenotypes of the parents, and the IR28 alleles at these loci had positive effect against rice FS, while the negative effects were found in Daguandao alleles at qFsr4. Both qFsr11 and qFsr12 should be useful in rice breeding for resistance to rice FS in marker-assisted selection program.

5.6.2 TRANSGENIC APPROACHES FOR FALSE SMUT DISEASE MANAGEMENT

Disease resistance genes could be sourced from plant pathogens themselves, as was possible with coat protein-mediated plant viral resistance and with toxin-inactivating protein-mediated bacterial resistance. Host plants also contribute an enormous number of disease resistance genes such as those encoding PR proteins, which have been used against fungal diseases (van Loon et al., 2006). Zhu et al. (2007) incorporated four different gene for the enhancement of disease resistances in super hybrid rice with four antifungal genes. The transgenic lines expressing four antifungal genes including RCH10, RAC22, -Glu and B-RIP showed not only high resistance to *M. grisea* but also enhanced resistance to rice FS (*U. virens*) and rice kernel smut (*Tilletia barclayana*). Therefore, an ingeniously planned genetic engineering involving a well-balanced expression of transgenes with different modes of action would ensure broad-spectrum and durable resistance against pathogens. Two transgenic rice lines, E122-1 and E127-1, which were overexpression of rice acidic chitinase gene (RAC22) and rice basic chitinase gene (RCH10) and transformed with alfalfa β-1,3-glucanase gene (β-Glu) and barley ribosome inactivating protein gene (B-RIP), were tested for their effects on the target fungal diseases (rice blast, rice FS, and rice brown spot) and the nontarget bacterial diseases (rice bacterial blight and rice bacterial leaf streak and insect pests (rice yellow stem borer and rice leaf roller) by means of inoculation in greenhouse and investigation in paddy fields, compared with their untransformed counterpart E32. Both E122-1 and E127-1 showed broad spectrum of resistance against rice blast isolates and were significantly lower in the percentages of infected plants and the disease index resulted from rice FS or rice brown spot, but had no significant difference in the percentages of susceptible plants and the disease index resulted from rice bacterial blight or rice bacterial leaf streak, the percentages of "dead heart" plants resulted from rice yellow stem borer and the percentages of leaf-rolled plants resulted from rice leaf roller in comparison with the counterpart E32. On the plants with resistance to rice bacterial blight, both E122-1 and E127-1 had significant higher number plants with no infected symptom and had significant lower number plants with the 3rd scale infected symptom than that of the counterpart E32. Both E122-1 and E127-1 exhibited a high resistance to rice blast, rice FS and rice brown spot and did not cause obvious population increase of rice bacterial blight, rice bacterial leaf streak, rice yellow stem borer and rice leaf roller as compared to their counterpart during growing period in paddy fields. PR-protein genes appear to be a very potential source for candidate genes for fungal resistance. These proteins may play a direct role in defense by attacking and degrading pathogen cell wall components. Typical candidate genes are that encoding chitinases and β-1, 3 glucanases (Leubner-Metzger & Meins, 1999). Increasing expression of individual and multiple PR proteins in various crops have demonstrated some success in enhancing disease resistance in particular pathogens. A recent research shows a chitinase gene from an antifungal biocontrol fungus species (*T. viride*) confers transgenic resistance against the rice sheath blight pathogen. A rice PR-5 protein gene in wheat delays onset of symptoms caused by the wheat scab pathogen (Han et al., 2015).

Andargie and Li (2016) studied a new fungal–plant pathosystem where *U. virens* (rice pathogen) was able to interact compatibly with the model plant *Arabidopsis thaliana*. After inoculation, disease symptoms were apparent on the leaves of the Arabidopsis plants after 6 days of post inoculation in the form of chlorosis. Further, cytological studies showed that *U. virens* caused a heavy infestation inside the cells of the chlorotic tissues. The study showed that *U. virens* isolates infect *Arabidopsis* and the plant subsequently activates different defense response mechanisms which are witnessed by the expression of PR genes, *PR-1*, *PR-2*, *PR-5*, *PDF1.1*, and *PDF1.2*.

5.6.3 THROUGH RNA SILENCING

Through RNA silencing, FS of rice can be controlled because RNA silencing is well recognized as a principal regulatory mechanism in eukaryotes to control gene expression. The key genes necessary for RNA silencing pathways are highly conserved in *U. virens* including three Argonaute-like proteins required for forming RNA-induced silencing complex, three RNA-dependent RNA polymerases, three ReqQ helicases, and two Dicer genes (Zhang et al., 2014). These genes are speculated to function in two fungus-specific RNA-silencing pathways: meiotic silencing of unpaired DNA and quelling. These genes are listed in Table 5.2.

TABLE 5.2 Potential Genes Involved in RNA Silencing in *U. virens* (Zhang et al., 2014).

RNA-silencing components	*U. virens*
Argonaute-like	UV_479, UV_1389, UV_7275
RNA-dependent RNA polymerase	UV_1782, UV_7050 , UV_7057
Dicer	UV_6162, UV_7267
ReqQ helicases	UV_4904, UV_7800, UV_7858

(Reprinted by permission from Macmillan Publishers Ltd: Zhang, Y.; et al. Specific Adaptation of Ustilaginoidea virens in Occupying Host Florets Revealed by Comparative and Functional Genomics. *Nature Commun.* **2014,** *5,* 3849. © 2014.)

Recently, adopting new strategies for designing bio-fungicides based on the concept of comparative genomics and drug-discovery docking approach, gives a set of bioactive compounds that are likely to effect the MAT 1-2-1 mating type gene protein which plays an essential role in sexual propagation of *U. virens*. MAT 1-2-1 protein can be major cause of infection in the host. Computational approaches like docking can be used for identification of inhibitors, which can bind to the targets with experimental or modeled structure. For docking approaches, ligand and receptor protein are necessary. Therefore, this project was aimed at designing the structure of MAT 1-2-1 protein and studying the binding of various bioactive compounds similar to podophyllin which has antifungal properties with this protein through docking techniques (Mukesh et al., 2013).

5.7 CONCLUSION

The controlling of FS disease through molecular approaches is a not a cake walk. Because effect of the pathogen on the host is visible only after flowering when myceliums grows into the ovary of the individual kernels and transform them into large, velvety, yellow-green balls. By using effective fungicides, bioagents like *Trichoderma* and *Pseudomonas*, these antagonists are usually able to overcome the fungistatic effect of soil that results from the presence of metabolites produced by other species, including plants, and to survive under very extreme competitive conditions. Through cultural practices like bunds and fields, cleaning reduce the incidence as the disease has been reported on some of the weeds. Nowadays, it has become routine to transfer genes from one organism to another, it is possible to introduce genes conferring disease resistance into crop plants. For that we have to introduce FS resistance gene in susceptible rice variety by using direct and indirect technique. Gene gun is direct gene-transfer technique, and through *Agrobacterium* mediated gene transfer is an indirect method. Through molecular technique by using ELISA is based on the ability of an antibody to recognize and bind to a specific antigen, a substance associated with a plant pathogen. Still biotechnologists are doing work on controlling of FS by using PCR, RNA silencing, and through bioinformatics. But the best way to control FS is by using FS resistant varieties.

KEYWORDS

- false smut
- destructive diseases
- quantitative trait loci
- molecular breeding
- booting stage

REFERENCES

Abbas, H. K.; Sciumbato, G.; Keeking, B. First Report of False Smut of Corn (*Zea mays*) in the Mississippi Delta. *Plant Dis.* **2002,** *86,* 1179.

Andargie, M.; Li, J. *Arabidopsis thaliana*: A Model Host Plant to Study Plant–Pathogen Interaction Using Rice False Smut Isolates of *Ustilaginoidea virens*. *Front. Plant Sci.* **2016**, *7*, 192.

Areygunawardena, D. V. W. Fungicidal Control of Plant Diseases. In: An Invitational Paper to the First International Congress in Plant Pathology, London, 1968; pp 40–42.

Ashizawa, T.; Takahashi, M.; Arai, M.; Arie, T. Rice False Smut Pathogen, *Ustilaginoidea virens*, Invades through Small Gap at the Apex of a Rice Spikelet before Heading. *J. Gen. Plant Pathol.* **2012**, *78*, 255–259.

Ashizawa, T.; Takahashi, M.; Moriwaki, J.; Kazuyuki, H. Quantification of the Rice False Smut Pathogen *Ustilaginoidea virens* from Soil in Japan Using Real-time PCR. *Eur. J. Plant Pathol.* **2010**, *128*, 221–232.

Atia, M. M. M. Rice False Smut (*Ustilaginoidea virens*) in Egypt. *J. Plant Dis. Prot.* **2004**, *111*, 71–82.

Bagnaresi, P.; Biselli, C.; Orru, L.; Urso, S; Crispino, L.; Abbruscato, P.; Piffanelli, P.; Lupotto, E.; Cattivelli, L.; Vale, G. Comparative Effector-mediated Suppression of Chitin-triggered Immunity by *Magnaporthe oryzae* is Necessary for Rice Blast Disease. *Plant Cell* **2012**, *24*, 322–335.

Bailey, A. M.; Mitchell, D. J.; Manjunath, K. L.; Nolasco, G.; Niblett, C. L. Identification to the Species Level of the Plant Pathogens *Phytophthora* and *Pythium* by Using Unique Sequences of the ITS1 Region of Ribosomal DNA as Capture Probes for PCR ELISA. *FEMS Microb. Lett.* **2002**, *201*, 153–158.

Bari, R.; Jones, J. D. Role of Plant Hormones in Plant Defence Responses. *Plant Mol. Biol.* **2009**, *69*, 473–488.

Bischoff, J. F.; Sullivan, R. F.; Kjer, K. M.; White, J. F. Phylogenetic Placement of the Anamorphic Tribe *Ustilaginoideae* (*Hypocreales, Ascomycota*). *Mycologia* **2004**, *96*, 1088–1094.

Biswas, A. False Smut Disease of Rice: A Review. *Environ. Ecol.* **2001**, *19*, 67–83.

Braun, J.; Bos, M. S. The Changing Economics and Politics of Rice: Implications for Food Security, Globalization, and Environmental Sustainability. In *Rice Is Life: Scientific Perspectives for the 21st Century*; Toriyama, K., Heong, K. L., Hardy, B., Eds.; International Rice Research Institute, Japan International Research Center for Agricultural Sciences: Tsukuba, Japan, Los Banos, Philippines, 2004; pp 7–20.

Brooks, S. A.; Anders, M. M.; Yeater, K. M. Effect of Cultural Management Practices on the Severity of False Smut and Kernel Smut of Rice. *Plant Dis.* **2009**, *93*, 1202–1208.

Cartwright, R. D.; Lee, F. N.; Beaty, T.; Sutton, E. A.; Parsons, C. E. Reaction of Rice Cultivars/lines to False Smut, Stem Rot, and Black Sheath Rot Disease. *Univ. Ark. Agric. Exp. Stat. Res. Ser.* **2000**, *476*, 158–168.

Chao, J. Q.; Jin, J.; Wang, D.; Han, R.; Zhu, R. S.; Zhu, Y. G.; Li, S. Q. Cytological and Transcriptional Dynamics Analysis of Host Plant Revealed Stage-specific Biological Processes Related to Compatible Rice–*Ustilaginoidea virens* Interaction. *PLoS One* **2014**, *9*, e91391.

Cooke, M. C. Some Extra-European Fungi. *Grevillea* **1878**, *7*, 13–15.

Deng, G. S. Present Status of Research on False Smut in China. *Plant Protect.* **1989**, *15*, 39–40.

Dodan, D. S.; Ram, S.; Sunder, S. False Smut of Rice. Present Status. *Agric. Rev.* **1996**, *17*, 227–240.

Fan, J.; Guo, X.; Huang, F.; Li, Y.; Liu, Y.; Li, L.; Xu, Y.; Zhao, J.; Xiong, H.; Yu, J.; Wang, W. Epiphytic Colonization of *Ustilaginoidea virens* on Biotic and Abiotic Surfaces Implies the Widespread Presence of Primary Inoculum for Rice False Smut Disease. *Plant Pathol.* **2014**, *63*, 937–945.

Fan, R. H.; Wang, Y. Q.; Liu, B.; Zhang, J. Z.; Hu, D. W. The Process of Asexual Spore Formation and Examination of Chalmydospore Germination of *Ustilaginoidea virens*. *Mycosystema* **2010**, *29*, 188–192.

FAO. *FAOSTAT Database*, 2009. FAO: Rome, Available at: www.faostat.fao.org (last accessed June 2010).

FAO. *World Rice Information, Issue No.1.* FAO: Rome, Italy, 1995.

Faure, J.; & Mazaud, F. Rice Quality Criteria and the European Market. In: Proceedings of the 18th Session of the International Rice Commission, 5–9 September 1996, Rome, Italy, 1996; pp 121–131.

Fu, R.; Ding, L.; Zhu, J.; Li, P.; Zheng, A. P. Morphological Structure of Propagules and Electrophoretic Karyotype Analysis of False Smut *Villosiclava virens* in Rice. *J. Microbiol.* **2012**, *50*, 263–269.

Fujita, Y.; Sonoda, R.; Yaegashi, H. Inoculation with Conidiospores of False Smut Fungus to Rice Panicles at the Booting Stage. *Ann. Phytopathol. Soc. Jpn.* **1989**, *55*, 629–634.

Goud, J. C.; &Termorshuizen, A. J. Quality of Methods to Quantify Microsclerotia of *Verticillium dahliae* in Soil. *Eur. J. Plant Pathol.* **2003**, *109*, 523–534.

Grant, M.; Jones, J. G. Hormone (Dis)harmony Moulds Plant Health and Disease. *Science*, **2009**, *324*, 750–752.

Guo, X., Li, Y.; Fan, J.; Li, L.; Huang, F.; Wang, W. Progress in the Study of False Smut Disease in Rice. *J. Agric. Sci. Technol.* **2012**, *2*, 1211–1217.

Han, Y.; Zhang, K.; Yang, J.; Zhang, N.; Fang, A.; Zhang, Y.; Liu, Y.; Chen, Z.; Hsiang, T.; Sun, W. Differential Expression Profiling of the Early Response to *Ustilaginoidea virens* between False Smut Resistant and Susceptible Rice Varieties. *BMC Genomics* **2015**, *16*, 955–957.

Ikegami, H. Seedling Inoculation with the Chlamydospores of the False Smut Fungus. *Ann. Phytopathol. Soc. Jpn.* **1962**, *27*, 16–23.

Ikegami, H. Studies on the False Smut of Rice X, Invasion of Chlamydospores and Hyphae of the False Smut Fungus into Rice Plants. *Res. Bull. Fac. Agric. Gifu Univ.* **1963**, *18*, 54–60.

Kepler, R. M.; Sung, G. H.; Harada, Y.; Tanaka, K.; Tanaka, E.; Hosoya, T.; Bischoff, J. F.; Spatafora, J. W. Host Jumping onto Close Relatives and Across Kingdoms by *Tyrannicordyceps* (Clavicipitaceae) gen. nov. and *Ustilaginoidea* (Clavicipitaceae). *Am. J. Bot.* **2012**, *99*, 552–561.

Khush, G. S. Harnessing Science and Technology for Sustainable Rice-based Production Systems. In: Proceedings of FAO Rice Conference "Rice is life." *Int. Rice Comm. Newslett.* **2004**, *53*, 17–23.

Kim, K. W.; Park, E. W. Ultrastructure of Spined Conidia and Hyphae of the Rice False Smut Fungus *Ustilaginoidea virens*. *Microbiology* **2007**, *38*, 626–631.

Koiso, Y.; Li, Y.; Iwasaki, S.; Hanaoka, K.; Kobayashi, T.; Sonoda, R. *Ustiloxins, antimitotic Cyclic Peptides from False Smut Balls on Rice Panicles caused by Ustilaginoidea virens. J. Antibiot.* **1994**, *47*, 765–773.

Koiso, Y.; Morisaki, N.; Yamashita, Y.; Mitsui, Y.; Shirai, R.; Hashimoto, Y.; Iwasaki, S. Isolation and Structure of an Antimitotic Cyclic Peptide, Ustiloxin F: Chemical Interrelation with a Homologous Peptide, Ustiloxin B. *J. Antibiot.* **1998**, *51*, 418–422.

Koyama, K.; Natori, S. Further Characterization of Seven Bis(naphtho-gpyrone) Congeners of Ustilaginoidins, Coloring Matters of *Claviceps virens* (*Ustilaginoidea virens*). *Chem. Pharmacol. Bull.* **1988**, *36*, 146–152.

Koyama, K.; Ominato, K.; Natori, S.; Tashiro, T.; Tsuruo, T. Cytotoxicity and Antitumor Activities of Fungal Bis(naphtho-gamma-pyrone) Derivatives. *J. Pharm.* **1998**, *11*, 630–635.

Kumar, V.; Ladha, J. K. Direct Seeding of Rice: Recent Developments and Future Research Needs. *Adv. Agron.* **2011**, *111*, 297–413.

Landgraf, A.; Reckmann, B.; Pingoud, A. Direct Analysis of Polymerase Chain Reaction Products Using Enzyme-linked-immunosorbent-assay Techniques. *Anal. Biochem.* **1991**, *198*, 86–91.

Leubner-Metzger, G.; Meins, F. Functions and Regulation of Plant β-1,3-glucanases (PR-2). In: *Pathogenesis Related Proteins in Plants*; Datta, S. K., Muthukrishnan, S., Eds.; CRC Press: Boca Raton, FL, 1999; pp 49–76.

Li, W.; Li, L.; Feng, A.; Zhu, X.; Li, J. Rice False Smut Fungus, *Ustilaginoidea virens*, Inhibits Pollen Germination and Degrades the Integuments of Rice Ovule. *Am. J. Plant Sci.* **2013**, *4*, 2295–2304.

Li, Y. S.; Zhu, Z.; Zhang, Y. D.; Zhao, L.; Wang, C. L. Genetic Analysis of Rice False Smut Resistance Using Mixed Major Genes and Polygenes Inheritance Model. *Acta Agron. Sci.* **2008**, *34*, 1728–1733.

Li, Y.; Koiso, Y.; Kobayashi, H.; Hashimoto, Y.; Iwasaki, S. Ustiloxins, New Antimitotic Cyclic Peptides: Interaction with Porcine Brain Tubulin. *Biochem. Pharmacol.* **1995**, *49*, 1367–1372.

Ling, K. C. Studies on Rice Diseases. In: *Rice Improvement in China and Other Asian Countries*; International Rice Research Institute and Chinese Academy of Agricultural Science, 1980; pp 135–148.

Liu, X.; Bai, X.; Wang, X.; Chu, C. OsWRKY71, a Rice Transcription Factor, is Involved in Rice Defense Response. *J. Plant Physiol.* **2007**, *164*, 969–979.

Mei, C.; Qi, M.; Sheng, G.; Yang, Y. Inducible Overexpression of a Rice Allene Oxide Synthase Gene Increases the Endogenous Jasmonic Acid Level, PR Gene Expression, and Host Resistance to Fungal Infection. *Mol. Plant Microb. Interact.* **2006**, *19*, 1127–1137.

Mentlak, T. A.; Kombrink, A.; Shinya, T.; Ryder, L. S.; Otomo, I.; Saitoh, H.; Terauchi, R.; Nishizawa, Y.; Shibuya, N; Thomma, B. P. H. J.; Talbot, N. J. Effector-mediated Suppression of Chitin-triggered Immunity *by Magnaporthe oryzae* is Necessary for Rice Blast Disease. Plant Cell **2012**, *24*, 322–335.

Miao, Q. M. Studies on Infection Route of Rice False Smut. *J. Yunnan Agric. Univ.* **1992**, *7*, 40–42.

Mukesh, N.; Shamshad, A.; Anantha, M. S.; Kumar, Y.; Maiti, D. Comparative Homology Modelling and Docking Study of MAT 1-2-1 Protein for Designing Bio-active Molecules. *J. Adv. Bioinform. Appl. Res.* **2013**, *4*, 72–79.

Nakamura, K. I.; Izumiyama, N.; Ohtsubo, K. I.; Koiso, Y.; Iwasaki, S.; Sonoda, R.; Fujita, Y.; Yaegashi, H.; Sato, Z. "Lupinosis"-like Lesions in Mice Caused by Ustiloxin, Produced by *Ustilaginoieda virens*: A Morphological Study. *Nat. Toxins* **1994**, *2*, 2–28.

Qu, S. H. *Rice Diseases*. CMI Kew Survey: England, 1972; p 230.

Rush, M. C.; Shahjahan, A. K. M.; Jones, J. P. Outbreak of False Smut of Rice in Louisiana. *Plant Dis.* **2000**, *84*, 100.

Schroud, P.; TeBeest, D. O. Germination and Infection of Rice Roots by Spores of *Ustilaginoidea virens*. *AAES Res. Ser.* **2005**, *540*, 143–151.

Shan, T. Determination and Analysis of Ustiloxins A and B by LC-ESI-MS and HPLC in False Smut Balls of Rice. *Int. J. Mol. Sci.* **2012**, *13*, 11275–11287.

Shetty, S. A.; Shetty, H. S. An Alternative Host for *Ustilaginodea virens* (Cooke) Takahashi. *IRRI Newslett.* **1985,** *10,* 11.

Shetty, S. A.; Shetty, H. S. Role of *Panicum trypheron* in Annual Recurrence of False Smut of Rice. *Bryol. Mycol. Soc.* **1987,** *88,* 409–411.

Singh, A. K.; Pophaly, D. J. An Unusual Rice False Smut Epidemic Reported in Raigarh District, Chhattisgarh. *Int. Rice Res. Notes* **2005,** *35,* 1–3.

Singh, G. P.; Singh, R. N.; Singh, A. Status of False Smut (FS) of Rice in Eastern Uttar Pradesh, India. *IRRI Newslett.* **1987,** *12,* 28.

Singh, R. A.; Dube, K. S. Assessment of Loss in Seven Rice Cultivars Due to False Smut. *Ind. Phytopathol.* **1978,** *31,* 186–188.

Singh, R. A.; Dubey, K. S. Sclerotial Germination and Ascospore Formation of *Claviceps oryzae-sativae* in India. *Ind. Phytopathol.* **1984,** *37,* 168–170.

Srivastava, R.; Srivastava, K. K.; Sarkar, J. D.; Srivastava, P. Analysis of Problems Faced by the Farmers in Adoption of Control Measures of Diseases of Rice. *J. Interacad.* **2010,** *14,* 260–266.

Sutton, D. M.; Mac Hardy, W. E.; Lord, W. G. Effects of Shredding or Treating Apple Leaf Litter with Urea on Ascospore Dose of *Venturia inaequalis* and Disease Buildup. *Plant Dis.* **2000,** *84,* 1319–1326.

Tanaka, E.; Ashizawa, T.; Sonoda, R.; Tanaka, C. *Villosiclava virens* Gen. Nov., Comb. Nov., the Teleomorph of *Ustilaginoidea virens*, the Causal Agent of Rice False Smut. *Mycotaxonomy* **2008,** *106,* 491–501.

Tanaka, E.; Kumagawa, T.; Chihiro, T.; Hironori, K. Simple Transformation of Rice False Smut Fungus *Villosiclava virens* by Electroporation of Intact Conidia. *Mycoscience* **2011,** *52,* 344–348.

Tanaka, E.; Tanaka, C. Phylogenetic Study of Clavicipitaceous Fungi Using Acetaldehyde Dehydrogenase Gene Sequences. *Mycoscience* **2008,** *49,* 115–125.

Tanaka, T.; Ashizawa, T.; Sonoda, R.; Tanaka, C. *Villosiclava virens* gen. nov., com. nov., Teleomorph of *Ustilaginoidea virens*, the Causal Agent of Rice False Smut. *Mycotaxonomy* **2008,** *106,* 491–501.

Tang, Y. X.; Jin, J.; Hu, D. W.; Yong, M. L.; Xu, Y.; He, L. P. Elucidation of the Infection Process of *Ustilaginoidea virens* (teleomorph: *Villosiclava virens*) in Rice Spikelets. *Plant Pathol.* **2013,** *62,* 1–8.

Teng, P. S. Integrated Pest Management in Rice: An Analysis of Status Quo with Recommendation for Action. In: *Project Report Submitted to the Multiagency Integrated Pest Management Task Force*; 1990, Vol. 2; pp 79–84.

van Loon, L. C.; Rep, M.; Pieterse, C. Significance of Inducible Defense-related Proteins in Infected Plants. *Ann. Rev. Phytopathol.* **2006,** *44,* 35–62.

Wang, G. L. Studies on the Infection Period and the Infection Gate of the Chlamydospores of *Ustilaginoidea virens* (Cooke) Tak. on Rice. *Acta Phytophyl. Sin.* **1992,** *19,* 97–100.

Wang, Z.; Gerstein, M.; Snyder, M. RNA-Seq: A Revolutionary Tool for Transcriptomics. *Nat. Rev. Genet.* **2009,** *10,* 57–63.

Yu-Sheng, L.; Sheng-Dong, H.; Juan, Y.; Cai-Lin, Y. Analysis of Quantitative Trait Loci for Resistance to Rice False Smut. *Chin. J. Rice Sci.* **2011,** *37,* 778–783.

Zhang, Y.; Zhang, K.; Fang, A.; Han, Y.; Yang, J.; Xue, M.; Bao, J.; Hu, D.; Zhou, B.; Sun, X.; Li, S.; Wen, M.; Yao, N.; Ma, L. J.; Liu, Y.; Zhang, M.; Huang, F.; Luo, C.; Zhou, L.; Li, J.; Chen, Z.; Miao, M.; Wang, S.; Lai, J.; Xu, J. R.; Hsiang, T.; Peng, Y. L.; Sun, W. Specific Adaptation of *Ustilaginoidea virens* in Occupying Host Florets Revealed by Comparative and Functional Genomics. *Nature Commun.* **2014,** *5,* 3849.

Zhou, L. In: *Mycotoxins: Properties, Applications and Hazards of False Smut*; Melborn, B., Greene, J. C., Eds.; Nova Science Publishers: Hauppauge, NY, 2003; pp 109–130.

Zhou, Y. L.; Izumitsu, K.; Sonoda, R.; Nakazaki, T.; Tanaka, E.; Tsuda, M.; Tanaka, C. PCR-based Specific Detection of *Ustilaginoidea virens* and *Ephelis japonica*. *J. Phytopathol.* **2003**, *151*, 513–518.

Zhu, H.; Xu, X.; Xiao, G.; Yuan, L.; Li, B. Enhancing Disease Resistance of Super Hybrid Rice with Four Antifungal Genes. *Sci. China Life Sci.* **2007**, *50*, 31–39.

CHAPTER 6

BIOLOGICAL AND BIOTECHNOLOGICAL APPROACHES TO MANAGE BROWN SPOT (*HELMINTHOSPORIUM ORYZAE*) DISEASE OF RICE

SANJEEV KUMAR[1], NISHANT PRAKASH[2], KAHKASHAN ARZOO[3], and ERAYYA[1*]

[1]*Department of Plant Pathology, Bihar Agricultural University, Sabour, Bhagalpur 813210, Bihar, India*

[2]*Krishi Vigyan Kendra, Arwal, Bihar Agricultural University, Sabour, Bhagalpur 813210, Bihar, India*

[3]*Department of Plant Pathology, G. B. Pant University of Agriculture and Technology, Pantnagar 263145, Uttarakhand, India*

[*]Corresponding author. E-mail: erayyapath@gmail.com

CONTENTS

ABSTRACT

Brown spot of rice is caused by fungus, *Helminthosporium oryzae*. It is one of the most deadly diseases of rice. Many strategies are available for management of the disease, but the instability of pathogen threaten the rice crop. The present chapter outlines the application of biological and molecular approaches for the management of brown spot disease. In general, chemical control method is most effective; however, the use of chemicals is not generally desired due to the serious environmental threat it possesses, although biocontrol agents for brown spot have been successfully deployed to combat the disease under in vitro and in vivo conditions. For effective management of brown spot disease, breeding program should be focused on utilizing the broad spectrum of resistance genes and pyramiding of genes and quantitative trait loci (QTL). The availability of rice and *H. oryzae* genome sequence data are facilitating blast resistance management program to new paradigms which includes isolation and characterization of R and Avr genes. With the identification, isolation, and characterization of brown spot blast resistance genes in rice. In the present chapter, role of biological control including bioagents, botanicals, and resistance-inducing substances; molecular breeding; transgenic; and few new methods in controlling rice's brown spot disease is discussed in detail. The updated information will be helpful guidance for pathologists and rice breeders to develop durable brown spot resistant rice varieties through advanced molecular techniques and also be helpful in sustainable and environmental friendly rice crop production.

6.1   INTRODUCTION

Rice (*Oryza sativa* L.) is a plant belonging to the family of grasses, Gramineae (Poaceae). It is one of the three major food crops of the world and forms the staple diet of about half of the world's population. The global production of rice has been estimated to be at the level of 650 million tons and the area under rice cultivation is estimated to be around 156 million hectares (FAOSTAT, 2009). Asia is the leader in rice production, accounting for about 90% of the world's production. Over 75% of the world's supply is consumed by people in Asian countries and thus rice is of immense importance to food security of Asia. The demand for rice is expected to increase further in view of expected increase in the population. India has a long history of rice cultivation. Globally, it stands first in rice area and second in rice production after China. It contributes 21.5% of global rice production. Within the country, rice occupies one quarter of the total cropped area, contributes about 43% of total food grain production and continues to play a vital role in the national food and livelihood security system. India is one of the leading exporters of rice, particularly basmati rice.

Rice belongs to the genus *Oryza* and the tribe Oryzeae of the family Gramineae (Poaceae). The genus *Oryza* contains 25 recognized species, of which 23 are wild species and 2, *O. sativa* and *O. glaberrima*, are cultivated (Brar & Khush, 2003; Takahashi, 1984; Vaughan, 1994). *O. sativa* is the most widely grown of the two cultivated species. It is grown worldwide including in Asian, North and South American, European Union, Middle Eastern, and African countries. However, *O. glaberrima* is grown solely in West African countries.

The center of origin and centers of diversity of two cultivated species *O. sativa* and *O. glaberrima* have been identified using genetic diversity, historical and archaeological evidences, and geographical distribution. It is generally agreed that river valleys of Yangtze, Mekon Rivers could be the primary centers of origin of *O. sativa* while Delta of Niger River in Africa as the primary center of origin of *O. glaberrima* (Porteres, 1956; OECD, 1999). The foothills of the Himalayas, Chhattisgarh, Jeypore Tract of Orissa, northeastern India, northern parts of Myanmar and Thailand, Yunnan Province of China, etc. are some of the centers. The Inner delta of Niger River and some areas around Guinean coast of the Africa are considered to be center of diversity of the African species of *O. glaberrima* (Chang, 1976; Oka, 1988). De Condelle (1886) and Watt (1892) thought that south India was the place where cultivated rice originated. Vavilov (1926) suggested that India and Burma should be regarded as origin of cultivated rice.

In India, rice is grown in an area of 43.95 m ha with production and productivity of 106.31 million tons and 24.24 q/ha, respectively. West Bengal, Uttar Pradesh, Andhra Pradesh, Punjab, and Odisha are the major rice-growing states. West Bengal ranks first in rice production and productivity of 15.31 million ton and 27.86 q/ha, respectively. However, rice-growing area is maximum in Uttar Pradesh (5.98 m ha). Nutritional value of rice is also high; 100 g of milled rice grains contains calories (kcal) 345.0, calcium (mg) 10.0, moisture (g) 13.7, iron (mg) 0.7, carbohydrates (g) 78.2, magnesium (mg) 90.0, protein (g) 6.8, riboflavin (mg) 0.06, fat (g) 0.5, thiamine (mg) 0.06, fiber (g) 0.2, niacin (mg) 1.9, phosphorus (mg) 160.0, folic acid (mg) 8.0, minerals (g) 0.6, copper (mg) 0.14, and essential amino acids (mg) 1.09 (Gopalan et al., 2007).

The pest and diseases are the major production constraints in rice production. It suffers from several diseases caused by fungi, bacteria, viruses, nematodes, and mycoplasma. Out of these, blast (*Pyricularia oryzae*), brown spot (*Cochliobolus miyabeanus*), sheath blight (*Rhizoctonia solani*), false smut (*Claviceps oryzae sativa*), bacterial leaf blight (*Xanthomonas oryzae*), and narrow brown leaf spot (*Cercospora oryzae*) cause considerable damage to crop in terms of yield. Hence, in the present chapter, the biological and biotechnological strategies pertaining to management of brown spot of rice is discussed in detail.

6.2 BROWN SPOT (SESAME LEAF SPOT) OF RICE

6.2.1 CAUSAL ORGANISM

Brown spot disease is caused by the fungal pathogen *Helminthosporium oryzae* (Syn.: *Drechslera oryzae*; sexual stage: *C. miyabeanus*), which causes significant yield loss in rice production.

6.2.2 SYMPTOMS

Fungus attacks the crop from seedling in nursery to milk stage in main field. Symptoms appear as lesions (spots) on the coleoptile, leaf blade, leaf sheath, and glume, being most prominent on the leaf blade and glumes. The disease appears first as minute brown dots, later becoming cylindrical or oval to circular. The several spots coalesce and the leaf dries up. The seedlings die and affected nurseries can be often recognized from a distance by their brownish scorched appearance. Dark brown or black spots also appear on glumes which contain large number of conidiophores and conidia of the fungus. It causes failure of seed germination, seedling mortality and reduces the grain quality and weight.

6.2.3 PATHOGEN

H. oryzae produces greyish-brown to dark-brown septate mycelium. Conidiophores may arise singly or in small groups. They are straight, sometime geniculate, pale to brown in color. Conidia are usually curved with a bulge in the center and tapering toward the ends occasionally almost straight, pale olive green to golden brown color and are 6–14 septate. The perfect stage of the fungus is *C. miyabeanus*. It produces perithecia with asci containing 6–15 septate, filamentous or long cylinderical, hyaline to pale olive green ascospores. It produces C25 terpenoid phytotoxins called ophiobolin A (or Cochliobolin A), ophiobolin B (or cochliobolin B), and ophiobolin I. Ophiobolin A is most toxic. This breakdown the protein fragment of cell wall resulting in partial disruption of integrity of cell.

6.2.4 HISTORY AND ECONOMIC IMPORTANCE

Brown spot of rice caused by *Bipolaris oryzae* Subr. and Jain (=*H. oryzae* Breda de Haan telemorph = *C. miyabeanus*) is known to occur in Japan since 1900. It is also called as "nai-yake", that is, seedling blight, sesame leaf spot, and Helminthosporiosis. The disease has been reported to occur in all the rice-growing countries including India, Japan, China, Burma, Sri Lanka, Bangladesh, Iran, Africa, South America, Russia, North America, Philippines, Saudi Arabia, Australia, Malaya, and Thailand (Khalili et al., 2012). In India, it is known to occur in all the rice-growing states (Ou, 1985), since its first report from Madras in 1919 by Sundraraman. The disease is more severe in dry/direct seeded rice in the states of Bihar, Chhatisgarh, Madhya Pradesh, Orissa, Assam, Jharkhand, and West Bengal. The disease is of great importance in several countries and has been reported to cause enormous losses in grain yield (up to 90%) particularly when leaf spotting phase assumes epiphytotic proportions as observed in Great Bengal Famine during 1942 (Ghose et al., 1960). The disease especially occurs in environment where water supply is scarce combined with nutritional imbalance particularly lack of nitrogen (Baranwal et al., 2013).

The loss in grain yield has been reported to vary with rice cultivars and stage of infection (Kulkarni et al., 1980). Yield losses in rice cv. Tilakkachery (susceptible) and Bhasamanik (moderately resistant) at heavily infected sites in West Bengal, India were 5.6–11.7% and 2.2%, respectively (Chattopadhyay et al., 1975). Heavy infection significantly reduces the number of tillers & grains and lowers the quality and weight of individual grains resulting in a loss of 30–43%, while it was only 12% under moderate and nonsignificant (negligible) at lower infection ratings (Ou, 1985). Average losses of 8.2–23.0% in grain yield were reported from North-western Sierra Leone during 1983–1985 (Fomba & Singh, 1990). Glume blotch phase of the disease has been reported to cause more damage (Kulkarni et al., 1986) and reduce seed germinability (Hiremath et al., 1983).

6.2.5 FAVORABLE CONDITIONS

Temperature of 25–30°C with relative humidity above 80% is highly favorable. Excess of nitrogen aggravates the disease incidence.

6.2.6 MODE OF SPREAD AND SURVIVAL

The infected seeds are the most common source of primary infection. The conidia present on infected grain and mycelium in the infected tissue may viable for 2–3 years. The fungus may survive in the soil for 28 months at 30°C and five months at 35°C. Air-borne conidia infect the plants both in nursery and in main field. Maximum flight of conidia takes place at a wind velocity of 4.0–8.8 h. Minimum temperature of 27–28°C, relative humidity of 90–99%, and rainfall of 0.4–14.4 mm favors the dispersal of the conidia to maximum extent. The fungus also survives on collateral hosts like *Leersia hexandra*, *Arundo donux*, and *Echinochlora colonum*.

6.3 MANAGEMENT OF BROWN SPOT IN RICE

Field sanitation-removal of collateral hosts and infected debris in the field. Crop rotation, adjustment of planting time, and proper fertilization are suggested. Use of slow release nitrogenous fertilizers is advisable. Grow disease-tolerant varieties, namely, Co44, Cauvery, Bala Bhavani. Use disease-free seeds. Treat the seeds with Thiram or Captan at 4 g/kg. Spray the nursery with Edifenphos 40 ml or Mancozeb 80 g or Captafol 40 g for 8 cent nursery. Spray the crop in the main field with Edifenphos 500 ml or Mancozeb 1 kg or Captafol 625 g/ha.

Brown spot disease caused by *Drechslera oryza* is the most destructive disease occurring in almost all the rice-growing areas of world. In some regions, it is of no consequence, while in certain states of India, Pakistan, Malaysia, Indonesia, Philippines, Indo-China, Japan, and United States of America, it has been reported to cause heavy losses (Gangopadhyay, 1983). In India, it occurs more or less every year in mild or severe form, occasionally as an epidemic. The damage to the crop results from poor germination of seed, leaf spot causing general weakening of the plant and poor grain setting and infection of grain making them unsuitable for seed. This disease was very much in news in India, when the Famine Enquiry Commission, 1945, concluded that this disease was one of principal causes of the famous Bengal famine of 1942, and it caused 50–90% yield losses (Ghose et al.,

1948; Padmanabhan, 1973). In north Bihar, the disease occurred in epidemic form during 1979–1982 (Misra, 1985).

It is believed that this disease was present for many years but the description of the fungus was made by Breda de Haan only in 1922. Ocfemia (1924) reported on the occurrence of the disease in USA, Japan, Philippines, and other countries. In India, the first report on this disease was made by Sunderraman from Madras in 1919 (Singh, 1983); Mitra (1931) made detailed studies of the causal fungus. In India, Butler (1905, 1918) and Mitra (1931) were the pioneer workers on the genus. Misra and Singh (1966) added to the knowledge of physiology, variability, and pathogenicity of *Helminthosporium* species, but there had been no detailed and organized study of *H. oryzae* Breda de Haan occurring throughout the country.

Continuous, inappropriate, and non-discriminative use of chemicals is known to cause undesirable effects such as residual toxicity, development of pathogen resistance to fungicides, environmental pollution, health hazards to humans and animals, and increased expenditure for plant protection. Instead, plant pathologists have focused their attention to develop environmentally safe, long-lasting, and effective biocontrol and biotechnological methods for the management of brown spot of rice.

6.3.1 BIOLOGICAL CONTROL

Research on biological control of rice pathogens started recently, mainly in the 1980s. Research is still concentrated on the identification, evaluation, and formulation of potential biocontrol agents for deployment. A number of fungus, bacteria, virus, nematode, and mycoplasma-like organisms cause disease to rice plants.

Rice disease management strategies mainly aim at prevention of outbreak or epidemics through the use of host plant resistance (HPR) and chemical pesticides. The persistent, injudicious use of chemicals has toxic effects on nontarget organisms and can cause undesirable changes in the environment. Most of these chemicals are too expensive for the resource poor farmers of Asia, where 90% of the world's rice is grown. Large-scale and long-term use of resistant cultivars is likely to result in significant shifts in the virulence characteristics of pathogens, culminating in resistance breakdown. However, research during the previous two decades indicates another potential option for rice disease management, that is, biological control of rice diseases. Biocontrol assumes special significance being an eco-friendly and cost-effective strategy which can be used in integration with other strategies for a greater level of protection with sustained rice yields (Campbell, 1989).

6.3.2 BIOLOGICAL CONTROL AGENTS

A diverse group of biocontrol agents such as bacteria, fungi, and viruses exist in nature. Among them, bacterial antagonist is considered ideal candidates because of their rapid growth, ease of handling, and aggressive colonizing character. Bacterial antagonists *Pseudomonas*, and *Bacillus* in particular, are good candidates for biological control. *Bacilli* are germ-positive endospore-producing bacteria that are tolerant to heat and desiccation; a very good feature required for field application. The pseudomonads are germ-negative rods and have simple nutritional requirements; they are excellent colonizers and widely prevalent in rice rhizosphere.

The fluorescent and non-fluorescent strains of a number of antagonistic bacteria associated with upland and lowland rice rhizosphere soils have been found effective *in vitro*, greenhouse and the field against *R. solani* (sheath blight). As many as 23 bacterial antagonists belonging to the genera *Bacillus*, *Pseudomonas*, *Serratia*, and *Erwinia* have been found to inhibit myce-lial growth of *R. solani*, while a few of them also inhibit growth of other fungal pathogens like *Sclerotium oryzae* (stem rot), *B. oryzae* (brown spot), *P. grisea* (blast), *Sarocladium oryzae* (sheath rot) and *Fusarium fuijkuroi* (bakanae). Laboratory studies also revealed that a large number of bacterial strains possess the ability to protect rice plants from diseases such as brown spot, blast, sheath blight, sheath rot, and stem rot. About 40 bacterial isolates antagonistic to the rice sheath blight pathogen have been identified.

Important fungal antagonists include *Trichoderma* spp., *Penicillium*, *Myrothecium verrucaria*, *Chaetomium globosum*, and *Laerisaria arvalis*. The important biocontrol agents of major rice diseases are listed (Table 6.1).

Among various fungal and bacterial biocontrol agents, *Trichoderma* spp. was most frequently used against various plant diseases. Research during the previous two decades has led to the possibility of biological control as an increasingly realistic option for rice disease management (Tsahouridou & Thanassoulopouloh, 2002). *Trichoderma* spp. has been shown to be effec-tive for the control of brown spot disease and the increase of plant growth on rice (Harish et al., 2007a). Rice plants sprayed with spore suspension of *T. harzianum* obtained a significant reduction in the severity of disease under greenhouse conditions (Abdel-Fattah et al., 2007) Also, *Trichoderma* species are able to colonize the root surface and rhizosphere from the treated seeds, protecting them from fungal diseases and stimulate plant growth and productivity (Baker, 2004). This study was accomplished to obtain indige-nous *Trichoderma* isolates from paddy fields and to examine their biocontrol activities against *B. oryzae* in vitro as well as in vivo and also to evaluate their effects on rice growth parameters.

TABLE 6.1 Bio-control agents used against major diseases of rice.

Disease	Disease causal organism	Bio-control agents
Blast	*Pyricularia grisea*	*Trichoderma harzianum*
		T. viride
		T. reesei
		Pseudomonas fluorescens
		P. aeruginosa
		Bacillus subtilis
Brown spot	*Bipolaris oryzae*	*Trichoderma harzianum*
		T. virens
		T. atroviride
		P. fluorescens
		P. aeruginosa
		P. putida
		B. subtilis
		B. pumilus
		Serratia marcescens
Bakane disease	*Gibberella fujikuroi*	*Trichoderma harzianum*
		T. viride
		P. fluorescens
		Bacillus oryzicola (endophytic bacteria)
Sheath blight	*Rhizoctonia solani*	*T. harzianum*
		T. viride
		P. fluorescens
		P. putida
		P. aeruginosa
		B. subtilis
		B. laterosporus
		B. pumilus
		B. megaterium
		Serratia marcescens
		Enterobacter spp.
Sheath rot	*Sarocladium oryzae*	*P. fluorescens*
		P. aeruginosa
		B. subtilis
Stem rot	*Sclerotium oryzae*	*P. fluorescens*
		P. aeruginosa
		B. subtilis
Bacterial blight	*Xanthomonas oryzae* pv. *oryzae*	*P. fluorescens*
		Bacillus sp.
		Lysobacter antibioticus

There have been only a few reports on the improvement of brown spot disease control involving biological control agents. However, the use of antagonistic microbes for plant health management has emerged as a viable technology in the recent past. Commercially available antagonistic microbes, mostly belonging to the genera *Pseudomonas* and *Trichoderma*, can reduce the damage by direct effects on the pathogens (mycoparasitism, antibiosis, competition for iron) or by improving plant immunity (induced resistance [IR]) (Singh et al., 2005). Direct antagonism has been the key factor in suppression of many soil-borne pathogens, while IR is active against diverse foliar pathogens including both bacteria and fungi (Shoresh et al., 2010). Seed treatments with *Trichoderma viride* or *T. harzianum* have reduced disease by 70% (Biswas et al., 2010). Over 70% disease reduction has been achieved too from the use of selected *Pseudomonas* spp. isolates (Joshi et al., 2007; Ludwig et al., 2009). Direct foliar application of *T. harzianum* has also been reported to reduce the disease intensity and significantly improve grain yield, total grain carbohydrate and protein, in addition to a significant improvement in the total photosynthetic pigments in rice leaves (Abdel-Fattah et al., 2007). Such an alternative mode of brown spot disease management is largely untapped, but holds great promise. *Trichoderma* spp., well-known mycoparasites, can colonize roots internally and help improving nutrient uptake and mobilization, enhance nitrogen use efficiency, promote root growth and plant biomass, and improve tolerance to various physiological stresses, including soil salinity and drought through the reduction of oxidative damage that stresses cause (Harman, 2011; Shoresh et al., 2010). Use of these microbes could suppress disease through direct antagonism against the pathogen because imbalanced plant nutrition and drought stresses are predisposing factors for BS development. This effect would be combined with improved plant nutrient supply and delayed onset of water deficit in plant tissues (Bae et al., 2009) altering plant physiology to the disadvantage of the pathogen. Further, these microorganisms could contribute to the buildup of active plant immunity through the activation predisposition, the effect of drought (in interaction with the former component) and the interplay of the first two factors with soil characteristics on the hydromineral plant nutrition.

Commercially available antagonistic *Pseudomonas* and *Trichoderma* species can suppress diseases by direct effect on the pathogen through mycoparasitism, antibiosis, and competition for iron/nutrients or by improving plant immunity through IR (Singh et al., 2005). Isolates of fluorescent *Pseudomonas* from soil reduced the fungal growth and brown spot incidence (Ray et al., 1990). Spray application of talc based *P. fluorescens* has been found effective in reducing brown spot severity (Joshi et al., 2007). *Bacillus mega-terium* completely inhibited the growth of *B. oryzae* at 1×10^4 bacterial cells/ml with an ED_{50} value of 1×10^3 cells/ml and greatly reduced the disease incidence in field experiment (Islam & Nandi, 1985). *B. subtilis* (Sarala et al., 2004) and *T. viride* (Kumar & Mishra, 1994) showed strong antagonism against *B. oryzae*. In another study, *T. pseudokoningii* was found to be the most useful antagonist in terms of reduction in disease incidence and enhancement of seed germination and seedling growth (Krishnamurthy et al., 2001). *Trichoderma* spp. are known to improve nutrient intake and mobilization and to enhance nitrogen use efficiency promoting root growth and plant biomass and improving tolerance to soil salinity and drought (Shoresh et al., 2010). Among six phylloplane microorganisms, *Cladosporium* spp. was very effective in inhibiting the mycelial growth and spore germination of *B. oryzae* followed by *Penicillium* spp. and *Aspergillus flavus* (Harish et al., 2007b). They further found that *T. viride* inhibited the mycelial growth and spore germination by 63% and 77% followed by *T. harzianum* and *T. reesei* (Harish et al., 2008). A significant reduction in mycelial growth (55–58%) and seed-borne infection of *B. oryzae* with a bioformulation of *T. harzianum* has also been recorded (Biswas et al., 2008). Seed treatment with *T. viride* and *T. harzianum* (Biswas et al., 2010) and *Pseudomonas* species (Ludwig et al., 2009) has yielded 70% or more reduction in brown spot. Foliar application of *T. harzianum* has also been shown to reduce the disease intensity and significantly improve grain yield, total grain carbohydrates and proteins in addition to a significant improvement in the total photosynthetic pigment in the rice leaves (Abdel-Fattah et al., 2007). Kumawat et al. (2008) found that pre-application of spore suspension of bioagents (*T. harzianum* and *T. viride*) protected paddy plants against challenge infection of *B. oryzae*, which was attributed to increased level of total soluble protein and total phenol content. In Iran, two strains of *T. harzianum* controlled the disease and one strain of *T. atroviride* increased the seedling growth (Khalili et al., 2012). Spraying F 402 fraction of antifungal substances produced by *Fusarium graminearum* completely inhibited the growth of *C. miyabeanus* at 50 μg/ml and provided >80% control of brown spot (Kim et al., 1995). Inoculation of susceptible rice cultivars with avirulent strain of the pathogen induced resistance in host and reduced the disease index by 83–85% (Sinha & Trivedi, 1969). The same effect was obtained by pre-treating plants with liquid in which germinating spores had been incubated for 24 h (Sinha & Das, 1972). Beside these, sheep urine and goat dung (20%) have also been found promising against brown spot (Selvi et al., 2008).

6.3.3 BOTANICALS

Plant extracts and botanicals have also been found to be effective against brown spot disease. Leaf extracts from *Artabotrys hexapetalus* (Grainge & Alvarez, 1987) and peppermint, *Piper nigrum* and garlic extract (Alice & Rao, 1987) were most inhibitory to *B. oryzae*. Leaf extracts of *Juglans regia* reduced mycelial growth of *B. oryzae* by 64% (Bisht & Khulbe, 1995), while aqueous extracts of *Acorus calamus* by 80% along with 45.3% reduction in brown spot incidence (Jitendiya Devi & Chhetry, 2013). Ganesan and Krishnaraju (1995) observed that out of antifungal leaf extracts of 23 plant species, extracts of *Leucas aspera*, *Polygonum chinense*, and *Spermacoce articularis* (*S. hispida*) inhibited spore germination, while remaining extracts inhibited the growth of germinated spores. Similarly, aqueous leaf extracts of *Anacardium occidentale*, *Bixa orellana*, *Ichnocarpus frutescens*, *Leea* species, *Macaranga peltata*, and *Uvaria navum* completely inhibited the conidial germination. Germ-tube elongation was inhibited by extracts of *Cleome aspera*, *Delonix regia*, *Gliricidia sepium*, *Hibscus surattensis*, *Quisqualis indica*, and *Zornia gibbosa* (Ganesan, 1994). The extract of *Agave americana* at 0.1% (Kumar, 2006) and *A. sativum* and *Pithecellobium dulce* at 10% (Raju et al., 2004) provided >50% and 90% inhibition of spore germination and mycelial growth of *B. oryzae*, respectively. Aqueous leaf extract of *Thuja orientalis* proved better than *Azadirachta indica*, *Clerodendron inermae*, *Catharanthus roseus*, *Tridax procumbens*, *Colens aromaticus*, *Ruta graviolens*, and *L. aspera* for minimizing the incidence of *B. oryzae* and enhancing seed germination and seeding growth (Krishnamurthy et al., 2001). Methanol extract of *Prosopis juliflora* has been reported to provide complete inhibition of mycelial growth of the pathogen at 800 ppm (Raghavendra et al., 2002). Water and ethanol extracts from leaves and oil extract from seeds of *A. indica* were found effective in reducing the radial growth of *C. miyabeanus* in culture and controlling the spread of brown spot in rice (Amadioha, 2002). Similarly, oils from palmarosa (0.1%), palmarosa + neem 1:2 at 0.1% (Sarala et al., 2004), lemon grass (*Cymbopogon flexuous*) at 1%, and cinnamon (*Cinnamomum zeylanicum*) at 1.0% (Raji, 2004), *Hedychium spicatum* at 1.0×10^3 ml/l and *Acorus calamus* at 0.5×10^3 ml/l (Mishra et al., 2003) provided 90–100% growth inhibition of *B. oryzae*. Oil from seeds of *Aphanamixis polystachy* inhibited the growth of *B. oryzae* by 40% and 50% at 20 and 40 μl/disc, respectively (Bhuyan et al., 2000), Harish et al. (2004) indicated that rhizome extract of turmeric (*Curcuma longa*), seed extract of sundavathal (*Solanum indicum*), and vedpalai (*Wrightia tinctoria*) inhibited the mycelia growth and spore germination of the pathogen. They further observed that leaf extracts of *Nerium oleander* and *P. dulce* provided >75% and >80% inhibition of mycelial growth and spore germination of *B. oryzae*.

In field experiments, two sprays of neem cake extract, *N. oleander* leaf extract, and *T. viride* reduced brown spot incidence by 70%, 53%, and 48%, respectively with significant increase in grain yield (Harish et al., 2008). Among various biopesticides, Tricure 5 ml/l, Biotos 2.5 ml/l, and Achook 5 ml/l (Kumar & Rai, 2008) and Neemazal 3 ml/l and Wanis 5 ml/l (Sunder et al., 2010) reduced brown spot severity and increased the yield significantly. Natarajan and Lalithakumari (1987) reported that the leaf extract of *Lawsonia inermis* possessed antifungal activity against *B. oryzae* due to presence of 2-hydroxy-1,4-naphthoquinone (lawsone). Among leaf extracts of some angiosperms, *Adenocalymma allicea*, known to contain a volatile oil, completely inhibited the mycelia growth of *B. oryzae*. The volatile oil was not only fungicidal at its minimum inhibitory concentration of 500 ppm but also reduced the disease incidence (Chaturvedi et al., 1987). Mimosine extracted from seeds of *Leucaena leucocephala* (Kandasamy & Arunachalam, 1994) and essential oils from *Citronella winterianus* (Kole et al., 1993) exhibited antifungal activity against *B. oryzae*. Salamargine extracted from *Solanum nigrum* inhibited spore germination of *B. oryzae* by 83% and germ-tube elongation by 100% at 500 μl/ml (Chelvan & Sumathi, 1994). Methanolic extracts of medicinal plants, namely, *Bergia capensis*, *Marselia quadrifolia*, *Lippia nodiflora*, *Eclipta prostrata*, and *Commleina clavata* (Maninegalai et al., 2011) and essential plant oils from basil (*Ocimum basilicum*) and sweet fennel (*Ocimum gratissimum*) showed good inhibitory activity against *B. oryzae* (Piyo et al., 2009).

6.3.4   RESISTANCE INDUCING CHEMICALS

Spraying with nonconventional chemicals has also proved to be an effective indirect approach of disease management. Foliar spray of ferric chloride (Vidhyasekaran et al., 1986), calcium chloride, and α-amino-*n*-butyric acid (Bala et al., 2007) induced resistance in rice plants against *B. oryzae*, while chitosan provided complete inhibition of fungal growth at 1000 ppm (Rivero et al., 2008). Amongst 15 phytoalexin inducers, spray application of sodium malonate, sodium molybdate, cadmium chloride, ferric chloride, and DL-methionine proved to be the most promising in reducing brown spot severity on rice leaves (Giri & Sinha, 1979). They further observed that seed treatment with ferric chloride, nickel nitrate, sodium molybdate, and

DL-methionine gave best results in pot experiments, while seed treatment with ferric chloride, sodium molybdate, and mercuric chloride proved better in field. In root dip test, cadmium chloride, barium chloride, sodium malonate, and sodium molybdate gave the best results. In spray treatments, the maximum protection against *C. miyabeanus* was recorded with DL-methionine, indole-3-acetic acid, and sodium malonate when sprayed on 3-week-old potted seedlings (Giri & Sinha, 1979). Among five resistance inducers evaluated, seed treatment with digionin followed by its foliar spray at 75 days after transplanting proved significantly superior in reducing disease severity (Satija et al., 2005b). Seed soaking in dilute concentration of phytoalexin inducers (metabolic inhibitors, growth regulators, and amino acid) for 24 h was more effective than pre-inoculation foliar spray and root dip for 24 h. The treated inoculated plants exhibited moderate to high fungitoxicity up to 5–7 weeks depending on the degree of resistance induced (Giri & Sinha, 1983). Trivedi and Sinha (1980) observed that among three heavy metal salts and four amino acids tested, silver nitrate at 10-2 M, and DL-methionine and DL-norleucine at 10-1 M provided 23–82% disease control in pot experiments. Resorcinol and phloroglucinol (phenolics) have been reported to inhibit the growth of *B. oryzae* at 10 mM (Kalaichelvan & Elangovan, 1995). Sunder et al. (2010) reported that among nine nonconventional chemicals, ferric chloride, sodium salenate, and nickel nitrate reduced both the leaf spot and stalk rot phases of the disease. However, the latter two chemicals were slightly phytotoxic on rice leaves. de Vleesschauwer et al. (2010) observed that the treatment of rice plants with abscisic acid (ABA) significantly enhanced resistance to *B. oryzae* owing to ABA-induced suppression of the ethylene response. Benzoic acid and salicylic acid have not only been found inhibitory to fungus in vitro but also observed to reduce the disease severity significantly along with enhanced total photosynthetic pigments, total carbohydrates, and protein content of rice grains (Shabana et al., 2008).

6.4 BIOTECHNOLOGICAL APPROACHES

6.4.1 *HOST PLANT RESISTANCE*

In modern terms "biotechnology" is defined as the manipulation, genetic modification, and multiplication of living organisms through novel technologies, such as tissue culture and genetic engineering, resulting in the production of improved or new organisms and products that can be used in a variety of ways. Biotechnology offers various tools which are proven very useful in management of brown spot of rice.

Study of microbial population of a particular pathogen is essential for its management. Population structure of a pathogen over a large geographical area portrays a picture of pathogen and is helpful in preparation of strategy for its management. Population structure over a large geographical area provide us information about the nature of variability of a pathogen, that is, highly variable or less variable and number of races of a pathogen in a particular area. With the help of degree of variation of a pathogen in a partic-ular area, we can screen different resistant source of host against pathogen.

6.4.2 MOLECULAR MARKER FOR THE STUDY OF POPULATION STRUCTURE OF A BROWN SPOT OF RICE PATHOGEN

Today various molecular markers are used for the population study of a pathogen. RFLP (restricted fragment length polymorphism), RAPD (randomly amplified polymorphic DNA), SSR (single sequence repeats), amplified fragment length polymorphism, etc. are widely used in micro-bial population study. *Helminthosporium* pathogen causes disease in rice, maize, and wheat. *Helminthosporium* pathogens showed interspecific varia-tion. Different isolates of *Bipolaris sorokiniana* from wheat, *B. oryzae* from rice and *B. maydis* from maize was collected and subjected to PCR-RFLP (Weikert-Oliveira et al., 2002). On PCR-RFLP, three groups were character-ized: first group include all isolates from *B. sorokiniana*, second include all isolates from *B. oryzae*, and third group include all isolates from *B. maydis*. Isolates of *B. oryzae* also have huge intra-specific variation. In Iran, finger-printing analysis of *B. oryzae* by rep-PCR using BOX and REP primer, 15 clonal lineage and 54 haplotype were reported. The largest and most common lineage contained 36 haplotype (Nazari et al., 2015). RAPD-PCR show considerable variation among Indian and Iranian isolates of *B. oryzae* (Archana et al., 2014; Motlagh & Anvari, 2010). URP primers also consid-erable variation among Indian isolates of *B. oryzae* (Kandan et al., 2013). Because of non-reproducibility of RAPD marker, SSR and ISSR markers are used for genetic variability and population study. ISSR markers produce reproducible banding pattern.

The analysis of ISSR polymorphism among populations of *B. oryzae* from different regions of India revealed the occurrence of high level of poly-morphism, indicating a wide and diverse genetic base. ISSR markers also indicate that *B. oryzae* isolates from diverse geographical regions of India may be genetically heterogeneous and the interrelationship amongst the different isolates can be easily, precisely, and reliably explained by ISSR-polymerase chain reaction technology (Archana et al., 2014).

6.4.3 PARTIAL RESISTANCE AND QUANTITATIVE TRAIT LOCI FOR BROWN SPOT

Goel et al. (2006) analyzed the inheritance of resistance to brown spot from crosses involving *O. nivara* germplasm and hypothesized that additive, dominant, and gene interactions were involved. Three QTL were detected in cultivar Tadukan (qBS2, qBS9, and qBS11) on chromosomes 2, 9, and 11, respectively (Sato et al., 2008) qBS11 being considered of the defense pathways, similar to the effects of *P. fluorescens* on rice, which induce resistance to sheath blight (Nandakumar et al., 2001).

Understanding variation and diversity in pathogen population and the mechanisms that influence the genotypic changes in the pathogen population is an essential step in disease management strategies. The information on pathogen variation is also useful in identifying and characterizing resistant germplasm, however, this study is limited in brown spot pathogen. HPR is an effective way to manage brown spot (Savary et al., 2011). Effective breeding for resistance must target a population instead of an individual (McDonald, 1997), and hence, knowledge of the pathogen population structure and virulence characteristics that represent this population is necessary. Although tests for resistance have continued in many countries, the variability of the fungus has not been studied in great detail (Ou, 1985). A few studies have addressed the molecular variation of *B. oryzae* (Burgos et al., 2013). Recently, genetic diversity analysis of *B. oryzae* has been done using URP markers (Kandan et al., 2013). Using of gene sequence data to clarify evolutionary relationships and determine taxonomic status of organisms, including fungi, is common nowadays. *Cochliobolus* has been segregated into two groups and *B. oryzae* is placed with *C. sativus* and *C. heterostrophus* (Tazick & Tajik, 2013). Recently, the full genome sequence of *C. miyabeanus* strain (WK1C) has been done and is available at the Joint Genome Institute. Besides, this four other *Cochlibolus* species, genome sequences are also available (Condon et al., 2013). The comparative genome analysis of *Cochlibolus* species has led to understanding the role of for secondary metabolism and small secreted proteins in pathogenesis and diversity existing at inter and intra specific level (Condon et al., 2013).

B. oryzae produces melanin which is a polymer of 1,8-dihydroxynaphthalene (DHN). DHN-melanin biosynthesis starts with a polyketide synthase (PKS) using acetate as a precursor. A number of genes involved in melanin biosynthetic pathway have been isolated and characterized in *B. oryzae*. SCD1, a gene encoding scytalone dehydratase, has been cloned and sequence analysis showed that SCD1 encodes a putative protein that has 185 amino acids, a molecular weight of 21 kDa and 51–75% sequence identity to other fungal scytalone dehydratases. Targeted disruption of this gene (SCD1) showed that it is necessary for melanin biosynthesis in *B. oryzae*. Northern blot analysis further revealed that SCD1 transcripts are specifically enhanced by near ultraviolet (300–400 nm) radiation (Kihara et al., 2004b). Understanding the role of the photomorphogenetic response in conidial development of *B. oryzae* has potential to aid the development of control methods. It was recently reported that expressions of three melanin biosynthesis genes (Moriwaki et al., 2004), and the photolyase gene (Kihara et al., 2004a) are specifically enhanced by NUV radiation. A number of many possible transformation systems and functional genomics strategies have been developed in filamentous fungi (Weld et al., 2006). In order to characterize the gene function of *B. oryzae*, the gene knockout strategy was carried out, which is to replace a wild-type gene with a version of that gene that has been disrupted by an antibiotic agent (Moriwaki et al., 2004). A melanin regulation 1 gene (BMR1) encoding a transcription factor for melanin biosynthesis genes has been isolated and characterized. Sequence analysis showed that the BMR1 gene encodes a putative protein of 1012 amino acids that has 99% sequence similarity to transcription factor Cmr1 of *C. heterostrophus*. The predicted *B. oryzae* Bmr1 protein has two DNA-binding motifs, two Cys2His2 zinc finger domains, and a Zn(II)2Cys6 binuclear cluster domain at the N-terminal region of Bmr1. Targeted disruption of the BMR1 gene showed that BMR1 is essential for melanin biosynthesis in *B. oryzae*. The overexpression of the BMR1 gene led to more dark colonies than in the wild-type strain under dark conditions. Real-time PCR analysis showed that the BMR1 expression of the over expression transformant was about 10-fold in transformants than that of the wild type under dark conditions (Kihara et al., 2008).

Recently, RNA-mediated gene silencing (RNA silencing) has emerged as a powerful tool for gene targeting in fungi, plants, and animals. This strategy exploits an endogenous gene-regulatory mechanism of eukaryotic cells, in which regulatory double-stranded RNAs (dsRNAs) interfere with homologous mRNA either by triggering its degradation or inhibiting its transcription or translation (Almeida & Allshire, 2005). For gene targeting, dsRNA homologous to the target gene is introduced into the organism either directly or indirectly as a construct leading to its endogenous expression. Effective gene silencing in *B. oryzae* using this method has been demonstrated. An endogenous (PKS1 gene was used to demonstrate gene silencing as a marker. The PKS1 is involved in fungal DHN melanin biosynthesis pathways, and targeted gene disruption (knockout) for the PKS1 gene showed a melanin-deficient (albino) phenotype (Moriwaki et al., 2004). Silencing of the PKS1 gene resulted in an albino phenotype and reduction of PKS1 mRNA expression. These results demonstrate the applicability of targeted gene silencing as a useful reverse-genetics approach in *B. oryzae* (Moriwaki et al., 2008).

In order to reveal the photomorphogenic response and to identify new genes up regulated by NUV irradiation, suppression subtractive hybridization (SSH) was carried out in *B. oryzae*. To confirm the differential gene expression in NUV-irradiated mycelia, quantitative realtime PCR (qRT-PCR) analysis has been performed among 301 genes arbitrarily chosen from 1170 cDNA clones. The expression of 46 genes (named NUV01 to NUV46) was found to be significantly enhanced (>4-fold) by NUV irradiation. Sequence analysis revealed that 23 out of the 46 sequences (50%) showed significant matches to known fungal genes. The 46 genes were categorized as either BLR1-dependent or BLR1- independent expression groups using the BLR1-deficient mutant, which presumably lacks the blue/UVA-absorbing photoreceptor. This finding demonstrates that NUV irradiation can induce gene regulation, and that this response may be mediated by both a blue/UV absorbing photoreceptor and an as-yet-unidentified photoreceptor in *B. oryzae* (Kihara et al., 2014). The interaction between rice and *C. miyabeanus* is inadequately understood from the perspective of genetic and molecular mechanisms, although it has been reported that, like other *Cochliobolus* species, the fungus utilizes phytotoxins to trigger host cell death (Ahn et al., 2005). Other necrotrophic *Cochliobolus* spp. and related taxa (e.g., *Pyrenophora tritici repentis*, *Stagonospora nodorum*, *Alternaria alternata*), are notorious for their ability to evolve novel, highly virulent, races producing host selective toxins (HSTs). Differential response of Tetep (resistant) and Nakdong (susceptible) was observed to *Magnaporthe grisea* and *C. miyabeanus*. The expression pattern of the pathogenesis-related (PR) and JAmyb genes in response to *C. miyabeanus* was nearly identical between cvs. Nakdong and Tetep, and neither BTH nor MeJA treatment significantly modified their expression patterns in response to *C. miyabeanus* infection. This suggests that rice employs distinct mechanisms for its defense against *M. grisea* and *C. miyabeanus* (Ahn et al., 2005).

Exogenously administered ABA enhances basal resistance of rice against the brown spot pathogen *B. oryzae*. Microscopic analysis of early infection events in control and ABA-treated plants revealed that this ABA inducible resistance (ABA-IR) is based on restriction of fungal progression in the mesophyll. ABA-IR requires a functional G-alpha-protein. In addition, several lines of evidence suggest that ABA steers its positive effect on brown spot resistance through antagonistic cross talk with the ethylene (ET) response pathway. Exogenous ethephon application enhances susceptibility, whereas genetic disruption of ET signaling renders plants less vulnerable to *C. miyabeanus* attack, thereby inducing a level of resistance similar to that observed on ABA-treated wild-type plants. Moreover, ABA treatment alleviates *C. miyabeanus* induced activation of the ET reporter gene EBP89, while derepression of pathogen triggered EBP89 transcription via RNA interference mediated Knock down of OsMPK5, an ABA-primed mitogen-activated protein kinase gene, compromises ABA-IR. Collectively, these data favor a model, whereby exogenous ABA enhances resistance against *C. miyabeanus* at least in part by suppressing pathogen induced ET action in an OsMPK5 dependent. Recently, for better understanding of the rice *C. miyabeanus* interaction, proteomic approach has been used. 2-DE approach after PEG-fractionation of total proteins coupled with MS (MALDI-TOF/TOF and nESILC–MS/MS) analyses led to identification of 49 unique proteins out of 63 differential spots. SDS-PAGE in combination with nESI-LC-MS/MS shotgun approach resulted in cataloging of 501 unique proteins, of which 470 and 31 proteins were secreted from rice and C. miyabeanus, respectively. The enzymes involved in Calvin cycle and glycolysis decreased in their protein abundance, whereas enzymes in the TCA cycle, amino acids, and ethylene biosynthesis increased. Differential proteomes also generated distribution of identified proteins in the intracellular and extracellular spaces, providing a better insight into defense responses of proteins in rice against *C. miyabeanus*. Established proteome of the rice *C. miyabeanus* interaction serves not only a good resource for the scientific community but also highlight its significance from biological aspects (Kim et al., 2014).

6.4.4 THROUGH BIOTECHNOLOGICAL APPROACHES

Tissue culture has been used for development brown spot resistant rice. *Helminthosporium* phytotoxin incorporated media with concentration of 25%. Rice plants were grown in phytotoxin incorporated media. These rice plants found resistance against brown spot disease. This resistance also inherits second generation (Ling et al., 1985). With the help of Particle bombardment method antifungal gene, namely, Chitinases, β-1, 3-glucanases, and maize ribosome-inactivating proteins are subjected for gene pyramiding in a rice host plant. This genetic engineered rice host plant confers resistance against *R. solani*, *Bipolaris oryzae*, and *M. grisea* (Kim et al., 2003; Zhang et al., 2009). Two resveratrol rice lines, Iksan515 and Iksan526, were obtained by transforming rice lines. Binary vectors pSB22 was used to transform rice line with *resveratrol synthase* gene (*AhSTS1*, *RS3*) (Baek et al., 2013). But these rice lines showed susceptibility against brown spot and blast of rice (Qin et al., 2013). Exogenous application of chemicals also activates defense genes in rice plant. Exogenous application of ABA induces resistance in rice against brown spot disease altering the expression of OsMPK-5 gene. ABA enhances the accumulation of OsMPK transcript and confers resistance.

HPR to disease is an effective and economical way to manage brown spot disease. However, breeding efforts have emphasized acute diseases such as leaf blast and bacterial blight rather than chronic diseases such as brown spot (Savary et al., 2011), despite the importance of brown spot. The search for sources of resistance to brown spot has been a long-standing effort (Chakrabarti, 2001; Nagai & Hara, 1930). Satija et al. (2005a) identified 15 *O. sativa* entries out of 124 that were classified as resistant (less than 5% severity). Conversely, Hossain et al. (2004) identified one resistant variety out of 29 entries. Screening of upland rice germplasm, exotic and indigenous to eastern India, has revealed that partial and complete resistance to the brown spot pathogen is expressed by several genotypes under field conditions (Shukla et al., 1995). It seems that the sources of resistance amongst *O. sativa* entries are few and recent research (e.g., Goel et al., 2006) has been exploring other pools, especially *O. nivara*. Major in genes Adair (1941) suggested that resistance was recessive, involving several genes. Later studies showed that resistance, or susceptibility, could be associated with a limited number of genes. Balal et al. (1979) found two dominant genes were associated with resistance, while one gene was associated with susceptibility.

6.5 KNOWLEDGE GAP STILL PERSISTS IN ADDRESSING ISSUES

(1) Why are brown spot epidemics so slow to initiate and increase?

(2) Why is brown spot such a persistent disease in some rice producing areas and less so in others?

(3) What are the sources of primary inoculum and their quantitative role in the course of epidemics?

(4) How can HPR be best characterized screened and deployed?

(5) What are the physical factors (and the physiology of the host plant) that must be accounted for in so doing?

(6) What genes (including QTL) are involved in both susceptibility (resistance) to brown spot and tolerance to physiological stress (e.g., drought) and how could they interact?

(7) What interactions of microorganisms with the host plant and/or the pathogen trigger could cascade reactions that hamper, or favors, resistance to disease and tolerance to physiological stresses?

6.6   CONCLUSION

Many farmers in South Asia, particularly India, are practicing biological control of brown spot and other fungal diseases of rice. Seed treatment with biocontrol agents is becoming increasingly popular. Research into the mechanisms underpinning biocontrol agents interactions with HPR and plant physiology and the contribution of biocontrol agents to nutrient uptake, better growth, tolerance to environmental stresses, could pave the way toward novel understandings of plant–pathogen interactions, as well as providing means to sustain the adoption of durable disease management practices by farmers. Present review highlights that brown spot remains a poorly defined rice crop health problem being a reflection of the large number of factors determining the course of epidemics and their outcomes. While a lack of understanding of the pathogen's life cycle and how it is influenced by environmental factors contributes to this poor definition, understanding the effects of global change and of shifts in socioeconomic contexts, which determine crop management and inputs, make brown spot a marker of global and climate changes. Such interactions are not unique. Brown spot may thus be seen as a model system to characterize the behavior of many complex pathosystems responding to unfolding climate and global changes.

A number of naturally occurring fungal isolates can inhibit growth of *B. oryzae*. Furthermore, this research raises some interesting possibilities for future research. These include testing whether the antagonist can show the same level of efficacy under natural field conditions; testing whether mixtures of biocontrol agents are more effective than a single strain on control of disease severity and testing for preparing the best formulation of promising *Trichoderma* isolates. HPR against this disease has not been given due attention as has been given to other diseases such as blast and bacterial blight. There is a need for concerted efforts on identification of sources of resistance and their utilization in breeding program. Quantitative disease resistance and understanding the underlying resistance mechanisms need to be addressed. Standard methodologies and most valuable stage of the crop at which assessment is to be done need to be developed for quantitative disease measurements. Besides, host resistance with advances made in genomics, mechanisms of resistance, and mechanisms involved in Si uptake, their distribution within the plant need to be explored.

KEYWORDS

- *Oryza*
- pest
- disease
- pathogen
- fungicide

REFERENCES

Abdel-Fattah, G. M.; Shabana, Y. M.; Ismail, A. E.; Rashad, Y. M. *Trichoderma harzianum*: A Biocontrol Agent against *Bipolaris oryzae*. *Mycopathologia* **2007**, *164*, 81–89.

Adair, C. R. Inheritance in Rice of Reaction to *Helminthosporium oryzae* and *Cercospora oryzae*. Technical Bulletin, United States Department of Agriculture, Washington, D.C. No., 1941; pp 1–18.

Ahn, I. P.; Kim, S.; Kang, S.; Suh, S. C.; Lee, Y. H. Rice Defense Mechanisms against *Cochliobolus miyabeanus* and *Magnaporthe grisea* are Distinct. *Phytopathology* **2005**, *95*, 1248–1255.

Alice, D.; Rao, A. V. Antifungal Effects of Plant Extracts on *Drechslera oryzae* in Rice. *Int. Rice Res. Newsl.* **1987**, *12*, 28.

Almeida, R.; Allshire, R. C. RNA Silencing and Genome Regulation. *Trends Cell Biol.* **2005**, *15*, 251–258.

Amadioha, A. C. Fungitoxic Effects of Extracts of *Azadirachta indica* against *Cochliobolus miyabeanus* Causing Brown Spot Disease of Rice. *Arch. Phytopathol. Pl. Prot.* **2002**, *35*, 37–42.

Archana, B.; Kini K. R.; Prakash, H. S. Genetic Diversity and Population Structure among Isolates of the Brown Spot Fungus, *Bipolaris oryzae*, as Revealed by Inter-simple Sequence Repeats (ISSR). *AfrBiotech.* **2014**, *13*, 238–244.

Bae, H.; Sicher, R. C.; Kim, M. S.; Kim, S.; Strem, M. D.; Melnick, R. L.; Bailey, B. A. The Beneficial Endophyte *Trichoderma hamatum* isolate DIS 219b Promotes Growth and Delays the Onset of the Drought Response in *Theobroma cacao*. *J. Exp. Bot.* **2009**, *60*, 3279–3295.

Baek, S. H.; Shin, W. C.; Ryu, H. S.; Lee, D. W.; Moon, E.; Seo, C. S.; Hwang, E.; Lee, H. S.; Ahn, M. H.; Jeon, Y.; Kang, H. J.; Lee, S. W.; Kim, S. Y.; D'Souza, R.; Kim, H. J.; Hong, S. T.; Jeon, J. S. Creation of Resveratrol-Enriched Rice for the Treatment of Metabolic Syndrome and Related Diseases. *PLoS One* **2013**, *8*, 1–10.

Baker, R. *Trichoderma* spp. as Plant-growth Stimulants. *Biotechnology* **2004**, *7*, 97–106.

Bala, J.; Chahal, S. S.; Pannu, P. P. S. Induction of Resistance in Rice by Chemicals Compounds against Brown Spot Disease under Nitrogen and Water Stress Conditions. *J. Mycol. Pl. Pathol.* **2007**, *37*, 588–589.

Balal, M. S.; Omar, R. A.; El-Khadem, M. M.; Aidy, I. R. Inheritance of Resistance to the Brown Spot Disease of Rice, *Cochliobolus miyabeanus*. *Agric. Res. Rev.* **1979**, *57*, 119–133.

Baranwal, M. K.; Kotasthane, A.; Magculia, N.; Mukherjee, P. K.; Savary, S.; Sharma, A. K.; Singh, H. B.; Singh, U. S.; Sparks, A. H.; Variar, M.; Zaidi, N. A Review on Crop Losses, Epidemiology and Disease Management of Rice Brown Spot to Identify Research Priorities and Knowledge Gaps. *Eur. J. Pl. Pathol.* **2013**, *136*, 443–457.

Bhuyan, M. A. K.; Begum, J.; Chowdhury, J. U.; Ahmed, K.; Anwar, M. N. Antimicrobial Activity of Oil and Crude Alkaloids from Seeds of *Aphanamixis polystachya* (Wall.) R. N. Parker. *Bangl. J. Bot.* **2000**, *29*(1), 1–5.

Bisht, G. S.; Khulbe, R. D. *In vitro* Efficacy of Leaf Extracts of Certain Indigenous Medicinal Plants against Brown Leaf Spot Pathogen of Rice. *Indian Phytopathol.* **1995**, *48*, 480–482.

Biswas, C.; Srivastava, S. S. L.; Biswas, S. K. Effect of Biotic, Abiotic and Botanical Inducers on Crop Growth and Severity of Brown Spot in Rice. *Indian Phytopathol.* **2010**, *63*, 187–191.

Biswas, S. K.; Ved, R.; Srivastava, S. S. L.; Singh, R. Influence of Seed Treatment with Biocides and Foliar Spray with Fungicides for Management of Brown Leaf Spot and Sheath Blight of Paddy. *Ind. Phytopathol.* **2008**, *61*, 55–59.

Brar, D. S.; Khush, G. S. Utilization of Wild Species of Genus *Oryza* in Rice Improvement. In: *Monograph on Genus Oryza*; Nanda, J. S., Sharma, S. D., Eds.; Science Publishers, Inc.: Enfield, NH, 2003; pp 283–309.

Burgos, M. R. G.; Katimbang, M. L. B.; Dela Paz, M. A. G.; Beligan, G. A.; Goodwin, P. H.; Ona, I. P.; Mauleon, R. P.; Ardales, E. Y.; Vera Cruz, C. M. Genotypic Variability and Aggressiveness of *Bipolaris oryzae* in the Philippines. *Eur. J. Pl. Pathol.* **2013**, *137*, 415–429.

Butler, E. J. Fungal Disease of Rice in Bengal. *Mem. Dep. Agr. India Bot.* 1905, *1*, 1–53.

Butler, E. J. *Fungi and Diseases in Plants.* Thacker Spink & Co.: Calcutta, 1918.

Campbell, R. *Biological Control of Microbial Plant Pathogens.* Cambridge University Press: Cambridge, 1989; p 219.

Chakrabarti, N. K. Epidemiology and Disease Management of Brown Spot of Rice in India. In: *Major Fungal Disease of Rice: Recent Advances*; Chakrabarti, N. K., Ed.; Kluwer Academic Publishers: Berlin, 2001, 293–306.

Chang, T. T. The Origin, Evolution, Cultivation, Dissemination and Diversification of Asian and African Rices. *Euphytica* **1976**, *25*, 425–441.

Chattopadhyay, S. B.; Chakrabarti, N. K.; Ghosh, A. K. Estimation of Loss in Yield of Rice due to Infection of Brown Spot Incited by *Helminthosporium oryzae*. FAO *Intern. Rice Comm. Newsl.* **1975**, *24*, 67–70.

Chaturvedi, R.; Dikshit, A.; Dixit, S. N. *Adenocalymma allicea*, a New Source of Natural Fungi Toxicant. *Trop. Agric.* **1987**, *64*, 318–322.

Chelvan, P. T. K.; Sumathi, L. A Fungitoxic Substance from *Solanum nigrum. Indian Phytopathol.* **1994**, *47*, 424–426.

Condon, B. J.; Leng, Y.; Wu, D.; Bushley, K. E.; Ohm, R. A.; Otillar, R.; Martin, J.; Schackwitz, W.; Grimwood, J.; Mohd. Zainudin, N.; Xue, C.; Wang, R.; Manning, V. A.; Dhillon, B.; Tu, Z. J.; Steffenson, B. J.; Salamov, A.; Sun, H.; Lowry, S.; LaButti, K.; Han, J.; Copeland, A.; Lindquist, E.; Barry, K.; Schmutz, J.; Baker, S. E.; Ciuffetti, L. M.; Grigoriev, I. V.; Zhong, S.; Turgeon, B. G. Comparative Genome Structure, Secondary Metabolite, and Effector Coding Capacity across Cochliobolus Pathogens. *PLoS Genet.* **2013**, *9*, e1003233.

De Condelle, A. P. *Origin of Cultivated Plants*. Keganpaul Trench and Co.: London, 1886.

De Vleesschauwer, D. D.; Yang, Y.; Cruz, C. V.; Hofte, M. Abscisic Acid-induced Resistance against the Brown Spot Pathogen *Cochliobolus miyabeanus* in Rice involves MAP Kinase-mediated Repression of Ethylene Signaling. *Plant Physiol.* **2010**, *152*, 2036–2052.

FAOSTAT. Rice Market Monitor. Trade Markets, 2009; p 12.

Fomba, N. N.; Singh, N. Crop Losses Caused by Rice Brown Spot Disease in Mangrove Swamps of North Western Sierra Leone. *Trop. Pest Mgm.* **1990**, *36*, 387–393.

Ganesan, T. Antifungal Properties of Wild Plants. *Adv. Pl. Sci.* **1994**, *7*, 185–187.

Ganesan, T.; Krishnaraju, J. Antifungal Properties of Wild Plants. *Adv. Pl. Sci.* **1995**, *8*, 194–196.

Gangopadhyay, S. *Current Concept on Fungal Diseases of Rice*; Today and Tomorrow's Printers & Publishers: 24-B/5, Desh Bandhu Gupta Road, Karol Bagh, New Delhi, 1983; p 114.

Ghose, R. L. M.; Ghatge, M. B.; Subrahmanayan, V. *Rice in India (Revised Edition)*; I.C.A.R.; New Delhi, 1948; p 474.

Ghose, R. L. M.; Ghatge, M. B.; Subramanian, V. *Rice in India (Revised Edition)*; ICAR: New Delhi, 1960; p 474.

Giri, D. N.; Sinha, A. K. Control of Brown Spot Disease of Rice Seedlings by Treatment with a Select Group of Chemicals. *Z. Pflanzenkrankh. Pflanzensch.* **1983**, *90*, 479–487.

Giri, D. N.; Sinha, A. K. Effect of Non-toxic Chemicals on Brown Spot Disease in Rice Seedlings. *Int. Rice Res. Newsl.* **1979**, *4*, 10–11.

Goel, R. K.; Bala, R.; Singh, K. Genetic Characterization of Resistance to Brown Leaf Spot caused by *Drechslera oryzae* in Some Wild Rice (*Oryza sativa*) Lines. *Ind. J. Agric. Sci.* **2006**, *76*, 705–707.

Gopalan, C.; Rama Sastri, B. V.; Balasubramanian, S. *Nutritive Value of Indian Foods*, National Institute of Nutrition (NIN), ICMR, 2007.

Grainge, M. D.; Alvarez, A. M. Antibacterial and Antifungal Activity of *Artabotrys hexapetalus* Leaf Extracts. *Int. J. Trop. Pl. Dis.* **1987**, *5*, 173–179.

Harish, S.; Duraiswamy, S.; Ramalingam, R.; Ebenezar, E. G.; Seetharaman, K. Use of Plant Extracts and Biocontrol Agents for the Management of Brown Spot Disease in Rice. *Biocontrol.* **2008**, *53*, 555–567.

Harish, S.; Saravanakumar, D.; Kamalakannan, A.; Vivekananthan, R.; Ebenezar, E. G.; Seetharaman, K. Phylloplane Microorganisms as a Potential Biocontrol Agent against *Helminthosporium oryzae* Bredade Haan, the Incident of Rice Brown Spot. *Arch. Phytopathol. Pl. Prot.* **2007a**, *40*, 148–157.

Harish, S.; Saravanan, T.; Radjacommare, R.; Ebenezar, E. G.; Seetharaman, K. Mycotoxic Effect of Seed Extracts against *Helminthosporium oryzae* Breda de Haan, the Incitant of Rice Brown Spot. *J. Biol. Sci.* **2004**, *4*, 366–369.

Harish, S.; Saravavakumar, D.; Radjacommar, R.; Ebenezar, E. G.; Seetharaman, K. Use of Plant Extracts and Biocontrol Agents for the Management of Brown Spot Disease in Rice. *Biocontrol.* **2007b**, *53*(3), 555–567.

Harman, G. E. *Trichoderma*-not just for Biocontrol Anymore. *Phytoparasitica* **2011**, *39*(2), 103–108.

Hiremath, P. C.; Hegde, R. K.; Kulkarni, B. G. Effect of Seed Infection by Brown Leaf Spot on Rice Seed Germination. *Curr. Res.* **1983**, *12*, 74–76.

Hossain, M.; Khalequzzaman, K. M.; Mollah, M. R. A.; Hussain, M. A.; Rahim, M. A. Reaction of Breeding Lines/Cultivars of Rice against Brown Spot and Blast under Field Condition. *Asian J. Plant Sci.* **2004**, *3*, 614–617.

Islam, K. Z.; Nandi, B. Control of Brown Spot of Rice by *Bacillus megaterium. Z. Pflanzenkrankh. Pflanzensch.* **1985**, *92*, 241–246.

Jitendiya Devi, O.; Chhetry, G. K. N. Evaluation of Antifungal Properties of Certain Plants against *Drechslera oryzae* Brown Spot of Rice in Manipur Valley. *Int. J. Sci. Res. Publ.* **2013**, *3*, 1–3.

Joshi, N.; Brar, K. S.; Pannu, P. P. S.; Singh, P. Field Efficacy of Fungal and Bacterial Antagonists against Brown Spot of Rice. *J. Biol. Contr.* **2007**, *21*, 159–162.

Kalaichelvan, P. T.; Elangovan, N.; Effect of Phenolics on *Drechslera oryzae. Indian Phytopathol.* **1995**, *48*, 271–274.

Kandan, A.; Akhtar, J.; Singh, B.; Dixit, D.; Chand, D.; Agarwal, P. C.; Roy, A.; Rajkumar, S. Population Genetic Diversity Analysis of *Bipolaris oryzae* Fungi Infecting *Oryza sativa* in India Using URP Markers. *Ecoscan* **2013**, *7*, 123–128.

Kandasamy, M.; Arunachalam, R. Biochemical Mechanism of Mimosine Toxicity to Fungi. *Int. J. Trop. Pl. Dis.* **1994**, *12*, 171–176.

Khalili, E.; Sadravi, M.; Naeimi, S.; Khosravi, V. Biological Control of Rice Brown Spot with Native Isolates of Three *Trichoderma* Species. *Braz. J. Microbiol.* **2012**, *43*, 297–305.

Kihara, J, Moriwaki, A.; Tanaka, N.; Tanaka, C.; Ueno, M.; Arase, S. Characterization of the BMR1 Gene Encoding a Transcription Factor for Melanin Biosynthesis Genes in the Phytopathogenic Fungus *Bipolaris oryzae. FEMS Microbiol. Lett.* **2008**, *281*, 221–227.

Kihara, J.; Moriwaki, A.; Matsuo, N.; Arase, S.; Honda, Y. Cloning, Functional Characterization, and Near Violet Radiation-enhanced Expression of a Photolyase Gene (PHR1) from the Phytopathogenic Fungus *Bipolaris oryzae. Curr. Genet.* **2004a**, *46*, 37–46.

Kihara, J.; Moriwaki, A.; Ueno, M.; Tokunaga, T.; Arase, S.; Honda, Y. Cloning, Functional Analysis and Expression of a Scytalone Dehydratase Gene (SCD1) Involved in Melanin Biosynthesis of the Phytopathogenic Fungus *Bipolaris oryzae. Curr. Genet.* **2004b**, *45*, 197.

Kihara, J.; Tanaka, N.; Ueno, M.; Arase, S. Identification and Expression Analysis of Regulatory Genes Induced by Near-ultraviolet Irradiation in *Bipolaris oryzae. Adv. Microbiol.* **2014**, *4*, 233–241.

Kim, B. S.; Kim, K. W.; Lee, J. K.; Lee, Y. W.; Cho, K. Y. Isolation and Purification of Several Substances Produced by *Fusarium graminearum* and their Antimicrobial Activities. *Korean J. Pl. Pathol.* **1995**, *11*, 158–164.

Kim, J. K.; Jang, I. C.; Wu, R.; Zuo, W. N.; Boston, R. S.; Lee, Y. H.; Ahn, I. P.; Nahm, B. H. Co-expression of a Modified Maize Ribosome-inactivating Protein and a Rice Basic

Chitinase Gene in Transgenic Rice Plants Confers Enhanced Resistance to Sheath Blight. *Transgen. Res.* **2003**, *12*, 475–484.

Kim, J. Y.; Wu, J.; Kwon, S. J.; Oh, H. L.; So, E. K.; Sang, G.; Wang, Y. A.; Ganesh, K.; Rakwal, R.; Kang; Kyu, Y.; Ahn, P.; Kim; B. G; Kim, S. T. Proteomics of Rice and *Cochliobolus miyabeanus* Fungal Interaction: Insight into Proteins at Intracellular and Extracellular Spaces. *Proteomics* **2014**, *14*, 2307.

Kole, C.; Pattanaik, S.; Subramanyam, V. R.; Narain, A. Antifungal Efficacy of Oil and its Genetic Variability in *Citronella. Crop Res. Hisar* **1993**, *6*, 509–512.

Krishnamurthy, C. D.; Lokesh, S.; Shetty, H. S. Occurrence, Transmission and Remedial Aspects of *Drechslera oryzae* in Paddy (*Oryza sativa* L.). *Seed Res.* **2001**, *29*, 63–70.

Kulkarni, S.; Ramakrishnan, K.; Hegde, R. K. Demonstration on Supervisory Control of Brown Leaf Spot of Rice Caused by *Drechslera oryzae* (Breda de Haan) Subram. and Jain ex. M. B. Ellis. *Pl. Pathol. Newsl.* **1986**, *4*, 22.

Kulkarni, S.; Ramakrishnan, K.; Hegde, R. K. Incidence of Brown Leaf Spot of Rice Caused by *Drechslera oryzae* (Breda de Haan) Subram. & Jain under Different Agroclimatic Conditions of Karnataka. *Mysore J. Agric. Sci.* **1980**, *14*, 321–322.

Kumar, A. Evaluation of Botanicals against Major Pathogens of Rice. *Ind. Phytopathol.* **2006**, *59*, 509–511.

Kumar, R. N.; Mishra, R. R. Interaction Studies in vitro Between Brown Spot Pathogen of Paddy and Certain Epiphytic Phylloplane Fungi. *Plant Cell Incom. Newsl.* **1994**, *26*, 40–47.

Kumar, S.; Rai, B. Evaluation of New Fungicides and Biopesticides against Brown Spot of Rice. *Ind. Agric.* **2008**, *52*, 117–119.

Kumawat, G. T.; Biswas, S. K.; Srivastava, S. S. L. Biochemical Evidence of Defence Response in Paddy Induced by Bio-agents against Brown Leaf Spot Pathogen. *Indian Phytopathol.* **2008**, *61*, 197–203.

Ling, D. H.; Vidhyaseharan, P.; Borromeo, E. S.; Mew, T. W. In Vitro Screening of Rice Germplasm for Resistance to Brown Spot Disease Using Phytotoxin. *Theor. Appl. Genet.* **1985**, *71*, 133–135.

Ludwig, J.; Moura, A. B.; dos Santos, A. S.; Ribeiro, A. S. Seed Microbiolization for the Control of Rice Brown Spot and Leaf Scald. *Trop. Pl. Pathol.* **2009**, *34*, 322–328.

Ludwig, J.; Moura, A. B.; dos Santos, A. S.; Ribeiro, A. S. Seed Microbiolization for the Control of Rice Brown Spot and Leaf Scald. *Trop. Plant Pathol.* **2009**, *34*, 322–328.

Maninegalai, V.; Ambikapathy, V.; Panneerselvam, A. Antifungal Potentiality of Some Medicinal Plants Extracts against *Bipolaris oryzae* (Breda de Haan). *Asian J. Plant. Sci. Res.* **2011**, *1*, 77–80.

McDonald, B. A. The Population Genetics of Fungi: Tools and Techniques. *Phytopathology* **1997**, *87*, 448–453.

Mishra, D.; Samuel, C. O.; Tripathi, S. C. Evaluation of Some Essential Oils against Seed-borne Pathogen of Rice. *Indian Phytopathol.* **2003**, *56*, 212–213.

Misra, A. P.; Singh, T. B. Efficacy of Carbon Copper and Organic Fungicides against Seedling Blight and Least Spot of Paddy *Proc. Bihar Acad. Agril. Sci.* 1966, *15*, 41–46.

Misra, A. K. Variability in *Drechslera oryzae*—The Causal Organism of Brown Spot Disease of Rice. *Ind. Phytopathol.* **1985**, *38*, 168–169.

Mitra, M. Saltation in the Genus *Helminthosporium. Trans. Br. Mycol. Soc.* 1931, *16*, 115–127.

Moriwaki, A.; Kihara, J.; Kobayashi, T.; Tokunaga, T.; Arase, S.; Honda, Y. Insertional Mutagenesis and Characterization of a Polyketide Synthase Gene (PKS1) required for Melanin Biosynthesis in *Bipolaris oryzae. FEMS Microbiol. Lett.* **2004**, *238*, 1–8.

Moriwaki, A.; Ueno, M.; Arase, S.; Kihara, J. RNA-mediated Gene Silencing in the Phyto-pathogenic Fungus *Bipolaris oryzae*. *FEMS Microbiol. Lett.* **2008**, *269*, 85–89.

Motlagh, M. R. S.; Anvari, M. Genetic Variation in a Population of *Bipolaris oryzae* based on RAPD-PCR in North of Iran. *Afr. J. Biotechnol.* **2010**, *9*(36), 5800–5804.

Nagai, I.; Hara, S. On the Inheritance of Variegation Disease in a Strain of Rice Plant. *Jpn. J. Genet.* **1930**, *5*, 140–144.

Nandakumar, R.; Babu, S.; Viswanathan, R.; Raguchander, T.; Samiyappan, R. Induction of Systemic Resistance in Rice against Sheath Blight Disease by Pseudomonas fluorescens. *Soil Biol. Biochem.* **2001**, *33*, 603–612.

Natrajan, M. R.; Lalithakumari, D. Antifungal Activity of the Leaf Extract of *Lawsonia inermis* on *Drechslera oryzae*. *Indian Phytopathol.* **1987**, *40*, 390–395.

Nazari, S.; Javan-Nikkhah, M.; Fotouhifar, K. B.; Khosravi, V.; Alizadeh, A. *Bipolaris* Species Associated with Rice Plant: Pathogenicity and Genetic Diversity of *Bipolaris oryzae* Using rep-PCR in Mazandaran Province of Iran. *J. Crop Prot.* **2015**, *4*, 497–508.

Ocfemia, G. C. The *Helminthosporium* Disease of Rice Occurring in the Southern United States and in the Philippines. *Am. J. Bot.* 1924, *11*, 385–408.

OECD. *Consensus Document on the Biology of Oryza sativa (Rice)*. ENV/JM/MOMO (99) 26. Organization for Economic Co-operation and Development, 1999.

Oka, H. I. *Origin of Cultivated Rice*. Elsevier: Amsterdam, 1988.

Ou, S. H. *Rice Diseases*, 2nd ed.; CMI: Kew, England, 1985; p 370.

Padmanabhan, S. Y. The Great Bengal Famine. *Ann. Rev. Phytopathol.* 1973, *11*, 11–26.

Piyo, A.; Udomsilp, J.; Khang-Khun, P.; Thobunluepop, P. Antifungal Activity of Essential Oils from Basil (*Ocimum basilicum* Linn.) and Sweet Fennel (*Ocimum gratissimum* Linn.): Alternative Strategies to Control Pathogenic Fungi in Organic Rice. *Asian J. Food Agroind.* **2009**, (*Special issue*), S2–S9.

Porteres, R. Taxonomic agrobotanique der riz cultives *O. sativa* Linne. et *O. glaberrima* Steudelo, *J. Agric. Trop. Bot. Appl.* **1956**, *3*, 341–384.

Qin, Y.; Kim, S. M.; Ahn, H.; Lee, J. H.; Baek, S. H.; Shin, K. S.; Woo, H. J.; Cho, H. S.; Kweon, S. J.; Lim, M. H. Bioassay for the Response of Resveratrol Transgenic Rice Lines to Bacterial and Fungal Diseases. *Plant Breed. Biotechnol.* **2013**, *1*, 253–261.

Raghavendra, M. P.; Satish, S.; Raveesha, K. A. *Prosopis juliflora* Swartz: A Potential Plant for the Management of Fungal Diseases of Crops. *J. Mycol. Pl. Pathol.* **2002**, *32*, 392–393.

Raji, P. Inhibitory Effect of Plant Oils on *Helminthosporium oryzae* Causing Brown Spot of Rice. *Ind. Phytopathol.* **2004**, *57*, 342.

Raju, K.; Manian, S.; Anbuganapathi, G. Effects of Certain Herbal Extracts on the Rice Brown Leaf Spot Pathogen. *J. Ecotoxicol. Environ. Monitor.* **2004**, *14*, 31–37.

Ray, S.; Ghosh, M.; Mukharjee, N. Fluorescent Pseudomonads for Plant Disease Control. *J. Mycopathol. Res.* **1990**, *28*, 135–140.

Rivero, D.; Cruz, A.; Martinez, B.; Rodriguez, A. T.; Ramirez, M. A. In vitro Antifungal Activity of K 1 and SIGMA Chitosans against *Bipolaris oryzae* (Breda de Haan) Shoem. *Riv. Prot. Veg.* **2008**, *23*, 43–47.

Sarala, L.; Muthusamy, M.; Karunanithi, K. Management of Grain Discolouration of Rice with Antagonistic Organisms. *Res. Crops* **2004**, *5*, 165–167.

Satija, A.; Chahal, S. S.; Pannu, P. P. S. Effect of Some Chemical Compounds in Induction of Resistance against Brown Leaf Spot of Rice. *Pl. Dis. Res.* **2005b**, *20*, 115–118.

Satija, A.; Chahal, S. S.; Pannu, P. P. S. Evaluation of Rice Genotypes against Brown Leaf Spot Disease. *Plant Dis. Res. (Ludhiana)* **2005a**, *20*, 163–164.

Sato, H.; Ando, I.; Hirabayashi, H.; Takeuchi, Y.; Arase, S.; Kihara, J.; Mizobuchi, M. QTL Analysis of Brown Spot Resistance in Rice (*Oryza sativa* L.). *Breed. Sci.* **2008**, *58*, 93–96.

Savary, S.; Nelson, A.; Sparks, A. H.; Willocquet, L.; Duveiller, E.; Mahuku, G. International Agricultural Research Tackling the Effects of Global and Climate Changes on Plant Diseases in the Developing World. *Plant Dis.* **2011**, *48*, 1–40.

Selvi, M. T.; Balabaskar, P.; Kurucheve, V. Field Evaluation of Certain Animal Excreta against Brown Leaf Spot of Rice (*Helminthosporium oryzae* Breda de Haan) Subram. and Jain. *Adv. Pl. Sci.* **2008**, *21*, 51–53.

Shabana, Y. M.; Abdel-Fattah; G. M.; Ismail, A. E.; Rashad, Y. M. Control of Brown Spot Pathogen on Rice (*Bipolaris oryzae*) Using Some Phenolic Antioxidants. *Braz. J. Microbiol.* **2008**, *39*, 438–444.

Shoresh, M.; Harman, G. E.; Mastouri, F. Induced Systemic Resistance and Plant Responses to Fungal Biocontrol Agents. *Ann. Rev. Phytopathol.* **2010**, *48*, 21–43.

Shoresh, M.; Harman, G. E.; Mastouri, F. Induced Systemic Resistance and Plant Responses to Fungal Biocontrol Agents. *Annu. Rev. Phytopathol.* **2010**, *48*, 21–43.

Shukla, V. D.; Chauhan, J. S.; Variar, M.; Maiti, D.; Chauhan, V. S.; Tomar, J. B. Reaction of Traditional Rainfed Rice Accessions to Brown Spot, Blast and Sheath Rot Diseases. *Indian Phytopathol.* **1995**, *48*, 433–435.

Singh, R. K.; Singh, C. V.; Shukla, V. D. Phosphorus Nutrition Reduces Brown Spot Incidence in Rainfed Upland Rice. *Int. Rice Res. Not.* **2005**, *30*(2), 31–32.

Singh, R. S. *Plant Diseases*; Oxford and IBH Publishing Co.: New Delhi, 1983; pp 387.

Singh, R.; Dabur, K. R.; Malik, R. K. Long-term Response of Zero-tillage: Soil Fungi, Nematodes and Diseases of Rice–Wheat System, Technical Bulletin (7), CCS HAU Rice Research Station, Kaul, Department of Nematology and Directorate of Extension Education, CCS HAU: Hisar, 2005; pp 16.

Sinha, A. K.; Das, N. C. Induced Resistance in Rice Plants to *Helminthosporium oryzae*. *Physiol. Pl. Pathol.* **1972**, *2*, 401–410.

Sinha, A. K.; Trivedi, N. Immunization of Rice Plants against *Helminthosporium* Infection. *Nature* **1969**, *223*, 963–967.

Sunder, S.; Singh, R.; Dodan, D. S. Evaluation of Fungicides, Botanicals and Non-conventional Chemicals against Brown Spot of Rice. *Ind. Phytopathol.* **2010**, *63*, 192–194.

Takahashi, N., Ed. *Biology of Rice*; Japan Scientific Societies Press/Elsevier: Tokyo/Amsterdam, **1984**; pp 3–30.

Tazick, Z.; Tajik, M. A. Taxonomic Position of *Bipolaris oryzae* among Other *Cochliobolus* Species using Ribosomal Region and Some Protein Coding Genes. *Res. J. Recent Sci.* **2013**, *2*, 212–216.

Trivedi, N.; Sinha, A. K. Effect of Pre-inoculation Treatments with Some Heavy Metal Salts and Amino Acids on Brown Spot Disease in Rice Seedlings. *Proc. Indian Acad. Sci. Pl. Sci.* **1980**, *89*, 283–289.

Tsahouridou, P. C.; Thanassoulopoulosh, C. C. Proliferation of *Trichoderma koningii* in the Tomato Rhizosphere and the Suppression of Damping off by *Sclerotium rolfsii*. *Soil Biol. Biochem.* **2002**, *34*, 767–776.

Vasudevan, P.; Kavitha, S.; Priyadarsini, V. B.; Babuje, L.; Gnanamanickam, S. S. Biological Control of Rice Diseases. In: *Biological Control of Crop Diseases*; Gnanamanickam, S. S., Ed.; Marcel Dekker, Inc.: New York, Basel, 2002; pp 11–32.

Vaughan, D. A. The Wild Relatives of Rice. International Rice Research Institute: Manila, 1994.

Vavilov, N. I. Studies on the Origin of Cultivated Plants. *Bull. Appl. Bot. Plant Breed.* 1926, *16*(2), 1–248.

Vidhyasekaran, P.; Borromeo, E. S.; Mew, T. W. Host Specific Toxin Production by *Helminthosporium oryzae*. *Phytopathology* **1986**, *76*, 261–266.

Vleesschauwer, D. D.; Yang, Y.; Cruz, C. V.; Hofte, M. Abscisic Acid-induced Resistance against the Brown Spot Pathogen *Cochliobolus miyabeanus* in Rice involves MAP Kinase-mediated Repression of Ethylene Signaling. *Plant Physiol.* **2010**, *152*, 2036–2052.

Watt, G. *Dictionary of the Economic Products of India.* 1892, Vol. *5*, 502–654.

Weikert-Oliveira, R. C. B.; Resende, M. A.; Valerio, H. M.; Caligiorne, R. B.; Paiva, E. Genetic Variation among Pathogens causing "*Helminthosporium*" Diseases of Rice, Maize and Wheat. *Fitopatol. Bras.* **2002**, *27*, 639–643.

Weld, R. J.; Plummer, K. M.; Carpenter, M. A.; Ridway, H. J. Approaches to Functional Genomics in Filamentous Fungi. *Cell Res.* **2006**, *16*, 31–44.

Zhang, H.; Li, G.; Li, W.; Song, F. Transgenic Strategies for Improving Rice Disease Resistance. *Afr. J. Biotech.* **2009**, *8*(9), 1750–1757.

CHAPTER 7

MOLECULAR AND BIOLOGICAL APPROACHES FOR MANAGEMENT OF ROOT-KNOT DISEASE OF RICE CAUSED BY *MELOIDOGYNE GRAMINICOLA*

NEELAM MAURYA[1], PINTOO KUMAR[2], and DHARMENDRA KUMAR[1*]

[1]Department of Plant Pathology, N. D. University of Agriculture and Technology, Kumarganj, Faizabad 224229, Uttar Pradesh, India

[2]Department of Nematology, N. D. University of Agriculture and Technology, Kumarganj, Faizabad 224229, Uttar Pradesh, India

[*]Corresponding author. E-mail: dkumar_nduat@yahoo.in

CONTENTS

ABSTRACT

Rice is an important source of calories for more than one-third of the human population living on this planet. Several biotic, mesobiotic, and abiotic stresses limit the rice productivity. Among the biotic stresses, plant–parasitic nematodes are quite important. Root-knot disease of rice caused by *Meloidogyne graminicola* (Golden and Birchfield) is one of the most devastating and destructive disease of upland and rain-fed lowland rice (*Oryza sativa* L.) in Asia and Africa. In India, the disease is widespread in rice–wheat cropping system with significant yield losses. In deepwater rice, root-knot nematode infected seedlings remain stunted; unable to grow above flood water and perish due to continuous submergence the juveniles cause disruption, hypertrophy, and hyperplasia of cortical cells by intra-cellular migration and releasing esophageal gland secretions. This chapter reviews research on root-knot disease of rice and its pathogen *M. graminicola* during the past years in relation to etiology, host range, host–pathogen relationship and management through cultural and biological and biotechnological approaches.

7.1 INTRODUCTION

Rice (*Oryza sativa* L.) is one of the most widely consumed staple foods for a large part of the world's human population, especially in Asia and African countries. In these countries, there is a very high consumption of rice that annually exceeds 100 kg/capita (Seck et al., 2012). Among the biotic stresses, plant–parasitic nematodes are one of the major constraints in the rice production (Soriano et al., 1999) causing 10–25% yield losses (Bridge et al., 2005). Among the different plant–parasitic nematodes, *Meloidogyne* spp., belongs to a group of root-knot nematodes (RKNs), is represented by over 90 species that have been described so far (Moens et al., 2009). *M. graminicola* is one of the most important species of RKN associated with root of rice crop (De Waele & Elsen, 2007). This nematode species is an obligate sedentary endoparasite that settles in roots and completes their life cycle taking nutrition by feeding on host cells (Williamson & Gleason, 2003) and causes extensive damage to plant growth and yield of rice. The RKN causes the formation of galls on the rice roots. After penetrating the root elongation zone and migrating intercellularly toward the root tip, RKNs enter the vascular cylinder, where they puncture the cell wall with their stylet and inject secretions from their pharyngeal glands into the plant cell to induce a permanent feeding site known as giant cells (Davis et al., 2000; Gheysen & Mitchum, 2009).

The RKN, *M. graminicola*, is the most widely distributed serious pest of rice (*O. sativa* L.) in the subtropics and tropics and is considered economically important in all rice ecosystems. It has been reported from all rice-growing regions in India, Burma, Bangladesh, Thailand, Vietnam, China, Philippines, and USA both on upland and lowland deepwater rice (Pankaj et al., 2010). *M. graminicola* was first reported in 1965 on grasses (*Echinochloa colonum, Poa annua, Alopecurus carolinianus, Eleusine indica*) and oats (Golden & Birchfield, 1965) from USA in the states of Louisiana Georgia, and recently in Florida on sandbur (*Cenchrus* spp.) (Brito et al., 2004; Handoo et al., 2003; Minton et al., 1987; Power et al., 2005). Later, this nematode was described as *M. graminicola* (Golden & Birchfield, 1968). In India, RKN *M. graminicola* for the first time was reported from Orissa (Patnaik, 1969). *M. graminicola* is equally prevalent on upland (rainfed) or lowland (irrigated) as on deep-water rice. According to Bridge et al. (2005), annual yield losses, worldwide, of rice crop due to plant parasitic nematodes is estimated to range from 10% to 25%. Yield reduction by *M. graminicola* becomes more severe when the soil is alternatingly dry and flooded under rain-fed conditions; hence, water-management practice affects the damaging potential of *M. graminicola* (Prot & Matias, 1995; Tandingan et al., 1996). In India, *M. graminicola* is the dominant species infecting rice. Sudden outbreak of *M. graminicola* infestation in 1500 ha area in Mandya (Karnataka, India) during kharif, 2001 stands as an example for our limited understanding of this nematode (Prasad & Varaprasad, 2001).

7.2   THE PATHOGEN

The male and female of RKNs are easily distinguishable morphologically. The males are wormlike and about 1.2–1.5-mm long by 30–36 μm *in diameter.* Adult females appear to be pear-shaped to spheroid with elongated neck and about 0.40–1.30-mm long by 0.27–0.75-mm wide, which is usually embedded in root tissue (Agrios, 2005). Their body does not transform into a cyst-like structure. Females have six large unicellular rectal glands in the posterior part of the body, which excrete a gelatinous matrix to form an egg sac, in which many eggs are deposited. The stylet is mostly 9–18-μm long with three small, prominent, dorsally curved basal knobs (Dutta et al, 2012). The esophageal glands overlap the anterior end of the intestine. The females have two ovaries that fill most of the swollen body cavity. The vulva is typically terminal with the anus, flush with or slightly raised from the body contour and surrounded by cuticular striae, which form a pattern of fine lines resembling human fingerprints called the perineal pattern. Infective second-stage juveniles are short (0.3–0.5 mm) and have a weak cephalic framework. The esophageal gland lobe overlaps the intestine ventrally. The tail tip tapers to a long, fine point with a long hyaline region (Dutta et al., 2012).

7.2.1 SYMPTOMS

Aboveground symptoms are retard plant growth, newly emerged leaves distorted and crinkled along the margin. The nematode cause unfilled spikelets, reduce tiller development, and cause chlorosis, stunting, and wilting symptoms (Agrios, 2005; Jaiswal et al., 2012) under upland and intermittently flooded conditions. The underground infectious symptoms include the development of hook, spindle, or club shaped galls on the root system of rice plants.

7.2.2 ETIOLOGY

M. graminicola is a damaging parasite on upland, lowland and deepwater rice. It is well adapted to flooded conditions and can survive in waterlogged soil as eggs in egg masses or as juveniles for long periods. Numbers of *M. graminicola* decline rapidly after 4 months, but some egg masses can remain viable for at least 14 months in waterlogged soil (Rao & Israel, 1973; Singh et al., 2003). *M. graminicola* can also survive in soil flooded to a depth of 1 m for at least 5 months. It cannot invade rice in flooded conditions but quickly invades when infested soils are drained. It can survive in roots of infected plants. It prefers soil moisture of 32%. It develops best in moisture of 20–30% and soil dryness at rice tillering and panicle initiation (Dutta et al., 2012). Its population increases with the growth of susceptible rice plants. The presence of relatively broad host range and many of the alternative vegetable crops that are grown during dry season are favorable for this nematode. A temperature of 22–29°C is suitable for the prevalence of the nematode (Rao & Israel, 1973; Singh et al., 2003).

7.2.3 HOST RANGE

M. graminicola, like many other species of *Meloidogyne*, has a wide host range. *M. graminicola* generally prefer cereal hosts but can also infect some dicotyledonous plants. Tomato cv. Rutgers was an experimental host for most of the root-knot species, but did not allow multiplication of root-knot species attacking cereals, like *M. graminicola*, *M. oryzae*, and *M. graminis* (Sasser & Triantaphyllou, 1977). However, *M. oryzae* multiplied well in Tomato cv. Money Maker (Maas et al., 1978) in contrast to *M. graminicola* that did not multiply in any of the tomato cultivars tested (Manser, 1971). *M. graminicola* galled roots and reproduced at higher rates than *M. incognita* on most of the *Trifolium* species evaluated. Although the two root-knot species varied in their ability to gall *Trifolium* species, the level of root galling indicated that most of the *Trifolium* species evaluated were susceptible to both RKN species (Windham & Pederson, 1992). Aside from the rice plant, it also prefers *E. colonum* (Golden & Birchfield, 1965). *Ranunculus pusillus*, *Cyperus compressus* L. (Yik & Birchfield, 1979), *Panicum miliaceum* L., *Pennisetum typhoides* (Burm. F) Stapf and C. E. Hubb and *Glycine max* (L.) Merr (Roy, 1978), *Echinochloa crusgalli*, *E. colona*, *E. indica*, *Paspalum sanguinola*, *Eclipta alba*, *Grangea madraspatensis*, *Phyllanthus urinaria*, *Fimbristylis miliacea*, *Blumea* sp., *Vandellia* sp., *Jussieua repens*, *Andropogon* sp., chilies, tomato, wheat, *Panicum* spp. (Rao et al., 1970), *Cyperus deformis* (Bajaj & Dabur, 2000), banana (Reversat & Soriano, 2002), and onion (Gergon et al., 2002).

7.2.4 LIFE CYCLE

The life cycle of *M. graminicola* varies considerably in different environments, ranging from a very short life cycle of 19 days at temperatures ranging from 22 to 29°C in Bangladesh (Bridge & Page, 1982) to up to 51 days in some regions in India (Rao & Israel, 1973). The nematode experiences four molts throughout its life cycle. The first molt takes place inside the egg and the motile second-stage juveniles (J_2) of the *M. graminicola* invades the rice root in the root elongation zone and moves toward the root tip, where it punctures some selected vascular cells with its stylet and injects pharyngeal secretions into it, ultimately leading to the reorganization of these cells into typical feeding structures called giant cells, from which the nematode feeds for the remainder of its sedentary (sessile and swollen females) life cycle

(Gheysen & Mitchum, 2011; Williamson & Hussey, 1996). At 1st day after infection, swelling of infected root tips can be observed. At 3rd days, hyperplasia and hypertrophy of the cells surrounding the giant cells result in the formation of terminal hook-like galls, a typical characteristic of the rice RKN (Bridge et al., 2005). After three molts, the females lay their eggs inside the galls, whereas most other RKNs deposit egg masses at the gall surface, and hatched juveniles can reinfect the same or adjacent roots. Up to 50 females can be found in a single gall, indicating that the level of infestation can be very high (Bridge et al., 2005). The post penetration incompatibility in resistant crops is usually associated with the suboptimal development of giant cells that fail to develop or develop only partially with limited hyperplasia and hypertrophy (Orion et al., 1980).

7.2.5 DIVERSITY AMONG THE ROOT-KNOT NEMATODE

RKN are worldwide in distribution and morphologically and genetically diverse. Using *SSU rDNA* analysis, De Ley et al. (2002) placed *M. incognita* a mitotic parthenogenetic species in clade I and *M. graminicola* a meiotic parthenogenetic species in clade III. These species of RKN have different modes of reproduction and are evolutionarily distant from each other (Triantaphyllou, 1985). *M. graminicola*, unlike other *Meloidogyne* spp., is remarkably well adapted to flooded conditions, enabling it to continue multiplying in the host tissues even when the roots are deep in water (De Waele & Elsen, 2007). Nevertheless, little is known about the molecular mechanisms that make a particular plant a host or a nonhost for a given plant parasitic nematode species. Some nonhost plants react to invading nematodes with an active hypersensitivity response possibly indicating the presence of a resistance gene, but in other plants preformed nematicidal metabolites such as alkaloids, phenolics, and sesquiterpenes as well as phytoalexins produced in response to invasion appear to play a role in nematode rejection. Nonhost plants may lack the genes for susceptibility required for production and maintenance of feeding site. Development of fundamental studies to understand what causes a plant to be a nonhost could contribute to the development of novel strategies to control plant's parasitic nematodes based on disruption of nematode behavior or adjustment of host response (Dutta et al., 2011).

7.3 MANAGEMENT OF ROOT-KNOT NEMATODE

7.3.1 CULTURAL MANAGEMENT

Root-knot can be controlled effectively in the green house with steam sterilization of the soil. In the field, the best control is obtained by continuous flooding, raising the rice seedlings in flooded soils and crop rotation. These practices will help prevent root invasion by the nematodes. Soil solarization, bare fallow period, and crop rotation with nonhost crops, namely, sweet potato, cowpea, sesamum, castor, sunflower, soybean, turnip, and cauliflower inhibit nematode development (Rao et al., 1984; Rao, 1985). Green manuring with marigold (*Tagetes* sp.) is also effective in lowering RKN populations because of its nematicidal properties (Gergon et al., 2001; Polthanee & Yamazaki, 1996). Prot et al. (1994) positively correlated the nitrogen concentration in roots with initial population and the number of juveniles of *M. graminicola* recovered from the roots. They observed that nitrogen application increased growth and yield whether plants were infested by the nematode or not. However, since the percent of yield loss remained approximately constant for a given initial population across the range of nitrogen quantities applied, nitrogen applications do not reduce the relative nematode effect. Soil amendments with decaffeinated tea waste or water hyacinth compost (300 or 600 g/4.5 kg soil) reduced RKN infestation and increased plant growth (Roy, 1976).

7.3.2 BIOLOGICAL MANAGEMENT

Antagonistic microorganisms have been studied to develop a biological alternative to chemical pesticides for the control of nematodes. The RKN can be controlled by treating nematode infested soil with endospores of the bacterium *Pasteuria penetrans*, species of *Bacillus* and *Pseudomonas* or *Burkholderia cepacia*. The spores of the fungus, *Dactylella oviparasitica*, *Trichoderma* spp., which parasitize the eggs of *M. graminicola* and in some experiments by treating infested soils with spores of the vesicular-arbuscular mycorrhizal fungi *Gigaspora* and *Glomus*. Application of colony forming units of nematode-trapping fungi *Arthrobotrys dactyloides*, *Dactylaria brochopaga* (Singh et al., 2007), and *Arthrobotrys oligospora* (Singh et al., 2012) in root-knot infested soil was found to reduce the number of galls, number of females, J_2 and number of egg sacs and improved the plant growth. The efficient strain of *P. fluorescens* in India was formulated as bionematicide for the Indian farmers (Rao et al., 2011).

7.4   RESISTANCE SOURCES AGAINST FOR ROOT-KNOT NEMATODES

The major genes for resistance already known to breeders are prime candidates for genetic breeding. However, lack of resistance to the nematode has been a major factor hindering the genetic improvement of cultivated rice. Several high-yielding cultivars of *O. sativa* have been screened, but not enough genetic variability has been found for resistance to the rice RKN (Soriano, 1995; Tandingan et al., 1996). *Oryza glaberrima*, a cultivated rice, exhibited a hypersensitive resistant reaction to *Meloidogyne jawmica* by forming necrotic tissues in invaded roots and consequently suppressing nematode development (Di Vito et al., 1996). Two individuals of the *O. longistaminata* accession WLO2 (WLO2-2 and WLO2-15) exhibited a strong resistance to the RKN, considering the significantly low nematode population density obtained compared with the susceptible check, UPLRi5 (Soriano et al., 1999).

Specific resistances to *Meloidogyne* spp. were identified in the African relative species like *O. glaberrima* (Diomande, 1984; Soriano et al., 1999) and progenies derived from interspecific crosses are currently being tested for nematode resistance.

Accessions with good resistance to *M. graminicola*, showing consistently low nematode reproduction compared with susceptible reference genotypes, have been found in *O. longistaminata* A. Chev. & Roehrich and *O. glaberrima* Steud. (Cabasan et al., 2012; Plowright et al., 1999; Soriano et al., 1999) but so far introgression of this resistance into Asian rice has not been very successful because the interspecific progenies do not express the same degree of resistance observed in African rice. In Asian rice, some resistance to *M. graminicola* has been reported (Jena & Rao, 1976; Prasad et al., 2006; Sharma-Poudyal et al., 2004; Yik & Birchfield, 1979); however, according to Bridge et al. (2005), only a few of these accessions are truly resistant and the majorities are in fact susceptible to *M. graminicola*. Srivastava et al. (2011) reported two rice cultivars Achhoo and Naggardhan against *M. graminicola*. Das et al. (2011) reported that wild rice *O. glaberrima* accessions CG 14 and TOG 5674 behaved as true resistant references and rice cultivars WAB 638-1 and IRAT 216, IR 81426-B-B-186-4, and IR81449-B-B-51-4 showed significant resistant reaction against *M. graminicola*.

In cultivated rice varieties two accessions Khao Pahk Maw (an *aus* from Thailand) and LD 24 (an *indica* from Sri Lanka) appeared to be resistant, which was confirmed in large pot experiments where no galls were observed. Detailed observations on these two accessions revealed no nematodes inside the roots 2 days after inoculation and very few females after 17 days (5 in Khao Pahk Maw and <1 in LD 24, in comparison with >100 in the susceptible controls). These two cultivars appear ideal donors for breeding RKN resistance (Dimkpa et al., 2016).

7.4.1 MORPHOLOGICAL BASIS OF RESISTANCE AS PER ROOT HISTOLOGY

Jena and Rao (1977) hypothesized that differences in root morphology may also explain differences in host response upon *M. graminicola* infection between susceptible and resistant rice genotypes. Further other also showed significant differences in some root morphology parameters among the rice genotypes (Cabasan et al., 2014). In the susceptible rice genotypes, root and stele diameters, and thickness of the cortex were greater than in the resistant rice genotypes as also observed (Jena & Rao, 1977). A greater root diameter may facilitate the penetration and migration of a larger number of nematodes. Furthermore, roots with a wider stele diameter will have more conditions for the establishment of more feeding sites and formation of giant cells near the vascular cylinder. The cell wall ingrowths of the giant cells usually come in contact with the xylem vessels to transport water and nutrients (Abad et al., 2009) but no difference was observed in xylem diameter between susceptible and resistant rice genotypes. The outer portion of the root is the first obstacle for the invading J2 but, in our study, the thickness of the outer part of the root was similar in the susceptible and resistant rice genotypes. Suberin and structural proteins in maize root cells were considered a good indication of the presence of a mechanical obstacle to the penetration of pathogens (Schreiber et al., 1999) but differences in suberization between susceptible and resistant rice genotypes were not observed.

7.4.2 RESISTANCE PROTEIN ISOLATED FROM RICE

Some cysteine proteinase inhibitor isolated from rice and their activity also reported in the RKNs. A cysteine proteinase inhibitor from rice, oryzacystatin I (OC-I), completely inhibited the proteolytic activity of all stages of

M. hapla. The tighter the enzyme–inhibitor complex, more effective is the inhibitor; however, OC-I did not bind with high affinity to the proteinases of *M. incognita* and *M. javanica.* Rice also produces OC-II, another cystatin which proved effective against these two nematode species, demonstrating a great specificity between particular plant inhibitors and specific nematode proteinases (Michaud et al., 1996; Vrain, 1999). Root-knot and sugar-beet cyst nematodes did not develop normally in transgenic *Arabidopsis* roots expressing the wild-rice inhibitor OC-I or the variant protein OC-IDD86. Female nematodes in transformed and untransformed roots develop initially at the same rate, but females in roots expressing the mutated protein remained smaller and produced very few eggs (Urwin et al., 1997).

7.4.3 BIOCHEMICAL AND MOLECULAR PATTERNS OF RICE AND THEIR RESPONSES AGAINST M. GRAMINICOLA

The *M. graminicola* life cycle in rice roots was recently investigated by histopathological analysis in several *O. sativa* and *O. glaberrima* rice varieties (Cabasan et al., 2012). Analysis of the molecular rice responses to *M. graminicola* infection showed that the hormone-mediated resistance signaling pathways controlled by salicylic acid (SA), jasmonic acid (JA), and ethylene (ET) are repressed soon after infection by *M. graminicola* (Nahar et al., 2011). The expression of HR-like reaction and/or late defense response in infected roots could indicate that multiple genes are likely to be involved in the resistance of *O. glaberrima* genotypes to *M. graminicola* such as in pepper with Me3 gene controlling HR-like response and Me1 gene respon-sible for blocking nematode development (Castagnone-Sereno et al., 1996). RKN species are able to suppress rice basal defense genes expression at early stages after infection and hypothesized that RKN repress the transcrip-tion of key immune regulators in order to lower the basal defense.

Nahar et al. (2011) and Ji et al. (2013) who reported that rice cv. Nippon-bare defense genes expression was repressed in roots and giant cells early upon infection with *M. graminicola.* The mRNA levels of four immune-related genes tested (*OsWRKY45, OsPR1b, OsEin2b,* and *JiOsPR10*) were significantly attenuated in gall tissues (Nahar et al., 2011). The JA/ET signaling pathway seems to play a major role in basal defense to nema-todes (Bhattarai et al. 2008; Nahar et al., 2011). When exogenous ethephon and methyl jasmonate were supplied to rice plants, *M. graminicola* was less effective in counteracting root defense pathways. Here, the OsAOS2 gene, which codes the first enzyme for JA biosynthesis pathway, was not induced in response to nematode infection at the time studied. This suggests that the plant did not perceive nematode attack or that proper signalization was impaired (Nguyen et al., 2014). Ji et al. (2015) reported that the nonprotein amino acid beta aminobutyric acid can induce defense against RKNs in rice independent of the JA and ET pathways, but rather acts through activation of lignin and callose production.

Ji et al. (2013) studied transcriptional changes in rice after at 7 days of inoculation in genes encoding histone-modifying and small-RNA-processing enzymes confirm a role for epigenetic processes in transcriptional reprogramming of the root cells to form nematode feeding sites. Nguyen et al. (2014) reported that *M. incognita* expressed the calreticulin gene (*Mi-crt*) in infected rice roots and that several rice defense genes expression are downregulated at an early stage of infection when the nematode starts feeding from root cells. A series of genes involved in the rice immune responses were selected, including those from signaling, SA- and JA-dependent resistance signaling pathways (Delteil et al., 2010). Nguyen et al. (2014) studied different gene to study the rice and *Meloidogyne incognita*, that is, *OsMAPK6*, *OsMAPK5a*, and *OsMAPK20* for early signaling (phosphorylation cascades in PTI and ETI), *OsAOS2* (JA pathway), *OsEDS1*, and *OsPAD4* (SA-dependent resistance), *OsRAC1* (oxidative burst), *OsNIH1*, and *OsWRKY13* (positive transcriptional regulators of defense genes). During the infection process, plant–pathogenic nematodes cause major changes in plant gene expression; nematode-induced transcriptome changes have been identified by microarray analysis. At 7 days, the galls have matured, and giant cells within the gall tissue are fully developed. Nematodes developing within these galls are now at the J3 or J4 stage. In 7 days after inoculation galls, genes involved in "transcription," "posttranslational" protein modification (Engler et al., 1999).

7.5 BIOTECHNOLOGICAL APPROACHES FOR ROOT-KNOT NEMATODES

7.5.1 IMPORTANT QTLS IDENTIFIED IN RICE AGAINST ROOT-KNOT NEMATODES

Applying plant molecular genetics has resulted in the identification of several nematode resistance (R) genes or quantitative trait loci (QTLs) for resistance to sedentary endoparasitic nematodes, and some genes mapped to chromosomal locations or linkage maps (Ganal et al., 1995; Messeguer et al., 1991). A few have been cloned (Thurau et al., 2010; Veremis & Roberts, 2000). One of the best characterized and commercially used RKN resistance genes is *Mi-1.2*, which is found in the wild relative of tomato (*Lycopersicon peruvianum* complex) and confers resistance to several *Meloidogyne* species (Veremis & Roberts, 2000). In African rice, the major gene *Hsa-1Og* confers resistance to the cyst nematode *Heterodera sacchari* and has been mapped on chromosome 11 of an *O. sativa* × *O. glaberrima* interspecific cross (Lorieux et al., 2003). In Asian rice, QTLs for partial resistance to *M. graminicola* were identified on chromosomes 1, 2, 6, 7, 9, and 11 using F6 recombinant inbred lines of a Bala × Azucena *O. sativa* mapping population (Shrestha et al., 2007).

Dimkpa et al. (2016) examined the root galling of 332 accessions of the RDP1 panel 2 weeks after inoculation with J_2 of *M. graminicola* to evaluate the host response of these genotypes, to identify resistant genotypes, and to highlight potential QTLs and candidate genes worthy of further characterization. Additional experiments with a small number of accessions were carried out to examine if observations were reproducible across differing experimental conditions. A genome-wide association study revealed 11 QTLs, 2 of which are close to epistatic loci detected in the Bala × Azucena population. The discussion highlights a small number of candidate genes worth exploring further, in particular many genes with lectin domains and genes on chromosome 11 with homology to the *Hordeum Mla* locus.

Shrestha et al. (2007) reported QTLs for partial resistance to *M. graminicola* in *O. sativa* and resistance in the *O. glaberrima* accession CG14. Lorieux et al. (2003) reported resistance to *Meloidogyne* spp. and the cyst nematode *Heterodera sacchari* in varieties of the African rice cultivated species *O. glaberrima*. Effort has been made in the past to transfer the *Hsa-1Og* gene from rice *O. glaberrima* chromosome 11 to *O. sativa* through interspecific crossing but the interspecific progenies tested so far have not been able to express the same degree of resistance (Lorieux et al., 2003; Plowright et al., 1999).

The *Hsa-1*^{Og} gene is linked to the markers located on the long arm of chromosome 11 (Lorieux *et al.,* 2003), whereas our QTL is located on the short arm of the chromosome. The ability of the gene to confer resistance to RKNs has not been tested, but reliance on major resistances can select for virulent populations, ultimately leading to loss of resistance and has previously been seen for nematode pests (Roberts, 2002).

7.5.2   RNAI AND TRANSGENIC STUDY AGAINST ROOT-KNOT NEMATODE

There are several reports of in-vitro RNAi strategies for the down-regulation of targeted specific genes with corresponding phenotypic changes such as decline in parasitic success, reduction in fecundity, motility inhibition, reduced host location ability, and decline in penetration and reproduction in host roots against the nematodes. These successful results have been demonstrated successfully in root-knot, cyst, lesion, pine wilt, burrowing, and white-tip nematodes (Tan et al., 2013). Huang et al. (2006) utilized in-vitro RNAi approaches to silence the parasitism gene *16D10* in RKN and validate that the parasitism gene has an essential role in RKN parasitism of plants. Ingestion of *16D10* dsRNA *in vitro* silenced *16D10* in RKN and resulted in reduced nematode infectivity. Banakar et al. (2015) validated RNAi construct against *Meloidogyne incognita* with two FMRFamide-like peptide genes, flp-14 and flp-18, and a subventral pharyngeal gland specific gene, 16D10. RNAi-silencing construct further reduced the attraction of *M. incognita* at different time intervals both in combination under in-vitro condition. Silencing of the genes reduced nematode infection by 23–30% and development as indicated by a reduction in the number of females by 26–62%. Reproduction was decreased by 27–73% and fecundity was decreased by 19–51%. In-situ hybridization revealed the expression of flp-18 in cells associated with the ventral and retrovesicular ganglia of the central nervous system. qRT-PCR supported the correlation between phenotypic effects of silencing with that of transcript quantification. Niu et al. (2012) validated the potential of *Mi-Rpn7* as a target for controlling RKN *Meloidogyne incognita* and evaluated the feasibility of modified platform for the assessment of silencing phenotypes. They knocked down the *Rpn7* gene of *M. incognita* using RNAi *in vitro* and *in vivo*. After soaking with 408-bp *Rpn7* dsRNA, pre-parasitic second-stage juvenile (J2) nematodes showed specific transcript knockdown, resulting in an interrupted locomotion in an attraction assay with Pluronic gel medium, and consequently in a reduction of nematode infection ranging from 55.2% to 66.5%. With in-vivo expression of *Rpn7*dsRNA in transformed composite plants, the amount of egg mass per gram root tissue was reduced by 34% ($P < 0.05$) and the number of eggs per gram root tissue was reduced by 50.8% ($P < 0.05$). Our results demonstrated that the silencing of the *Rpn7* gene in *M. incognita* J_2s significantly reduced motility and infectivity. Although it does not confer complete resistance, *Mi-Rpn7* RNAi in hairy roots produced significant negative impacts on reproduction and motility of *M. incognita*. Papolu et al. (2013) demonstrated the significance of two FMRFamide like peptide genes (*flp-14* and *flp-18*) for infection and development of resistance to *M. incognita* through host-derived RNAi. The study demonstrated both in-vitro and in-planta validation of RNAi-induced silencing of the two genes cloned from J$_2$ stage of *M. incognita*. In-vitro silencing of both the genes interfered with nematode migration toward the host roots and subsequent invasion into the roots. Transgenic tobacco lines were developed with RNAi constructs of *flp-14* and *flp-18* and evaluated against *M. incognita*. The transformed plants did not show any visible phenotypic variations suggesting the absence of any off-target effects. Bioefficacy studies with deliberate challenging of *M. incognita* resulted in 50–80% reduction in infection and multiplication confirming the silencing effect. We have provided evidence for in-vitro and in-planta silencing of the genes by expression analysis using qRT-PCR.

Vain et al. (1998) have engineered rice with a gene-encoding oryzacys-tatin (OCIDD86). Transformed plant expressed with a proteinase inhibitor up to 0.2% of the total soluble protein. This resulted in a significant 55% reduction in egg production by *M. incognita*. Cystine PIs are likely to pose a valuable class of antinematode genes, especially since cystine-proteinase inhibitor expressed in transgenic potato, provided resistance against *Globodera pallida* under field conditions. The advantage of these strategies is that they are preformed defense like α-terthienyl and do not require any nematode infection for induced resistance.

7.6 CONCLUSION

M. graminicola causes considerable reduction in the yield of rice crop. The lack of resistance to this disease has been a major factor hindering the genetic improvement of cultivated rice. There is no true sources of resistance has been identified among cultivated rice spp.; however, wild rice sp. *O. glaberrima* accessions CG 14 and TOG 5674 behaved as high level of resistance (true resistance). Cultivated rice varieties Khao Pahk Maw, LD 24 appeared to be resistant and WAB 638-1, IRAT 216, IR 81426-B-B-186-4, and IR81449-B-B-51-4 showed significant resistant reaction against *M. graminicola*. Very little resistance to *M. graminicola* in Asian rice has been reported. Due to lack of highly resistance in cultivated rice in Asia, other methods could be used for management of root-knot disease. For the management of this disease, one of the most common methods is used of nematicides. These nematicides cause serious health hazards to human being and also killed beneficial earth worms, microbes, and fungi which suppress the nematodes from the soil. Hence, nowadays more emphasis are given on other methods for management of this disease like growing resistant varieties, agronomical practices, and bio-agents because they are eco-friendly and safe.

KEYWORDS

- **nematodes**
- **disease**
- **yield loss**
- **endoparasite**
- **biological control**

REFERENCES

Abad, P.; Castagnone-Sereno, P.; Rosso, M. N.; de Almeida Engler, J.; Favery, B. Invasion, Feeding and Development. In: *Root-knot Nematodes*; Perry, R. N., Moens, M., Starr, J. L., Eds.; CABI Publishing: Wallingford, UK, 2009; pp 162–181.

Agrios, G. N. *Plant Pathology*, 5th ed.; Elsevier, Academic Press: UK, 2005; pp 838–841.

Bajaj, H. K.; & Dabur, K. R. *Cyperus deformis*, a New Host Record of Rice Root-knot Nematode, *Meloidogyne graminicola. Ind. J. Nematol.* **2000**, *30*, 256.

Banakar, P.; Sharma, A.; Lilley, C. J.; Gantasala, N. P.; Kumar, M.; Rao, U. Bottom Combinatorial In Vitro RNAi of Two Neuropeptide Genes and a Pharyngeal Gland Gene on *Meloidogyne incognita. Nematology* **2015**, *17*(2), 155–167.

Bhattarai, K. K.; Xie, Q. G.; Mantelin, S.; Bishnoi, U.; Girke, T.; Navarre, D. A.; Kaloshian, I. Tomato Susceptibility to Root-knot Nematodes Requires an Intact Jasmonic Acid Signaling Pathway. *Mol. Pl.–Microbe Interact.* **2008**, *21*(9), 1205–1214.

Bridge, J.; Page, S. L. J. The Rice Root-knot Nematode, *Meloidogyne graminicola*, on Deep Water Rice (*Oryza sativa* subsp. *indica*). *Rev. Dev. Nematol.* **1982**, *5*, 225–232.

Bridge, J.; Plowright, R. A.; Peng, D. Nematode Parasites of Rice. In: *Plant–parasitic Nematodes in Subtropical and Tropical Agriculture*; Luc, M., Sikora, R. A., Bridge, J., Eds.; CAB International: Wallingford, 2005; pp 87–130.

Brito, J. A.; Centinas, R.; Powers, T. O.; Inserra, R. N.; McAvoy, E. J.; Mendes, M. L.; Crow, T. W.; Dickson, D. W. Identification and Host Preference of *Meloidogyne mayaguensis* and Other Root-knot Nematodes from Florida and their Susceptibility to Pasteuria penetrants. *J. Nematol.* **2004**, *36*, 308–309.

Cabasan, M. T. N.; Kumar, A.; De Waele, D. Comparison of Migration, Penetration, Development and Reproduction of *Meloidogyne graminicola* on Susceptible and Resistant Rice Genotypes. *Nematology* **2012**, *14*, 405–415.

Cabasan, T. N. M.; Kumar, A.; Bellafiore, S.; Waele, D. D. Histopathology of the Rice Root-knot Nematode, *Meloidogyne graminicola*, on *Oryza sativa* and *O. glaberrima*. *Nematology* **2014**, *16*, 73–81.

Castagnone-Sereno, E.; Bongiovanni, M.; Palloix, A.; Dalmasso, A. Selection for Meloidogyne Incognita Virulence against Resistance Genes from Tomato and Pepper and Specificity of the Virulence/Resistance Determinants. *Eur. J. Pl. Pathol.* **1996**, *102*, 585–590.

Das, K.; Zhao, D.; Waele, D. D.; Tiwari, R. K. S.; Shrivastava, D. K.; Kumar, A. Reactions of Traditional Upland and Aerobic Rice Genotypes to Rice Root Knot Nematode (*Meloidogyne graminicola*). *J. Pl. Breed. Crop Sci.* **2011**, *3*, 131–137.

Davis, E. L.; Hussey, R. S.; Baum, T. J.; Bakker, J.; Schots, A.; Rosso, M. N.; Abad, P. Nematode Parasitism Genes. *Ann. Rev. Phytopathol.* **2000**, *38*, 365–396.

De Ley, I. T.; De Ley, P.; Vierstraete, A.; Karssen, G.; Moens, M.; Vanfleteren, J. Phylogenetic Analyses of *Meloidogyne* Small Subunit rDNA. *J. Nematol.* **2002**, *34*, 319–327.

De Waele, D.; Elsen, A. Challenges in Tropical Plant Nematology. *Annu. Rev. Phytopathol.* **2007**, *45*, 457–485.

Delteil, A.; Zhang, J.; Lessard, P.; Morel, J. B. Potential Candidate Genes for Improving Rice Disease Resistance. *Rice* **2010**, *3*, 56–71.

Di Vito, M.; Vovlsan, N.; Lamberti, F.; Zaccheog, G.; Catalano, F.; Di Vito, M. Pathogenicity of *Meloidogyne javanica* on Asian and African rice. *Nematol. Mediterr.* **1996**, *24*, 95–99.

Dimkpa, S. O. N.; Lahari, Z.; Shrestha, R.; Douglas, A.; Gheysen, G.; Price, A. H. A Genome-wide Association Study of a Global Rice Panel Reveals Resistance in *Oryza sativa* to Root-knot Nematodes. *J. Exp. Bot.* **2016**, *67(4), 1191–200.*

Diomande, M. Response of Upland Rice Cultivars to *Meloidogyne* Species. *Rev. Nematol.* **1984**, *7*, 57–63.

Dutta, T. K.; Ganguly, A. K.; Gaur, H. S. Global Status of Rice Root-knot Nematode, *Meloidogyne graminicola*. *Afr. J. Microbiol. Res.* **2012**, *6*, 6016–6021.

Dutta, T. K.; Powers, S. J.; Kerry, B. R.; Gaur, H. S.; Curtis, R. H. C. Comparison of Host Recognition, Invasion, Development and Reproduction of *Meloidogyne graminicola* and *M. incognita* on Rice and Tomato. *J. Nematol.* **2011**, *13*, 509–520.

Engler, J. D.; De-Vleesschauwer, V.; Burssens, S.; Celenza, J. L.; Inze, D.; Van-Montagu, M.; Engler, G.; Gheysen, G. Molecular Markers and Cell Cycle Inhibitors Show the Importance of Cell Cycle Progression in Nematode-induced Galls and Syncytia. *Pl. Cell* **1999**, *11*, 793–807.

Ganal, M. W.; Simon, R.; Brommonschenkel, S.; Arndt, M.; Phillips, M. S.; Tanksley, S. D.; Kumar, A. Genetic Mapping of a Wide Spectrum Nematode Resistance Gene (Hero) against *Globodera rostochiensis* in Tomato. *Mol. Pl. Microb.* **1995**, *8*, 886–891.

Gergon, E. B.; Miller, S. A.; Davide, R. G.; Opina, O. S.; Obien, S. R. Evaluation of Cultural Practices (Surface Burning, Deep Ploughing, Organic Amendments) for Management of Rice Root-knot Nematode in Rice—Onion Cropping System and their Effect on Onion (*Allium cepa* L.) Yield. *Int. J. Pest Manage.* **2001**, *47*, 265–272.

Gergon, E. B.; Miller, S. A.; Halbrendt, J. M.; Davide, R. G. Effect of Rice Root-knot Nematode on Growth and Yield of Yellow Granex Onion. *Pl. Dis.* **2002**, *86*(12), 1339–1344.

Gheysen, G.; Mitchum, M. G. How Nematodes Manipulate Plant Development Pathways for Infection. *Curr. Opin. Pl. Biol.* **2011**, *14*, 415–421.

Gheysen, G.; Mitchum, M. G. Molecular Insights in the Susceptible Plant Response to Nematode Infection. *Pl. Cell Monogram.* **2009**, *15*, 45–81.

Golden, A. M.; Birchfield, W. *Meloidogyne graminicola* (Heteroderidae), a New Species of Root-knot Nematode from Grass. *Proc. Helminthol. Soc. Wash.* **1995**, *32*, 228–231.

Golden, A. M.; Birchfield, W. Rice Rootknot Nematode *(Meloidogyne graminicola)* as a New Pest of Rice. *Pl. Dis. Reptr.* **1968,** *2*, 423.

Handoo, Z. A.; Klassen, W.; Abdul-Baki, A.; Bryan, H. H.; Wang, Q. First Record of Rice-root Nematode *(Meloidogyne graminicola)* in Florida. *J. Nematol.* **2003,** *35*(3), 342.

Huang, G.; Allen, R.; Davis, E. L.; Baum, T. J.; Hussey, R. S. Engineering Broad Root-knot Resistance in Transgenic Plants by RNAi Silencing of a Conserved and Essential Root-knot Nematode Parasitism Gene. *Proc. Natl. Acad. Sci. U.S.A.* **2006,** *103*(39), 14302–14306.

Jaiswal, R. K.; Kumar, D.; Singh, K. P. Relationship between Growth of Rice Seedlings and Time of Infection with *Meloidogyne graminicola. Libyan Agric. Res. Center J. Int.* **2012,** *3*, 13–17.

Jena, R. N.; Rao, Y. S. Nature of Resistance in Rice *(Oryza sativa* L.) to the Root Knot Nematode *(Meloidogyne graminicola).* II. Histopathology of Nematode Infection in Rice Varieties. *Proc. Ind. Acad. Sci.* **1977,** *86*, 87–91.

Jena, R. N.; Rao, Y. S. Nature of Root-knot *(Meloidogyne graminicola)* Resistance in Rice *(Oryza sativa).* I. Isolation of Resistant Genotypes. *Proc. Indian Acad. Sci.* **1976,** *83*, 177–184.

Ji, H.; Gheysen, G.; Denil, S.; Lindsey, K.; Topping, J. F.; Nahar, K.; Kyndt, T. Transcriptional Analysis through RNA Sequencing of Giant Cells Induced by *Meloidogyne graminicola* in Rice Roots. *J. Exp. Bot.* **2013,** *64*(12), 3885–3898.

Ji, H.; Kyndt, T.; He, W.; Vanholme, B.; Gheysen, G. β-Aminobutyric Acid-induced Resistance against Root-knot Nematodes in Rice is Based on Increased Basal Defense. *Mol. Pl.–Microbe Interact.* **2015,** *28*, 519–533.

Lorieux, M.; Reversat, G.; Diaz, S. X. G.; Denance, C.; Jouvenet, N.; Orieux, Y.; Bourger, N.; Pando-Bahuon, A.; Ghesquiere, A. Linkage Mapping of *Hsa-1Og*, a Resistance Gene of African Rice to the Cyst Nematode, *Heterodera sacchari. Theor. Appl. Genet.* **2003,** *107*, 691–696.

Maas, P. W.; Sanders, H.; Dede, J. *Meloidogyne oryzae* n. sp. (Nematoda, Meloidogynidae) Infesting Irrigated Rice in Surinam (South America). *Nematology* **1978,** *24*, 305–312.

Manser, D. P. Notes on the Rice Root-knot Nematode in Laos. *FAO Pl. Protect. Bull.* **1971,** *19*, 136–139.

Messeguer, R.; Ganal, M.; de Vicente, M. C.; Young, N. D.; Bolkan, H.; Tanksley, S. D. High Resolution RFLP Map Around the Root Knot Nematode Resistance Gene *(Mi)* in Tomato. *Theor. Appl. Genet.* **1991,** *82*, 529–536.

Michaud, D.; Nguyen-Quoc, B.; Vrain, T. C.; Fong, D.; Yelle, S. Response of Digestive Cysteine Proteinases from the Colorado Potato Beetle *(Leptinotarsa decemlineata)* and the Black Vine Weevil *(Otiorynchus sulcatus)* to a Recombinant Form of Human Stefin A. *Arch. Insect Biochem. Physiol.* **1996,** *31*, 451–464.

Minton, N. A.; Tukcker, E. T.; Golden, A. M. First Report of *Meloidogyne graminicola* in Georgia. *Pl. Dis.* **1987,** *71*, 376.

Moens, M.; Perry, R. N.; Starr, J. L. *Meloidogyne* species—A Diverse Group of Novel and Important Plant Parasites. In: *Root-knot Nematodes*; Perry, R. N., Moens M., Starr J. L., Eds.; CABI International: Cambridge, MA **2009**; pp 1–17.

Nahar, K.; Kyndt, T.; De Vleesschauwer, D.; Höfte, M.; Gheysen, G. The Jasmonate Pathway is a Key Player in Systemically Induced Defense against Root Knot Nematodes in Rice. *Pl. Physiol.* **2011,** *157*(1), 305–316.

Nguyen, P. U.; Bellafiore, S.; Petitot, A. S.; Haidar, R.; Bak, A.; Abed, A.; Gantet, P.; Mezzalira, I.; Engler, J. A.; Fernandez, D. *Meloidogyne incognita*-rice *(Oryza sativa)* Interaction:

A New Model System to Study Plant–Root-knot Nematode Interactions in Monocotyledons. *Rice* **2014**, *7*, 23.

Niu, J.; Jian, H.; Xu, J.; Chen, C.; Guo, Q.; Liu, Q.; Guo, Y. RNAi Silencing of the *Meloidogyne incognita Rpn7* Gene Reduces Nematode Parasitic Success. *Eur. J. Pl. Pathol.* **2012**, *134*(1), 131–144.

Orion, D.; Wergin, W. P.; Endo, B. Y. Inhibition of Syncytia Formation and Root-knot Nematode Development on Cultures of Excised Tomato Roots. *J. Nematol.* **1980**, *12*, 196–203.

Pankaj, Sharma, H. K.; Prasad, J. S. The Rice Root-knot Nematode, *Meloidogyne graminicola*: An Emerging Problem in Rice–Wheat Cropping System. *Ind. J. Nematol.* **2010**, *40*, 1–11.

Papolu, P. K.; Gantasala, N. P.; Kamaraju, D.; Banakar, P.; Sreevathsa, R.; Rao, U. Utility of Host Delivered RNAi of Two FMRF Amide Like Peptides, *flp-14* and *flp-18*, for the Management of Root Knot Nematode, *Meloidogyne incognita. PLoS One*, **2013**, *8*(11), e80603.

Patnaik, N. C. *All India Nematology Symposium*, August, 21–21, 1967, New Delhi. *Phytopathology* **1969**, *26*, 348–349.

Plowright, R. A.; Coyne, D. L.; Nash, P.; Jones, M. P. Resistance to the Rice Nematodes *Heterodera sacchari, Meloidogyne graminicola* and *M. incognita* in *Oryza glaberrima* and *O. glaberrima* × *O. sativa* Hybrids. *Nematology* **1999**, *1*, 745–751.

Polthanee, A.; Yamazaki, K. Effect of Marigold (*Tagetes patula* L.) on Parasitic Nematodes of Rice in Northeast Thailand. *Kaen. Kaset. Khon. Kaen. Agr. J.* **1996**, *24*(3), 105–107.

Power, T. O.; Mullin, P. G.; Harris, T. S.; Sutton, L. A.; Higgins, R. S. Incorporating Molecular Identification of *Meloidogyne* spp. Into a Large-scale Regional Survey. *J. Nematol.* **2005**, *37*(2), 226–235.

Prasad, J. S.; Varaprasad, K. S. *Ufra* Nematode, *Ditylenchus angustus* is Seed Borne. *Crop Protec.* **2001**, *21*(1), 75–76.

Prasad, J. S.; Vijayakumar, C. H. M.; Sankar, M.; Varaprasad, K. S.; Srinivasa, P. M.; Kondala, R. Y. Root-knot Nematode Resistance in Advanced Back Cross Populations of Rice Developed for Water Stress Conditions. *Nematol. Medit.* **2006**, *34*, 3–8.

Prot, J. C.; Matias, D. Effects of Water Regime on the Distribution of *Meloidogyne graminicola* and Other Root-parasitic Nematodes in a Rice Field Toposequence and Pathogenicity of *M. graminicola* on Rice Cultivar UPL R15. *Nematologica* **1995**, *41*, 219–228.

Prot, J. C. The Combination of Nematodes, *Sesbania rostrata*, and Rice: The Two Sides of the Coin. *Int. Rice Res. Notes* **1994**, *19*(3), 30–31.

Rao, M. S.; Reddy, M. S.; Wang, Q.; Li, Y.; Zhang, L.; Du, B.; Yellareddygari, S. K. R. Potential of PGPR in the Management of Nematodes and as Bionematicides Research Initiatives—National and International. In: *Proceedings of the 2nd Asian PGPR Conference*, 21–24 August, 2011, Beijing, China, 2011; pp 216–224.

Rao, Y. S.; Israel, P. Life History and Bionomics of *Meloidogyne graminicola*, the Rice Root-knot Nematode. *Ind. Phytopathol.* **1973**, *26*, 333–340.

Rao, Y. S.; Israel, P.; Biswas, H. Weed and Rotation Crop Plants as Hosts for the Rice Root-knot Nematode, *Meliodogyne graminicola* (Golden and Birchfield). *Oryza* **1970**, *7*(2), 137–142.

Rao, Y. S.; Prasad, J. S.; Rao, A. V. S. Interaction of Cyst and Root-knot Nematodes in Roots of Rice. *Rev. Nematol.* **1984**, *7*(2), 117–120.

Rao, Y. S. *Research on Rice Nematodes*. In: *Rice in India. ICAR Monograph*; Padmanabhan, S. Y., Ed.; 1985, Chapter 21; pp 591–561.

Reversat, G.; Soriano, I. The Potential Role of Bananas in Spreading Rice Root-knot Nematode, *Meloidogyne graminicola. Int. Rice Res. Notes* **2002**, *27*(2), 23–24.

Roberts, P. A. Concepts and Consequences of Resistance. In: *Plant Resistance to Parasitic Nematodes*; Starr, J. L., Cook, R., Bridge, J., Eds.; CABI Publishing: Wallingford, UK. 2002; pp 23–41.

Roy, A. K. Effect of Decaffeinated Tea Waste and Water Hyacinth Compost on the Control of *Meloidogyne graminicola* on Rice. *Ind. J. Nematol.* **1976**, *6*, 73–77.

Roy, A. K. Host Suitability of Some Crops to *Meloidogyne graminicola. Ind. Phytopathol.* **1978**, *30*, 483–485.

Sasser, J. N.; Triantaphyllou, A. C. Identification of *Meloidogyne* Species and Races. *J. Nematol.* **1977**, *9*, 283.

Schreiber, L.; Hartmann, K.; Skrabs, M.; Jurgen, Z. Apoplastic Barriers in Roots: Chemical Composition of Endodermal and Hypodermal Cell Walls. *J. Exp. Bot.* **1999**, *50*, 1267–1280.

Seck, P. A.; Diagne, A.; Mohanty, S.; Wopereis, M. C. S. Crops that Feed the World 7: Rice. *Food Sec.* **2012**, *4*, 7–24.

Sharma-Poudyal, D.; Pokharel, R. R.; Shrestha, S. M.; Khatri-Chhetri, G. B. Evaluation of Common Nepalese Rice Cultivars against Rice Root Knot Nematode. *Nepal Agric. Res. J.* **2004**, *5*, 33–36.

Shrestha, R.; Uzzo, F.; Wilson, M. J.; Price, A. H. Physiological and Genetic Mapping Study of Tolerance to Root-knot Nematode in Rice. *New Phytologist* **2007**, *176*, 665–672.

Singh, I.; Gaur, H. S.; Briar. S. K.; Sharma, S. K.; Sakhuja, P. K. Role of Wheat in Sustaining *Meloidogyne graminicola* in Rice–Wheat Cropping System. *Int. J. Nematol.* **2003**, *13*, 79–86.

Singh, K. P.; Jaiswal, R. K.; Kumar, N.; Kumar, D. Nematophagous Fungi Associated with Root Galls of Rice Caused by *Meloidogyne graminicola* and its Control by *Arthrobotrys dactyloides* and *Dactylaria brochopaga. J. Phytopathol.* **2007**, *155*, 193–197.

Singh, U. B.; Sahu, A.; Singh, R. K.; Singh, D. P.; Meena, K. K.; Srivastava, J. S.; Renu, Manna, M. C. Evaluation of Biocontrol Potential of *Arthrobotrys oligospora* against *Meloidogyne graminicola* and *Rhizoctonia solani* in Rice (*Oryza sativa* L.). *Biol. Control.* **2012**, *60*, 262–270.

Soriano, I. R.; Schmidt, V.; Brar, D.; Prot, J. C.; Reversat, G. Resistance to Rice Rootknot Nematode *Meloidogyne graminicola* Identified in *Oryza longistaminata* and *O. glaberrima. Nematology* **1999**, *1*(4), 395–398.

Sorianoi, R. Identification of Five Root-knot Nematode (*Meloidogyne* spp.) Lines and Comparison of their Effects on the Growth and Yield of Selected Rice Cultivars Grown under Different Water Regimes and Soil Types. Master of Science Thesis, University of the Philippines at Los Banos: College, Laguna, 1995; p 189.

Srivastava, A.; Rana, V.; Rana, S.; Singh, D.; Singh, V. Screening of Rice and Wheat Cultivars for Resistance against Root-knot Nematode, *Meloidogyne graminicola* (Golden and Birchfield) in Rice–Wheat Cropping System. *J. Rice Res.* **2011**, *4*(2), 8–19.

Tan, J. C. H.; Jones, M. G. K.; Fuso-Nyarko, J. Gene Silencing in Root Lesion Nematodes (*Pratylenchus* spp.) Significantly Reduces Reproduction in a Plant Host. *Exp. Parasitol.* **2013**, *133*, 166–178.

Tandingan, I.; Prot, J. C.; Davide, R. Influence of Water Management on Tolerance of Rice Cultivars for *Meloidogyne graminicola. Fund. Appl. Nematol.* **1996**, *19*, 189–192.

Thurau, T.; Ye, W.; Cai, D. Insect and Nematode Resistance. In: *Modification of Plants: Genetic, Biotechnology in Agriculture and Forestry*; Kempken, F., Jung, C., Eds.; Springer-Verlag: Berlin-Heidelberg, 2010, pp 177–197.

Triantaphyllou, A. C. Cytogenetics, Cytotaxonomy and Phylogeny of Root-knot Nematodes. In: *An Advanced Treatise on Meloidogyne: Biology and control*; Carter, K. R., Sasser, J. N., Eds.; North Carolina State University Graphics: Raleigh, USA, 1985; pp 107–114.

Urwin, P. E.; Lilley, C. J.; McPherson, M. J.; Atkinson, H. J. Resistance to Both Cyst and Root-knot Nematodes Conferred by Transgenic *Arabidopsis* Expressing a Modified Plant Cystatin. *Pl. J.* **1997,** *12,* 455–461.

Vain, P. B.; Borland, B. M. C.; Clarke, M. C. G.; Richard, G. M.; Beavis, M. H.; Liu, H. A.; Kohli, A. M.; Leech, M.; Snape, J.; Christou, P.; Atkinson, H. Expression of an Engineered Cysteine Proteinase Inhibitor (Oryzacystatin IDD86) for Nematode Resistance in Transgenic Rice Plants. *Theor. Appl. Genet.* **1998,** *96,* 266–271.

Veremis, J. C.; Roberts, P. A. Diversity of Heat-stable Genotype Specific Resistance to *Meloidogyne* in Maranon Races of *Lycopersicon peruvianum* Complex. *Euphytica* **2000,** *111,* 9–16.

Vrain, T. C. Engineering Natural and Synthetic Resistance for Nematode Management. *J. Nematol.* **1999,** *31*(4), 424–436.

Williamson, V. M.; Gleason, C. A. Plant–nematode Interactions. *Curr. Opin. Pl. Biol.* **2003,** *6*(4), 327–333.

Williamson, V. M.; Hussey, R. S. Nematode Pathogenesis and Resistance in Plants. *Pl. Cell* **1996,** *8,* 1735–1745.

Windham, G. L.; Pederson, G. A. Comparison of Reproduction by *Meloidogyne graminicola* and *M. incognita* on *Trifolium* Species. *J. Nematol.* **1992,** *24*(2), 257–261.

Yik, C. P.; Birchfield, W. Host Studies and Reactions of Cultivars to *Meloidogyne graminicola*. *J. Phytopathol.* **1979,** *69,* 497–499.

CHAPTER 8

CONTROLLING YELLOW STEM BORER IN RICE BY USING MOLECULAR TOOLS

MD. SHAMIM[1,2], PRASHANT YADAV[1,3], N. A. KHAN[1], POONAM[1], MAHESH KUMAR[2], TUSHAR RANJAN[2], RAVI RANJAN KUMAR[2], ANAND KUMAR[4], and K. N. SINGH[1*]

[1]*Department of Plant Molecular Biology and Genetic Engineering, N. D. University of Agriculture and Technology, Kumarganj, Faizabad 224229, Uttar Pradesh, India*

[2]*Department of Molecular Biology and Genetic Engineering, Bihar Agricultural University, Sabour, Bhagalpur 813210, Bihar, India*

[3]*Division of Crop Improvement, National Research Centre on Rapeseed-Mustard, Sewar, Bharatpur 321303, Rajasthan, India*

[4]*Department of Plant Breeding and Genetics, Bihar Agricultural University, Sabour, Bhagalpur 813210, Bihar, India*

Corresponding author. E-mail: kapildeos@hotmail.com

CONTENTS

ABSTRACT

Rice (*Oryza sativa* L.) is the one most important food crop of world's population. Among the major insect pests of rice, yellow stem borer (YSB) (*Scirpophaga incertulas*), a monophagous pest is considered as one of the most important pest of rain-fed low land and flood-prone rice ecosystems. YSB larvae feeds inside the rice stem, causes "dead heart" in the vegetative stage ultimately leading to "white head" in the reproductive stage. Breeding of YSB resistance in rice is difficult due to the complex genetic traits, inherent difficulties in screening and poor understanding of the genetics of resistance in the cultivated rice. The uses of chemical pesticides were effective for the control of this pest in rice field. Though, for an extended time, chemical control has shown to be ineffective because the insect larvae feed inside the stem pith and remain out of the reach to the sprayed pesticide. Resistance rice varieties will be useful for the managing against YSB. However, good level of resistance against the YSB has been rare in the cultivated rice germplasms. Transgenic rice developed by cry gene from *Bacillus thuringiensis* showed very good level of resistance in laboratory as well as in field conditions.

8.1 INTRODUCTION

Rice (*Oryza sativa* L.) is cultivated worldwide over an area of about 153.51 m ha with an annual production of about 614.65 m t with an average productivity of 4.00 t/ha. India has the largest area (43.0 m ha) constituting 28.01% of the land under rice in the world and ranks second in total production (89.81 m t) next to China (184.25 m t) with an average productivity of 2.10 t/ha. Among various constraints of rice production, the insect-pest is of prime importance. Disease and insect-pest are estimated to cause yield losses to the tune of 30–40%. About 100 species of insect-pest infest and damage the rice plant. Of these 15 insect-pests are of major significance and are generally regular in occurrences. The stem borer, leaf hopper, plant hopper, ghandhi bug, rice leaf folder, gall midge, rice hispa, cut warms, and army worms are considered to be most destructive (Pathak & Dyck, 1973). Out of 50 different stem borer species attacking rice crop from seedling to maturity stage three stem borer namely Yellow stem borer (YSB) (*Scirpophaga incertulas*), dark headed stem borer (*Chilo polychrysus*), and pink stem borer (PSB) (*Sesamia infrans*) are of major significance in Asia (Banerjee, 1971; Kapur, 1967; Pathak & Khan, 1994; Rao, 1964). Among these stem borers, YSB is very serious pest of rice throughout and Asia (Table 8.1) and creating hurdles in achieving the expected high-yield potential (Chaudhary et al., 1984; Patanakanjorn & Pathak, 1967; Sunio et al., 1986). The extent of borer induced yield losses in rice have been estimated to range from 30% to 70% in outbreak per years and from 2% to 20% in non-outbreak per years in Bangladesh (Catling et al., 1987) and in India (Chelliah et al., 1989; Imayavaramban et al., 2004; Satpathi et al., 2012), respectively. Chemical fertilizers and adoption of integrated pest control methodology under modern IPM mostly relies for the control of YSB in rice. Thus, achieving this aim, pest control will have to rely on integrated pest management (IPM) practices which comprise different crop planting techniques and insect resistant plants (Baloch & Abdullah, 2011) to improve productivity and sustainability. Rice breeding programs often highlighted selection for insect resistant rice varieties and made much progress. Various researches confirmed that certain rice cultivars had less borers' damage than more susceptible rice varieties (Khan et al., 2005). However, lack of a high level of resistance against the YSB in cultivated rice varieties had delayed the development of suitable rice varieties in the past (Bentur, 2006).

TABLE 8.1 Major Insect Pests (Lepidoptera, Diptera, and Coleoptera) of Rice Crop and Extent of Losses Caused by Them.

Major Insect pests of rice		Insect Order	Yield loss	Reference
Common name	**Scientific name**			
Yellow stem borer	*Scirpophaga incertulas*	Lepidoptera	10–48%	AICRIP (1988)
Rice leaf folder	*Cnaphalocrocis medinalis*	Lepidoptera	10–50%	Nair (2001)
Whorl maggot	*Hydrellia philippira*	Diptera	20–30%	Nair (2001)
Gall midge	*Orseolia oryzae*	Diptera	8–50%	Nair (2001)
Rice hispa	*Dicladispa armigera*	Coleoptera	6–65%	Nair (2001)

Though some resistance to stem borers reported appears to be under polygenic control (Khush, 1984), many morphological, anatomical, physiological, and biochemical factors have been reported to be associated with resistance, each controlled by different sets of genes (Chaudhary et al., 1984). With advances in biotechnology, breeding of horizontal resistance, whereby resistance is based on many genes, along with genetically enhanced sustainable pest resistance with fusion genes, is becoming more popular (Wan, 2006). Recent progress in rice transformation technologies has made it possible to produce genetically modified (GM) new rice cultivars with improved resistance to insect pests by genetic engineering. Thus, development of knowledge of damage mechanisms associated with YSB feeding and resistance mechanism can be of great value to scientist and farmers in enabling strategies to tackle this most important pest. There are reports on transgenic insect resistant plants by the different trail confirmation of accomplishment that has become increasingly more difficult for opponents of genetic engineering to disregard (Bhattacharya et al., 2006).

8.2   YELLOW STEM BORER

The YSB *S.* (tryoryza) *incertulas* (Walker) belong to the family *Pyralidiae* of order *Lepidoptera*. The insect is dimorphic and the moth of stem borer is nocturnal in habit. The male moth is light brown with numerous brownish dots (Fig. 8.1a–c). The female moths are straw colored and have a very distinct black spot in the center of each forewing. The hind wing is pale and straw colored. In northern region of India, the pest is active from April to October and hibernates from November to March as full-grown larvae in rice stubbles. The larvae of *S. incertulas* bores inside the stem during vegetative stage and causes death and dried up of the youngest leaf a symptom (Fig. 8.1d) referred to as "dead heart" (DH). Infestation during the later plant growth stages results in the formation of empty grains or panicle termed as "white ear heads" (WE). Larvae feeding also cause plant reduction in plant vigor fewer litter, stunt growth, and unfilled grains (Pathak, 1968). The reduction in yield has related with an increase in the levels of YSB population. The avoidable losses caused by insect has been reported to be range of 10–80% in different agro-climatic regions with a mean yield loss of 40% (Manwan & Vega, 1975) for every percent of WEs, 1–2% loss in yield may be expected (Pathak, 1977).

The adult of *S. incertulas* is a nocturnal, positively phototrophic and strong flier. The moths lay eggs in masses near the tip of leaf blade of rice, usually containing 50–80 eggs. The eggs are covered with pale orange-brown hairs from anal tufts of the female moths. The eggs laid on rice within a field are generally randomly distributed (Pathak, 1977). Islam and Catling (1991) reported that the female YSB laid on an average 197 eggs within 3 days of emergence. The larval and pupal stage and total life cycle lasted 25.40, 9.1, and 45.80 days, respectively. The female of *S. incertulas* laid on an average of 133.40 ± 29.90 eggs, which hatched within 6–8 days. There were five larval instars and larval stage lasted for 27–30 days.

FIGURE 8.1 (a) Female yellow stem borer moths on rice plant, (b) yellow stem borer larvae on rice leaf, (c) IIIrd instar yellow stem borer larva on rice culm, and (d) dead heart symptom on rice by yellow stem borer infestation (photograph: Md. Shamim).

Malhi and Brar (1998) studied the biology of *S. incertulas* on Basmati rice cv. Basmati 370 and reported that the freshly laid egg masses covered within hair of female tuft varied in shape from oval to somewhat rectangular. The incubation period lasted from 6.70 and 6.83 days during July and August, respectively. Larval stage passed through six instars with total larval duration of 28.85 and 32.96 days during July and August–September, respectively. The prepupal and pupal periods varied from 1.15 to 1.25 and 6.90 to 6.96 days, respectively.

8.2.1 DAMAGE AND YIELD LOSSES BY YSB

About a week after hatching the larvae from the leaf sheaths bore into the stem and staying in pith, feed on the inner surface of the walls. Such feeding often results in a severing of the topical parts of the plant from the base, resulting the central leaf whorl does not unfold turns brownish and dried off, although the lower leaves remain green and healthy. This condition is known as DHs and the affected tillers dry out without bearing panicles. After panicle initiation, severing of growing plants parts from base results in drying of the panicles, which may not have emerged at all, and these that have already emerged do not produce grains. This condition is known as "white ear" (WE). Although the damage becomes evident only as DH and WE. Such damage results in reduction of plant vigor, fewer tillers and many unfilled grains (Pathak, 1968, 1977). According to Israel and Abraham (1967), 1.0% increase in stem borer infestation resulted in reduction of grain yield by 0.28% in young plants and by 0.62% in plants at reproductive stage.

In India, it was estimated that 21.60% yield loss was due to damage caused by *S. incertulas* at early stage of plant growth and 26.90% yield loss due to damage caused at heading stage. Estimation of yield loss caused by *S. incertulas* revealed that the influence of WE was much greater than that of DH on yield of rice cv. GR11, and losses ranged from 1.98% to 2.10% and 1.74% to 2.47% unit damage, respectively (Pandya et al., 1994).

8.2.2 SEASONAL OCCURRENCE OF YSB

In general, stem borers are polyvoltine but the number of generation in a year depends on the environmental factors, primarily temperature and crop availability. According to the area, the insect hibernate, aestivate, or remain active throughout the year and occur in different seasonal patterns (Pathak, 1968, 1977). Tripathi et al. (1997) reported that the incidence of *S. incertulas* varied considerably between season and years. Based on the seasonal incidence of YSB of rice in semi deep water situation of Bhubaneswar, it was observed that the YSB produced two broods, with the first peaking during the last week of September and the seasonal peaking during the second week of November, with coincided with the dough stage of rice. Egg mosses were more abundant in the second brood ($1.33/m^2$) than the first ($1.00/m^2$).

8.3   YELLOW STEM BORER MANAGEMENT

8.3.1   CHEMICAL CONTROL OF YELLOW STEM BORER

Out of large number of insecticides tested on YSB, more than 30 were reported to be effective in controlling this pest. The more common recommended insecticides were monocrotophos 36 EC, endosulfan 35 EC, quinalphos 25 EC, chlorpyriphos 20 EC, carbonyl 50 EC, melathion 50 EC, phosolone 35 EC, diazinon 30 EC, and phosphamidon 85 EC, phorat 10 G, aldicarb 10 G, and disulfoton 5 G (Kulshreshtha & Nigam, 1987). Use of monocarotophos and chlorpyriphos were tested at the rate of 0.02% and 0.03% on *S. incertulas* on Basmati rice under field condition for the control. Chlorpyriphos at the rate of 0.03% proved most effective insecticides as it delayed the first application of granular insecticide by a maximum of 20 days. Significantly lower number of WE heads and higher grain yield over recorded due to application of chlorpyriphos (Brar et al., 1994).

The crop was not protected at any stage by insecticidal application so as to see occurrence of YSB in natural condition. However, indiscriminate use of insecticide has led to serious consequences like harmful residues to insecticides, pest resistance of insecticides, pest resurgence, outbreak of secondary pest, and considerable harmful effect on nontarget species including parasite and predator and insect pollinator. Many pesticides, particularly, those based on organophosphates are also toxic to humans. Further, to clearly demonstrate that overreliance on pesticide is non-suitable, many insects have resistant to pesticides, as has been the case with rice brown plant hopper (*Nilopavata lugens*) through much of SE Asia. Controlling YSB with insecticides is not only costly but usually achieves poor results particularly in the wet season as timing of application is highly critical (Sunio et al., 1986). But till recently, use of insecticides was considered to be most effective tool to overcome this problem. However, insecticides, pest resurgence, outbreaks of secondary pests are considerable harmful effect on nontarget species including parasites, predators, and insect pollinators.

8.3.2   MICROBIAL AND BACTERIAL CONTROL OF YELLOW STEM BORER

The biopesticides such as *Bacillus thuringiensis* (*Bt*) is least harmful to natural enemies because of its selectivity. The use of *Bt* formulation has increased, particularly in IPM program. Rath (1999) reported that four *Bt* formulation (Dalfin 85%, Biolep, BT II, and Dipel 3.5%) and four neem formulations (Nimbicide 2%, Neemax 2%, Achook 2%, and Neemgold 2%) were evaluated against YSB of rice. All four neem formulation and two *Bt* formulation (BT II and Dipel 3.5%) protected the crop from YSB damage to a limited extent but proved less effective than the insecticide chloropyriphos. Panda et al. (2006) reported that five commercial formulation of *B. thuringiensis* var. *kustaki* were tested against the rice stem borer (*S. incertulas*) Bioasp, Biolep, and Biotex sprayed at the rate of 2.1 kg/ha were found moderately effective against stem borer. But the foliar application of *Bt* break down quickly under field conditions due to UV sensitivity and rainfall. With the advent of recombinant DNA technologies, insecticide proteins present in *Bt* have been expressed in crop plants to ensure durable insect resistance, thus this technology is now frequently used. The insecticidal bacterium *B. thuringiensis* is the most successful and widely used biological control against to insect-pest in the world for 40 years. But foliar application of *Bt* breaks down quickly under field conditions due to UV sensitivity and rainfall (Gopalswamy et al., 2003).

Rani et al. (2006) studied host location and acceptance by egg parasitoids can be mediated by close-range host stimuli. In this study, we tested the response of *Trichogramma japonicum* Ashmead to cuticular extracts of adult and larval rice YSB, *Scripophaga incertulas* Walker. We also studied the wasps' response to extracts from YSB larval frass. Laboratory bioassays revealed that hexane extracts of the adult host body stimulate ovipositor probing of *T. japonicum*. Extracts of larval frass also stimulated parasitization. In contrast, host larval cuticular extracts had no effect on parasitization rates. Fractionation of the crude extracts of adult YSB cuticular extracts was performed using silica gel chromatography, followed by bioassays of the individual fractions to test their effects on wasp behavior. Analyses of the most active fractions by gas chromatography–mass spectrometry revealed that the extract contained saturated long-chain alkanes and alkenes, with carbon numbers ranging from C_{20} to C_{32}. Hydrocarbons were applied onto host eggs to test their effects on parasitization rates. Treatments of eggs with docosane, tetracosane, pentacosane, and eicosane enhanced host egg parasitization, while pentadecane, hexadecane, and nonadecane deterred oviposition.

Shamim et al. (2011) isolated and purified protease inhibitor (PI) from the mature seeds of jackfruit (*Artocarpus heterophyllus*). The isolated PI strongly inhibited papain and midgut proteases of YSB (*S. incertulas*) larvae, as seen by in vitro assay. The purified PI was active over a wide range of pH with the maximum activity between pH 4 and 10. This protein was also stable up to 80°C, but the retained activity was lost at 100°C, when heated for 30 min. The molecular mass of the purified cysteine-like PI is to be 14.50 kDa as determined by SDS-PAGE. Significant reduction in larval weight and mortality was observed, when fresh rice culms with PI was fed to the YSB larvae. These results may provide important information to control the YSB in rice with respect to naturally occurring insecticidal proteins. The observed differences would potentially translate into reductions in population growth of YSB, indicating a potential value of using jackfruit PI for protecting rice plants against damage by the YSB.

Yadav et al. (2012) evaluated the biological activities of jackfruit PI and *Bt* toxins against *S. incertulas* (YSB) and *Sesamia inferens* (PSB) under laboratory. Lethal dose (LD_{50}) was quantified against the IInd instar larvae of YSB and PSB. The LD_{50} of jackfruit PI was 25.6 ± 0.14 and 22.5 ± 0.42 µg/g against the YSB and PSB, respectively. However, the LD_{50} of *Bt* toxin was 4±0.32 µg/g in case of YSB and 2.5±0.38 µg/g for PSB.

Alcantara et al. (2004) studied the receptor binding step in the molecular mode of action of five δ-endotoxins (*Cry1Ab*, *Cry1Ac*, *Cry1C*, *Cry2A*, and *Cry9C*) from *B. thuringiensis* was examined to find toxins with different receptor sites in the midgut of the YSB *S. incertulas* (Walker) (Lepidoptera: Pyralidae) and striped stem borer (SSB) *Chilo suppressalis* (Walker), and homologous competition assays were used to estimate binding affinities (K_{com}) of ^{125}I-labeled toxins to brush border membrane vesicles (BBMV). The SSB BBMV affinities in decreasing order were reported as *Cry1Ab* = *Cry1Ac* > *Cry9C* > *Cry2A* > *Cry1C*. In YSB, the order of decreasing similarity was *Cry1Ac* > *Cry1Ab* > *Cry9C* = *Cry2A* > *Cry1C*. The number of binding sites (B_{max}) expected by homologous competition binding. Results of the heterologous competition binding assays suggest that *Cry1Ab* and *Cry1Ac* contend for the similar binding sites in SSB and YSB. Other toxins bind with weak (*Cry1C*, *Cry2A*) or no affinity (*Cry9C*) to Cry1Ab and *Cry1Ac* binding sites in both species. *Cry2A* was found to lowest toxicity against 10 days old SSB, and *Cry1Ab* and *Cry1Ac* were the most toxic than other *Cry* protein. The finding of the study showed that *Cry1Ab* or *Cry1Ac* could be combined with either *Cry1C*, *Cry2A*, or *Cry9C* for more durable resistance in transgenic rice. *Cry1Ab* should not be used together with *Cry1Ac* because a mutation in one receptor site could diminish binding of both toxins.

Gayen et al. (2015) studied the analogous form of the *B. thuringiensis vip3Aa* insecticidal toxin gene, named *vip3BR*, was identified and characterized, and exhibited similar attributes to the well-known Vip3Aa toxin. Vip3BR possessed broad-spectrum lepidopteran-specific insecticidal properties effective against most major crop pests of the Indian subcontinent.

A Vip3BR toxin protein N-terminal deletion mutant, Ndv200, showed increased insecticidal potency relative to the native toxin, which conferred efficacy against four major crop pests, including cotton boll worm (*Helicoverpa armigera*), black cut worm (*Agrotis ipsilon*), cotton leaf worm (*Spodoptera littoralis*), and rice YSB (*S. incertulas*). Ligand blot analysis indicated the Ndv200 toxin recognized the same larval midgut receptors as the native Vip3BR toxin, but differed from receptors recognized by Cry1A toxins. In the present study, we tested the prospect of the *vip3BR* and *ndv200* toxin gene as candidate in development of insect-resistant genetically engineered crop plants by generating transgenic tobacco plant. The study revealed that the *ndv200* mutant of *vip3BR* insecticidal toxin gene is a strong and prospective candidate for the next generation of GM crop plants resistant to lepidopteran insects.

8.3.3 DIVERSITY IN YELLOW STEM BORER

Diversity assessment in the any pest population is very important for the control of particular pest. Kumar et al. (2001) examined the genetic variation among populations of *S. incertulas* collected from 28 hot spot locations in India using the Random amplified polymorphic DNA (RAPD)-PCR. All in all, 32 primers were used and 354 amplification products were observed. However, no RAPD-PCR bands diagnostic to the pest population were identified from any specific region. Panda et al. (2012) studied the genetic variability among the geographically isolated populations of YSB from forty locations of Orissa using RAPD markers. Twenty five different RAPD primers were used to amplify genomes of forty YSB populations. A total of 126 bands were amplified by 10 primers, of which 125 are polymorphic. Thirteen unique bands were identified which will be useful for developing diagnostic markers. Genetic similarity among YSB populations varied from 0.098 to 0.753, with an average of 0.423 indicating that wide genetic variation exists among YSB populations at molecular level. Most of the populations could be uniquely distinguished from each other and grouped into three major clusters at 30% level of genetic similarity. Further study with host differentials can ascertain their biotype status.

Sutrisno (2015) investigated genetic variation among the different *S. incertulas* population in Indonesia by mitochondrial sequence variation. A 685-bp segment of mitochondrial DNA, COII, was amplified from 42 YSB samples from five locations in Java (Madiun, Ngawi, Wonogiri, Tasikmalaya, and Indramayu). Six different haplotypes (YSB1, YSB2, YSB3, YSB4,

YSB5, and YSB6) were identified in the sequenced YSB populations, with haplotype YSB2 being dominant. Finally, based on the mitochondrial DNA CO II sequence from 42 different samples of YSB, it was concluded that YSB within populations and among populations in Java exhibited very low genetic diversity.

8.4 MOLECULAR APPROACHES FOR THE DEVELOPING RESISTANCE IN RICE AGAINST YELLOW STEM BORER

The mechanism of rice varietals resistance to the YSB, *S. incertulas*, was investigated and analyzed by earlier workers. The resistant cultivars had the lower interval between vascular bundles and larger sheath bridge in the outer leaf sheath than the susceptible ones (Fang et al., 2002). Mann and Shukla (2005) studied different rice genotypes for mechanisms of resistance against stem borer. Four genotypes including Pakistan Basmati were scored resistant/moderately resistant both at DHs and white heads stages in early and normal transplanting dates. These four genotypes possessed better anti-xenosis for egg laying and antibiosis for larval survival and larval and pupal weights against borers compared to the susceptible variety Basmati-370.

Chen and Romena (2006) studied the feeding pattern of *S. incertulas* on different on wild and cultivated rice accessions was determined where larvae feed and if feeding route influences larval survival and development. Three cultivated (Taichung Native 1, IR64, and IR72) and three wild rice accessions (two accessions of *Oryza nivara* and one accession of *O. rufipogon*) were selected and larvae were introduced onto booting plants and sampled after five different time internal, that is, 6 h, 1 day, 2 days, 4 days, and 7 days. Approximately, 25% more YSB larvae survived on cultivated accessions than on wild accessions. Larvae were also 15% more likely to feed on the panicle of cultivated accessions than wild accessions, and panicle-feeding improved larval survival and development. Study concluded that YSB feeding route depended on plant phenology; larvae were more likely to feed on the panicle than on vegetative structures on booting, heading, and flowering tillers. Because all rice stems were cut by the seventh day if larvae fed on the panicle, resistance during the booting phase may be effective if it reduces the likelihood of panicle feeding or if strong antibiotic resistance can be found in the panicle.

Padhi (2006) studied that the susceptible cultivar showed high amounts of amino acids and nucleic acids (RNA and DNA), causing profuse tillering than resistant to borer attacks. Xu et al. (2006) monitored the relationship

among resistance to stem borer and morphological and anatomical characteristics of rice plants: the highest number of tillers, higher plant height, leaf
width, leaf angle of rice plant, the narrowest space between vascular bundles
in leaf sheath of the tillers than the broadest, higher number of silica cells
in leaf sheath, and the diameter of vascular bundles may influence the resistance of rice cultivars.

8.4.1   RESISTANCE LINKED MARKER AND QTLS AGAINST YELLOW STEM BORER

There are very few reports on the resistance rice cultivars against the YSB.
And a good level of resistance against the widespread YSB has been rare in
the rice germplasm. However, moderate resistance against YSB in few cultivated and wild rice accessions were reported (Table 8.2). Selvi et al. (2002a)
described for the first time, the identification of RAPD markers associated
with YSB resistance and susceptibility, their sufficiently closer linkage to
the genes regulating the trait and their unambiguous scorability in the resistant and susceptible cultivars that were tested. Identification of molecular
markers linked to the trait would enhance phenotypic evaluation for the trait.
An F_2 population was developed using parents contrasting in their reaction
to YSB resistance. RAPD analysis, in conjunction with bulk segregant analysis (BSA), enabled to identify four phenotype specific RAPD markers. The
markers C1320 and K6695 were linked with resistance and AH5660 and
C41300 with susceptibility. The markers K6695 and AH5660 were linked
to the gene(s) at distances of 12.8 and 14.9 cm, respectively. Scoring of
these markers in a set of germplasm confirmed their reproducibility and the
association with the trait. The same authors (Selvi et al., 2002b) investigated the possibility to tag genes for YSB resistance with micro satellite
markers. An F_2-mapping population was derived from a cross between the
moderately resistant variety W1263 and a highly susceptible variety Co43.
Parental survey using 48 microsatellite primer pairs revealed that 25 primers
were polymorphic between the parents. BSA with the F_2 mapping population
showing extreme cases of resistance and susceptibility could identify two
primers RM241 and RM219 segregating with the resistant and susceptible
bulk. Further analysis with F_2 individuals revealed that the marker RM241
is associated with the trait.

Selvi et al. (2003) also reported that YSB is governed by polygenes.
RAPD analysis in conjunction with BSA identified OPK6 695 as a marker
specifically amplified from the DNA of the resistant parent, progenies,

and other resistant cultivars and OPHA5 660 that specifically amplified in the susceptible parent, progenies, and susceptible cultivars. The identified markers showed linkage with YSB resistance. RM241, located on chromosome 4, was also found to be associated with the trait. Further marker analysis is required to place more markers closer to the gene(s) for YSB resistance. Mohankumar et al. (2001) screened different rice line and reported that rice cultivar W1263 shoed resistance against YSB. Further W1263 was crossed with the susceptible rice cultivar CO43. The F_1 and F_2 populations were screened against YSB under field and green house conditions for both DHs and WE disease. Further scoring of 250 different RILs (F_8) lines for various morphological traits showing wide range of variation indicating the suitability for QTL mapping against YSB.

TABLE 8.2   Tolerance/Resistance Rice Accessions and Cultivars against Yellow Stem Borer.

S. No.	Cultivated rice/Wild rice	Reference
1	W1236	Mohankumar et al. (2001)
2	*Oryza brachyantha*, *Oryza officinalis*, *Oryza ridleyi*, and *Porteesia coartate* (saline resistance wild rice)	Behura et al. (2011)
3	Basmati 15-13, Jajai 25/A, Basmati 15-13, and CR-33	Sarwa (2013)
4	*Oryza brachyantha*	Narain et al. (2016)

Behura et al. (2011) screened different wild rice germplasm against YSB (*S. incertulas* L.). Wild rice *Oryza brachyantha*, *Oryza officinalis*, *Oryza ridleyi*, and *Porteresia coarctata* were found to be resistant/tolerant against YSB. The resistance parents were crossed to the susceptible for the development of resistance against YSB. Further, BC_1F_1 interspecific hybrid of *O. sativa* cv. "Savitri"/*O. brachyantha* was backcrossed with recurrent parent "Savitri" with the help of embryo rescue technique. The morphological characters of the embryo rescued plants were studied, and they were observed to be phenotypically different types, namely grassy, bushy, erect, dwarf, sterile, and pseudonormal, etc. The cytological analysis of the variants was done for the identification of the different chromosome variants for YSB resistance genes to cultivated rice.

Sarwa (2013) evaluated different rice and reported some showed resistance reactions, namely, Basmati-15-15, Jajai-25/A, Basmati-15-13, and CR-33 against YSB. These four genotypes possessed better antixenosis against adult egg laying and antibiosis for larval survival of stem borer as compared to the susceptible genome. These genotypes can be used in resistance development breeding program and should be admired in stem borer endemic areas.

Narain et al. (2016) studied the resistance in the different wild rice against YSB. The African wild rice *O. brachyantha*, showed resistance to YSB, was exploited as a resistance source for introgression of YSB resistance trait into cultivated rice (*O. sativa*), through development of monosomic alien addition lines (MAALs). Different BC_2F_1 backcross hybrids (*O. sativa* cv. Savitri/*O. brachyantha*//*O. sativa* cv Savitri///*O. sativa* cv. Savitri) were developed employing embryo rescue. Different hybrid embryos that were collected at 10 and 12 days after pollination showed highest percentage of survival in culture with a germination percentage of 35.8%. Cytological analysis of pollen mother cells (PMCs) of the BC_2F_1 hybrids revealed 15 hybrids with $2n + 1$ ($2n = 25$) chromosome number that exhibited typical trisomic chromosomal configurations. These aneuploids were putatively designated as MAALs and characterized morphologically and cytologically. Based on their morphological similarity to primary trisomics of the cultivated rice (*O. sativa*), 8 MAALs, namely, MAALs-4 (Sterile), 5 (Twisted leaf), 7 (Narrow leaf), 8 (Rolled leaf), 9 (Stout), 10 (Short grain), 11 (Pseudonormal), and 12 (Tall) were isolated, each of which exhibited several distinct morphological features. Among the 505 PMCs analyzed, 79.80% had 12_{II}, 5.45% had 11_{II}, and 1.78% PMCs had 1_{III} configurations. The size of the extra chromosome was found to be smaller than those of *O. sativa* chromosomes and in 97.25% of PMCs it remained unpaired and at a distance from the rest of the chromosomes. Of the eight MAALs screened for YSB resistance, MAAL 11 was found to be moderately resistant. Undesirable morphological traits of *O. brachyantha* were found to be eliminated in the MAALs.

8.4.2   TRANSGENIC APPROACHES AGAINST YELLOW STEM BORER

Insect resistance in crop plants has been one of the major "success stories" of the application of plant genetic engineering to agriculture, and genetically engineered insect tolerant corn, potato, and cotton plants expressing a gene encoding the bacterial endotoxin from *B. thuringiensis* is now a commercial reality, at least in the USA. It is also considered advantageous to generate *Bt*-transgenics with multiple toxin systems to control rapid development of pest resistance to the insecticidal cry protein (ICP). Larvae of YSB, *S. incertulas*, a major lepidopteran insect pest of rice, and cause massive losses of rice yield. Studies on insect feeding and on the binding properties of ICP to brush border membrane receptors in the midgut of YSB larvae revealed that *cryI Ab* and *cryI Ac* are two individually suitable candidate genes for developing YSB-resistant rice (Nayak et al., 1997). The production of transgenic crops has been rapidly advanced during the last decades with the commercial introduction of *Bt.* transgenic (Table 8.3).

TABLE 8.3   Transgenic Rice Cultivars Development against Yellow Stem Borer by *Bt* and Other Genes.

Sl. No.	Gene used for transformation	Transformed rice varieties	Reference
1	Cry1Ab	Indica, IR58	Wunn et al. (1996)
2	IR64 Cry1Ac	Indica,	Nayak et al. (1997)
4	Cry1Ab	Aromatic rice, Tarom molaii	Ghareyazie et al. (1997)
5	Cry1Aa, Cry1Ac, Cry2A, Cry1C	Indica, Japonica	Lee et al. (1997)
6	Cry1Ab	Japonica, Taipei309	Wu et al. (1997)
7	Cry1Ab, Cry1Ac	Japonica rice	Cheng et al. (1998)
8	Cry1Ab	Indica and Japonica lines	Datta et al. (1998)
9	Cry1Ab	Deep water indica variety, Vaidehi	Alam et al. (1998)
10	Cry2A	Indica, Basmati 370, and M7	Maqbool et al. (1998)
11	Cry1Ab	Maintainer line IR68899B	Alam et al. (1999)
12	Cry1Ab, Xa21	Indica, Pusa Basmati 1	Gosal et al. (2000)
13	cry1A, cry1Ab, cry1Ac, cry1C, and cry2A	Indica	Intikhab et al.(2000)
14	Cry1Ab	PR16, PR18, PR57, PR58	Ye et al. (2000)
15	Cry1Ab, Cry1Ac	Indica cv. Minghui 63, Shanyou 63	Tu et al. (2000)
16	Cry1Ab	Japonica Elite line KMD1	Shu et al. (2000)
17	Cry1Ab	Indica, IR64	Maiti et al. (2001)
18	Cry1Ac	Pusa Basmati 1	Gosal et al. (2001)
19	Cry1Ac, Cry2A Snowdrop lectin gene	Indica, M7 and Basmati 370	Maqbool et al. (2001)
20	Cry1Ab	Indica	Aguda et al. (2001)
21	Cry1Ac	Indica Pusa Basmati 1, IR 64 and Karnal local	Khanna and Raina (2002)
22	Chimeric Bt gene, Cry1Ab, Hybrid Bt gene Cry1Ab/Cry1Ac	Maintainer lines—IR 68899B and IR68897B Restorer lines—MH63 and BR827-35R	Balachandran et al. (2002)

TABLE 8.3   *(Continued)*

Sl. No.	Gene used for transformation	Transformed rice varieties	Reference
23	Cry1Ac	Indica, R3 progeny of basmati	Gosal et al. (2003)
24	Cry1Ab, gna	Javanica, progenies Rajalele	Slamet et al. (2003)
25	Cry1Ac	Indica, IR64, Pusa Basmati 1, Karnal local	Raina et al. (2003)
26	Cry1Ac, cry2A	Indica, basmati rice	Husnain et al. (2003)
27	Xa21, Bt fusion gene, Chitinase gene	Indica, IR 72	Datta et al. (2003)
28	Cry1Ab, Cry1Ac, gna	Chaitanya, Phalguna and Swarna, besides three parents, viz. IR58025A (CMS line), IR58025B (maintainer line) and Vajram (restorer)	Ramesh et al. (2004)
29	Cry1Ab, Cry1Ac, Cry1C, Cry2A, Cry9C	Indica	Alcantara et al. (2004)
30	Cry1Ac, Cry2A	Indica, Basmati	Bashir et al. (2005)
31	Cry2A	Indica, restorer line Minghuli 63	Chen et al. (2005)
32	Translationally fused gene, Cry1Ab-1B and hybrid Bt gene, Cry1A/Cry1Ac	Elite Vietnamese	Ho et al. (2006)
33	Japonica, Tainung 67 Potato proteinase inhibitor 2 (Pin 2)	Indica, Pusa Basmati 1	Bhutani et al. (2006)
34	Cry1Ac, Cry2A	Indica, Basmati 370	Riaz et al. (2006)
35	P-I, P-II, P-III Cry1Ab	Korean varieties	Kim et al. (2008)
36	Cry1AC, Cry2A, and Cry9C	Munghui	Chen et al. (2008)
37	Cry1Ab and Vip2H	Xiushui 110	Chen et al. (2010)
38	Cr1B and Cry1Ab	Pusa Basmati 1	Kumar et al. (2010)
39	Cy1Ab	IR64	Dandapat et al. (2014)
40	Cry1C	KR022 restorer line	Wan et al. (2014)
45	Vip3BR		Gayen et al. (2015)
46	Protease inhibitor gene		
47	Potato chymotrpsin inhibitor	Swarna	Rao et al. (2009)

Adapted with modification from Deka and Barthakur (2009).

Nayak et al. (1997) transformed an *indica* rice IR64 plant by a *crylAc* gene under the control of the maize ubiquitin 1 promoter, along with the first intron of the maize *ubiquitin 1* gene, and the *nos* terminator. The gene construct was transferred by using the particle bombardment method on embryogenic calli. Six highly expressive independent transgenic plants with insecticidal proteins were identified. Molecular analyses and insect-feeding assays of two transgenic lines revealed that the transferred synthetic *CrylAc* gene was expressed stably in the transgenic T_2 were highly toxic and lessened the damage after infestation with YSB larvae. Since then, several groups have transformed rice varieties like IR64, Karnal Local, etc. with *Bt* genes such as *Cry1Ab*, *Cry1Ac*, and others to obtain resistance against YSB (Khanna & Raina, 2002).

Cheng et al. (1998) generated up to 2600 transgenic rice plants after *Agrobacterium*-mediated transformation. The transformed plants contained fully modified (plant codon optimized) versions of two synthetic crystal proteins, namely, *crylA(b)* and *crylA(c)* coding sequences from *B. thuringiensis*. These sequences were placed under control of the maize *ubiquitin* promoter, the CaMV35S promoter, and the *Brassica Bp10* gene promoter to achieve high and tissue-specific expression of the lepidopteran-specific d-endotoxins. Accumulation of high levels of *CrylA(b)* and *CrylA(c)* proteins was detected in R0 plants (up to 3% of soluble proteins). Bioassays with R_1 transgenic plants showed that the transgenic rice were highly toxic to two major pests namely, SSB (*C. suppressalis*) and YSB (*S. incertulas*), with mortalities of 97–100% within 5 days after infestation. According to Sehgal et al. (2001), the use of pest resistant crop varieties is the easiest, most effective, compatible, economical, and practical method among all the pest management practices. Such crop varieties are extensively used in pest-prone areas as a principal method of IPM or as a supplement to other pest management strategies.

Shu et al. (2000) reported that transgenic rice transformed with a synthetic *cry1Ab* gene was found significantly tolerant to eight lepidopteran insects, including SSB (*C. suppressalis*) and YSB (*S. incertulas*). Furthermore, two lines from *Bt* rice plants were highly resistant to lepidopteran pests under field conditions (Deka & Barthakur, 2010; Kumar et al., 2008; Wang et al., 2014).

Husnain et al. (2002) studied expression of insecticidal gene cry1Ab, under three different promoters, was studied in leaves, stem, and panicles to determine organ-specificity in Basmati rice. Enhanced resistance against two Lepidopteran insects, stem bore (*S. incertulas*) and leaf folder (*Cnaphalocrocis medinalis*) was observed. The result of western hybridization and insect bioassays demonstrated that all these promoters express the cry1Ab gene at similar levels in leaves and panicles. The *Cry1Ab* gene was expressed in stems at 0.05% of the total protein under the control of the PEPC promoter alone, or in combination with the pollen-specific promoter. On the other hand, it was expressed at 0.15% under the control of the ubiquitin promoter. Southern blot hybridization of these plants indicated integration of the complete plant transcriptional unit at multiple insertion sites. These results demonstrated that a specific promoter could be used to limit the expression of *Cry1Ab* gene in the desired parts of Basmati rice plants.

Ramesh et al. (2004) developed indica rice (*O. sativa* L.) varieties are known to be recalcitrant to *Agrobacterium*-mediated genetic transformation. We have used the pSB111 super-binary vectors containing *B. thuringiensis* δ-endotoxin synthetic *Cry1Ab* and *Cry1Ac* genes driven by maize ubiquitin promoter, and snowdrop lectin gene gna driven by rice sucrose synthase promoter along with the herbicide resistance gene bar driven by cauliflower mosaic virus 35S promoter, to transform various indica rice lines susceptible to major insect pests, namely, YSB and three sap-sucking insects. The present transformation protocol has been optimized to enhance the frequency of T-DNA transfer into the rice callus cells during cocultivation with *Agrobacterium*. Transformation studies with the gusA containing pTOK233 vector revealed substantial increase (61.54–133.33%) in the number of calli showing transient GUS expression when the calli were treated with 100 mM acetosyringone before cocultivation. After cocultivation with *Agrobacterium*, embryogenic calli were selected on the medium-containing phosphinothricin. Southern blot analyses of primary transformants revealed the stable integration of bar, gna, and cry coding sequences into rice genome with a predominant single copy integration and without any rearrangement of T-DNA. Northern blot analyses revealed the expression of cry and gna genes in different transformants. Molecular analyses of T_1 transgenics showed a monogenic (3:1) pattern of transgene segregation. Furthermore, co-segregation of bar–cry and bar–gna in T_1 lines confirms that bar–cry and bar–gna are integrated at single sites in the rice genome. Transgenic lines expressing cry and gna exhibited substantial resistance against YSB as well as three major sap-sucking insects of rice. This is the first report dealing with the successful introduction of three exotic resistant genes into diverse indica rice lines using pSB111 super-binary vectors of *Agrobacterium tumefaciens*.

Ramesha et al. (2004) developed YSB-resistant transgenic parental lines, involved in hybrid rice, were produced by *Agrobacterium*-mediated gene transfer method. Two pSB111 super-binary vectors containing modified

Cry1Ab/Cry1Ac genes driven by maize ubiquitin promoter and herbicide resistance gene bar driven by cauliflower mosaic virus 35S promoter were used in this study. Embryogenic calli after cocultivation with *Agrobacterium* were selected on the medium-containing phosphinothricin. Southern blot analyses of primary transformants revealed the stable integration of bar, *Cry1Ab*, and *Cry1Ac* coding sequences into the genomes of three parental lines with a predominant single copy integration and without any rearrangement of T-DNA. T_1 progeny plants disclosed a monogenic pattern (3:1) of transgene segregation as confirmed by molecular analyses. Furthermore, the co-segregation of bar and cry genes in T_1 progenies suggested that the transgenes are integrated at a single site in the rice genome. In different primary transformants with alien inbuilt resistance, the levels of *Cry* proteins varied between 0.03% and 0.13% of total soluble proteins. These transgenic lines expressing insecticidal proteins afforded substantial resistance against stem borers. This is the first report of its kind dealing with the introduction of *B. thuringiensis* (*Bt*) cry genes into the elite parental lines involved in the development of hybrid rice.

Kim et al. (2008) developed transgenic Korean rice plants containing the *Cry1Ab* gene for resistance against YSB (*S. incertulas*, YSB). More than 100 independent transgenic lines from three Korean varieties (P-I, P-II, and P-III) were generated. The amount of Cry1Ab in transgenic T_0 plants was as high as 2.88% of total soluble proteins. These levels were sufficient to cause 100% mortality of YSB larvae. The majority of T_1 transgenic lines originated from the varieties P-I and P-II followed a Mendelian fashion of segregation. Deviation from the expected segregation ratio was observed in a small number of the transgenic lines of P-I and P-II origins. However, this deviation was primarily observed in the P-III originated lines. Segregation analysis of the T_1 generation indicated that 1–3 copies of the *Cry1Ab* gene were integrated into the genome of the majority of the transgenic lines originating from varieties P-I and P-II. Stunted and semi-fertile mutants were observed in some transgenic lines. These aberrations were either independent or closely linked to the introduced *Cry1Ab* gene loci in different transgenic lines. Reduction in GUS expression levels and loss of toxicity against YSB larvae were found in some transgenic lines. The transgenic T_3 and T_4 lines causing 100% mortality of third instar YSB larvae in the lab were completely protected in the field. Analysis of important yield components on nine selected transgenic lines indicated that stem length, panicle length, grain number per panicle, and seed setting rates were reduced in transgenic plants compared to those in non-transgenic parental rice lines. Number of panicles per cluster, however, was significantly higher in transgenic plants.

The numerical value of the average yield was in general greater in the controls than in all the transgenic lines, indicating some "yield drag". Since some selected lines were highly resistant to the YSB with good yielding potential, they offer effective potential for use in insect resistance management programs.

Chen et al. (2008) constructed 10 transgenic *B. thuringiensis Bt* rice, *O. sativa* L., lines with different *Bt* genes (two *Cry1Ac* lines, three *Cry2A* lines, and five *Cry9C* lines) derived from Minghui 63 cultivar. The transformed rice was evaluated in both the laboratory and the field against YSB. The transformed rice was screened by using the first instars of two main rice lepidopteran insect species: YSB, *S. incertula*s (Walker) and Asiatic rice borer, *C. suppressalis* (Walker). Bioassay exhibit all transgenic lines have high toxicity to these two rice borers. Field condition evaluation results also showed that all transgenic lines were highly resistant against neonate larvae of *S. incertula*s compared with the non-transformed Minghui63. The highest expression level was found with *Cry9C* gene followed by *Cry2A* gene and the *Cry1Ac* gene by the sandwich enzymelinked immunosorbent assay. When *Bt* transgenic rice culm cuttings with three classes of different Bt proteins (*Cry1Ac*, *Cry2A*, and *Cry9C*) were feeded to the stem borers, significant reduction in weight and mortality were recorded.

Chen et al. (2010) transformed six transgenic rice, *O. sativa* L. lines (G6H1, G6H2, G6H3, G6H4, G6H5, and G6H6) expressing a fused *Cry1Ab/Vip3H* protein. The transgenic rice lines were evaluated for resistance against the Asiatic rice borer, *C. suppressalis* (Walker) (Lepidoptera: Crambidae), and the stem borer *Sesamia inferens* (Walker) (Lepidoptera: Noctuidae) in the laboratory and field. The bioassay findings showed 100% mortality of Asiatic rice borer and *S. inferens* neonate larvae on six transgenic lines at168 h after infestation. The cumulative feeding area of rice by Asiatic rice borer neonate larvae on all transgenic lines was significantly reduced in comparison to untransformed parental "Xiushui 110" rice. Field evaluation showed that damage during the vegetative stage (DH) or during the reproductive stage (white head) caused by Asiatic rice borer and *S. inferens* for transgenic lines was much lower than the control. For three lines (G6H1, G6H2, and G6H6), no damage was found during the entire growing period. Estimation of fused *Cry1Ab/Vip3H* protein concentrations using PathoScreen kit for *Bt Cry1Ab/1Ac* protein indicated that the expression levels of *Cry1Ab* protein both in main stems (within the average range of 0.006–0.073% of total soluble protein) and their flag leaves (within the average range of 0.001–0.038% of total soluble protein) were significantly different among six transgenic lines at different developmental stages. Both laboratory and field researches suggested that the transgenic rice lines have considerable potential for protecting rice from attack by both stem borers.

Kumar et al. (2010) reported marker-free (clean DNA) transgenic rice plant (*O. sativa*), carrying minimal gene expression cassettes. The transgenic *indica* rice contains a translational fusion of two different *B. thuringiensis* (*Bt*) genes, namely *cry1B-1Aa*, motivated by the green tissue specific phosphoenol pyruvate carboxylase (*PEPC*) promoter. Marker-free transgenic Pusa Basmati 1 rice plants were regenerated by mature seed derived calli after co-bombarded with gene expression cassettes of the *Bt* gene and the marker *hpt* gene. The clean DNA fragments for bombardment were obtained by restriction digestion and gel extraction. Through biolistic transformation, 67 independent Pusa Basmati 1 transformants rice were generated. Transformation frequency reached 3.3%, and 81% of the transgenic rice plants were co-transformants. Further, stable integration of the *Bt* gene was confirmed, and the insert copy number was determined by Southern blotting. Western analysis and ELISA revealed a high level of Bit *potein* expression in transgenic plants. Progeny analysis confirmed stable inheritance of the *Bt* gene according to the Mendelian (3:1) ratio. Insect bioassays confirmed complete protection of transgenic plants from YSB infestation. PCR analysis of T_2 progeny of Pusa Basmati 1 plants resulted in the recovery of up to 4% marker-free transgenic rice plants (Dandapat et al., 2014).

The primary technical constraint plant scientists face in generating insect resistant transgenic crops with insecticidal *B. thuringiensis* (*Bt*) crystal protein (*Cry*) genes remains failing to generate sufficiently large numbers of effective resistant transgenic plant lines. One possible means to overcome this challenge is through deployment of a *cry* toxin gene that contains high levels of insecticidal specific activity for target insect pests. In the present study, we tested this hypothesis using a natural variant of the *Cry1Ab* toxin under laboratory conditions that possessed increased insecticidal potency against the YSB (YSB, *Scirpophaga incertulus*), one of the most damaging rice insect pests. Following adoption of a stringent selection strategy for YSB resistant transgenic rice lines under field conditions, results showed recovery of a significantly higher number of YSB resistant independent transgenic plant lines with the variant *cry1Ab* gene relative to transgenic plant lines harboring *cry1Ab berliner* gene. Structural homology modeling of the variant toxin peptide with the *Cry1Aa* toxin molecule, circular dichroism spectral analysis, and hydropathy plot analysis indicated that serine substitution by phenylalanine at amino acid position 223 of the Cry1Ab toxin molecule resulted in a changed role for α-helix 7 in domain I of *Cry1Ab* for enhanced toxicity (Dandapat et al., 2014).

Wan et al. (2014) studied breeding of a new restorer line KR022 containing stacked BPH-resistance genes *Bph14* and *Bph15*, *Bt* gene *cry1C*. A rice restorer line KR022 with BPH-resistance genes *Bph14* and *Bph15* was used as a recurrent parent to cross with the transgenic rice cultivar T1C-19 of *cry1C* and *bar* genes during the breeding process. The restorer line KR022 was developed from the backcross populations of R022 and T1C-19 through molecular marker assisted selection and glufosinate-resistance selection process. The *cry1C* and *bar* genes were found to integrate on chromosome 11 of rice var. KR022, and the genome recovery of KR022 was up to 95.8% of the R022 genome. The quantification of *cry1C* protein expression showed different level of expression of *cry1C* protein at different levels in the leaf, stem, panicle, endosperm and root of KR022 and its hybrid rice. The insect resistance assessment showed that KR022 and its hybrid rice had good resistance to rice leaf folder and stem borer under laboratory and field condition. The field trial showed there was no significant difference in the different agronomic traits between KR022 and its recurrent parent R022. The different four hybrids from KR022 showed higher than the control II-You 838. Transgenic rice KR022 and its hybrid were also found resistance to the herbicide glufosinate.

8.4.3 PROTEASE INHIBITOR

There are several natural toxins proteins derived from different plant sources were used against the insect pest and in disease control. Among the natural plant toxins PIs are one of the parts of the natural plant defense coordination against insect predation (Johnson et al., 1989). PIs play a task in the plant defense response against the insects via the inhibition of digestive protease present in the insect mid-gut. PI binds to the active site of the insect mid-gut enzyme to form a complex with a very low dissociation constant, thus effectively blocking the active site. The mechanism of bindings of the plant PI to the insect protease appears to be similar with all the four classes (serine, cysteine, aspartic, and metallo) of inhibitor (Walker et al., 1998). Insecticidal effects of PI, especially cysteine and serine PIs have been studied by in vitro inhibition or diet incorporation assay studies against the different pests (Oppert et al., 2003). Different PI causes delayed growth and development, reduced fecundity, and sometimes increased mortality to the insects. Thus use of genes encoding PIs to transform crop plants for resistance to insect pests has been well documented in the different cereal crops (Schuler et al., 1998).

Bhutani et al. (2006) transformed two Basmati rice varieties (Pusa Basmati 1 and Taraori Basmati) and a Japonica rice variety TNG67 (Tainung 67) by using LBA4404 (pSB1 + pRKJ1 and pSB1 + pRKJ2) strains harboring potato proteinase inhibitor 2 gene (Pin2) driven by Pin2 wound inducible promoter. B-glucuronidase gene with an intron (pRKJ2) was used for transient Gus (uidA) expression and production of stable uidA expressing calluses. In the transgenic rice presence and copy number of the transgene was confirmed by a simple, rapid and low-cost polymerase chain reaction method. Out of 108 independent regenerated transgenic T_0 plants, 96 plants were confirmed the presence of transgene integration. More than 90% of the transgenic plants were fertile and roots of most of the plants harboring plasmid pSB1 + pRKJ2 were found to be Gus positive (88.5%). Cut-leaf, cut-stem, and whole plant bioassay of 10 T_2 transgenic rice lines of three cultivars during vegetative and earhead stage of plants under greenhouse and laboratory conditions showed significantly higher larval mortality and lower damaged leaf area as compared to the control (non-transgenic) plants. The positive transgenic plants indicating increased resistance against YSB (*S. incertulas* Walker) in field and under greenhouse condition.

Rao et al. (2009) developed transgenic rice Swarna, the most popular indica rice cultivar (*O. sativa* L.) in South-East Asia, with a potato chymotrypsin inhibitor gene (*pin2*) through *Agrobacterium*-mediated transformation. Four out of nine primary transgenic plants had a single-copy T-DNA insertion while other five plants had two copies. Mendelian pattern of inheritance of the transgene (*pin2*) was observed in the T_1 generation progeny plants. Whole plant bioassays conducted at both vegetative and reproductive stages and cut stem assays showed enhanced levels of resistance of transgenic rice against YSB. The transgenic rice lines with plant derived proteinase inhibitor genes would develop into resistant cultivars to fit into resistance breeding strategies as an important component of IPM in rice.

8.4.4 RNAI APPROACH

Kola et al. (2016) attempted to determine the effect of dsRNA designed from two genes Cytochrome P450 derivative (CYP6) and Aminopeptidase N (APN) of rice YSB on growth and development of insect. The bioassay was determined by the injecting of chemically synthesized 5′ FAM labeled 21-nt dsRNA into rice cut stems. The injected stem was further allowed to feed the larvae, which resulted in increased mortality and observed growth and development changes in larval length and weight compared with its untreated control at 12–15 days. The results were further supported by the reduction in transcripts expression of the above said genes in treated larvae. The finding of study clearly showed that YSB larvae fed on dsRNA designed from Cytochrome P450 and Aminopeptidase N has detrimental effect on larval growth and development of YSB resistance in rice using RNAi approach.

8.5 CONCLUSION

There is urgent need to develop new rice varieties against YSB resistance. Many research works have been conducted on the natural resistance system(s) against different insect-pest of rice, that is, brown plant hopper, green leafhopper, white-backed plant hopper, and rice gall midge. Similarly, there is an urgent need to identification and characterization of novel resistance gene(s) from the existing cultivated rice varieties or wild rice accessions to confer resistance against YSB. The resistance rice varieties used as a core of IPM in rice production. For the development of effective IPM, insect resistance genes are projected to participate a crucial on role. Thus, here is an insistent need to identify and clone the possible resistance genes and utilize them by introgressing into the elite rice germplasms.

KEYWORDS

- **dead heart**
- **white ear head**
- **transgenic**
- **protease inhibitor**
- **insect resistance**

REFERENCES

Aguda, R. M.; Datta, K.; Tu, J.; Datta, S. K.; Cohen, M. B. Expression of *Bt* Genes under Control of Different Promoters in Rice at Vegetative and Flowering Stages. *Int. Rice Res. Not.* **2001,** *26,* 26–27.

Alam, M. F.; Datta, K.; Abrigo, E.; Oliva, N.; Tu, J.; Virmani, S. S.; Datta, S. K. Transgenic Insect-resistant Maintainer Line (IR68899B) for Improvement of Hybrid Rice. *Plant Cell Rep.* **1999,** *18,* 572–575.

Alam, M. F.; Datta, K.; Abrigo, E.; Vasquez, A.; Senadhira, D. Production of Transgenic Deep Water Indica Rice Plants Expressing a Synthetic *Bacillus thuringiensis Cry1Ab* Gene with Enhanced Resistance to Yellow Stem Borer. *Plant Sci.* **1998,** *35,* 25–30.

Alcantara, E. P.; Aguda, R. M.; Curtiss, A.; Dean, D. H.; Cohen, M. B. *Bacillus thuringiensis* δ-endotoxin binding to Brush Border Membrane Vesicles of Rice Stem Borers. *Arch. Insect Biochem. Physiol.* **2004,** *559,* 169–177.

All India Coordinated Rice Improvement Programme (AICRIP), Indian Councul of Agricultural Research (NCIPM), State Agricultural Universities, and DPPQ & S, Faridabad, 1988.

Balachandran, S.; Chandel, G.; Alam, M. F.; Tu, J.; Virmani, S. S.; Datta, S. K. Improving Hybrid Rice through Another Culture and Transgenic Approaches. In: 4th International Symposium on Hybrid Rice, Hanoi, Vietnam, 2002; pp 105–118.

Baloch, S. M.; Abdullah, K. Effect of Planting Techniques on Incidence of Stem Borers (*Scirpophaga* spp.) in Transplanted and Direct Wet-seeded Rice. *Pak. J. Zool.* **2011,** *43,* 9–4.

Banerjee, S. N. In: Symposium on Rice Insects. *Trop. Agric. Res. Ser.* **1971,** *5,* 83–90.

Bashir, K.; Husnain, T.; Fatima, T.; Riaz, N.; Makhdoom, R.; Riazuddin, S. Novel Indica Basmati Line (B-370) Expressing Two Unrelated Genes of *Bacillus thuringiensis* is Highly Resistant to Two Lepidopteran Insects in the Field. *Crop Prot.* **2005,** *24,* 870–879.

Behura, N.; Sen, P.; Kar, M. A. Introgression of Yellow Stem Borer (*Scirpophaga incertulus*) Resistance Genes into Cultivated Rice (*Oryza* sp.) from Wild Species. *Ind. J. Agric. Sci.* **2011,** *81,* 359–362.

Bentur, J. S. *Host Plant Resistance to Insects as a Core of Rice IPM*; Science, Technology and Trade for Peace and Prosperity (IRRI, ICAR), McMillan India Ltd., 2006; pp 419–435.

Bhattacharya, J.; Mukherjee, R.; Banga, A.; Dandapat, A.; Mandal, C. C.; Hossain, M. A. *A Transgenic Approach for Developing Insect Resistant Rice Plant Types*; Science, Technology and Trade for Peace And Prosperity (IRRI, ICAR). McMillan India Ltd. 2006; pp 245–264.

Bhutani, S.; Kumar, R.; Chauhan, R.; Singh, R.; Chowdhury, V. K.; Chowdhury, J. B.; Jain, R. K. Development of Transgenic Indica Rice Plants Containing Potato Proteinase Inhibitor 2 (Pin2) Gene with improved Defense against Yellow Stem Borer. *Physiol. Mol. Biol. Plant.* **2006,** *12,* 43–52.

Brar, D. S.; Singh, R. J.; Mahal, M. S.; Singh, B. Effectiveness of Seedling Root Tip Treatment with Some Insecticides for the Control of Stem Borer on Basmati Rice, Pest Management. *Ecol. Zool.* **1994,** *2,* 119–122.

Catling, H. D.; Islam, Z.; Patrasudhi, R. Assessing Yield Losses in Deepwater Rice Due to Yellow Stem Borer *Scirpophaga incertulas* (Walker) in Bangladesh and Thailand. *Crop Prot.* **1987,** *6,* 20–27.

Chaudhary, R. C.; Khush, G. S.; Heinrichs, E. A. Varietal Resistance to Rice Stem Borers in Asia. *Insect Sci. Appl.* **1984,** *5,* 447–463.

Chelliah, A.; Benthur, J. S.; Prakasa, R. P. S. Approaches to Rice Management—Achievements and Opportunities. *Oryza* **1989,** *26,* 12–26.

Chen, H.; Tang, W.; Xu, C.; Li, X.; Lin, Y.; Zhang, Q. Transgenic Indica Rice Plants Harboring a Synthetic *cry2A* Gene of *Bacillus thuringiensis* Exhibit Enhanced Resistance against Lepidopteran Rice Pests. *Theor. Appl. Genet.* **2005,** *111,* 330–337.

Chen, H.; Zhang, G.; Zhang, O.; Lin, Y. Effect of Transgenic *Bacillus thuringiensis* Rice Lines on Mortality and Feeding Behavior of Rice Stem Borers (Lepidoptera: Crambidae). *J. Econ. Entomol.* **2008,** *101,* 182–189.

Chen, Y. H.; Romena, A. Feeding Patterns of *Scirpophaga incertulas* (Lepidoptera: Crambidae) on Wild and Cultivated Rice during the Booting Stage. *Environ. Entomol.* **2006,** *35,* 1094–1102.

Chen, Y.; Tian, J.; Shen, Z.; Peng, Y.; Hu, C.; Guo, Y.; Ye, G. Transgenic Rice Plants Expressing a Fused Protein of *Cry1Ab/Vip3H* Has Resistance to Rice Stem Borers under Laboratory and Field Conditions. *J. Econ. Entom.* **2010**, *103*, 1444–1453.

Cheng, X.; Sardana, R.; Kaplan, H.; Altosaar, I. *Agrobacterium*-transformed Rice Plants Expressing Synthetic *cryIA(b)* and *cryIA(c)* Genes are Highly Toxic to Striped Stem Borer and Yellow Stem Borer. *Proc. Natl. Acad. Sci. U. S. A.* **1998**, *95*, 2767–2772.

Dandapat, A.; Bhattacharyya, J.; Gayen, S.; Chakraborty, A.; Banga, A.; Mukherjee, R.; Mandal, C. C.; Hossain, M. A.; Roy, S.; Basu, A.; Sen, S. K. Variant *cry1Ab* entomocidal *Bacillus thuringiensis* Toxin Gene Facilitates the Recovery of an Increased Number of Lepidopteran Insect Resistant Independent Rice Transformants against Yellow Stem Borer (*Scirpophaga incertulus*) Inflicted Damage. *J. Plant Biochem. Biotech.* **2014**, *23*, 81–92.

Datta, K.; Baisakh, N.; Thet, K. M.; Tu, J.; Arboleda, M.; Oliva, N.; Datta, S. K. Transgenesis-breeding for Multiple Plant Protection. *Philipp. J. Crop. Sci.* **2003**, *27*, 29.

Datta, K.; Vasquez, A.; Tu, J.; Torrizo, L.; Alam, M. F.; Oliva, N.; Abrigo, E.; Khus, G. S.; Datta, S. K. Constitutive and Tissue Specific Differential Expression of *Cry1Ab* Gene in Transgenic Rice Plants Conferring Resistance of Rice Insect Pest. *Theor. Appl. Genet.* **1998**, *97*, 20–30.

Deka, S.; Barthakur, S. Overview on Current Status of Biotechnological Interventions on Yellow Stem Borer *Scirpophaga incertulas* (Lepidoptera: Crambidae) Resistance in Rice. *Biotech. Adv.* **2010**, *28*, 70–81.

Fang, J. C.; Guo, H. F.; Wang, J. P. Mechanism of Rice Variety Resistance to the Yellow Stem Borer, *Scirpophaga incertulas* (Walker). *Chin. Rice Res. Newslett.* **2002**, *10*, 18–19.

Gayen, S.; Samanta, M. K.; Hossain, M. A.; Mandal, C. C.; Sen, S. K. A Deletion Mutant *ndv200* of the *Bacillus thuringiensis vip3BR* Insecticidal Toxin Gene is a Prospective Candidate for the Next Generation of Genetically Modified Crop Plants Resistant to Lepidopteran Insect Damage. *Planta* **2015**, *242*, 269–281.

Ghareyazie, B.; Alinia, F.; Menguito, C. A.; Rubia, L. G.; de Palma, J. M.; Liwanag, E. A.; Cohen, M. B.; Khush, G. S.; Bennett, J. Enhanced Resistance to Two Stem Borers in an Aromatic Rice Containing a Synthetic *cryIA(b)* Gene. *Mol. Breed.* **1997**, *3*, 401–414.

Gopalswamy, S. V. S.; Subbaratanam, G. V.; Sharma, H. C. Developmentof Resistance in Insects to Transgenic Plant with *Bacillus thuringiensis* Gene. Current Status and Management Strategies. *Res. Pest Manage. News Lett.* **2003**, 230–235.

Gosal, S. S.; Gill, I. M. S.; Gill, R.; Sindhu, A. S. Tissue Culture and Transformation of Some Field Crops. Food Security and Environment Protection in the New Millennium. In *Proc. Asian Agril. Cong.* Manila, 2001; p 314.

Gosal, S. S.; Gill, R.; Sindhu, A. S.; Navraj, K.; Dhaliwal, H. S.; Christou, P. Introducing the *Cry*1Ac Gene into Basmati Rice and Transmitting Transgenes to R3 Progeny. In: Proceedings of the International Rice Research Institute, 2003; pp 558–560.

Ho, N. H.; Baisakh, N.; Oliva, N.; Datta, K.; Frutos, R.; Datta, S. K. Translational Fusion Hybrid Bt Genes Confer Resistance against Yellow Stem Borer in Transgenic Elite Vietnamese Rice Cultivars. *Crop Sci.* **2006**, *46*, 781–789.

Husnain, T.; Asad, J.; Maqbool, S. B.; Datta, S. K.; Riazuddin, S. Variability in Expression of Insecticidal Cry1Ab Gene in Indica Basmati Rice. *Euphytica* **2002**, *128*, 121–128.

Husnain, T.; Bokhari, S. M.; Riaz, N.; Fatima, T.; Shahid, A. A.; Bashir, K.; Jan, A.; Riazuddin, S. Pesticidal Genes of *Bacillus thuringiensis* in Transgenic Rice Technology to Breed Insect Resistance. *Pak. J. Biochem. Mol. Biol.* **2003**, *36*, 133–142.

Imayavaramban, V.; Thanunathan, K.; Thiruppathi, M.; Singaravel, R.; Dandapani, A.; Selva-kumar, P. Effect of Combining Organic and Inorganic Fertilizers for Sustained Productivity of Traditional Rice cv. Kambanchamba. *Res. Crop.* **2004,** *5,* 11–13.

Intikhab, S.; Karim, S.; Riazuddin, S. Natural Variation among Rice Yellow Stem Borer and Rice Leaf Folder Populations to C-delta Endotoxins. *Pak. J. Biol. Sci.* **2000,** *3,* 1285–1289.

Islam, Z.; Catling, H. D. Biology and Behaviour of Rice YSB in Deep Water Rice. *J. Plant Prot. Trop.* **1991,** *8,* 85–86.

Israel, P.; Abraham, T. P. Technique for Assessing Crop Losses Caused by Stem Borers in Tropical Areas. In: Proc. Symp. Major Insect Pests of Rice Plant, IRRI, John Hopkins Press: Baltimore, MD, 1967; pp 266–275.

Johnson, R.; Narvaez, J.; An, G.; Ryan, C. Expression of Proteinase Inhibitors I and II in Transgenic Tobacco Plants: Effects on Natural Defense against *Manduca sexta* Larvae. *Proc. Natl. Acad. Sci. U.S.A.* **1989,** *86,* 871–875.

Kapur, A. P. Taxonomy of the Rice Stem Borers. In: The Major Insect Pest of Rice Plant. Proc. Symposium at International Rice Research Institute, Philippines, Johns Hopkins Press: Baltimore, MD, 1967; pp 3–43.

Khan, R. A.; Junaid, A. K.; Jamil, F. F.; Hamed, M. Resistance of Different Basmati Rice Varieties to Stem Borers under Different Control Tactics of IPM and Evaluation of Yield. *Pak. J. Bot.* **2005,** *37,* 319–324.

Khanna, H.; & Raina, S.; Elite Indica Transgenic Rice Plants Expressing Modified *Cry1Ac* Endotoxin of *Bacillus thuringiensis* Show Enhanced Resistance to Yellow Stem Borer. *Transgen. Res.* **2002,** *11,* 411–423.

Khush, G. S. Breeding Rice for Resistance to Insects. *Protect. Ecol.* **1984,** *7,* 147–165.

Kim, S.; Kim, C.; Li, W.; Kim, T.; Li, Y.; Zaidi, M. A.; Altosaar, I. Inheritance and Field Performance of Transgenic Korean *Bt* Rice Lines Resistant to Rice Yellow Stem Borer. *Euphytica* **2008,** *164,* 829–839.

Kola, V. S.; Renuka, P.; Padmakumari, A. P.; Mangrauthia, S. K.; Balachandran, S. M.; Ravindra, B. V.; Madhav, M. S. Silencing of CYP6 and APN Genes affects the Growth and Development of Rice Yellow Stem Borer, *Scirpophaga incertulas. Front. Physiol.* **2016,** *7,* 20.

Kulshreshtha, J. P.; Nigam, P. M. Integrated Management of Key Pest of Paddy. In: *Recent Advances in Entomology*; Mathur, Y. K., Bhattacharya, Upadhyay, N. D., Srivartava, J. P., Eds.; New Gopal Printing Press Parade: Kanpur, 1987; pp 186–211.

Kumar, L. S.; Sawant, A. S.; Gupta, V. S.; Ranjekar, P. K. Genetic Variation in Indian Popu-lations of *Scirpophaga incertulas* as Revealed by RAPD-PCR Analysis. *Biochem. Genet.* **2001,** *39,* 43–57.

Kumar, S.; Arul, L.; Talwar, D. Generation of Marker-free *Bt* Transgenic *indica* Rice and Evaluation of its Yellow Stem Borer Resistance. *J. Appl. Genet.* **2010,** *51,* 243–257.

Kumar, S.; Chandra, A.; Pandey, K. C. *Bacillus thuringiensis* (*Bt*) Transgenic Crop: An Envi-ronment Friendly Insect-pest Management Strategy. *J. Environ. Biol.* **2008,** *29,* 641–653.

Lee, M.; Aguda, R. M.; Cohen, M. B.; Gould, F. L.; Dean, D. H. Determination of Binding of *Bacillus thuringiensis* Delta Endotoxin Receptors to Rice Stem Borer Midguts. *Appl. Environ Biol.* **1997,** *63,* 1453–1459.

Lee, J.; Bricker, T. M.; Lefevre, M.; Pinson, S. R. M.; Oard, H. J. Proteomic and genetic approaches to identifying defence-related proteins in rice challenged with the fungal pathogen *Rhizoctonia solani. Mol. Plant Pathol.* **2006,** *7,* 405–416.

Maiti, M. K.; Nayak, P.; Basu, A.; Dandapat, A.; Mandal, C.; Ghose, D.; and Sen, S. K. Performance of *Bt* IR64 Rice Plants Resistant Against Yellow Stem Borer in their Advanced Generations. Food Security and Environment Protection in the New Millennium. In Proc. Asian Agril. Congress. Manila, 2001; pp 314.

Malhi, B. S.; Brar, D. S. Biology of Yellow Stem Borer on Basmati Rice. *J. Insect Sci.* **1998,** *11*, 127–129.

Mann, R. S.; Shukla, K. K.; Screening and Mechanisms of Resistance in Rice against Yellow Stem Borer, *Scirpophaga incertulas* (Walker). *Crop Res.* **2005,** *29*, 300–305.

Manwan, I.; Vega, C. R. Resistance of Rice varieties to Yellow Stem Borer, *Tryporyza incertulas* (Walker). Saturday Seminar, International Rice Research Institute, Los Banos, Laguna, Philippines, 1975; p 37.

Maqbool, S. B.; Riazuddin, S.; Loc, N. T.; Gatehouse, A. M. R.; Christou, P. Expression of Multiple Insecticidal Genesconfers Broad Resistance against a Range of Different Rice Pests. *Mol. Breed.* **2001,** *7*, 85–93.

Maqbool, S. B.; Husnain, T.; Riazuddin, S.; Masson, L.; Christou, P. Effective Control of Yellow Stem Borer and Rice Leaf Folder in Transgenic Rice Indica Varieties Basmati 370 andM7 using the Novel Endotoxin *cryIIA Bacillus thuringinensis* Gene. *Mol. Breed.* **1998,** *6*, 1–7.

Mohankumar, S.; Thiruvengadam, V.; Samiayyan, K.; Shanmugasundaram, P. Generation and Screening of Recombinant Inbred Lines of Rice for Yellow Stem Borer Resistance. *Ind. J. Exp. Biol.* **2001,** *41*, 348–351.

Nair, M. R. G. K. Insects and Mites of Crops in India. NASS USDA Report 2001, Acerage, 2001.

Narain, A.; Kar, M. K.; Kaliaperumal, V.; Sen, P. Development of Monosomic Alien Addition Lines from the Wild Rice (*Oryza brachyantha* A. Chev. et Roehr.) for Introgression of Yellow Stem Borer (*Scirpophaga incertulas* Walker.) Resistance into Cultivated Rice (*Oryza sativa* L.). *Euphytica* **2016,** *19*, 1–11.

Nayak, P.; Basu, D.; Das, S.; Basu, A.; Ghosh, D.; Ramakrishnan, N. A.; Ghosh, M.; Sen, S. K. Transgenic Elite Indica Rice Plants Expressing *Cry*IAc δ-endotoxin of *Bacillus thuringiensis* are Resistant against Yellow Stem Borer (*Scirpophaga incertulas*). *Proc. Natl. Acad. Sci. U.S.A.* **1997,** *94*, 2111–2116.

Oppert, B.; Morgan, T. D.; Hartzer, K.; Lenarcic, B.; Galesa, K.; Brzin, J.; Turk, V.; Yoza, K.; Ohtsubo, K.; Kramer, K. J. Effects of Proteinase Inhibitors on Digestive Proteinases and Growth of the Red Flour Beetle, *Tribolium castaneum* (Herbst) (Coleoptera: Tenebrionidae). *Comp. Biochem. Physiol.* **2003,** *134*, 481–490.

Padhi, G. Role of Amino Acids and Nucleic Acids in Host Plant Resistance of Rice to Yellow Stem Borer, *Scirpophaga incertulas* Wlk. *Environ. Ecol.* **2006,** *24*, 742–745.

Panda, R. S.; Mohanty, S. K.; Behera, L.; Sasmal, S.; Sahu, S. C. Diversity Analysis of Rice Yellow Stem Borer Populations of Orissa Using RAPD Markers. *J. Insect Sci.* **2012,** *25*, 373–379.

Panda, S. K.; Nayak, S. K.; Behera, V. K. Five Efficacy of Five Commercial Formulation of *Bacillus thurinsiensis* var. Karastaki against the Rice Stem Borer and Leaf Folder. *Pest Manage. Ecol. Zool.* **2006,** *7*, 143–146.

Panda, S. K.; Nayak, S. K.; Behera, V. K. Five Efficacy of Five Commercial Formulation of *Bacillus thurinsiensis* var. Karastaki against the Rice Stem Borer and Leaf Folder. *Pest Manage. Ecol. Zool.* **1999,** *7*, 143–146.

Pandya, H. V.; Shah, A. H.; Purohit, M. S.; Patel, C. B. Estimation of Losses due to Rice Stem Borer, *Scirpophaga incertulas* (Walker). *Gujrat Agric. Univ. Res. J.* **1994,** *20,* 164–166.

Patanakanjorn, S.; Pathak, M. D. Varietal Resistance of Rice to Asiatic Rice Stem Borer, Chilo Supprenalis (*Lepidoptera crambidae*), and Its Association with Various Plant Characters. *Ann. Entom. Soc. Am.* **1967,** *60,* 287–292.

Pathak, M. D.; Dyck, V. A. Developing an Integrated Method of Rice Insect Pest Control, *PANS* **1973,** *12,* 534–544.

Pathak, M. D.; Khan, Z. R. *Insect-pests of Rice*; International Rice Research Institute: Manila, Philippines, 1994; p 89.

Pathak, M. D. Ecology of Common Insect-pest of Rice. *Ann. Rev. Entomol.* **1968,** *13,* 257–273.

Pathak, M. D. *Insect-pest of Rice*; International Rice Research Institute: Manila, Philippines, 1977; p 68.

Pinson, S. R. M.; Capdevielle, F. M.; Oard, J. H. Confirming QTLs and finding additional loci conditioning sheath blight resistance in rice using recombinant inbred lines. *Crop Sci.* **2005,** *45,* 503–510.

Raina, S. K.; Khanna, H. K.; Talwar, D.; Tiwari, A.; Kumar, U. Insect Bioassays of Transgenic Indica Ice carrying a Synthetic Bt Toxin Gene, cry 1Ac. Advances in Rice Genetics. In: Proc. 4th Int. Rice Res. Inst. 2003; pp 567–569.

Ramesh. S.; Nagadhara, D.; Reddy, V. D.; Rao, K. V. Production of Transgenic Indica Rice Resistant to Yellow Stem Borer and Sapsucking Insects, using Super-binary Vectors of *Agrobacterium tumefaciens*. *Plant Sci.* **2004,** *166,* 1077–1085.

Ramesha, S.; Nagadhara, D.; Pasalu, I. C.; Kumari, A. P.; Sarma, N. P.; Reddya, V. D.; Raoa, K. V. Development of Stem Borer Resistant Transgenic Parental Lines involved in the Production of Hybrid Rice. *J. Biotechnol.* **2004,** *111,* 131–141.

Rani, P. U.; Kumari, S. I.; Sriramakrishna, T.; Sudhakar, T. R. Kairomones Extracted from Rice Yellow Stem Borer and their Influence on Egg Parasitization by *Trichogramma japonicum* Ashmead. *J. Chem. Ecol.* **2006,** *31,* 59–73.

Rao, M. V. R.; Behera, K. S.; Baisakh, N.; Datta, S. K.; Rao, G. J. N. Transgenic *Indica* Rice Cultivar 'Swarna' Expressing a Potato Chymotrypsin Inhibitor *pin2* Gene Show Enhanced Levels of Resistance to Yellow Stem Borer. *Plant Cell: Tissue Organ Cult.* **2009,** *99,* 277–285.

Rao, V. P. Surveys for Natural Enemies of Pests of Paddy. Final Technical Report of Project; CIBC: Bangalore, India, **1964**; p 85.

Rath, P. C. Evaluation of Some *Bacillus thuringiensis* and Neem Formulation against Yellow Stem Borer of Rice, Under Rainfed Lowlands. *Oryza* **1999,** *36,* 398–399.

Riaz, N.; Husnain, T.; Fatima, T.; Makhdoom, R.; Bashir, K.; Masson, L.; Altosaar, I.; Riazuddin, S. Development of Indica Basmati Rice Harboring Two Insecticidal Genes for Sustainable Resistance against Lepidopteran Insects. *S. Afr. J. Bot.* **2006,** *72,* 217–223.

Sarwar, M. Valuation of Some Aromatic Rice (*Oryza sativa* L.) Genetic Materials to Achieve Tolerant Resources for Rice Stem Borers (Lepidoptera: Pyralidae). *Int. J. Sci. Res. Environ. Sci.* **2013,** *1,* 285–290.

Satpathi, C. R.; Chakraborty, K.; Shikari, D.; Acharjee, P. Consequences of Feeding by Yellow Stem Borer (*Scirpophaga incertulas* Walk.) on Rice Cultivar Swarna Mashuri (MTU 7029). *World Appl. Sci. J.* **2012,** *17,* 532–539.

Schuler, T. H.; Poppy, G. M.; Kerry, B. R.; Denholm, I. Insect Resistant Transgenic Plants. *Trends Biotechnol.* **1998**, *16*, 168–175.

Sehgal, M.; Jeswani, M. D.; Kalra, N. Management of Insect, Disease and Nematode Pests of Rice–Wheat in the Indo-Gangetic Plains. *J. Crop Prod.* **2001**, *4*, 167–226.

Selvi, A.; Shanmugasundaram, P.; Mohankumar, S.; Raja, J. A. J. Molecular Marker Association for Yellow Stem Borer Resistance in Rice. *Mol. Biol.* **2003**, *3*, 117–124.

Selvi, A.; Shanmugasundaram, P.; Mohankumar, S.; Raja, J. A. J.; Sadasivam, S. Microsatellite Markers for Yellow Stem Borer, *Scirpophaga incertulas* (Walker) Resistance in Rice. *Plant Cell Biotechnol. Mol. Biol.* **2002b**, *3*, 117–124.

Selvi, A.; Shanmugasundaram, P.; Mohankumar, S.; Raja, J. A. J. Molecular Markers for Yellow Stem Borer, *Scirpophaga incertulas* (Walker) Resistance in Rice. *Euphytica* **2002a**, *124*, 371–377.

Shamim, M.; Khan, N. A.; Singh, K. N. Inhibition of Midgut Protease of Yellow Stem Borer (*Scirpophaga incertulas*) by Cysteine Protease-like Inhibitor from Mature Jackfruit (*Artocarpus heterophyllus*) Seed. *Acta Physiol. Plant.* **2011**, *33*, 2249–2257.

Shu, Q. U.; Ye, G. Y.; Cui, H. R.; Cheng, X. Y.; Xiang, Y. B;, Wu, D. X.; Gao, M. W.; Xia, Y. W.; Hu, C.; Sardana, R.; Altosaar, I. Transgenic Rice Plants with a Synthetic cry1Ab Gene from *Bacillus thuringiensis* were Highly Resistant to Eight Lepidopteran Rice Pest Species. *Mol. Breed.* **2000**, *6*, 433–439.

Slamet, L. I. H.; Novalina, S.; Damayanti, D.; Sutrisno, Christou, P.; Aswidinoor, H. Inheritance of *cry1Ab* and Snowdrop lectin gna Genes in Transgenic Javanica Rice Progenies and Bioassay for Resistance to Brown Plant Hopper and Yellow Stem Borer; IRRI, 2003; pp 565–566.

Sunio, L. M.; Heinrichs, E. A.; Tryon, E. H. Evolution of Rice Cultivars for Resistance to Yellow Stem Borer, *Scirpophaga incertulas* (walker). In: Saturday Seminar, International Rice Research Institute: Los Bonos, Leguna, Philippines, **1986**; p 15.

Sutrisno, H. Mitochondrial DNA Variation of the Rice Yellow Stem Borer, *Scirpophaga incertulas* (Lepidoptera: Crambidae) in Java, Indonesia. *Treubia* **2015**, *42*, 9–12.

Tripathi, M. K.; Senapati, B.; Dash, S. K. Pest Status and Seasonal Incidence of Stem Borer Complex of Rice in Semi Deep Water Situation at Bhubaneswar. *J. Appl. Biol.* **1997**, *7*, 71–74.

Tu, J.; Zhang, G.; Datta, K.; Xu, C.; He, Y.; Zhang, Q.; Khus, G. S.; Datta, S. K. Field Performance of Transgenic Elite Commercial Hybrid Rice Expressing *Bacilus thuringiensis* Delta Endotoxin. *Nat. Biotechnol.* **2000**, *18*, 1101–1104.

Walker, A. J.; Ford, L.; Majerus, M. E. N.; Eoghegan, I. E.; Birch, A. N. E.; Gatehouse, J. A.; Gatehouse, A. M. R. Characterisation of the Midgut Digestive Proteinase Activity of the Two-spot Ladybird (*Adalia bipunctata* L.) and its Sensitivity to Proteinase Inhibitors. *Insect. Biochem. Mol. Biol.* **1998**, *28*, 173–180.

Wan, B.; Zha, Z.; Li, M.; Xia, M.; Du, X.; Lin, Y.; Yin, D. Development of Elite Rice Restorer Lines in the Genetic Background of R022 Possessing Tolerance to Brown Planthopper, Stem Borer, Leaf Folder and Herbicide through Marker-assisted Breeding. *Euphytica* **2014**, *195*, 129–142.

Wan, J. M. Perspectives of Molecular Design Breeding in Crops. *Acta Agron. Sin.* **2006**, *32*, 455–462.

Wang, Y.; Zhang, L.; Li, H.; Han, L,; Liu, Y.; Zhu, Z.; Wang, F.; Peng, Y. Expression of Cry1Ab Protein in a Marker-free Transgenic Bt Rice Line and its Efficacy in Controlling a Target Pest *Chilo suppressalis* (Lepidoptera: Crambidae). *Environ. Entomol.* **2014**, *43*, 528–536.

Wu, C.; Fan, Y.; Zhang, C.; Oliva, N.; Datta, S. K. Transgenic Rice Plants Resistant to Yellow Stem Borer. *Plant Cell Rep.* **1997,** *17,* 129–132.

Wunn, J.; Kloti, A.; Burkhardt, P. K.; Biswas, G. C. G.; Launis, K.; Iglesias, V. A.; Potrykus, I. Transgenic Indica Rice Breeding Line IR58 Expressing a Synthetic cryIA(b) Gene from *Bacillus thuringiensis* Provides Effective Insect Pest Control. *Biotechnology* **1996,** *14,* 171–176.

Xu, H. X.; Lu, Z. X.; Chen, J. M.; Zhengand, X. S.; Yu, X. P. Resistance of Different Rice Varieties to the Striped Stem Borer, *Chilo suppressalis,* and its Relationship with the Morphological and Anatomic Characteristics of Rice. *Acta Phytol. Sin.* **2006,** *33,* 241–245.

Yadav, P.; Khan, N. A.; Shamim, M.; Srivastava, D.; Singh, K. N. Efficacy of Jackfruit Protease Inhibitor and *Bt* Protein against Yellow and Pink Stem Borers of Rice. *Curr. Adv. Agric. Sci.* **2012,** *4,* 152–155.

Ye, G.; Hu, C.; Shu, Q.; Cui, H.; Gao, M. The Application of Detached Leaf Bioassay for Evaluating the Resistance of Bt Transgenic Rice to Stem Borers. *Acta Phytoph. Sin.* **2000,** *27,* 1–6.

CHAPTER 9

MOLECULAR TOOLS FOR CONTROLLING RICE LEAF FOLDER (*CNAPHALOCROCIS MEDINALIS*)

MAHESH KUMAR[1*], JUHI[2], NAGATEJA NATRA[3], TUSHAR RANJAN[1], and RAVI RANJAN KUMAR[1]

[1]*Department of Molecular Biology and Genetic Engineering, Bihar Agricultural University, Sabour, Bhagalpur 813210, Bihar, India*

[2]*Department of Ethnobiology, Jiwaji University, Gwalior 474011, Madhya Pradesh, India*

[3]*Department of Plant Pathology, Irrigated Agriculture Research and Extension Center, Washington State University, 24106 N. Bunn Road, Prosser, WA 99350, USA*

Corresponding author. E-mail: maheshkumara2z@gmail.com

CONTENTS

ABSTRACT

Rice (*Oryza sativa* L.) is an important staple food crop providing a major source of nutritional calories for nearly one-third of world population. One of the major constraints in rice production is the damage caused by biotic stresses such as pest and disease. Among the insects, the major target pests of rice in Asia are the striped stem borer (SSB; *Chilo suppressalis* Walker), yellow stem borer (YSB; *Scirpophaga incertulas* Walker) (Lepidoptera; Pyralidae), and leaf folder (*Marasmia* spp. and *Cnaphalocrocis medinalis*). Insect pest management by chemicals has brought about considerable protection. Though plant breeders have been successful in developing insect resistant crop plants, their efforts are limited by the availability of insect resistance genes in germplasm and breakdown of resistance. The recent availability of the cloned resistance/ defense genes provides additional tools for genetic engineering of improved cultivars by transformation. Among different transformation methods of rice, particle/microprojectile (biolistic) bombardment is now successfully used in many laboratories. The different insecticidal genes used for the control of insect pests include protease inhibitors, lectins, amylase inhibitors, and δ-endotoxins (*Bt* gene) produced by the soil bacterium, *Bacillus thuringiensis*. Among them, *Bt* gene offers a great scope for controlling insect pests.

9.1 INTRODUCTION

Rice (*Oryza sativa* L.) is cultivated and consumed by 2500 million people in almost 112 countries. About 90% of the world's rice is grown and consumed in Asia (Ahmad et al., 2005). There are large numbers of insect pests, which damage rice crop right from nursery sowing to the harvest causing considerably high yield losses. About 128, different species of insect pests have been reported to attack rice crop. Insect pests damage rice crop at different stages of its growth. Among that, leaves feeding insect pests are of major importance because of their ability to defoliate or to remove the chlorophyll content of the leaves leading to considerable reduction in yield. Rice leaf folder [*C. medinalis* (Guen. Pyralidae Lepidoptera)] considered as pests of minor importance have increased in abundance in the late 1980s and have become major pests in many parts world (Ahmed et al., 2010). Paddy leaf folder is one of the most important insect pests (Gunathilagaraj & Gopalan, 1986). Out of the eight species of leaf folder, the most widespread and important one is *C. medinalis* (Guenee) (Bhatti et al., 1995). *C. medinalis* (LF) has been reported to attain the major pest status in some important paddy growing areas (Maragesan & Chellish, 1987). Second instars leaf folder larvae glues the growing paddy leaves longitudinally for accommodation and feeds on green foliage voraciously that results in papery dry leaves (Khan et al., 1989). Loss incurred to the growing paddy crop is insurmountable (Ahmed et al., 2010). Feeding often results in stunting, curling, or yellowing of plant green foliage (Alvi et al., 2003). The extent of loss may extend up to 63–80% depending on agro-ecological situations (Rajendran et al., 1986).

9.2 RICE LEAF FOLDER

Leaf folder of rice is also one of the most important insect pests in Indian subcontinent (Gunathilagaraj & Gopalan, 1986). Out of the different eight species of leaf folder, the most widespread and important one is *C. medinalis* in rice field (Bhatti et al., 1995). Leaf folder has been reported to attain the major pest status in some important rice growing areas of India (Maragesan & Chellish, 1987). Second instar leaf folder larvae glues the growing rice leaves longitudinally for accommodation and feeds on green foliage voraciously which results in papery dry leaves of rice (Khan et al., 1989). Loss incurred to the growing rice crop is intractable (Ahmed et al., 2010). Feeding often results in stunting, curling, or yellowing of rice green foliage of leaves (Alvi et al., 2003). Bhanu and Reddy (2008) reported that in favorable conditions leaf folder affected the crop adversely resulting in severe losses in rice production (Fig. 9.1).

FIGURE 9.1 Leaf folder symptom in rice (adapted from Rice Knowledge Portal Management, www.rkmp.co.in).

9.2.1 DAMAGE SYMPTOMS

The caterpillar folds the leaves longitudinally and remains inside. It scrapes the green tissues of the leaves and makes them white and dry. During severe infestation, the whole field exhibits scorched appearance.

9.2.2 LIFE CYCLE OF RICE LEAF FOLDER

The adult moth is often seen in the field during daytime. The moth is brownish with many dark wavy lines in center and dark band on margin of wings. The female moth lays eggs in batches of 10–12, which are arranged in linear row in the lower surface of leaves. The eggs are flat, oval in shape, and yellowish white in color. The egg period is 4–7 days. Larva is 15–20-mm long, pale green, transparent, actively moving caterpillar. The larval period is 15–20 days. It pupates inside the leaf fold. The pupa is greenish brown. The pupal period is 6–8 days. Total life cycle of leaf folder is completed in 25–35 days (Fig. 9.2).

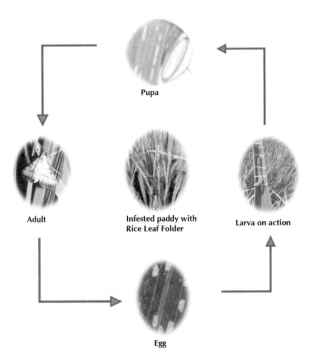

FIGURE 9.2   Life cycle of leaf folder (adapted and modified from Rice Knowledge Portal Management, www.rkmp.co.in).

9.3 MANAGEMENT OF LEAF FOLDER

9.3.1 CULTURAL PRACTICES

- Early planting may help to avoid greater degrees of leaf damage.
- Wider spacing (22.5 × 20 cm and 30 × 20 cm) and low usage of nitrogenous fertilizers decreases leaf damage.
- Excess fertilized plots seem to attract females for oviposition. Therefore, it is advisable to avoid the over fertilization.
- Egg predators (Crickets) inhabit surrounding grass habitats might be worthwhile.
- Higher damages will occur in shaded areas. Therefore, remove the causes of shading within the field.

9.3.2 MECHANICAL PRACTICES

There are several mechanical practices for the leaf folder management in rice:

- Collection of the egg masses and larvae of pest to be placed in bamboo cages for conservation of biocontrol agents.
- Removal and destruction of leaf folder infested part of the plant.
- Use of coir rope in rice for dislodging of leaf folder, cut worm, and case worm larvae, etc. onto kerosinized water (1 L of kerosene mixed on 25 kg soil and broadcast in 1 ha).

9.3.3 CHEMICAL CONTROL

The control of rice insect pests has often relied on extensive use of insecticides, which disrupt the beneficial insects and other insect fauna and also cause environmental contamination. The heavy use of insecticides and high fertilizer rates seem to favor leaf-folder population outbreaks (Gottfried & Fallil, 1980). For the control of this pest in Asia, more than 25% of applied pesticides are aimed to this pest in a year (Heong & Escalanda, 1997). Application of Lorsban, Sumithion, Methyl Parathion, Denital, and Thiodan gave more than 90% mortality of the insect larvae and were statistically at par in controlling the rice leaf folder.

Ramasubbaiah et al. (1980) tested fenthion, phosphamidin, fenitrothion, endosulphon dimethoate, quinalphos, diazinon, and carbaryl against *C. medinalis* and reported that all insecticides gave effective control. Effect of some foliar insecticides against leaf folder and found that cypermethrin and fanvalerate provided effect control and increased the yield of rice significantly. Hence, this study was taken up and the results of which will be useful for the scientists to work out for an eco-friendly integrated pest management; thus, it will support the ultimate beneficiary to the farming community (Ahmed et al., 2010).

9.3.4 BIOLOGICAL CONTROL PRACTICES

The control of the leaf folder used of different biological practices that are given below

Augmentation and conservation

- *Trichogramma japonocum* and *T. chilonis* may be released @ 1 lakh/ ha on appearance of moth of leaf folder/egg masses.
- Natural biocontrol agents such as dragon flies, damsal flies, spiders, drynids, mired bugs, Bracon, carabids, grasshopper, Tetrastichus, coccinellids, etc. should be conserved.
- Collection of egg masses of Rice leaf folder and putting them in a bamboo cage-cum-percher till flowering which will permit the escape of egg parasites and trap and kill the hatching larvae. Besides, these would allow perching of predatory birds.

Generally, the threshold limit of leaf folder in rice field is "Early to tillering stage two fully damaged leaves (FDL) with larva per hill and Panicle initiation to booting stage two FDL per hill". The release of the parasitoid and predators of the rice leaf folder in the field of rice is important to control the leaf folder different stages, namely, egg, larval, pupal, and adult. Major parasitoids and predators are mentioned in Table 9.1.

TABLE 9.1 Major Parasitoids and Predators of Rice Leaf Folder.

Natural enemy category	Natural enemy	Pest attacked and feeding potential
I Parasitoid		
Egg parasitoid	*Trichogramma chilonis*	Egg parasitoid of leaf folder
Larval parasitoids	*Stenobracon nicevillei*	Larval parasitoid of leaf folder
Larval and pupal parasitoids	*Brachymerta lasus* and *Brachymerta excarinata*	Larval and pupal parasitoid of leaf folder
II Predators		
Coccinellid beetles	*Micrapis hirashimai,* ladybird beetles	Preying on small hopper, small larvae, and exposed eggs
	Harmonia octamaculata	Preying on small hopper, small larvae, and exposed eggs
Spiders	*Pardosa psuedoannulata,* wolf spider	Prey leaf folder moth, leaf-and-plant hopper, and stem borer fly
	Tetragnatha maxillosa, long-jawed spider	Prey leaf folder moth, leaf-and-plant hopper, and stem borer
Carabid beetle	*Ophionea nigrofasciata,* ground beetle	Preying leaf folder larvae and plant hoppers
Rove beetle	*Paederus fuscipes*	Preying eggs, and small moths

9.4 MOLECULAR APPROACHES

Stem borer and leaf folder are the most serious and devastating insects of rice. When compared to all other insect pests of rice, stem borer occurs throughout Asia and cause chronic yield losses of 2–10% (Savary et al., 2000). Although the chemical pesticides are effective against some pest populations, they have broad host range and toxic to noninsect pests also. The worldwide expenditure on insecticide is estimated to be US $ 8.1 million every year (Shelton et al., 2000). Even though the agrochemical industry is investing in the production of safe and more environment friendly pesticides, much higher levels of protection are still required (Gatehouse, 1999).

The use of biological insecticides is simple and cost-effective among all the other approaches to control insect pests on crops. Biological control of insects is more popular as it has several advantages over the chemical pesticides. Among various biocontrol agents, *Bacillus thuringiensis* offers greater scope for controlling insect pests. A number of other strategies including the use of genes for protease inhibitors, lectins, amylase inhibitors, and δ-endotoxins are tried against these insects, among them the insecticidal δ-endotoxin genes present in a soil bacterium *B. thuringiensis* (*Bt*) offers great scope for controlling insect pests (Shelton et al., 2000).

9.4.1 BACILLUS THURINGIENSIS

B. thuringiensis is an endospore-forming soil bacterium characterized by the presence of a crystal protein within the cytoplasm of the sporulating cell. These proteins are known as δ-endotoxin because of their intracellular location and as Cry proteins because of crystalline nature of the inclusions, which exhibit highly specific insecticidal activities against larvae of lepidoptera, coleoptera, and/or diptera (Kostichka et al., 1996). *B. thuringiensis* was initially isolated in Japan by Ishiwata and formally described by Berliner in 1915 (Siegel, 2000). Since Schnepf and Whiteley (1981) cloned *cry* genes in 1981 for the first time, hundreds of *cry* genes have been successfully cloned (Wasano et al., 2001). Initially, *cry* genes were expressed in plant colonizing microorganisms to target the stem and root dwelling insect pests. Nowadays, the most efficient way to deliver *cry* genes seems to be the development of transgenic plants expressing those genes.

9.4.2 MODE OF ACTION OF BT ENDOTOXINS

Bt toxins are only active in insects when ingested orally, as their mode of action is in the midgut. When insects come into contact with and ingest the crystalline inclusions, these inclusions as well as the toxins in them (protoxin) first gets proteolytically processed and converted into active toxin, before binding to the receptors (Boucias & Pendland, 1994; Pientrantonio & Gill, 1996; Watson et al., 1992). So, the specificity of *Bt* strains may be caused by the presence or absence of specific proteases in the insect midgut. The target organ for *Bt* toxin is the midgut. There is an important difference in the mode of action of *Bt* toxins produced in the *Bt* plant. Transgenic *Bt* plants produce toxins instead of protoxins, as only the part of the gene that codes for the active toxin is inserted in the plant genome. As a result, after ingestion the δ-endotoxin may bind directly to specific receptors on the midgut membrane, without the involvement of insect proteolytic enzymes. In the case of *Bt* plants, specificity may be based on the presence of specific receptors. So far, the identified receptors to which δ-endotoxin may bind are general receptors, that is, amino-peptidase-N-proteins, cadherin-like proteins or glycolipids (Ferre et al., 1991; Gahan et al., 2001; Van Rie et al., 1990). Binding of the toxins to the receptors is a prerequisite for their insecticidal activity, although binding alone is not sufficient for toxicity (de Maagd et al., 1999a,b, 2001; Schnepf et al., 1998). The rate at which the ligand is inserted into the target membrane decisively influences its relative toxicity (Boucias & Pendland, 1994). However, only when binding is followed by pore formation in the midgut membrane, the midgut cells get disrupted and the insect dies (de Maagd et al., 1999a,b, 2001; Jenkins et al., 2000; Schnepf et al., 1998).

9.4.3 CLASSIFICATION OF CRYSTAL PROTEIN GENES

The first attempt to systematically classify *Bt*-crystal protein genes was undertaken by Hofte and Whiteley (1989). The principle toxicity spectrum was denoted by Roman letters (from I to IV) grouped under different classes such as *cryI*, *cryII*, *cryIII*, and *cryIV* depending on the host specificity as well as the degree of amino acid homology (Crickmore et al., 1998). Within the class, major and minor amino acid differences were denoted by upper and lower case letters, respectively (e.g., *cryIAa*). *Bt*-genes also encode cytolytic proteins, and they are totally different from *cry* genes. They are designated as *cyt* genes. Each crystal protein is characterized by its own particular insecticidal activity and spectrum (Peferoen, 1992). The crystal protein genes are grouped into different classes as described below.

9.4.3.1 CRYI

The *cryI* genes code for lepidopteran specific bipyramidal crystal proteins which are having a molecular weight range of 130–140 kDa. Three crystal protein genes *cryIAa*, *cryIAb*, and *cryIAc* are very closely related; the proteins encoded by these genes show 82–90% amino acid homology.

9.4.3.2 CRYII

The *cryII* class of genes encode about 65–71 kDa proteins which form cuboidal inclusions during sporulation (Sasaki et al., 1997). The *CryII* proteins are toxic to both lepidopteran and dipteran larvae (*Cry2Aa*) or to Lepidopteran larvae alone (*Cry2Ab, Cry2Ac, Cry2Ad*) (Sauka et al., 2005).

9.4.3.3 CRYIII

The *cryIII* class encodes 73 kDa coleopteran specific proteins (Herrn-stadt et al., 1987). CryIIIA and CryIIIB proteins show 66–67% amino acid homology. The CryIIIC are produced as protoxins of about 129 kDa, crystallizes as bipyramidal crystals unlike the CryII toxins which are stored as rhomboid crystals (McPherson et al., 1988).

9.4.3.4 CRYIV

The *cryIV* class of genes is composed of a mixed group of dipteran specific genes encoding polypeptides which are divided into four subclasses, namely, *cryIVA, cryIVB, cryIVC,* and *cryIVD* encoding about 135, 128, 78, and 73 kDa proteins, respectively.

9.4.3.5 CRYV

The *cryV* class of genes encodes a protein having molecular weight of about 80 kDa. They showed toxicity toward coleopteran and lepidopteran larvae (Tailor et al., 1992).

9.4.3.6 CRYVI

The *cryVI* class is reported to exhibit activity against nematodes (Feitelson et al., 1992).

Some *B. thuringiensis* strains have also been reported to be active against mites, nematodes, flatworms, protozoa (Feitelson et al., 1992), and aphids (Walters & English, 1995). Crickmore et al. (1998) revised the nomenclature for the *cry* and *cyt* genes. In this newly revised system, a broad definition was given to *Cry* protein: a parasporal inclusion (crystal) protein from *Bt* that exhibits some experimentally verifiable toxic effect to a target organism, or any protein that has obvious sequence similarity to a known *Cry* protein. Similarly *Cyt* protein denotes a parasporal inclusion protein from *Bt* that exhibit haemolytic activity, or any protein that has obvious sequence simi-larity to a known *Cyt* protein. The known *Cry* and *Cyt* proteins now fall into 32 sets including *Cyt1*, *Cyt2*, and *Cry1* through *Cry28* (Crickmore et al., 1998). Some of the important features of revised nomenclature are as follows:

The *cry* genes whose products share less than 45% amino acid sequence homology are characterized by different Arabic numbers, designated as primary ranks (e.g., *cry1*, *cry2*, etc.). The *cry* genes of the same primary rank whose products show less than 78% amino acid homology are differentiated by secondary ranks by using uppercase letters (e.g., *cry1A*, *cry1B*, etc.). The *cry* genes having same primary and secondary ranks whose products share less than 95% amino acid sequence homology receive separate tertiary rank, designated by lowercase letters (e.g., *cry1Aa*, *cry1Ab*, etc.). The *cry* genes whose products are different in amino acid sequence, but are more than 95% identical to each other are given separate quaternary ranks by another Arabic number (e.g., *cry1Aa1*, *cry1Aa2*, etc.).

9.4.4 APPLICATION OF CRYSTAL PROTEINS

9.4.4.1 BIOPESTICIDES

The use of chemical pesticides is being discouraged due to spread of insect resistance, environmental pollution, and rising costs (worldwide expendi-ture on insecticides $ 8.1 billion each year as reported by Krattiger, 1997) and is not recommended in recent years. Therefore, the use of bioinsecti-cides and an integrated approach to insect pest control has become necessary (Gupta, 1997). Biological pesticides based on the soil microbe, *B. thuringi-ensis* are becoming increasingly important for pest management. Commer-cial *Bt*-based bioinsecticides are usually the formulations of spores and crystalline inclusions that are released upon lysis of *Bt* during its stationary phase of growth. The products are applied at the rate of 10–50 g/acre. The molecular potency of *Bt* toxins is higher than synthetic pyrethroids and 80,000 times higher than organophosphates (Feitelson et al., 1992). The advantages of *Bt* over synthetic pesticides include lack of polluting resi-dues, high specificity to target insects, safety to nontarget organisms such as mammals, birds, amphibians, and reptiles. Biopesticides constitute only 1% of the pesticide market share; out of this, *Bt* products comprise of 95% of biopesticides. Most of the *Bt* products are based on subsp. *kurstaki* strain HD-1. It is effective over 200 crops and against more than 55 lepidopteran species (Justin et al., 1988; Rabindra & Jayaraj, 1998). The subsp. *tene-brionis* and/or *san diego* are effective against coleopteran insects, such as Colorado potato beetle (*Leptinotarsa decemlineata*). Some of the natural *Bt* products like Biobit, Dipet, Foray, Thuricide, Javelin, Spicturin, and geneti-cally modified *Bt* products like Agree, Design, Condor, Cutlass, Leptinox, MATTCH, MTRACK were registered for agricultural use (Navon, 2000).

9.4.5 TRANSGENIC PLANTS EXPRESSING CRYSTAL PROTEIN GENES

The first report on the use of the *Bt* gene expressed in plants for insect control occurred in 1987 (Barton et al., 1999; Fischhoff et al., 1987; Vaeck et al., 1987). After that, *B. thuringiensis* genes were successfully expressed in other economically important crops like rice, potato, maize, cotton, etc. The first field-test of transgenic upland cotton, *Gossypium hirsutum* L., containing δ-endotoxin gene from *Bt* was conducted in 1989 (Jenkins et al., 1991). The expression of a site-specific modification in the coding sequences (DNA), use of improved promoters in *Agrobacterium* Ti plasmid transfor-mation vectors, and the usage of a truncated gene (Perlak et al., 1990) have shown 100 fold increases in toxic protein expression in plant tissues than the wild type. In plants, the modified *cry1Ab* and *cry1Ac* genes expressed toxic protein up to 0.05–0.10% of the total soluble proteins in leaves. In 1996, *Bt* cotton (Bollgard™; Monsanto) was released to protect tobacco budworm and to a lesser extent cotton bollworm and pink bollworm. This *Bt* trans-genic cotton expressing a *cry1Ac* gene was able to increase average yield by 14% with a reduction in insecticide use to 300,000 ga in 1997.

Wunn et al. (1996) transformed the *indica* rice breeding line IR58, by particle bombardment with a truncated version of synthetic *cry1Ab* under the control of the *CaMV35S* promoter and allowed efficient production of the lepidopteron specific δ-endotoxin. The transgenic rice thus produced is resistant to major lepidopteran insect pests of rice and has the potential to significantly decrease yield losses, reduce the use of broad-spectrum chem-ical insecticides, and furthermore, reduce levels of mycotoxins, one of the unexpected benefits of reducing larval attacks (Cheng et al., 1998). Introduc-tion of stem borer resistance into the germplasm of an *indica* rice breeding line made it agronomically important trait available for conventional rice breeding program. The transgenic potato harboring *cry3A* gene is commer-cialized for its resistance against Colorado potato beetle (*L. decemlineata*; Perlak et al., 1993). Due to this, the insecticide usage was reduced by up to 40% for *Bt* plants in 1997 (de Maagd et al., 1999a,b). During 1996, US growers planted one million hectares of transgenic maize, cotton, and potatoes expressing *cry1Ab*, *cry1Ac*, and *cry3A* genes, respectively (Maqbool et al., 1998).

A modified *cry1Ab* gene has been inserted into a japonica rice cultivar, and its expression conferred resistance to two major rice pests, rice leaf folder and SSB (Fujimoto et al., 1993). Transformation of a synthetic *cry1Ab* gene in the IRRI breeding line IR58 resulted in effective control of two of the most destructive insect pests of rice in Asia, the YSB and the SSB (Wunn et al., 1996). Several reports describe rice transformation with *Bt* genes including *cry1Ab* and *cry1Ac* (Cheng et al., 1998; Fujimoto et al., 1993; Ghareyazie et al., 1997; Nayak et al., 1997).

Transgenic indica rice Basmati 370 and M7 expressing higher levels of Cry2A protein showed complete resistance to YSB and rice leaf folder (Maqbool et al., 1998). An elite rice line, KMD1 obtained from commercial Chinese japonica variety Xiushui11 containing a synthetic *cry1Ab* gene under the control of a maize ubiquitin promoter were resistant to eight lepidopteron pests (Shu et al., 2000). Expression of *Cry1Ba* endotoxins exhibits 10-fold lower lethal concentration (LC_{50}) than Cry1Ac in a SSB diet incorporation assay.

Simultaneous introduction of three insecticidal genes *cry1Ac*, *cry2A* and *gna* (*Galanthus nivalis agglutinin*, snowdrop lectin) into commercially important *indica* rice varieties (M7 and Basmati 370) showed protection against rice brown plant hopper, leaf folder and YSB. The triple transformants showed significantly higher resistance to these insects than plants expressing single genes (Maqbool et al., 2001). Expression of synthetic *cry1Ab* and *cry1Ac* genes in Basmati rice variety 370 through *Agrobacterium*-mediated transformation was found to control European corn borer (*Ostrinia nubialis*; Ahmad et al., 2002).

Alam et al. (1998) have successfully transformed a deep-water *indica* rice variety, Vaidehi with *cry1Ab* through biolistic-mediated transformation method. A synthetic *cry1Ac* gene introduced in rice (IR64, Pusa Basmati, Karnal Local) under the control of ubiquitin promoter exhibited total protection against neonate larvae of YSB (Khanna and Raina, 2002).

Wu et al. (2002) studied the inheritance and expression pattern of *cry1Ab* gene derived from different *Bt* transgenic *japonica* rice lines under field conditions. They found that *cry1Ab* gene driven by maize ubiquitin promoter displayed certain kinds of spatial and temporal expression pattern. The crosses between *japonica* and *indica* sub-species led to distortion segregation of *cry1Ab* gene in F_2 population. The content of the Cry1Ab protein in leaves peaked at the booting stage and it was lowest at the heading stage. Furthermore, the *Cry1Ab* expression of transgenic rice varied individually in different tissues based on temperature fluctuation. Jin et al. (2002) detected the *cry1Ab* transgene in transgenic rice plants developed by *Agrobacterium*-mediated and determined its chromosomal location by fluorescence in-situ hybridization, and the results showed that transferred DNA sequences remained linked in recipient genome.

Alam et al. (1999) introduced a new approach of using a transgenic maintainer line for developing insect resistant hybrid rice. Using the biolistic method of transformation, an elite IRRI maintainer line (IR68899B) has been transformed with the *cry1Ab* gene driven by the *CaMV35S* constitutive promoter. Since a cytoplasmic line is maintained by backcrossing to its isogenic maintainer line, the presence of resistance genes in the maintainer line will lead to the development of a CMS lines possessing these genes. Tu et al. (2000) reported the development of transgenic elite *indica* rice CMS restorer line of Minghui 63 and its derived hybrid rice Shangou 63. The lines expressed as *Bt* fusion gene derived from *cry1Ab* and *cry1Ac* under the control of rice actin promoter. The transgenic line (T51-1) and the hybrid showed protection against leaf folder and YSB.

9.4.6 *CLONING OF CRY GENES*

The main objective of cloning *cry* gene using recombinant DNA methodologies is to improve *Bt* strains available for better production and to improve toxicity. Kalman et al. (1993) cloned a *cry1C* gene (*cry1Cb*) from *Bt. gallariae*. The mosquitocidal gene, *cry11B*, encoding 81 kDa crystal protein in *Bt* subsp. *jegathesan* was cloned. The sequence of Cry2B protein showed significant homology toward Cry2A toxin (Cry4D) from *Bt* subsp. *israelensis*. A novel mosquitocidal protein gene, *cry20Aa* was cloned from *Bt* subsp. *fukuokaensis*. The gene product was naturally truncated and had a molecular weight of 86 kDa. The amino acid comparison showed *Cry20Aa* to be an entirely different protein (Lee and Gill, 1997). Two new crystal protein genes, *cry19A* and *orf2*, were isolated from *Bt* subsp. *jegathesan* encoding a 74.7-kDa protein and 60-kDa protein, respectively. Sasaki et al. (1997) cloned a *cry2A* type gene from *Bt* serovar *sotto* SKWOI-102-06 and designated as *cry2* (SKW). The cloned protein was toxic to *Bombyx mori*. The Cry2Aa of a new *Bt* strain (47–48) was effective against *Helicoverpa armigera*, cotton bollworm (Lenin et al., 2001).

9.4.7 INSECT RESISTANCE TO CRY PROTEINS

The continuous use of chemical insecticides led to the development of resistance among insect populations. Insects are highly adaptable to the changing environment and have evolved resistance to many chemical insecticides. de Maagd et al. (1999a,b) reported that even *Bt* toxins are not an exception for insects to develop resistance. *Bt* resistance has been documented for several insect pests that have been repeatedly selected for resistance in the laboratory. Tabashnik et al. (1990) reported resistance to *Bt* insecticides from field population of *Plutella xylostella*. In the specific case of Lepidoptera, laboratory selection has resulted in significant increases in LC_{50}s (up to 1000-fold) for diamondback moth, *P. xylostella* (L.), cabbage looper, *Tricoplusiani* Hubner (Estada & Ferre, 1992), Indian meal moth, *Plodia interpunctella* (H) (McGaughey & Whalon, 1992), tobacco budworm, *Heliothis virescens* (F) (Gould et al., 1992), beet armyworm, *Spodoptera exigua* (Hubner) (Moar et al., 1995), cotton leaf worm, *Spodoptera littoralis* Boisduval (Muller-Cohn et al., 1996) and European corn borer, *Ostrinia nubilalis* (Huang et al., 1997). The diamondback moth is notable as the only insect to evolve high levels of resistance in the field as a result of repeated use of *Bt* formulations (Tabashnik, 1994). However, the Indian meal moth probably evolved low levels of resistance in *Bt*-treated stored grains (McGaughey & Whalon, 1992).

The development of resistance to *P. xylostella* against *Bt* subsp. *kurstaki* was reported from Florida (Shelton et al., 1993), Japan (Hama et al., 1992), Philippines (Ferre et al., 1991), and China. A variety of studies conducted on different resistant populations shows that the primary reason of resistance lies in the lowering of the affinity of the toxin to the brush border membrane of midgut epithelial cells (Ferre et al., 1991). However, Gould et al. (1992) and Tabashnik (1994) have reported cross resistance to *Bt* toxins. The reason for this cross resistance was attributed to proteolysis of protoxins of decreased solublization of crystals in the midgut of the larvae. The laboratory studies revealed 76-fold increase in resistance toward *crylAc* in *H. armigera* by the end of the tenth generation (Kranthi et al., 2000). In the majority of the studies conducted to date, *Bt* resistance appears to be related to loss of membrane binding by Cry toxins (Ferre & Van Rie, 2002).

9.4.8 RESISTANCE MANAGEMENT STRATEGIES

Several strategies were proposed for management of resistance development in insects (Alstad & Andow, 1995; Gatehouse, 1999; Kumar et al.,

1996; McGaughey et al., 1998). The first step in resistance management is to establish a target pest's baseline susceptibility to insecticidal protein. Once a baseline has been established, regular monitoring of susceptible insects may indicate early stages of resistance among insect populations.

The use of multiple toxin genes at the same time (pyramiding genes) with different modes of action so that cross-resistance is unlikely to occur, that is, two *cry* genes for toxins with different receptors or a *cry* gene in combination with a different toxin gene. Second generation dual-Bt gene cottons Bollgard II[(R)] (*cry1Ac* + *cry2Ab*) and WideStrike™ (*cry1Ac* + *cry1F*) express two *Bt* endotoxins and were introduced in order to raise the level of control for *H. zea*, which was not satisfactorily controlled by the cry1Ac toxin alone (Gahan et al., 2001). The *Cry1Ac* and *Cry2Ab* toxins have different binding sites in the larval midgut and are considered to be a good combination to deploy in delaying resistance evolution.

Genetic engineering is now used to combine the toxicity of *cry* genes from *B. thuringiensis israelensis* and *B. sphaericus* (Bar et al., 1999; Poncet et al., 1994, 1997). Crop rotation with non-*Bt* crops, particularly if resistance is not stable in the insect population is another approach (de Maagd et al., 1999a,b).

Yet, another approach is high dose strategy. A high dose can be defined as a dose, which kills heterozygotes consistently and allows homozygotes to survive. Because homozygous resistant individuals are at very low frequency early in the evolution of resistance, suitable refuges may provide a continuous source of susceptible individuals (Denholm & Rowland, 1992). Another problem is the constitutive expression of *Bt* genes that led to significant selection pressure on pest population. The use of tissue specific promoters would decrease selection pressure by allowing pests to feed on unharmed economically less important parts of the plant (Vaeck et al., 1987; Wong et al., 1992). The spatial, temporal, and inducible expression of *Bt* genes in transgenic plants has thus become an attractive strategy of resistance management (Gujar et al., 2000).

9.4.9 OTHER INSECTICIDAL PROTEINS

Other genes for insect resistance such as proteinase inhibitors (PIs), amylase inhibitors, plant lectins, insect chitinases, and insect viruses are also being used to obtain transgenic insect resistance plants (Ranjekar et al., 2003). PIs are well studied plant defense proteins and are abundantly present in the storage organs (seeds and tubers). They were first shown as plant defense proteins in 1972 when the induction of PIs in potato and tomato was observed due to wounding and insect herbivory (Ryan, 1990). A direct proof of the protective role of PIs against insect herbivory was provided by Hilder et al. (1987) who showed the transgenic tobacco plants expressing cowpea trypsin inhibitor (CpTI) were resistant to the tobacco bud worm (*H. virescens*).

Hosoyama et al. (1995) and Irie et al. (1996) obtained transgenic rice lines through electroporation, harboring *Oryza* cystanin gene and corn cystatin gene, respectively. The potato proteinase inhibitor II (*pinII*) gene was introduced into several *japonica* rice varieties by Duan et al. (1996). Vain et al. (1998) reported the expression of an engineered cysteine PI for nematode resistance in transgenic rice plants. Several serine PIs have been expressed in transgenic plants for resistance against insect pests of the order lepidoptera while the cysteine PIs have been expressed against the coleopteran pests (Schuler et al., 1998).

Shade et al. (1994) reported that the introduction and expression of the bean amylase inhibitor gene (*αAl*) in pea conferred resistance to the bruchid beetles. Transgenic azuki bean expressing *αAl* gene was reported to be resistant to three species of bruchids (Ishimoto et al., 1996). Chitinase secreted from insect was used as an insecticidal protein. Schuler et al. (1998) reported that the plant metabolic enzymes such as tryptophan decarboxylase, polyphenol oxidase and lipoxygenase have been toxic to insects. Lectins are plant derived insecticidal protein genes, particularly for the control of insects belonging to Homoptera (Gatehouse, 1999). The expression of the gene encoding snowdrop lectin, *Galanthus nivalis agglutinin* (*gna*) in the phloem of transgenic rice plants conferred resistance to brown planthopper (Foissac et al., 2000; Rao et al., 1998; Sudhakar et al., 1998a,b).

Schuler et al. (1998) expressed wheat-germ agglutinin, pea lectin, jacalin, and rice lectin in the plants such as tobacco, maize, and potato mainly against aphids. Insecticidal virus, *H. armigera* stunt virus is specific to lepidopteran insects and its deployment in transgenic plants would not pose any risks (Gordon et al., 1996).

9.4.9.1 DIFFERENT STEPS INVOLVED IN RICE TRANSFORMATION FOR DEVELOPMENT OF TRANSGENIC

If efficient protocols for tissue culture and transformation are available, the production of fertile transgenic plants with any foreign gene of interest is possible (Sharma & Oritz, 2000). The basic requirements for genetic transformation are as follows:

(1) a candidate gene,
(2) a target genome,
(3) vector to carry the gene,
(4) tissue culture and regeneration system,
(5) efficient selection system to select transformed cells, and
(6) molecular and genetic characterization of putative transgenic plants.

9.4.9.2   EXPLANTS USED IN TRANSFORMATION

An ideal explant is essential for effective gene delivery and transformation system. Various explants like leaf bases (Dekeyser et al., 1990), anther culture-derived emryogenic cultures (Chair et al., 1996; Datta et al., 1990), immature embryos (Christou et al., 1991), embryogenic suspension cultures (Cao et al., 1992; Jain et al., 1996), mature-seed-derived calli (Hiei et al., 1994), germinating embryos (Chaudhury et al., 1995; Xu and Li, 1994), meristem disc (Christou & Ford, 1995), shoot apices, and shoot meristems (Park et al., 1996) were used for transformation with the help of suitable gene delivery techniques. Dong et al. (2001) reported the *Agrobacterium*-mediated transformation of rice inflorescence with *Agrobacterium* strain LBA4404-carrying plasmid pJD4.

9.4.9.3   GENE DELIVERY METHODS IN RICE

The first transgenic rice was obtained by polyethylene glycol (PEG)-mediated transfer of DNA into protoplasts in 1988 (Toriyama et al., 1988; Zhang & Wu, 1988). Datta et al. (1990) obtained transgenic *indica* plants for the first time in Chinsurah Boro II cultivar by PEG-mediated transformation. Alam et al. (1995) introduced the hygromycin phosphotransferase (*hpt*) gene into the protoplast of *indica* type rice cultivar Tepi Boro by PEG method. Using PEG-mediated method, useful genes have been delivered and rice plants expressing herbicide resistance (Datta et al., 1992), viral resistance (Peng et al., 1995), or bean *synthase* gene were developed. However, procedures for protoplast-mediated transformation of rice have certain problems. Although some successful results have been reported, it remains difficult to regenerate plants from protoplasts of elite *japonica* and the majority of *indica* varieties (Ayres & Park, 1994). Plants regenerated from protoplasts are often sterile and phenotypically abnormal (Abdullah et al., 1989; Battraw & Hall, 1992; Datta et al., 1992). Prolonged culture period as well as changes in ploidy level could be the reason for this problem (Yamagishi et al., 1996).

Electroporation was found to be a more efficient gene delivery method than PEG method. Electroporation is based on the use of short electric pulses to increase permeability of protoplasts. Since electroporation relies on protoplasts, this method also suffers from the same drawbacks as PEG-mediated method. Electroporation has been employed to obtain herbicide-resistant (Christou et al., 1991), virus-resistant (Hayakawa et al., 1996), and insect-resistant (Fujimoto et al., 1993; Wunn et al., 1996) transgenic rice plants. Electroporation has also been employed to produce rice expressing corn *cystatin* gene, which inhibits the insect pests (Irie et al., 1996).

Transformation of rice by macroinjection into floral tillers and through the pollen tube pathway (Luo & Wu, 1989) and the microinjection of DNA into microspore or proembryos (Potrykus, 1990) have also been attempted. More efficient and genotype-independent systems like particle bombardment and *Agrobacterium*-mediated gene transfer have become more popular than other plant transformation systems. Genetic transformation of rice through the use of *Agrobacterium* is preferred over bombardment as it appears to enable transfer of DNA with definite ends, minimal rearrangements, integration of small number of copies of the gene and more importantly, even large segments of DNA can be efficiently transferred (Agarwal et al., 2002; Komari et al., 1996). Other direct DNA transfer methods such as transformation of protoplast mediated by calcium phosphate (Datta et al., 1990; Jongsma et al., 1987) and transformation using silicon carbide whiskers have also been attempted.

Nowadays, particle bombardment mediated transformation and agrobacreium are commonly used and various explants like immature embryos, calli from mature embryos, suspension cultures, shoot or root axis and calli derived from germinating rice have been employed.

9.4.9.4 BIOLISTIC/PARTICLE BOMBARDMENT-MEDIATED TRANSFORMATION OF RICE

Sanford (1988) developed this method of introducing new genetic material into living plant cells, using a particle gun that can accelerate microscopic tungsten projectiles to initial velocities of about 1400 ft/s. Microprojectile bombardment is the most widely employed method for genotype independent plant transformation. This method was first used to deliver DNA and RNA into the epidermal cells of *Allium cepa* (Klein et al., 1987). This method involves the acceleration of DNA-coated metal particles through the cell wall and cell membrane (Christou, 1995).

Christou et al. (1991) successfully obtained first transgenic rice plant by using electric discharge particle bombardment. Following bombardment method, several important traits including herbicide resistance, pest and disease resistance have been introduced to rice. The cell suspension culture was used to impart herbicide resistance in plants by using the PDS 1000/He biolistic method (Cao et al., 1992). Kalla et al. (1994) generated transgenic rice plants expressing aleurone specific gene *Ltp2* from barley. A *cry1Ac* gene was constructed and placed under control of the maize ubiquitin 1 promoter, along with the first intron of the maize *ubiquitin 1* gene, and the *nos* terminator and was delivered to embryogenic calli of IR64, using the particle bombardment method. Molecular analyses and insect-feeding assays of two lines revealed that the transferred synthetic *cry1Ac* gene was expressed stably in the T_2 generation of these lines and that the transgenic rice plants were highly toxic to YSB larvae (Nayak et al., 1997).

Elite *indica* rice variety IR72 was reported to be transformed with a cloned gene *Xa21* through particle bombardment (Tu et al., 1998). Sudhakar et al. (1998a,b) developed a practical and efficient gene transfer system utilizing mature seed derived calli as explants by using a portable and inexpensive particle bombardment device. Co-transformation of *indica* rice with three plasmids carrying *cry1Ac, cry1Ab,* and *gna* with *hpt* genes showed more than 60% of transgenic plants carrying all the four genes (Maqbool & Christou, 1999). Nandadeva et al. (1999) reported that linear DNA is as effective as supercoiled plasmid DNA in particle bombardment mediated transformation. Fu et al. (2000) bombarded rice tissue with two constructs; a plasmid containing the *bar* gene and a linear DNA fragment isolated from the same plasmid, corresponding to the minimal *bar* gene expression cassette (promoter, open-reading frame, and terminator) and obtained phosphinothricin-resistant plants in both cases. By using minimal linear gene cassette they obtained simple integration events. Srivastava and Ow (2001) reported site-specific integration of transgene through the biolistic delivery of DNA using *cre-lox* mediated site specific integration.

Use of direct gene-transfer systems often results in integration into the genome fragmented and rearranged multiple copies of transgenes (Battraw & Hall, 1992; Cooley et al., 1995; Hayashimoto et al., 1990; Peng et al., 1992; Rathore et al., 1993; Tada et al., 1990). This feature of transgenic plants is undesirable since the number of copies of a transgene influences the co-expression of introduced genes (Finnegan & McElroy, 1994; Flavell, 1994). Increased copy number that resulted in reduced co-expression was observed in rice after gene transfer by biolistic methods (Christou, 1995; Cooley et al., 1995). Important traits which have been introduced through particle bombardment include herbicide resistance (Oard et al., 1996), tolerance to salt and water stress (Xu et al., 1996), disease resistance (Datta et al., 2001; Sivamani et al., 1999; Tu et al., 1998; Wang et al., 1998; Yuan et al., 2002; Zheng et al., 1997), and insect resistance (Maqbool et al., 2001; Sudhakar et al., 1998a,b).

9.4.9.5 AGROBACTERIUM-MEDIATED TRANSFORMATION

Plant transformation mediated by *A. tumefaciens* has become the most widely used method for introduction of foreign genes into plant cell and subsequent regeneration of transgenic plants. Since the first report on transgenic tobacco plant expressing foreign genes reported during 1988 by Herrera-Estrella, a great progress in understanding the *Agrobacterium*-mediated transformation in plants has been achieved (Dela Riva et al., 1998). *A. tumefaciens*, a Gram-negative soil bacterium, considered as a natural genetic tool for transformation, transfers a portion of its Ti (tumor inducing) plasmid called T-DNA into the host genomes through a well-developed system. Ti plasmids are on the order of 200–800 kbp in size (de Vos et al., 1981; Fortin et al., 1993; Gerard et al., 1992; Goodner et al., 2001; Hood et al., 1984; Otten et al., 1996; Suzuki et al., 2000). The T-DNA transfer is initiated in response to certain phenolics and sugar compounds from wounded plant cells. These phenolic compounds serve as inducers of the bacterial *vir* genes. Phenolic chemicals such as acetosyringone and related compounds are perceived via the VirA sensory protein which undergoes autophosphorylation followed by the transphosphorylation of VirG protein resulting in the activation of *vir* gene transcription. Most of the induced Vir proteins are directly involved in T-DNA processing from the Ti plasmid and subsequent transfer of T-DNA from bacterium to the plant.

A VirD2 protein along with VirD1 is directly involved in processing the T-DNA from the Ti plasmid. VirD2 nicks the Ti plasmid at 25-bp directly repeated sequences, called T-DNA borders that flank the T-DNA. Nicked T-DNA is then transferred to plants as a single stranded DNA molecule called T-strand. Thus, T-DNA enters the plant as a protein–nucleic acid complex composed of a single VirD2 molecule attached to a single-stranded T-DNA. Moreover, several other *vir* genes and host-coded proteins are also responsible for the T-DNA transfer process. Once the T-DNA enters the host nucleus, the production of plant growth regulators (auxin and cytokinins) occurs by the genes encoded by T-DNA which stimulates the enzymes involved in cell division leading to crown gall formation. Apart from growth regulators, T-DNA also codes for genes responsible for the production of opines which are consumed specifically by the infectious bacteria (Gelvin, 2000). To use this naturally occurring gene transfer mechanism for plant genetic engineering, all the genes that are naturally present in T-DNA are inactivated or replaced with foreign genes. A number of sophisticated plant transformation vectors were designed on the basis of this naturally occurring gene transfer mechanism and such vectors are widely employed in plant molecular biology and the genetic engineering of plants (Agarwal et al., 2002; Bevan, 1984; Hoekema et al., 1983; Hood et al., 1986, 1993).

The following two strategies are widely used for the introduction of engineered T-DNA into *A. tumefaciens*.

9.4.9.5.1 Binary Vector System

Two plasmids are involved in this gene transformation system and both are able to replicate separately inside the *Agrobacterium* cells. The first plasmid is a modified Ti plasmid (helper plasmid) which provides the *vir* functions in *trans* and the second contains gene of interest between the T-DNA borders. Disarmed helper Ti plasmids have been engineered by removing the oncogenic genes while still providing the necessary *vir* gene product required for transferring the T-DNA to the host plant cell (Hoekema et al., 1983).

9.4.9.5.2 Cointegrate Vector System

Gene to be transformed is integrated with in the T-DNA of a resident Ti plasmid forming an intermediate vector. The intermediate vector contains a small region of homology with the disarmed Ti plasmid and on mobilization into *Agrobacterium*; the intermediate vector combines with the disarmed Ti plasmid by single homologous recombination and forms cointegrated vector (Fraley et al., 1985).

9.4.9.6 AGROBACTERIUM-MEDIATED TRANSFORMATION OF RICE

Dicots are the natural hosts for *Agrobacterium*-mediated transformation. So systems of *Agrobacterium*-mediated transformation have been well established for many dicotyledonous plants (Arpaia et al., 1997; Ellul et al., 2003). Development of transgenic plants in monocots following *Agrobacterium*-mediated transformation seemed to be difficult. Yet, significant achievements have been made in the last decade (Hiei et al., 1994, 1997; Smith & Hood, 1995). Earlier success in this field came through the use of germinating embryos of *japonica* cultivars and employing a wide host range super virulent strain (Raineri et al., 1990). Chan et al. (1993) successfully transferred and expressed *uidA* (GUS) driven by a α-amylase promoter in a *japonica* rice cv. Tainung 62 employing *Agrobacterium*-mediated transformation system with immature embryos. Hiei et al. (1994) reported successful transformation of *japonica* rice through *Agrobacterium*-mediated gene transfer. They described several requirements for successful transformation such as the use of mature seed derived calli, acetosyringone (100 µM) and a temp of 22–28°C during cocultivation. Thereafter, several reports were made on *Agrobacterium*-mediated transformation of *indica* (Datta et al., 2000; Mohanty et al., 2002; Rashid et al., 1996), *japonica* (Aldemita & Hodges, 1996; Cheng et al., 1998; Dong et al., 2001; Toki, 1997; Yara et al., 2001) and *javanica* (Dong et al., 1996) cultivars. Datta et al. (2000, 2001) transferred three indica genotypes such as Pusa Basmati1, Tulsi, and Vaidehi with rice *chitinase* gene through *Agrobacterium*-mediated transformation method. Kumar et al. (1998) successfully transferred and expressed *uidA* (GUS) driven by a CaMV35S promoter in an *indica* rice cv. IR64 and IR72 employing *Agrobacterium*-mediated transformation system with mature embryos. Hiei et al. (1997) improved the efficiency of transformation by treating immature embryo with heat centrifugation before infection with *Agrobacterium tumefaciens* in rice.

9.4.9.7   EXPLANTS AND COCULTIVATION

Immature embryos were used for cocultivation with a strain of *A. tumefaciens* by Chan et al. (1993). Aldemita and Hodges (1996) reported successful transformation of *indica* rice using immature embryo. Datta et al. (1999, 2000) transferred rice *tlp* and *chitinase* genes into indica rice cultivars using immature embryos as explants using *Agrobacterium*-mediated method. Shoot apex was used as explant by Park et al. (1996). Two-to-five-week-old *japonica* rice calli were used by Nishizawa et al. (1999) for transforming *japonica* cultivars such as Koshihikari and Nipponbare. Khanna and Raina (1999) observed that the scutellar calli of IR64 and Karnal Local were found to be better explants due to their higher regeneration efficiency. Reproducible, efficient *Agrobacterium*-mediated transformation has been established for cultivars of *indica, japonica,* and *javanica* rice using embryogenic calli derived from mature seed scutella (Azhakanandam et al., 2000). Kumar et al. (2003) used mature seed derived calli as target tissues to deliver rice *chitinase* gene. Cocultivation medium composition is an important factor that governs the efficiency of transformation. Hiei et al. (1994) used simple modified N_6 medium containing acetosyringone, 2,4-D and casaminoacid for cocultivation. Usually, solid medium is better for cocultivation than liquid medium (Hiei et al., 1997). The acidic condition of cocultivation medium, culture temperature below 28°C, and high osmotic pressure have also been reported to be important for the expression of *vir* genes (Altmoerbe et al., 1988, 1989; Usami et al., 1988).

9.5 CONCLUSION

Rice, being the staple diet for nearly one-third of the world population, is a major target in crop improvement programs. Major limiting factors in rice production are the pest and diseases. Among insect pests, lepidopteron insects are the major ones in rice producing regions, for which about 50% of insecticidal sprays in Asian countries are targeted (Heong et al., 1994). Average yield losses caused by SSB, YSB, and leaf folders worldwide are estimated to be about 10 MT per annum (Gatehouse et al., 1992). Application of genetic engineering to transform rice plants with agronomically useful genes has become an important component of rice improvement programs. Genetic manipulation has several advantages over conventional plant breeding methods. It overcomes the incompatibility barriers among different crop species, a major limiting factor in any breeding program (Gatehouse et al., 1992). Transfer of a synthetic *Bt* (*cry1Ab*) gene into a IRRI breeding line, IR58 have resulted in an effective control of YSB and the SSB, the two most devastating insect pests of rice in Asia (Wunn et al., 1996). A synthetic *cry1Ac* gene introduced in rice (IR64, Pusa Basmati 1, and Karnal Local) under the control of ubiquitin promoter exhibited total protection against neonate larvae of YSB (Khanna & Raina, 2002). The introduction of crystal insecticidal protein genes from *Bt* into several crops through genetic engineering is often useful in controlling insect pest, namely, YSB and leaf folder.

KEYWORDS

- **leaf folder**
- **transgenic rice**
- **gene transfer**
- **insecticidal gene**
- ***Bt* gene**

REFERENCES

Abdullah, R.; Thompson, J.; Khush, G. S.; Kaushik, R. P.; Cocking, E. C. Protoclonal Variation in the Seed Progeny of Plants Regenerated from Rice Protoplasts. *Plant Sci.* **1989,** *65,* 97–101.

Agarwal, S. K.; Kapoor, A.; Grover, A. Binary Cloning Vectors for Efficient Genetic Transformation of Rice. *Curr. Sci.* **2002,** *82,* 873–876.

Ahmad, M.; Haq, I. U.; Wains, M. S.; Anwer M.; Ahmad, M. Screening of Advanced Breeding Materials for Resistance to Rice Leaf Folder under Field Conditions. In: Proceedings of the International Seminar on Rice Crop, Oct. 2–3, 2005, Rice Research Institute, Kalashahkaku, 2005; pp 293–296.

Ahmad, A.; Maqbool, S. B.; Riazuddin, S.; Sticklen, M. B. Expression of Synthetic *cry1Ab* and *cry1Ac* Genes in Basmati Rice (*Oryza sativa* L.) Variety 370 Via *Agrobacterium*-mediated Transformation for the Control of the European Corn Borer (*Ostrinia nubilalis*). *In Vitro Cell. Dev. Biol.* **2002,** *38,* 21.

Ahmed, H.; Khan, R. B.; Sharma, D.; Jamwal, V. V. S.; Gupta, S. Seasonal Incidence, Infestation and Trap Catches of *Cnaphalocrocis medinalis* (Guenee) in Rice. *Ann. Plant Prot. Sci.* **2010,** *18*(2), 380–383.

Alam, M. F.; Datta, K.; Abrigo, E.; Oliva, N.; Tu, J.; Virmani, S. S.; Datta, S. K. Transgenic Insect-resistant Maintainer Line (IR68899B) Improvement of Hybrid Rice. *Plant Cell Rep.* **1999,** *18,* 572–575.

Alam, M. F.; Datta, K.; Abrigo, E.; Vasquez, A.; Senadhira, D.; Datta, S. K. Production of Transgenic Deepwater *Indica* Rice Plants Expressing a Synthetic *Bacillus thuringiensis cry1A(b)* Gene with Enhanced Resistance to Yellow Stem Borer. *Plant Sci.* **1998,** *135,* 25–30.

Alam, M. F.; Oliva, N. I.; Zapata, F. J.; Datta, S. K. Fertile Transgenic *Indica* Rice Plants by PEG-mediated Protoplast Transformation. *J. Genet. Breed.* **1995,** *49,* 303–308.

Aldemita, P. R.; Hodges, T. K. *Agrobacterium tumefaciens*-mediated Transformation of *Japonica* and *Indica* Rice Varieties. *Planta* **1996,** *199,* 612–617.

Alstad, D. N.; Andow, D. A. Managing the Evolution of Insect Resistance to Transgenic Plants. *Science* **1995,** *268,* 1894–1896.

Altmoerbe, J.; Kulhmann, H.; Schroder, J. A. Differences in Induction of Ti-plasmid Virulence Genes *virG* and *virD* and Continued Control of *virD* Expression by Four External Factors. *Mol. Plant-Microbe Interact.* **1989,** *2,* 301–308.

Altmoerbe, J.; Neddermann, P.; von Lintig, J.; Schroder, J. Temperature-sensitive Step in Ti Plasmid *Vir*-region Induction and Correlation with Cytokinin Secretion by *Agrobacteria*. *Mol. Gen. Genet.* **1988**, *213*, 1–8.

Alvi, S. M.; Ali, M. A.; Chaudhary, S.; Iqbal, S. Population Trends and Chemical Control of Rice Leaf Folder, *Cnaphalocrocis medinalis* on Rice Crop. *Int. J. Agric. Biol.* **2003**, *5*, 615–617.

Arpaia, S.; Mennella, G.; Onofaro, V.; Perri, E.; Sunseri, F.; Rotino, G. L. Production of Transgenic Eggplant (*Solanum melongena* L.) Resistant to Colorado Potato Beetle (*Leptinotarsa decemlineata* Say). *Theor. Appl. Genet.* **1997**, *95*, 329–334.

Ayres, N. M.; Park, W. D. Genetic Transformation of Rice. *Crit. Rev. Plant Sci.* **1994**, *13*, 219–239.

Azhakanandam, K.; McCabe, M. S.; Power, J. B.; Lowe, K. C.; Cocking, E. C.; Davey, M. R. T-DNA Transfer, Integration, Expression and Inheritance in Rice: Effects of Plant Genotype and *Agrobacterium* super-virulence. *J. Plant Physiol.* **2000**, *157*, 429–439.

Bar, E.; Lieman-Hurwitz, J.; Rahamim, E.; Keynan, A.; Sandler, N. Cloning and expression of *Bacillus thuringiensis israelensis* δ-endotoxin DNA in *Bacillus* sp. *haericus*. *J. Invert. Pathol.* **1999**, *57*, 149–158.

Barton, K. A.; Whiteley, H.; Yang, N. S. *Bacillus thuringiensis* Delta-endotoxin in Transgenic *Nicotiana tabacum* Provides Resistance to Lepidopteran Insects. *Plant Physiol.* **1999**, *85*, 1103–1109.

Battraw, M.; Hall, T. C. Expression of a Chimeric Neomycin Phosphotransferase II Gene in First and Second Generation Transgenic Rice Plants. *Plant Sci.* **1992**, *86*, 191–202.

Bevan, M. Binary *Agrobacterium* Vectors for Plant Transformation. *Nucl. Acid Res.* **1984**, *12*, 8711–8721.

Bhanu, K. V.; Reddy, P. S. Field Evaluation of Certain Newer Insecticides against Rice Insect Pests. *J. Appl. Zool. Res.* **2008**, *19*(1), 11–14.

Bhatti, M. N. Rice Leaf Folder (*Cnaphalocrosis medinalis*): A Review. *Pak. Entomol.* **1995**, *17*, 126–131.

Boucias, D. G.; Pendland, J. C. *Principles of Insect Pathology*. Kluwer Academic Publishers: Dordrecht, The Netherlands, 1994.

Cao, J.; Duan, X.; McElroy, D.; Wu, R. Regeneration of Herbicide Resistant Transgenic Rice Plants Following Microprojectile-mediated Transformation of Suspension Culture Cells. *Plant Cell Rep.* **1992**, *11*, 586–591.

Chair, H.; Legavre, T.; Guiderdoni, E. Transformation of Haploid, Microspore-derived Cell Suspension Protoplasts of Rice (*Oryza sativa* L). *Plant Cell Rep.* **1996**, *15*, 766–770.

Chan, M. T.; Chang, H. H.; Ho, S. L.; Tong, W. F.; Yu, S. M. *Agrobacterium*-mediated Production of Transgenic Rice Plants Expressing a Chimeric δ-Amylase Promoter/δ-Glucuronidase Gene. *Plant Mol. Biol.* **1993**, *22*, 491–506.

Chaudhury, A.; Maheshwari, S. C.; Tyagi, A. K. Transient Expression of *gus* Gene in Intact Seed Embryos of *indica* Rice after Electroporation-mediated Gene Delivery. *Plant Cell Rep.* **1995**, *14*, 215–220.

Cheng, X. Y.; Sardana, R.; Kaplan, H.; Altosaar, I. *Agrobacterium*-transformed Rice Plants Expressing Synthetic *cry1A(b)* and *cry1A(c)* Genes are Highly Toxic to Striped Stem Borer and Yellow Stem Borer. *Proc. Natl. Acad. Sci. U.S.A.* **1998**, *95*, 2767–2772.

Christou, P. Particle Bombardment. *Meth. Cell Biol.* **1995**, *50*, 375–382.

Christou, P.; Ford, T. L. Recovery of Chimeric Rice Plants from Dry Seed Using Electric Discharge Particle Acceleration. *Ann. Bot.* **1995**, *75*, 449–454.

Christou, P.; Ford, T. L.; Kofron, M. Production of Transgenic Rice *(Oryza sativa* L.) Plants from Agronomically Important *indica* and *japonica* Varieties via Electric Discharge Particle Acceleration of Exogenous DNA into Immature Zygotic Embryos. *Biotechnol.* **1991,** *9,* 957–962.Cooley, J.; Ford, T.; Christou, P. Molecular and Genetic Characterization of Elite Transgenic Rice Plants Produced by Electric Discharge Particle Acceleration. *Theor. Appl. Genet.* **1995,** *90,* 97–104.

Crickmore, N.; Zeigler, D. R.; Feitelson, J.; Schnepf, E.; Vanrie, J; Lereclus, D.; Baum, J.; Dean, D. H. Revision of the Nomenclature for the *Bacillus thuringiensis* Pesticidal Crystal Proteins. *Microbiol. Mol. Biol. Rev.* **1998,** *62,* 807–813.

Datta, K.; Koukolikova-Nicola, Z.; Baisakh, N.; Oliva, N.; Datta, S. K. *Agrobacteium* Mediated Engineering for Sheath Blight Resistance of *indica* Rice Cultivars from Different Ecosystems. *Theor. Appl. Genet.* **2000,** *100,* 832–839.

Datta, K.; Tu, J.; Oliva, N.; Ona, I.; Velazhahan, R.; Mew, T. W.; Muthukrishnan, S.; Datta, S.K. Enhanced Resistance to Sheath Blight by Constitutive Expression of Infection-Related Rice Chitinase in Transgenic Elite *indica* Rice Cultivars. *Plant Sci.* **2001,** *60,* 405–414.

Datta, K.; Velazhahan, R.; Oliva, N.; Ona, I.; Mew, T.; Khush, G. S.; Muthukrishnan, S.; Datta, S. K. Over Expression of the Cloned Rice Thaumatin-like Protein (PR-5) Gene in Transgenic Rice Plants Enhances Environmental Friendly Resistance to *Rhizoctonia solani* Causing Sheath Blight Disease. *Theor. Appl. Genet.* **1999,** *98,* 1138–1145.

Datta, S. K.; Datta, K.; Soltanifar, N.; Donn, G.; Potrykus, I. Herbicide-resistant *Indica* Rice Plants from IRRI Breeding Line IR72 after PEG-mediated Transformation of Protoplasts. *Plant Mol. Biol.* **1992,** *20,* 619–629.

Datta, S. K.; Peterhans, A.; Datta, K.; Potrykus, I. Genetically Engineered Fertile *indica*-rice Recovered from Protoplasts. *Biotechnology* **1990,** *8,* 736–740.

De cosa, B.; Moar, W.; Lee, S. B.; Miller, M.; Danniel, H. Overexpression of the *Bt cry2Aa2* Operon in Chloroplasts Leads to Formation of Insecticidal Crystals. *Nat. Biotechnol.* **2001,** *19*(1), 71–74.

de Maagd, R. A.; Bakker, P. L.; Masson, L.; Adang, M. J.; Sangadala, S.; Stiekema, W.; Bosch, D. Domain III of the *Bacillus thuringiensis* delta-endotoxin CryIAc is Involved in Binding to *Manduca sexta* Brush Border Membranes and to Its Purified Aminopeptidase N. *Mol. Microbiol.* **1999a,** *31,* 463–471.

de Maagd, R. A.; Bosch, D.; Stiekema, W. *Bacillus thuringiensis* Toxin Mediated Insect Resistance in Plants. *Elsevier Sci.* **1999b,** *4,* 9–13.

de Maagd, R. A.; Bravo, A.; Crickmore, N. How *Bacillus thuringiensis* Has Evolved Specific Toxins to Colonize the Insect World. *Trend. Genet.* **2001,** *17,* 193–199.

de Vos, G.; DeBeuckeleer, M.; Van Montagu, M.; Schell, J. Restriction Endonuclease Mapping of the Octopine Tumor-inducing Plasmid pTiAch5 of *Agrobacterium tumefaciens* Addendum. *Plasmid* **1981,** *6,* 249–253.

Dekeyser, R. A.; Claes, B.; De Rycke, R. M. U.; Habets, M. E.; Van Montagu, M. C.; Caplan, A. B. Transient Gene Expression in Intact and Organised Rice Tissues. *Plant Cell* **1990,** *2,* 591–602.

Dela Riva, G. A.; Gonzalez-Cabrera, J.; Vazquez-Padron, R.; Ayra-Pardo, C. *Agrobacterium tumefaciens:* A Natural Tool for Plant Transformation. *Elect. J. Biotechnol.* **1998,** *1,* 1–16.

Delecluse, A.; Rosso, M. L.; Ragni, A. Cloning and Expression of a Novel Toxin Gene from *Bacillus thuringiensis* Subsp. *jegathesan* Encoding a Highly Mosquitocidal Protein. *Appl. Environ. Microbiol.* **1995,** *61,* 4230–4235.

Denholm, I.; Rowland, M. W. Tactics for Managing Pesticide Resistance in Arthropods. Theory and Practical. *Ann. Rev. Entomol.* **1992,** *37,* 91–112.

Dong, J.; Kharb, P.; Teng, W.; Hall, T. C. Characterization of Rice Transformed via an *Agrobacterium*-mediated Inflorescence Approach. *Mol. Breed.* **2001**, *7*, 187–194.

Dong, J.; Teng, W.; Buchholz, W. G.; Hall, T. C. *Agrobacterium*-mediated Transformation of Javanica Rice. *Mol. Breed.* **1996**, *2*, 267–276.

Duan, X.; Li, X.; Xue, Q.; Abo-El-Saad, M.; Xu, D.; Wu, R. Transgenic Rice Plants Harboring an Introduced Potato Proteinase Inhibitor II Gene are Insect Resistant. *Nature Biotechnol.* **1996**, *14*, 494–498.

Ellul, P.; Garcia-Sogo, B.; Pineda, B.; Rios, G.; Roig, L.A.; Moreno, V. The Ploidy Level of Transgenic Plants in *Agrobacterium*-mediated Transformation of Tomato Cotyledons (*Lycopersicon esculentum* L. Mill.) is Genotype and Procedure Dependent. *Theor. Appl. Genet.* **2003**, *106*, 231–238.

Estada, U.; Ferre, J. Laboratory Selection of *Tricoplusiani* for Resistance to *Bacillus thuringiensis* Cry1Ab δ-Endotoxins. In: 25th Annu. Meet. Soc. Invertebr. Pathol., Heidelberg, 1992.

Feitelson, J. S.; Payne, J.; Kim, L. *Bacillus thuringiensis*: Insect and Beyond. *Bio/Technology* **1992**, *10*, 271–275.

Ferre, J.; Van Rie, J. Biochemistry and Genetics of Insect Resistance to *Bacillus thuringiensis*. *Annu. Rev. Entomol.* **2002**, *47*, 501–533.

Ferre, J.; Real, M. D.; Van Rie, J.; Jansens, S.; Peferoen, M. Resistance to the *Bacillus thuringiensis* Bio-insecticide in a Field Population of *Plutella xylostella* is Due to a Change in a Midgut Membrane Receptor. *Proc. Natl. Acad. Sci. U.S.A.* **1991**, *88*, 5119–5123.

Finnegan, J.; McElroy, D. Transgenic Inactivation: Plants Fight Back. *Bio/Technology* **1994**, *12*, 883–888.

Fischhoff, D. A.; Bowdish, K. S.; Perlak, F. J.; Marrone, P. G.; McCormick, S. M.; Niedermeyer, J. G.; Dean, D. A.; Kusano-Knetzmer, K.; Mayer, E. J.; Rochester, D. E.; Rogers, S. G.; Fraley, R. T. Insect Tolerant Transgenic Tomato Plants. *Bio/Technology* **1987**, *5*, 807–813.

Flavell, R. B. Inactivation of Gene Expression in Plants as a Consequence of Specific Sequence Duplication. *Proc. Natl. Acad. Sci. U.S.A.* **1987**, *91*, 3490–3496.

Foissac, X.; Loc, N. T.; Christou, P.; Gatehouse, A. M. R.; Gatehouse, J. A. Linear Transgene Constructs Lacking Vector Backbone Sequences Generate Low-copy-number Transgenic Plants with Simple Integration Patterns. *Transgen. Res.* **2000**, *9*, 11–19.

Fortin, C.; Marquis, C.; Nester, E. W.; Dion, P. Dynamic Structure of *Agrobacterium tumefaciens* Ti Plasmids. *J. Bacteriol.* **1993**, *175*, 4790–4799.

Fraley, R.T.; Rogers, S. G.; Horsch, R. B.; Eichholtz, D. A.; Flick, J. S.; Fink, C. L.; Hoffmann, N. L.; Sanders, P. R. The SEV system: a New Disarmed Ti Plasmid Vector System for Plant Transformation. *Biotechnology* **1985**, *3*, 629–635.

Fu, X. D.; Duc, L.T.; Fontana, S.; Bong, B. B.; Tinjuangjun, P.; Sudhakar, D.; Twyman, R. M.; Christou, P.; Kohli, A. Linear Transgene Constructs Lacking Vector Backbone Sequences Generate Low-copy-number Transgenic Plants with Simple Integration Patterns. *Transgen. Res.* **2000**, *9*, 11–19.

Fujimoto, H.; Itoh, K.; Yamamoto, M.; Kyozuka, J.; Shimamoto, K. Insect Resistant Rice Generated by Introduction of a Modified Gene of *Bacillus thuringiensis*. *Biotechnol.* **1993**, *11*, 1151–1155.

Gahan, L. J.; Gould, F.; Heckel, D. G. Identification of a Gene Associated with *Bt* Resistance in *Heliothis virescens*. *Science* **2001**, *293*, 857–860.

Gatehouse, A. M. R. Biotechnological Applications of Plant Genes in the Production of Insect Resistant Crops. In: *Global Plant Genetic Resources for Insect Resistant Crops*; Clement, S. L., Quisenberry, S. S., Eds.; CRC Press LLC: Boca Raton, FL, 1999; pp 263–280.

Gatehouse, A. M. R.; Boulter, D.; Hilder, V. A. Potential of Plant-derived Genes in the Genetic Manipulation of Crops for Insect Resistance. In: *Plant Genetic Manipulation for Crop Protection. Biotechnology in Agriculture*; Gatehouse, A. M. R., Hilder, V. A., Boulter, D., Eds.; CAB International: Wallingford: UK, **1992**, 155–181.

Gelvin, S. B. *Agrobacterium* and Plant Genes Involved in T-DNA Transfer and Integration. *Annu. Rev. Plant. Physiol. Plant Mol. Biol.* **2000**, *51*, 223–256.

Gerard, J. C.; Canaday, J.; Szegedi, D.; de la Salle, H.; Otten, L. Physical Map of the Vitopine Ti Plasmid pTiS4. *Plasmid* **1992**, *28*, 146–156.

Ghareyazie, B.; Alinia, F.; Menguito, C. A.; Rubia, L. G.; Palma, J. M.; Liwanag, E. A.; Cohen, M. B.; Khush, G. S.; Bennett, J. Enhanced Resistance to Two Stem Borers in an Aromatic Rice Containing a Synthetic *cryIAb* Gene. *Mol. Breed.* **1997**, *3*, 401–414.

Goodner, B. W.; Markelz, B. P.; Flanagan, M. C.; Crowell, C. B.; Racette, J. L.; Schilling, A., Halfon, L. M.; Mellors, J. S.; Grabowski, G. Combined Genetic and Physical Map of the Complex Genome of *Agrobacterium tumefaciens. J. Bacteriol.* **2001**, *181*, 5160–5166.

Gordon, K. H. J.; Johnson, K. N.; Hanzlik, T. N. The Larger Genomic RNA of *Helicoverpa armigera* Stunt Tetra Virus Encodes the RNA Polymerase and Has a Novel 3′ Terminal tRNA Like Structure. *Virology* **1996**, *20*, 84–98.

Gottfried, F.; Fallil, F. The Spinning (Stitching) Behaviour of the Rice Leaf Folder, *Cnaphalocrocis medinalis. Ent. Exp. Appl.* **1980**, *29*(2), 138–146.

Gould, F; Martinez-Ramirez, A.; Anderson, A.; Ferre, J.; Silva, F. J.; Moar, W. J. Broad-spectrum Resistance to *Bacillus thuringiensis* Toxins in *Heliothis virescens. Proc. Natl. Acad. Sci. U.S.A.* **1992**, *89*, 7986–7988.

Gujar, G.; Archanakumari, T.; Kalia, V.; Chandrashekar, K. Spatial and Temporal Variation in Susceptibility of the American Bollworm *Helicoverpa armigera* (Hubner) to *Bacillus thuringiensis* var. *kurstaki* in India. *Curr. Sci.* **2000**, *78*, 995–1000.

Gunathilagaraj, K.; Gopalan, M. Rice Leaf Folder Complex in Madurai, TN, India. *Int. Rice Res. Notes* **1986**, *11*(6), 24.

Gupta, P. K. Use of Microbes in Industry and Agriculture. In: *Elements of Biotechnology*; Gupta, P. K., Ed.; Rastogi: Meerut, India, **1997**, 471–507.

Hama, H.; Suzuki, K.; Tanaka, H. Inheritance and Stability of Resistance to *Bacillus thuringiensis* formulations of the Diamondback Moth *Plutella xylostella* (L.) (Lepidoptera: Plutellidae). *Appl. Entomol. Zool.* **1992**, *27*, 355–362.

Hayakawa, T.; Zhu, Y.; Itoh, K.; Kimura, Y.; Izawa, T.; Shimamoto, K.; Toriyama, S. Genetically Engineered Rice Resistant to Rice Stripe Virus, an Insect-transmitted Virus. *Proc. Natl. Acad. Sci. U.S.A.* **1996**, *89*, 9865–9869.

Hayashimoto, A.; Li, Z.; Murai, N. A Polyethylene Glycol-mediated Protoplast Transformation System for Production of Fertile Transgenic Rice Plants. *Plant Physiol.* **1990**, *93*, 857–863.

Heong, K. L.; Escalanda, M. M. A Comparative Analysis of Pest Management Practices of Rice Farmers in Asia. In: *Pest Management of Rice Farmers of Asia*; Heong, K. L., Escalanda, M. M., Eds.; International Rice Research Institute: Philippines, 1997, pp 227–245.

Heong, K. L.; Escalada, M. M.; Mai, V. An Analysis of Insecticide Use in Rice: Case Studies in the Philippines and Vietnam. *Int. J. Pest Manage.* **1994**, *40*, 173–178.

Heong, K. L. Rice Leaf Folder: Are They Serious Pests? In: Research on Rice Leaf Folder Management in China, Proceeding of the International Workshop on Economic threshold level of for rice leaf folder in China, March 4–6, 1992, Beijing; Hu, G. W., Guo, Y. J., Li, S. W., Eds.; China Agricultural Science and Technology Publisher: Beijing, China, 1992; pp 8–11.

Herrnstadt, C.; Gilroy, T. E.; Sobieski, D. A.; Bennett, B. D.; Gaertner, F. H. Nucleotide Sequence and Deduced Amino Acid Sequence of a Coleopteran-active Delta-Endotoxin Gene from *Bacillus thuringiensis* subsp. *san diego. Gene* **1987**, *57*, 37–46.

Hiei, Y.; Komari, T.; Kubo, T. Transformation of Rice Mediated by *Agrobacterium tumefaciens. Plant Mol. Biol.* **1997**, *35*, 205–218.

Hiei, Y.; Ohta, S.; Komari, T.; Kumashiro, T. Efficient Transformation of Rice (*Oryza sativa* L.) Mediated by *Agrobacterium* and Sequence Analysis of the Boundaries of the T-DNA. *Plant J.* **1994**, *6*, 271–282.

Hilder, V. A.; Gatehouse, A. M. R.; Sheerman, S. E.; Barker, R. F.; Boulter, D. A Novel Mechanism of Resistance Engineered in Tobacco. *Nature* **1987**, *330*, 160–163.

Hoekema, A.; Hirsch, R. R.; Hooykaas, P. J. J.; Schilperoort, R. A. A Binary Plant Vector Strategy Based on Separation of *vir-* and T-region of *Agrobacterium tumefaciens* Ti Plasmid. *Nature* **1983**, *303*, 177–180.

Hofte, H.; Whiteley, H. R. Insecticidal Crystal Proteins of *Bacillus thuringiensis. Microbiol. Rev.* **1989**, *53*, 242–255.

Hood, E. E.; Gelvin, S. B.; Melchers, L. S.; Hoekema, A. New *Agrobacterium* Helper Plasmids for Gene Transfer to Plants. *Transgen. Res.* **1993**, *2*, 208–218.

Hood, E. E.; Helmer, G. L.; Fraley, R. T.; Chilton, M. D. The Hyper Virulence of *Agrobacterium tumefaciens* A281 is Encoded in a Region of pTiBo542 Outside of T-DNA. *J. Bacteriol.* **1986**, *168*, 1291–1301.

Hood, E. E.; Jen, G.; Kayes, L.; Kramer, J.; Fraley, R. T.; Chilton, M. D. Restriction Endonuclease Map of pTiBo542, a Potential Ti-plasmid Vector for Genetic Engineering of Plants. *Bio/Technology* **1984**, *2*, 702–709.

Hosoyama, H.; Irie, K.; Abe, K.; Arai, S. Introduction of a Chimeric Gene Encoding an Oryzacystatin-*b*-glucuronidase Fusion Protein into Rice Protoplasts and Regeneration of Transformed Plants. *Plant Cell Rep.* **1995**, *15*, 174–177.

Huang, F.; Higgins, R. A.; Buschman, L. T. Baseline Susceptibility and Changes in Susceptibility to *Bacillus thuringiensis* Subsp. *kurstaki* Under Selection Pressure in European Corn Borer (Lepidoptera: Pralidae). *J. Econ. Entomol.* **1997**, *90*, 1137–1143.

Irie, K.; Hosoyama, H.; Takeuchi, T.; Iwabuchi, K.; Waumabe, H.; Abe, M.; Abe, K.; Ami, S. Transgenic Rice Established to Express Corncystatin Exhibits Strong Inhibitory Activity against Insect Gut Proteinases. *Plant Mol. Biol.* **1996**, *30*, 149–157.

Ishimoto, M.; Sato, T.; Chrispeels, M. J.; Kitamura, K. Bruchid Resistance of Transgenic Azuki Bean Expressing Seed Alpha-amylase Inhibitor of Common Bean. *Entomol. Exp. Appl.* **1996**, *79*, 309–315.

Jain, R. K.; Jain, S.; Wang, B.; Wu, R. Optimization of Biolistic Method for Transient Gene Expression and Production of Agronomically Useful Basmati Rice Plants. *Plant Cell Rep.* **1996**, *15*, 963–968.

Jenkins, J. L.; Lee, M. K.; Valaitis, A. P.; Curtiss, A.; Dean, D. H. Bivalent Sequential Binding Model of a *Bacillus thuringiensis* Toxin to Gypsy Moth Aminopeptidase N Receptor. *J. Biol. Chem.* **2000**, *275*, 14423–14431.

Jenkins, J. N.; Parrott, W. L.; McCarty, Jr., J. C.; Barton, K. A.; Umbeck, P. F. Field Test of Transgenic Cotton Containing a *Bacillus thuringiensis* Gene. *Miss. Agri. For. Exp. Sta. Tech. Bull.* **1991**, *174*, 1–6.

Jin, W. W.; Li, Z. Y.; Fang, Q.; Altosaar, I.; Liu, L. H.; Song, Y. C. Fluorescence *in situ* Hybridization Analysis of Alien Genes in *Agrobacterium* Mediated *cry1Ab*-transformed Rice. *Ann. Bot.* **2002**, *90*, 31–36.

Jongsma, M.; Koornneef, M.; Zabel, P.; Hille, J. Tomato Protoplast DNA Transformation, Physical Linkage and Recombination of Exogenous DNA Sequences. *Plant Mol. Biol.* **1987**, *8*, 383–394.

Justin, G. L.; Rabindra, R. J.; Jayaraj, S.; Rangarajan, M. Laboratory Evaluation of Comparative Toxicity of *Bt* to Larvae of *P. xylostella* and *B. mori. J. Biol. Control.* **1988**, *2*, 137–138.

Kalla, R.; Shimamoto, K.; Potter, R.; Nielson, P. S.; Linnestad, C.; Olsen, O. The Promoter of the Barley Aleurone-specific Gene Encoding a Putative 7 kDa Lipid Transfer Protein Confers Aleurone Cell Specific Expression in Transgenic Rice. *Plant J.* **1994**, *6*, 849–860.

Kalman, S.; Kiehne, K. L.; Libs, J. L.; Yamamoto, T. Cloning of a Novel *cryIC*-type Gene from a Strain of *Bacillus thuringiensis* Subsp. *galleriae. Appl. Environ. Microbiol.* **1993**, *59*, 1131–1137.

Khan, M. R., Ahmad, M.; Ahmad, S. Some Studies on Biology, Chemical Control and Varietal Preference of Rice Leaf Folder, *Cnaphalocrocis medinalis. Pak. J. Agric. Sci.* **1989**, *26*, 253–263.

Khanna, H. K.; Raina, S. K. *Agrobacterium*-mediated Transformation of *indica* Rice Cultivars Using Binary and Superbinary Vectors. *Aust. J. Plant Physiol.* **1999**, *26*, 311–324.

Khanna, H. K.; Raina, S. K. Elite *Indica* Transgenic Rice Plants Expressing Modified Cry1Ac Endotoxins of *Bacillus thuringiensis* Show Enhanced Resistance to Yellow Stem Borer (*Scirpophaga incertulas*). *Transgen. Res.* **2002**, *11*, 411–423.

Klein, T. M.; Wolf, E. D.; Wu, K.; Sanford, J. C. High Velocity Microprojectiles for Delivering Nucleic Acids into Living Cells. *Nature* **1987**, *327*, 70–72.

Komari, T.; Hiei, Y.; Saito, Y.; Murai, N.; Kumashiro, T. Vectors Carrying Two Separate T-DNAs for Co-transformation of Higher Plants Mediated by *Agrobacterium tumefaciens* and Segregation of Transformants Free from Selection Markers. *Plant J.* **1996**, *10*, 165–174.

Kostichka, K.; Warren, G. W.; Mullins, M.; Mullins, A. D.; Craig, J. A.; Koziel, M. G.; Estruch, J. J. Cloning of a *CryV* Type Insecticidal Protein Gene from *Bacillus thuringiensis*: The *CryV* Encoded Protein is Expressed Early in Stationary Phase. *J. Bacteriol.* **1996**, *178*, 2141–2144.

Kranthi, K. R.; Kranthi, S.; Ali, S.; Banerjee, S. K. Resistance to Cry1Ac δ-Endotoxin of *Bacillus thuringiensis* in a Laboratory Selected Strain of *Helicoverpa armigera. Curr. Sci.* **2000**, *78*, 1001–1004.

Krattiger, A. F. Insect Resistance in Crops: A Case Study of *Bacillus thuringiensis* (Bt) and its Transfer to Developing Countries. *International Service for the Acquisition of Agri-Biotech Applications Briefs, No. 2*; ISAAA: Ithaca, NY, 1997; 42 pp.

Kumar, K. K.; Poovannan, K.; Nandakumar, R.; Thamilarasi, K.; Geetha, C.; Jayashree, N.; Kokiladevi, E.; Raja, J. A. J.; Samiyappan, R.; Sudhakar, D.; Balasubramanian, P. A High Throughput Functional Expression Assay System for a Defence Gene Conferring Transgenic Resistance on Rice against the Sheath Blight Pathogen, *Rhizoctonia solani. Plant Sci.* **2003**, *165*, 969–976.

Kumar, P. A.; Mandaokar, A.; Sreenivasa, K.; Chakrabarti, S. K.; Bisaria, S.; Sharma, S. R.; Kaur, S.; Sharma, R. P. Insect Resistant Transgenic Brinjal Plants. *Mol. Breed.* **1998**, *4*, 33–37.

Kumar, P. A.; Sharma, R. P.; Malik, V. S. The Insecticidal Proteins of *Bacillus thuringiensis. Adv. Appl. Microbiol.* **1996**, *42*, 1–43.

Lee, H. K.; Gill, S. S. Molecular Cloning and Characterization of a Novel Mosquitocidal Protein Gene from *Bacillus thuringiensis* Subsp. *fukuokaensis. Appl. Environ. Microbiol.* **1997**, *63*, 4664–4670.

Lenin, K.; Asia Mariam, M.; Udayasuriyan, V. Expression of a *cry2Aa* Gene in an Acrystalliferous *Bacillus thuringiensis* Strain and Toxicity of *cry2Aa* against *Helicoverpa armigera*. *World J. Microbiol. Biotechnol.* **2001**, *17*, 273–278.

Luo, Z.; Wu, R. A Simple Method for the Transformation of Rice via the Pollen Tube Pathway. *Plant Mol. Biol. Rep.* **1989**, *7*, 69–77.

Maqbool, S. B.; Christou, P. Multiple Traits of Agronomic Importance in Transgenic *indica* Rice Plants: Analysis of Transgene Integration Patterns, Expression Levels and Stability. *Mol. Breed.* **1999**, *5*, 471–480.

Maqbool, S. B.; Husnain, T.; Riazuddin, S.; Masson, L.; Christou, P. Effective Control of Yellow Stem Borer and Rice Leaf Folder in Transgenic Rice *Indica* Varieties Basmati 370 and M7 Using the Novel δ-Endotoxin *cry2ABacillus thuringiensis* Gene. *Mol. Breed.* **1998**, *4*, 501–507.

Maqbool, S. B.; Riazuddin, S.; Loc, N. T.; Gatehouse, A. M. R.; Gatehouse, J. A.; Christou, P. Expression of Multiple Insecticidal Genes Confers Broad Resistance Against a Range of Different Rice Pests. *Mol. Breed.* **2001**, *7*, 85–93.

Maragesan, S.; Chellish, S. Yield Losses and Economic Injury by Rice Leaf Folder. *Indian J. Agric. Sci.* **1987**, *56*, 282–285.

McGaughey, W. H.; Whalon, M. E. Managing Insect Resistance to *Bacillus thuringiensis* Toxins. *Science* **1992**, *258*, 1451–1455.

McGaughey, W. H.; Gould, F.; Gelertner, W. *Bt*-resistance Management. *Nat. Biotechnol.* **1998**, *16*, 144–146.

McPherson, S. A.; Perlak, F. J.; Fuchs, R. L.; Marrone, P. G.; Lavrik, P. B.; Fischhoff, D. A. Characterization of the Coleopteran-specific Protein Gene of *Bacillus thuringiensis* var. *tenebrionis*. *Biotechnology* **1988**, *6*, 61–66.

Moar, W. J.; Pusztai-Carey, Van Faassen, H.; Frutos, R.; Rang, C.; Luo, K.; Adang, M. J. Development of *Bacillus thuringiensis cry1C* Resistance by *Spodoptera exigua* (Lepidoptera: Noctuidae). *Appl. Environ. Microbiol.* **1995**, *61*, 2086–2092.

Mohanty, A.; Kathuria, H.; Ferjani, A.; Sakamoto, A.; Mohanty, P.; Murata, N.; Tyagi, A. K. Transgenics of an Elite *Indica* Rice Variety Pusa Basmati1 Harbouring the *codA* Genes are Highly Tolerant to Salt Stress. *Theor. Appl. Genet.* **2002**, *106*, 51–57.

Muller-Cohn, J.; Chaufaux, J.; Buisson, C.; Gilois, N.; Sanchis, V.; Lereclus, D. *Spodoptera littoralis* (Lepidoptera: Noctuidae) Resistance to *CryIC* and Cross-resistance to Other *Bacillus thuringiensis* Crystal Toxins. *J. Econ. Entomol.* **1996**, *89*, 791–797.

Nandadeva, Y. L.; Lupi, C. G.; Meyer, C. S.; Devi, P. S.; Potrykus, I.; Bilang, R. Microprojectile-mediated Transient and Integrative Transformation of Rice Embryogenic Suspension Cells: Effects of Osmotic Cell Conditioning and of the Physical Configuration of Plasmid DNA. *Plant Cell Rep.* **1999**, *18*, 500–504.

Navon, A. *Bacillus thuringiensis* Insecticides in Crop Protection Reality and Prospects. *Crop Protect.* **2000**, *19*, 676–699.

Nayak, K. P.; Basu, D.; Das, S.; Basu, A.; Ghosh, D.; Ramakrishna, N. A.; Ghosh, M.; Sen, S. K. Transgenic Elite *indica* Rice Plants Expressing *CryIAc* δ-Endotoxin of *Bacillus thuringiensis* are Resistant against Yellow Stem Borer (*Scirpophaga incertulas*). *Proc. Natl. Acad. Sci. U.S.A.* **1997**, *94*, 2111–2116.

Nishizawa, Y.; Nishio, Z.; Nakazono, K.; Soma, M.; Nakajima, E.; Ugaki, M.; Hibi, T. Enhanced Resistance to Blast (*Magnaporthe grisea*) in Transgenic *Japonica* Rice by Constitutive Expression of Rice Chitinase. *Theor. Appl. Genet.* **1999**, *99*, 383–390.

Oard, J. H.; Linscombe, S. D.; Braverman, M.P.; Jodari, F.; Blouin, D.C.; Leech, M.; Kohli, A.; Vain, P.; Cooley, J. C.; Christou, P. Development, Field Evaluation and Agronomic Performance of Transgenic Herbicide-resistant Rice. *Mol. Breed.* **1996**, *2*, 359–368.

Otten, L.; de Ruffray, P.; Momol, E. A.; Momol, M. T.; Burr, T. J. Phylogenetic Relationships between *Agrobacterium vitis* Isolates and their Ti Plasmids. *Mol. Plant-Microbe Interact.* **1996**, *9*, 782–786.

Park, S. H.; Pinson, S. R. M.; Smith, R. H. T-DNA Integration into Genomic DNA of Rice Following *Agrobacterium* Inoculation of Isolated Rice Shoot Apices. *Plant Mol. Biol.* **1996**, *32*, 1135–1148.

Peferoen, M. Insecticidal Promise of *Bacillus thuringiensis*. *BioScience* **1992**, *42*, 112–122.

Peng, J.; Kononowicz, H.; Hodges, T. K. Transgenic *Indica* Rice Plants. *Theor. Appl. Genet.* **1992**, *83*, 855–863.

Peng, J.; Wen, E.; Lister, R. L.; Hodges, T. K. Inheritance of *gusA* and *Neo* Genes in Transgenic Rice. *Plant Mol. Biol.* **1995**, *27*, 91–104.

Perlak, F. J.; Deaton, R. W.; Armstrong, T. A.; Fuchs, R. L.; Sims, S. R.; Greenplate, J. T.; Fischhoff, D. A. Insect Resistant Cotton Plant. *Bio/Technology* **1990**, *8*, 939–943.

Perlak, P. J.; Stone, T. B.; Muskopf, Y. M.; Peterson, L. J.; Parker, G. B.; McPherson, S. A. Genetically Improved Potatoes: Protection from Damage by Colorado Beetles. *Plant Mol. Biol.* **1993**, *1*, 1131–1145.

Pientrantonio, P. V.; Gill, S. S. *Bacillus thuringiensis* Endotoxins: Action on the Insect Midgut. In: *Biology of the Insect Midgut*; Lehane, M. J., Billingsley, P. F., Eds.; Chapman and Hall: London, 1996; pp 345–372.

Poncet, S.; Bernard, C.; Dervyn, E.; Cayley, J.; Klier, A.; Rapoport, G. Improvement of *Bacillus sphaericus* Toxicity against Dipteran Larvae by Integration, via Homologous Recombination, of the *cry11A* Toxin Gene from *Bacillus thuringiensis* Subsp. *israelensis*. *Appl. Environ. Microbiol.* **1997**, *63*, 4413–4420.

Poncet, S.; Delecluse, A.; Anello, G.; Klier, A.; Rapoport, G. Transfer and Expression of the *cryIVB* and *cryIVD* Genes of *Bacillus thuringiensis* Subsp. *israelensis* in *Bacillus sphaericus* 2297. *FEMS Microbiol. Lett.* **1994**, *117*, 91–96.

Potrykus, I. Gene Transfer to Plants: Assessment and Perspective. *Physiol. Plant.* **1990**, *79*, 125–134.

Rabindra, R. J.; Jayaraj, S. Effect of Granulosis Virus Infection on the Susceptibility of *H. armigera* to *Bt*, Endosulfan and Chlorpyriphos. *Pest Manage. Econ. Zool.* **1998**, *2*, 7–10.

Raineri, D. M.; Bottino, P.; Gordon, M. P.; Nester, E. W. *Agrobacterium*-mediated Transformation of Rice (*Oryza sativa* L.). *Bio/Technology* **1990**, *8*, 33–38.

Rajendran, R.; Rajendran, S.; Sandra, P.C. Varietals Resistance of Rice of Leaf Folder. *Int. Rice Res. News* **1986**, *11*, 17.

Ramasubbaiah, K.; Rao, P. S.; Rao, A. G. Nature of Damager and Control of Rice Leaf Folder. *Ind. J. Entomol.* **1980**, *42*, 214–217.

Ranjekar, P. K.; Patankar, A.; Gupta, V.; Bhatnagar, R.; Bentur, J.; Kumar, A. Genetic Engineering of Crop Plants for Insect Resistance. *Curr. Sci.* **2003**, *84*, 321–329.

Rao, K.V.; Rathore, K. S.; Hodges, T. K.; Fu, X.; Stoger, E.; Sudhakar, D.; Williams, S.; Christou, P.; Bharathi, M.; Bown, D. P.; Powell, K. S.; Spence, J.; Gatehouse, A.M.R.; Gatehouse, J. A. Expression of Snowdrop Lectin (*GNA*) in Transgenic Rice Plants Confers Resistance to Rice Brown Plant Hopper. *Plant J.* **1998**, *15*, 469–477.

Rashid, H.; Yokoi, S.; Toriyama, K.; Hinata, K. Transgenic Plant Production Mediated by *Agrobacterium* in *indica* rice. *Plant Cell Rep.* **1996**, *15*, 727–730.

Rathore, K. S.; Chowdhury, V. K.; Hodges, T. K. Use of *bar* as a Selectable Marker Gene and for the Production of Herbicide-resistant Rice Plants from Protoplasts. *Plant Mol. Biol.* **1993**, *21*, 871–884.

Ryan, C. A. Protease Inhibitors in Plants: Genes for Improving Defense against Insects and Pathogens. *Annu. Rev. Phytopathol.* **1990**, *28*, 425–449.

Sanford, J. C. The Biolistic Process. *Trends Biotechnol.* **1988**, *6*, 299–302.

Sasaki, J.; Asano, S.; Hashimoto, N.; Lay, B.; Hastowo, S.; Bando, H.; Iizuka, T. Characterization of a cry2A Gene Cloned from an Isolate of *Bacillus thuringiensis* Serovar Sotto. *Curr. Microbiol.* **1997**, *35*, 1–8.

Sauka, D, H.; Cozzi, J, G.; Benintende, G, B. Screening of cry2 Genes in *Bacillus thuringiensis* Isolates from Argentina. *Ant. Leeuw.* **2005**, *88*, 163–165.

Savary, S.; Willocquet, L.; Elazegui, F. A.; Castilla, N. P.; Teng, P. S. Rice Pest Constraints in Tropical Asia: Quantification of Yield Losses Due to Rice Pests in a Range of Production Situations. *Plant Dis.* **2000**, *84*, 357–369.

Schnepf, H, E.; Crickmore, N.; Vanrie, J.; Lereclus, D.; Baum, J.; Feitelson, J.; Jeigler, D. R.; Dean, D. H. *Bacillus thuringiensis* and Its Pesticidal Crystal Proteins. *Microbiol. Mol. Biol. Rev.* **1998**, *62*, 775–806.

Schnepf, H. E.; Whiteley, H. R. Cloning and Expression of the *Bacillus thuringiensis* Crystal Protein Gene in *Escherichia coli. Proc. Natl. Acad. Sci. U.S.A.* **1981**, *78*, 2893–2897.

Schuler, H. T.; Poppy, G. M.; Denholm, I. Insect Resistant Transgenic Plants. *Trends Biotechnol.* **1998**, *16*, 168–175.

Shade, R. E.; Schroeder, H. E.; Peuyo, J. J.; Murdock, L. M.; Higgins, T. J.; Chrispeels, M. J. Transgenic Pea Seeds Expressing the Alpha-amylase Inhibitor of the Common Bean are Resistant to Bruchid Beetles. *Biotechnology* **1994**, *12*, 793–796.

Sharma, K. K.; Ortiz, R. Program for the Application of Genetic Transformation for Crop Improvement in the Semiarid Tropics. *In vitro Cell Dev.* **2000**, *36*, 83–92.

Shelton, A. M.; Robertson, J. L.; Tang, J. D.; Perez, C.; Eigenbrode, S. D. Resistance of Diamond Back Moth to *Bacillus thuringiensis* Subspecies in the Field. *Econ. Entomol.* **1993**, *86*, 697–705.

Shelton, A. M.; Tang, J. D.; Roush, R. T.; Metz, T. D.; Earle, E. D. Field Tests on Managing Resistance to *Bt*-engineered Plants. *Nat. Biotechnol.* **2000**, *18*, 339–342.

Shu, Q.; Ye, G.; Cui, H.; Cheng, X.; Xiang, Y.; Wu, D.; Goa, M.; Xia, Y.; Hu, C.; Sardana, R.; Altosaar, I. Transgenic Rice Plants with a Synthetic *cry1Ab* Gene from *Bacillus thuringiensis* were Highly Resistant to Eight Lepidopteran Pest Species. *Transgen. Res.* **2000**, *9*, 433–439.

Siegel, J. P. Bacteria. In *Field Manual of Techniques in Invertebrate Pathology*; Lacey, L. L., Kaya, H. K., Eds.; Kluwer: Dordrecht, 2000; pp 209–230.

Sivamani, E.; Huet, H.; Shen, P.; Ong, C. A.; Kochko, A.; Fauquet, C. M.; Beachy, R. N. Rice Plants (*Oryza sativa* L.) Containing *Rice Tungro Spherical Virus* (RTSV) Coat Protein Transgenes Are Resistant to Virus Infection. *Mol. Breed.* **1999**, *5*, 177–185.

Smith, R. H.; Hood, E. E. *Agrobacterium tumefaciens* Transformation of Monocotyledons. *Crop Sci.* **1995**, *35*, 301–309.

Srivastava, V.; Ow, W. Biolistic Mediated Site-specific Integration in Rice. *Mol. Breed.* **2001**, *8*, 345–350.

Sudhakar, D.; Duc, L. T.; Bong, B. B.; Tinjuangjun, P.; Maqbool, S. B.; Valdez, M.; Jefferson, R.; Christou, P. An Efficient Rice Transformation System Utilising Mature Seed-derived Explants and a Portable, Inexpensive Particle Bombardment Device. *Transgen. Res.* **1998b**, *7*, 289–294.

Sudhakar, D.; Fu, X.; Stoger, E.; Williams, S.; Spence, J.; Brown, D. P.; Bharathi, M.; Gatehouse, J. A.; Christou, P. Expression and Immunolocalization of the Snowdrop Lectin, *GNA* in Transgenic Rice Plants. *Transgen. Res.* **1998a**, *7*, 371–378.

Suzuki, K.; Hattori, Y.; Uraji, M.; Ohta, N.; Iwata, K.; Murata, K.; Kato, A.; Yoshida, K. Complete Nucleotide Sequence of a Plant Tumor Inducing Ti Plasmid. *Gene* **2000**, *242*, 331–336.

Tabashnik, B. E.; Cushing, N. L.; Finson, N.; Johnson, M. W. Field Development of Resistance to *Bacillus thuringiensis* in Diamondback Moth (Lepidoptera: Plutellidae). *J. Econ. Entomol.* **1990**, *83*, 1671–1676.

Tabashnik, B. E. Evolution of Resistance to *Bacillus thuringiensis*. *Ann. Rev. Entomol.* **1994**, *39*, 47–79.

Tada, Y.; Sakamoto, M.; Fujimura, T. Efficient Gene Introduction into Rice by Electroporation and Analysis of Transgenic Plants: Use of Electroporation Buffer Lacking Chloride Ions. *Theor. Appl. Genet.* **1990**, *80*, 475–480.

Tailor, R.; Tippett, J.; Gibb, G.; Pells, S.; Pike, D.; Jordan, L.; Ely, S. Identification and Characterization of a Novel *Bacillus thuringiensis* δ-Endotoxin Entomocidal to Coleopteran and Lepidopteran Larvae. *Mol. Microbiol.* **1992**, *6*, 1211–1217.

Toki, S. Rapid and Efficient *Agrobacterium*-mediated Transformation in Rice. *Plant Mol. Biol. Rep.* **1997**, *15*, 16–21.

Toriyama, K.; Arimoto, Y.; Uchimiya, H.; Hinata, K. Transgenic Rice Plants after Direct Gene Transfer into Protoplasts. *Biotechnol.* **1988**, *6*, 1072–1074.

Tu, J.; Ona, I.; Zhang, Q.; Mew, T. W.; Khush, G. S.; Datta, S. K. Transgenic Rice Variety IR72 with *Xa21* is Resistant to Bacterial Blight. *Theor. Appl. Genet.* **1998**, *97*, 31–36.

Tu, J.; Zhang, G.; Datta, K.; Xu, C.; He, Y.; Zhang, G.; Khush, G. S.; Datta, S. K. Field Performance of Transgenic Elite Commercial Hybrid Line Expressing *Bacillus thuringiensis* δ-Endotoxin. *Nat. Biotechnol.* **2000**, *18*, 1101–1105.

Usami, S.; Okamoto, S.; Takebe, I.; Machida, Y. Factor Inducing *Agrobacterium tumefacians vir* Gene Expression Is Present in Monocotyledonous Plants. *Proc. Natl. Acad. Sci. U.S.A.* **1988**, *85*, 3748–3752.

Vaeck, M.; Reynaerts, A.; Hofte, H.; Jansens, S.; De Beuckeleer, M.; Dean, C.; Zabeau, M.; Van Montagu, M.; Leemans, J. Transgenic Plants Protected from Insect Attack. *Nature* **1987**, *328*, 33–37.

Vain, P.; Woriand, B.; Kohli, A.; Snape, J. W.; Christou, P. The Green Fluorescent Protein (GFP) as a Vital Screenable Marker in Rice Transformation. *Theor. Appl. Genet.* **1998**, *96*, 164–169.

Van Rie, J.; McGaughey, W. H.; Johnson, D. E.; Barnett, B. D.; Van Mellaert, H.; Mechanism of Insect Resistance to the Microbial Insecticide *Bacillus thuringiensis*. *Science* **1990**, *247*, 72–74.

Walters, F. S.; English, L. H.; Toxicity of *Bacillus thuringiensis* δ-Endotoxin Towards the Potato Aphid in an Artificial Diet Bioassay. *Entomol. Exp. Appl.* **1995**, *77*, 211–216.

Wang, G. L.; Ruan, D. L.; Song, W. Y.; Sideris, S.; Chen, L. L.; Pi, L.Y.; Zhang, S.; Zhang, Z.; Fauquet, C.; Gaut, B. S.; Whalen, M. C.; Ronald, P. C. The LRR Domain Encoded by the Rice Gene *Xa21D*, an *Xa21* Receptor-like Gene Family Member, Determines Race Specific Recognition and its Subject to Adaptive Evolution. *Plant Cell* **1998**, *10*, 765–779.

Wasano, N.; Ohba, M.; Miyamoto, K. Two Delta Endotoxin Genes, *cry9Da* and a Novel Related Gene, Commonly Occurring in Lepidoptera Specific *Bacillus thuringiensis* Japanese Isolates that Produce Spherical Parasporal Inclusions. *Curr. Microbiol.* **2001**, *42*(2), 129–133.

Watson, J. D.; Gilman, M.; Witkowski, J. Recombinant DNA, 2nd ed. Scientific American Books: New York, 1992.

Wong, E.Y.; Hironaka, C. M.; Fischhoff, D. A. *Arabidopsis thaliana* Small Subunit Leader and Transit Peptide Enhance the Expression of *Bacillus thuringiensis* Proteins in Transgenic Plants. *Plant Mol. Biol.* **1992,** *20,* 81–93.

Wu, A.; Sun, X.; Pang, Y.; Tang, K. Homozygous Transgenic Rice Lines Expressing *GNA* with Enhanced Resistance to the Rice Sap-sucking Pest *Laodelphax striatellus*. *Plant Breed.* **2002,** *121,* 93–95.

Wunn, J.; Kloti, A.; Burkhardt, P. K.; Biswas, G. C. G.; Launis, K.; Iglesias, K.; Potrykus, I. Transgenic *Indica* Rice Breeding Line IR58 Expressing a Synthetic *cry1Ab* Gene from *Bacillus thuringiensis* Provides Effective Insect Pest Control. *Bio/Technology* **1996,** *14,* 171–176.

Xu, D.; Xue, Q.; McElroy, D., Mawal, Y.; Hilder, V. A.; Wu, R. Constitutive Expression of a Cowpea Trypsin Inhibitor Gene, CpTi, in Transgenic Rice Plants Confers Resistance to Two Major Rice Insect Pests. *Mol. Breed.* **1996,** *2,* 167–173.

Xu, X.; Li, B. Fertile Transgenic *Indica* Rice Plants Obtained by Electroporation of the Seed Embryos Cells. *Plant Cell Rep.* **1994,** *13,* 237–242.

Yamagishi, M.; Otani, M.; Shimada, T. A Comparison of Somaclonal Variation in Rice Plants Derived and Not Derived from Protoplasts. *Plant Breed.* **1996,** *115,* 289–294.

Yara, A.; Otani, M.; Kusumi, K.; Matsuda, O.; Shimada, T.; Iba, K. Production of Transgenic *Japonica* Rice (*Oryza sativa* L.) Cultivar, Taichung 65, by *Agrobacterium*-mediated Method. *Plant Biotechnol.* **2001,** *18,* 305–310.

Yuan, H.; Ming, X.; Wang, L.; Hu, P.; An, C.; Chen, Z. Expressions of a Gene Encoding Trichosanthin in Transgenic Rice Plants Enhance Resistance to Fungus Blast Disease. *Plant Cell Rep.* **2002,** *20,* 992–998.

Zhang, W.; Wu, R. Efficient Regeneration of Transgenic Rice Plants from Rice Protoplasts and Correctly Regulated Expression of the Foreign Gene in the Plants. *Theor. Appl. Genet.* **1988,** *76,* 835–840.

Zheng, H. H.; Li, Y.; Yu, Z. H.; Li, W.; Chen, M. Y.; Ming, X. T.; Casper, R.; Chen, Z. L. Recovery of Transgenic Rice Plants Expressing the Rice Dwarf Virus Outer Coat Protein Gene (*S8*). *Theor. Appl. Genet.* **1997,** *94,* 522–527.

CHAPTER 10

BIOLOGICAL AND MOLECULAR APPROACHES FOR BIOTIC STRESS MANAGEMENT OF POSTHARVEST LOSSES IN RICE

DEEPAK KUMAR[1], MOHAMMED WASIM SIDDIQUI[2], and MD. SHAMIM[3*]

[1]*Research and Development Unit, Shri Ram Solvent Extractions Pvt. Ltd., Jaspur, Kashipur Road, U.S. Nagar, Jaspur 244712, Uttarakhand, India*

[2]*Department of Food Science and Post-Harvest Technology, Bihar Agricultural University, Sabour, Bhagalpur 813 210, Bihar, India*

[3]*Department of Molecular Biology and Genetic Engineering, Bihar Agricultural University, Sabour, Bhagalpur 813210, Bihar, India*

Corresponding author. E-mail: shamimnduat@gmail.com

CONTENTS

ABSTRACT

Rice is a staple food in several countries and serves as a good source of carbohydrates, B vitamins, and some minerals including the important trace elements like Selenium in vegetarian diets of majority of the population. But unfortunately, due to poor postharvest storage and pest and storage fungi attack estimated from up to 9% in developed countries to 20% or more in developing countries. Insect pest and storage fungi infestation have been reported to deteriorate the quality of cereals in terms of proteins, amino acids, starch, vitamins, etc. and is also responsible for creating unhygienic conditions due to mixing of insect fragments making it unfit for consumption. There is much interest in alternatives to conventional insecticides for controlling stored-product insects and fungi because of insecticide loss due to regulatory action and insect resistance and because of increasing consumer demand for product that is free of insects and insecticide residues. Some of the most promising molecular management tools for farm-stored grain are resistant varieties, transgenic cultivars, and ecofriendly approaches. Food safety is an area of concern as it has a direct impact on human health. In this background, the chapter demonstrates the critical analysis done by various researchers on quality parameters of infested rice along with a pragmatic solution to the serious problem related to food quality and safety.

10.1 INTRODUCTION

The environmental stresses, both biotic and abiotic, drastically affect the capacity of crop plants to produce their maximum yields. The various biotic factors like insects, fungi, virus, bacteria, and nematodes are a major threat to the productivity of a large number of crop plants (Rushton & Gurr, 2005). Despite the successes of the green revolution with substantial strides in food grains production, India is still classified by FAO as a low income, food deficit country, and nearly 26% of India's population are considered food insecure, consuming less than 80% of minimum energy requirements (Kumar & Bhatt, 2006).

The postharvest losses have been estimated at different stages in two major food grains, namely, rice and wheat in India. A recent estimate by the Ministry of Food and Civil Supplies, Government of India puts the total preventable postharvest losses of food grains at 10% of the total production or about 20 Mt, which is equivalent to the total food grains produced in Australia annually. In a country where 20% of the population is undernourished, postharvest losses of 20 Mt annually is a substantial avoidable waste. According to a World Bank study (1999), postharvest losses of food grains in India are 7–10% of the total production from farm to market level and 4–5% at market and distribution levels. Thus, the postharvest losses have impact at both the micro- and macro-levels of the economy. The harvested seeds of most of the orthodox crop seeds are usually dried and stored for at least one season until the commencement of the next growing season. The dry weather alters moisture content of the seed, thereby reducing the viability. Hence, seeds become practically worthless if they fail to give adequate plant stands in addition to healthy and vigorous plants. Good storage is therefore a basic requirement in seed production. The cereals, pulses, oilseeds, etc. are very important products for storage. A safe storage place must be provided for the grain produced until it is needed for consumption and multiplication purposes. Since grain production is seasonal, and consumption is continuous, safe storage must maintain grain quality and quantity. This means that grains have to be protected from weather, molds and other microorganisms, moisture, destructively huge temperatures, insects, rodents, birds, objectionable odors and contamination, and from unauthorized distribution.

In rural areas, different indigenous storage facilities are adopted by the farmers for storing the seeds. Seeds are stored in gunny sacks or vessels; mud pits; mud jar; and warehouses of mud, brick, or stone for small storage, whereas in large scale, seeds are stored for different storage purposes such as country elevator, seed processor, retail, research, and germplasm storage in stacking seed bags or containers with proper ventilation and temperature in cold store and warehouses. All practices of storage are not completely feasible to control the deterioration during storage.

The storage periods of more than 3 months make the commodity more vulnerable to insect attacks or storage fungi resulting in quantitative and qualitative losses in rice grain during storage (Ahmad, 1995; Singh et al., 1990). In India, the quantitative losses caused by storage insects are in the range of 5–25% per year (Prakash & Rao, 1995). For the management of storage pest, different fumigants such as methyl bromide, phosphine, or sulfuryl fluoride are used for storage insect controls that are not safe for the human health. Except that they induce resistance in storage pest. Additionally, storage fungi, namely, *Aspergillus* sp., *Penicillium* sp., and *Fusarium* sp. lead to the secretions of mycotoxins in food grain and deteriorate the grain quality which is important issue from consumer's health point of view (Gautam et al., 2012).

During storage, many physiological and biochemical changes occur in seeds during seed ageing. Deteriorative changes enhance when seed exposure to external challenges increases and decrease the ability of the seed to survive. It is an undesirable attribute of agriculture. Annual losses due to deterioration can be as much as 25% of the harvested crop. It is one of the basic reasons for low productivity (Shelar et al., 2008). The process has been described as cumulative, irreversible, degenerative, and inexorable process (Kapoor et al., 2011). These physiological and biochemical changes which include pasting properties, color, flavor, and composition affect the rice grain quality (Barber, 1972; Suzuki et al., 1999).

The biggest limitation of traditional breeding is its notoriously slow nature of transferring a desired trait that is often associated with linkage drag (undesirable traits) into an otherwise superior crop cultivar. The time needed to transfer a desired gene into a crop plant depends on the source of the gene and the evolutionary distance of that source to the recipient crop plant. If the gene source is a landrace or a related species, forming a primary gene pool with the crop species in question, the gene transfer may take 5–8 years. Less related wild species belonging to the secondary or even tertiary gene pool may be rich reservoirs of genes for agronomic traits like disease or pest resistance, but to transfer such genes into crop cultivars may take 10–15 years. Pre- and postfertilization barriers may impede sexual hybridization between the donor and the crop species with the problem of alien gene transfer (Jauhar, 2006). Great challenge of food security has directed plant scientists toward gene revolution that involves direct modification of qualitative and quantitative traits in an organism by transferring desired genes using "transgenic approach" or "genetic transformation." In contrast to classical breeding, genetic engineering offers an excellent tool for asexually inserting gene(s) of unrelated organisms into plant cells. This process may take less time thus accelerating the process of genetic improvement of crop plants. In addition, this exciting technology allows access to an unlimited gene pool without the constraint of sexual compatibility. Crop plants engineered to suit the environment better through incorporation of genes for tolerance to biotic and abiotic stresses have been suggested to represent an improvement in crop production (Ortiz, 1998). Over the past few decades, breeding possibilities have been broadened by genetic engineering and gene transfer technologies, as well as by gene mapping and identification of the genome sequences of model plants and crops which resulted in efficient transformation and generation of transgenic lines in a number of crop species (Gosal et al., 2009). Further, pyramiding of desirable genes with similar effects can also be achieved by using these approaches. Genetic transformation by *Agrobacterium*-mediated approaches (Sanghera et al., 2010a,b) has been successfully demonstrated for economically important traits like biotic resistance (Wani, 2010; Wani & Sanghera, 2010) and abiotic stress tolerance (Gosal et al., 2009; Wani & Sanghera, 2010) in different crop plants. Use of agrochemicals for the control of crop pests and diseases is however inefficient and environmentally and ethically unsound (Pimentel, 1995; Robinson, 1999). Genetic engineering could offer a remedy through more precise targeting of pest and viral diseases (Wani & Sanghera, 2010). Tremendous loss in yield of several economically important crops occurs due to biotic stresses (insect infestation and diseases). Significant researches have been made to produce plants with high resistance to insect pest and diseases following transgenic technology. The present chapter is an attempt to accomplish rapid crop improvements for biotic stresses using molecular technology.

10.2 POSTHARVEST INSECT PESTS OF RICE

The vast majority of insects that are pests of stored grain belong to just two orders: Coleoptera and Lepidoptera (Reed, 2010). According to their pest status, they can be divided into two groups: primary and secondary storage insect pests. Primary storage pests are able to destroy stored rice (paddy or milled grains) independently. They are responsible for severe losses of stored rice and other cereals in Africa. Secondary storage insect pests can only attack grains that have been damaged by primary insect pests (Table 10.1). Most of the storage pests have been dispersed across the world by international trade (Youm et al., 2011) and lead to quantitative and quali-tative losses, as well as price reduction in most African markets. A quan-titative loss of 18% was reported on farmers' stored rice in Benin over 4 months of storage (Togola et al., 2010). The quantitative loss was estimated on damaged and undamaged grains collected from farmers' stored paddy grains. The damage is higher when the storage exceeds 4 months, due to the high reproduction rate of these insects and their short life cycle of just 45 days (Togola, unpublished data). Moreover, these pests produce heat and moisture and contaminate rice grain with their waste products and secretions (Walker & Farrell, 2003). The damaged grains are inappropriate for trade, or for use, as food or seeds.

TABLE 10.1   Key Postharvest Insect Pests of Stored Paddy and Milled Rice.

Main species	Order (Family)	Status
Rice weevil, *Sitophilus oryzae* and Maize weevil, *Sitophilus zeamais*	Coleoptera (Curculionidae)	Primary pests of rice (paddy and milled grains)
Lesser grain borer, *Rhizopertha dominica*	Coleoptera (Bostrichidae)	Primary pest of rice (paddy and milled grains)
Angoumois grain moth, *Sitotroga cerealella*	Lepidoptera (Gelechiidae)	Primary pest of rice (paddy grain)
Red flour beetles, *Tribolium castaneum* and *T. confusum*	Coleoptera (Tenebrionidae)	Primary pest of rice (paddy and milled grains)
Rice moth, *Corcyra cephalonica*	Lepidoptera (Pyralidae)	Secondary insect pest of rice (paddy and milled grains)
Merchant grain beetle, *Oryzaephilus mercator*	Coleoptera (Sylvanidae)	Secondary insect pest of rice (paddy and milled grains)
Rust-red grain beetle, *Cryptolestes ferrugineus*	Coleoptera (Cucujidae)	Secondary insect pest of rice (paddy and milled grains)

Adapted and modified from Nwilene et al. (2013).

10.3   APPROACHES TO MANAGEMENT OF POSTHARVEST INSECT PESTS

10.3.1   POSTHARVEST INTEGRATED INSECT PEST MANAGEMENT

Postharvest insect pest management begins with good postharvest handling and management practices right from the field to the storage environment. The prerequisites and options for good storage pest management include (i) harvesting on time; (ii) maintenance and protection of the site and storage environment from birds, rodents, and the weather (controlling grain and air moisture), and basic hygiene using thermal disinfestations of the site by solar heat or treatment with traditional additives; and (iii) commodity management (cleaning and drying of appropriate packaging facilities) using hermetic storage (pits or metal drums) or treatment with synthetic insecticides/pesticides. Because of the danger of pesticides affecting the quality of rice stored for human consumption, research has made it possible to develop alternative measures to minimize pesticide risks to human and agroecosystem health. Plant products such as botanical extracts, essential oils, and vegetable oils are being explored as potential pest management tools because they are not toxic to plants, are systemic, biodegrade easily, and stimulate the host's metabolism (Dubey et al., 2008). These measures need to be explored more effectively in order to preserve the quantity and quality of stored rice in world. Moreover, technologies such as rendering males infertile by using ultraviolet rays, pheromone traps, and baits can be explored as future options to manage postharvest insects. Finally, quarantine measures should be rigorously applied in order to minimize the dispersion of invasive pests across national borders.

10.3.2 GRAIN PROTESTANTS

Application of insecticides directly to grain to prevent infestation may be warranted if grain is to be stored for more than 3–6 weeks at grain temperatures above 60–70°F. Summer-harvested grains that will be stored 1 month or longer and fall-harvested grains that will remain in storage beyond May or June of the year after harvest should be treated with a protectant insecticide. Incorporating a surface treatment is adequate for short-term protection. However, uniform application to all grain at the auger is necessary for long-term protection. If grain-protectant insecticides are applied at labeled rates, grain may be processed or fed to livestock with no waiting period. To protect against stored-grain beetles and weevils throughout the entire mass of grain within a bin, apply a protectant insecticide to grain as it is augured into the bin. Spray-on applicators may be mounted on the auger to apply liquid formulations (Yankanchi & Gadache, 2010). Dusts may be spread over a load of grain in a truck or wagon just before unloading. Protectant insecticides should not be applied to grain before high-temperature drying. A "top-dress" or "cap-off" treatment may be used to give some control of insects entering the top of the grain mass. If stirrators are used after a top-dress application, the surface of the grain mass will no longer be protected.

Insecticide Resistance in Stored Grain Insecticide resistance is an important worldwide problem that is especially common (on an international scale) in stored-product insects (Table 10.2). In Illinois, resistance to malathion is widespread among Indian meal moth populations (Benhalima et al., 2004). Some Illinois populations of the red flour beetle are resistant to malathion, but the range and intensity of this resistance problem are not well known. Populations of the hairy fungus beetle may be resistant to both actellic and malathion; the geographical range of resistant populations of this species is not known.

TABLE 10.2 Insecticides Registered for Use to Protect Stored Grain.

Insecticide	Insecticide for use-on	Rate (per 1000 bu)	Restrictions, comments
Bacillus thuringiensis (many trade names)	Rice, barley, corn, oats, rye, sorghum, soybean, sunflower, wheat	Rate depends upon the product formulation and concentration. Follow label directions for the product in use	These products control only the larval stage of Indian meal moths; they must be ingested by the larvae. Apply to the top 4–6 in. of grain as it is augured into the bin or incorporate by raking after the bin is filled
Deltamethrin plus chlorpyrifos-methyl (Storcide II)	Rice, barley, oats, rice, sorghum, wheat	6.6–12.4 fl oz in 5 gallons of water per 1000 bu. See product label for rates for individual commodities	Controls weevils, lesser grain borer, secondary beetles, and Indian meal moth. Dry grain to 12–13% moisture for residues to remain effective for 1 year or longer
Methoprene (Diacon II)	Rice, barley, corn, oats, wheat, sorghum	Apply 0.8–7.7 fl oz of formulated product in 5 gallons of water per 1000 bu. See product label for a tabular listing of dilutions	Do not apply to soybean. Methoprene prevents growth and development of immature insects but will not kill adults
Diatomaceous earth (several trade names)	Rice, barley, corn, oats, wheat, sorghum	Rate depends upon the product formulation and concentration. Follow label directions for the product in use	Do not apply to soybean. Diatomaceous earth prevents growth and development of immature insects but will not kill adults
Dichlorvos resin strips (DDVP, Vapona)	Rice, barley, corn, oats, wheat, sorghum	Hang one strip per 1000 cu ft of bin headspace	Dichlorvos strips release a vapor that kills adult Indian meal moths before they reproduce and lay eggs
Pirimiphos-methyl (Actellic 5E)	Rice, barley, corn, oats, wheat, sorghum	Apply 8.6–11.5 fl oz of actellic in 5 gallons of water per 1000 bu. Protects grain for up to 12 months at an application rate of 8.6 oz, and up to 18 months at the 11.5-oz rate	Do not apply to barley, oats, soybean, or wheat. Do not apply before high-temperature drying. Controls weevils, secondary beetles, and Indianmeal moth. Dry grain to 14–15% moisture for pirimiphos-methyl to persist for 1 year or longer. Cap-off treatments do not provide control of insects active beneath the treated layer
Pyrethrins plus piperonyl butoxide	Rice, barley, corn, oats, wheat, sorghum	Rate depends upon the product concentration. Follow label directions of the pro in use	Do not apply to soybean. Short-term residual activity. Useful mainly as a surface spray or aerosol to control larval and adult Indian meal moths, as well as other pests at the grain surface

Adapted and modified from Mason and Obermeyer (2010).

10.3.3 BOTANICAL INSECTICIDES

There is a plethora of studies on the use of plant extracts or whole plant materials for insect control, but few are used on a commercial scale (Rajandran & Sriranjini, 2008). Farmers often use homegrown or naturally occurring plant materials for insect control in developing countries. Problems with botanical insecticides are lack of consistency, safety concerns, and sometimes odor. It is often falsely assumed that because a plant material is used as a food flavoring or medicine that extracts from the material will be safe for human consumption. Various extracts from the neem tree, *Azadirachta indica*, collectively referred to as the insecticide Neem, are commercially available botanical insecticides, and local formulations have been widely used in some parts of the world for stored-product insect control (Koul et al., 1990). However, commercial formulations show only moderate levels of efficacy (Abate-Zeru, 2001; Kavallieratos et al., 2007). Crude pea flour and the protein-rich fraction of field peas, *Pisum* spp., as well as that of other food legumes (e.g., species of *Pissum*, *Phaseolus*, and *Vignia*) are toxic and repellent to stored-product insects (Bodnaryk et al., 1999; Fields, 2006). Direct application of protein-enriched pea flour to bulk grain at 0.1% by weight resulted in substantial reductions in stored-grain beetle populations (Hou & Fields, 2003), and broadscale application of pea flour to the inside of mills reportedly resulted in insect control, but such control was not at commercially acceptable levels like those achieved with synthetic fumigants. Pyrethrum, a commercial mixture of compounds derived from *Chrysanthemum cinerariifolium*, is perhaps the most successful botanical insecticide throughout all modern pest control, and this is certainly the case for stored products. The active ingredients from pyrethrum are called pyrethrins. Synergized pyrethrum commonly contains the synergist piperonyl butoxide, commonly referred to as PBO, which suppresses metabolic degradation of pyrethrins in the insect. Synergized pyrethrum is commonly used as an aerosol in flour mills (Toews et al., 2006) and is usually combined with another insecticide that has longer residual activity because the pyrethrum achieves only quick knockdown of insect pests at best, while the other insecticide with which it is combined provides longer activity (Arthur, 2008). Organically compliant pyrethrum, which lacks any synthetic synergist and is extracted from chrysanthemum flowers by methods approved by the USDA National Organic Program, has been registered in the United States in recent years and shows potential for managing stored-product insects (Campos-Figueroa, 2008), but registration of a stored-product use is pending and suitable efficacy has yet to be investigated.

10.3.4 MICROBIAL INSECTICIDES

A number of insect pathogens have been tested for control of stored-product insects, but none is in common use because of lack of sufficient, broad-spectrum efficacy. Many tests have been conducted to synergize pathogens with other control technologies, particularly those that might be expected to increase efficacy of pathogens, such as DE (51) by presumably abrading the cuticle, or grain varietal resistance by delaying larval development (Throne & Lord, 2004), both of which might make the insect more susceptible to the pathogen. Laboratory evaluations of the commercially available fungi *Beauveria bassiana* and *Metarhizium anisopliae* and the bacterium *Bacillus thuringiensis* (*Bt*), alone or in conjunction with another insecticidal material such as DE, generally result in complete control of only some stages of some species, while other stages or species are poorly controlled (Abdel-Razek, 2002; Kavallieratos et al., 2006; Throne & Lord, 2004). *Bt* generally has been most effective against Lepidoptera and Diptera, although some strains show increased efficacy for beetles (Abdel-Razek, 2002); however, efficacy is still poor compared with conventional insecticides. This lack of efficacy limits the use of pathogens in commercial applications. *Bt* has been registered for control of stored-product Lepidoptera for decades, but it has rarely been used because it does not control beetle pests. An effective granulosis virus specific for *P. interpunctella* was described and a method for low-cost mass-production was developed (Vail et al., 1991), but commercial adoption has been limited. Spinosad is an insecticide derived from metabolites in the fermentation of the actinomycete bacterium *Saccharopolyspora spinosa* Mertz and Yao (Actinomycetales: Actinomycetaceae) (Thompson et al., 2000). Spinosad is currently registered by the US EPA (2005) with a residue tolerance concentration of 1.5 ppm for use on stored grain in both conventional and organic formulations (Table 10.2). However, spinosad has not been released for use by the manufacturer as of this writing due to the lack of full approval for tolerance levels on stored grain by all international trading partners with the United States as called for under the Codex Alimentarius (internationally recognized standards or guidelines for food safety). There is much interest in the use of spinosad on stored grain because other residual insecticides registered in the United States and elsewhere have limited efficacy against the major pest of stored wheat, *Rhizopertha dominica*, either because of simple lack of efficacy or because of development of resistance. Spinosad is effective for season-long control of *R. dominica* in stored wheat; it is highly toxic to larvae of many stored-product insects and shows good compatibility with insect natural enemies (Daglish et al., 2008; Subramanyam et al., 2007; Toews & Subramanyam, 2004).

10.3.5   ENTOMOPATHOGENIC FUNGI

Although over 750 species of entomopathogenic fungi were reported to infect insects, few have received serious consideration as potential commercial candidates. The first registered mycoinsecticide was *Hirsutella thompsonii*, which has been known to cause dramatic epizootics in spider mites. The next mycoinsecticides are *Verticillium lecanii* and *Paecilomyces fumosoroseus*, which have been recently registered for control of whitefly, thrips, aphids spider mites, and storage insects. Insect fungi that have much broader host range are *B. bassiana* and *M. anisopliae*, which are effective against homopteran and lepidopteran greenhouse insects as well as coleopteran and lepidopteran field insects (Flexner & Belnavis, 1998). The broad host range of some insect fungi is an attractive characteristic for insect pests control. Nevertheless, there are numerous biotic and abiotic constraints on the ability of fungi to infect their hosts. These include desiccation, UV light, host behavior, temperature, pathogen vigor, and age. Certain aspects of the insecticidal efficacy of these fungi such as production, stability and application have been optimized by nongenetic means. For instance, advances in production and formulation technologies have contributed substantially to the cost-effectiveness and viability of mycoinsecticide as practical insect control agents.

Optimization of entomopathogenic fungi by genetic engineering is limited due to lack of knowledge of molecular and biochemical bases for fungal pathogenesis, and the unavailability of good cloning system for species other than deutromycete fungi. The molecular and biochemical bases of pathogenicity of *M. anisopliae* which cause green muscardine diseases have been well studied especially on host–cuticle penetration by the fungus. Various genes related to formation of the appressorium (a specialized structure involved in penetration of the insect cuticle by the fungus), virulence, and nutritional stress had been cloned from *M. anisopliae*. Additional copies of the *PrI* gene, which encodes a subtilisin-like protease involved in host cuticle penetration, were engineered into the genome of *M. anisopliae*. The larvae infected with recombinant strains died 25% sooner and feeding damage was reduced by 40% (St. Leger et al., 1996). The prospect of using recombinant fungi for insect control highlights the need for further research in identifying and manipulating genes involved in pathogenesis and monitoring of genetic exchange between strains by using isolate-specific molecular markers (Harrison & Bonning, 1998). Despite a potentially wide array of insecticidal proteins produced by entomopathogenic fungi, fungal genes have played little part in agricultural biotechnology to date.

10.3.6 ENTOMOPATHOGENIC NEMATODE

Insect nematodes have enormous potential for inoculative and inundative release and control of a wide range of insect pests. They are probably second only to bacteria (i.e., *Bt*) in terms of commercially important microbial insecticides. Commercially available species of nematode as bioinsecticide are in three families: Rhabditidae, Steinernematidae, and Heterorhabditidae. Nematodes parasitize their hosts by direct penetration either through the cuticle or natural opening in the host integument (i.e., spiracles, mouth, or anus). Insect death is not due to nematode itself but a symbiotic bacterium that is released upon entry into the host. The symbionts are specific with members of the genus *Xenorhabdus* associated with the steinernematids and *Photorhabdus* associated with the heterorhabditids (Lacey & Goettel, 1995). In general, both steinernematids and heterorhabditids tend to do best against soil-inhabiting insects and borers. There have been limited successes when applying to other insects. Strain selection and new formulations may be able to address this limitation. Molecular techniques such as RFLP, RAPD-PCR, AFLP, ribosomal internal transcribed spacer analysis, satellite DNA analysis have been applied to measure genetic diversity of the nematodes and provide an initial screen to identify useful strains. The development of large-scale in-vitro rearing systems and formulations that would allow for adequate shelf life and infectivity in the field are underway. Currently, nematodes are successfully grown in large-scale bioreactors similar to those used for the production of *Bt* or antibiotics. Formulation by chilling the produced nematodes prior to formulation and then mixing with materials that will enhance their handling, application, persistence, and storage will help to create a commercial venture. Another limitation of nematodes for insect control is their susceptibility to environmental stress, extreme temperature, solar radiation, and desiccation. The potential of genetic engineering to enhance these traits is being explored. In addition, genes that confer resistance to insecticide or fungicides could also be incorporated for protective purposes (Harrison & Bonning, 1998).

Efforts to engineer entomopathogenic nematodes have been immense and relied heavily on knowledge and techniques for manipulating of the

Caenorhabditis elegans genome (Poinar, 1991). The research has been focused mainly on enhancing the environmental stability with respect to heat tolerance. The nematode, *Heterorhabditis bacteriophora* was engineered to express C. *elegans* Hsp70A (heat-shock protein genes) to enhance tolerance of high temperatures (Hashmi et al., 1998). Research on genetic of heterorhabditid and steinernematid nematodes especially on pathogenicity traits are required for future genetic manipulation of more efficient strains of nematode for insect control.

10.3.7 RECOMBINANT BACULOVIRUSES FOR INSECT CONTROL

Entomopathogenic viruses have been employed as bioinsecticides for a wide range of situations from forest and field to food stores and greenhouses. Baculoviruses, particularly the nucleopolyhedro viruses (NPVs) are the most commonly used or considered for development as microbial insecticides mainly for the control of lepidopteran insects on field and vegetable crops. NPVs are formulated for application as sprays in the same fashion as chemical insecticide and *Bt* strains. However, only moderate success has been achieved due to several key limitations, which include a relatively slow speed of kill, a narrow spectrum of activity, less persistence in the field, and lack of a cost-effective system for mass production in vitro. Fermentation technology for their mass production on a large-scale commercial basis is extensively investigated to reduce the production cost.

Approaches to engineer NPVs as improved biological insecticide include deletion of genes that encode products prolonging host survival, and insertion of genes that express an insecticidal protein during viral replication. O'Reilly and Miller (1991) demonstrated that deletion of the ecdysteroid UDP-glucosyltransferase (*EGT*) gene of *Autographa californica* NPV caused infected fall armyworm, *Spodoptera frugiperda* to feed less and die about 30% sooner than larvae infected with wild-type AcNPV. Research to insert specific toxin genes or disrupters of larval development genes into baculovirus genome is progressing. The most common used strategy for engineering baculoviruses has exploited the polyhedrin or p10 promoters and the construction of recombinant baculovirus is achieved by allelic replacement of the polyhedrin gene by foreign genes. When the recombination is successful, the polyhedrin of p10 promoters drive the expression of the foreign gene to levels equivalent to those of polyhedrin or p10 in wild-type virus (Miller, 1995). Hundreds of proteins from viral, bacteria, animal, and plant origin have now been produced via such recombinant baculovirus expression vectors. Candidate genes for insertion into baculoviruses and potential to enhance pathogenicity and insecticidal activity. Introducing an insect-specific neurotoxin gene from the Algerian scorpion, *Androctonus australis*, and from the straw itch mite, *Pyemotes tritici* into insect genome using the baculovirus expression system have received the greatest attention to date (Treacy, 1999). There are several insect hormones that play vital role in the control of insect morphogenesis and reproduction and are focused for engineering into baculoviruses. These include eclosion hormone that initiates ecdysis, the process leading to the shedding of old cuticle, prothoracicotropic hormone, which is involved in triggering the molting process; allatostatins and allatotropins, which regulate the release of juvenile hormone; and diuretic hormone that regulates water balance and possibly blood pressure in insect. Another interesting gene for genetic manipulation of baculovirus is the enzyme gene, juvenile hormone esterase that caused the reduction in juvenile hormone (JH) level. A reduction in the titer of JH early in the last instar initiates metamorphosis and leads to cessation of feeding. Insertion of two or more toxin genes into baculoviruses has been studied and Hermann et al. (1995) found that binary mixtures of scorpion toxin, AaIT, and LqhIT injected into larval of *Helicoverpa virescens* induced 5–10-fold the levels of activity. The authors suggested that simultaneous expression in baculoviruses of synergistic combinations of insecticidal proteins could lead to even more potent, insect-selective bioinsecticides.

The development of baculovirus expression system and the accomplishment of insect cell-culture technology have broadened the utility of insect viruses as effective insecticides and as expression vector of foreign genes in eukaryote host for the production of useful proteins. Production, formulation, and application technology in conjunction with genetic engineering for fast kill and broader host range will be necessary to enable the development of more economic and efficacious viral products for insect control.

10.3.8   VARIETAL RESISTANCE

Resistant and tolerant rice cultivars play an important role in the reduction of yield losses due to insect pests and assessment of different rice varieties for insect resistance is an integral component of pest management. Because of its unique advantages (e.g., generally compatible with other control measures), host–plant resistance is a key component in the integrated control of rice insect pests in world. Success in identifying resistant material depends to a large extent on the ability to adequately evaluate germplasm and improved genotypes. Screening germplasm under artificial and natural pest infestations is essential for identifying better sources of resistance to major insect pests of rice. Knowledge of the mechanisms and factors contributing to host–plant resistance to insects is useful in selecting suitable criteria and breeding methodology for the genetic improvement of rice seeds for insect resistance. Some of the factors associated with resistance, such as silica content, hull integrity, and phenolic content in *Oryza sativa* varieties, can be used as "marker traits" to screen and select for resistance to pests. Considerable progress has been made by the Africa Rice Center (AfricaRice) in the development of NERICA varieties that combine the high yield potential of Asian rice (*O. sativa*) with many useful traits from the African cultivated species (*Oryza glaberrima*). Today, there are both upland and lowland NERICA varieties and traditional *O. sativa* varieties that are resistant or tolerant to some key pests of rice in Africa. Systematic evaluation of rice germplasm for AfRGM resulted in identification of over 50 primary sources of resistance among *O. glaberrima* and traditional *O. sativa* varieties (Nwilene et al., 2002). Several gall-midge-tolerant rice varieties (e.g., NERICA-L 19, NERICA-L 49, Cisadane, BW 348-1, Leizhung) have been developed and released for commercial cultivation in Africa. Nwilene et al. (2009) identified anti-xenotic and antibiotic traits associated with resistance to AfRGM in some of these rice varieties, but the traits have yet to be utilized in breeding. The quantitative trait loci or genes conferring resistance to AfRGM have also been identified from (*O. sativa* × *O. glaberrima*) crosses, ITA306 × TOS14519 and ITA306 × TOG7106. Some NERICA varieties (NERICA 5, 14, 18) are suitable for cultivation in termite-prone areas of north-central Nigeria (Agunbiade et al., 2009). Crop improvement programs need to place emphasis on developing germplasm with multiple resistance to key insect pests using biotechnological tools (e.g., marker-assisted selection), because there are often two or more stresses in most rice production environments.

10.3.9 BT-RICE FOR STORED INSECT-PEST

Rice transformation with a *Bt*-gene (delta-endotoxins from *B. thuringiensis*) was first targeted against Asian yellow stem borer (*Scirpophaga incertulas*), Asian striped stem borer (*Chilo suppressalis*), and leaffolder (*Cnaphalocrocis medinalis*) in Asia. Genetically modified rice plants (*Bt*-rice) resistant to striped stem borer, leaffolders, and other insects have been developed (Chen et al., 2005; Fujimoto et al., 1993), and two *Bt*-rice varieties (Huahui 1 and

Xianyou 63) were authorized for marketing, especially in China, in 2009 (Chen et al., 2011). Review of the literature on the entomological situation of *Bt*-rice in Asia and the current rice insect pest status in Africa led to the conclusion that, in the current state of knowledge, it seems inappropriate to introduce *Bt*-rice to control the diversity of rice insect pests in Africa (Silvie et al., 2012).

Two *Bt*-transgenic rice lines developed for control of the Asiatic rice borer, *C. suppressalis* Walker, incorporate cry1Aa and cry1B genes and had mixed nontarget effects on storage insects (Riudavets et al., 2006). *P. interpunctella* did not survive on semolina produced from the two lines, while *S. oryzae* progeny production was reduced on one of the lines and progeny production of the psocid *Liposcelis bostrychophila* (Badonnel) (Psocoptera: Liposcelididae) was reduced on the other line.

Moreover, the high diversity of insects found on rice in Africa (including Madagascar) militates against the use of *Bt*-rice due to the problem of resistance and that the toxins only affect a small number of stem borer species (van Rensburg, 2007). Response to the toxins may be quite different from one species to another. The introduction of *Bt*-rice in Africa is currently inappropriate because most African countries have no containment facilities to test transgenic rice for insect resistance. Regulatory frameworks are needed that promote quality control and informed use of transgenic crops and plant protection products.

10.4 POSTHARVEST DISEASES OF RICE AND THEIR MANAGEMENT

The postharvest diseases that cause spoilage of both durable and perishable commodities are widespread. Greater losses occur in developing countries due to nonavailability of proper storage and transportation facilities and improper handling methods, resulting in greater levels of injuries or wounds during harvesting and transit. Durable commodities are generally stored in a dry state with moisture level below 12%, whereas perishable products have higher moisture levels (about 50% or more) at the time of storage. The harvested produce might have been infected by pathogens prior to harvest under field conditions or they may get infected during transit and storage. It is estimated that, in the tropics, about 25% of all perishable food crops harvested are lost between harvest and consumption. Losses in durable commodities, such as cereals, oilseeds, and pulses, may be about 10% on a worldwide basis (Waller, 2002). The losses may be both quantitative and qualitative.

Under conditions favoring pathogens, loss caused by postharvest diseases may be greater than the economic gains achieved by improvements in primary production. Studies on postharvest diseases are primarily directed at preventing economic loss from spoilage of harvested commodities during transit and storage, and at eliminating the adverse effects of mycotoxins produced by fungal pathogens contaminating both durables and perishables. The mycotoxins are known to be carcinogenic, causing several serious ailments in humans and animals (Narayanasamy, 2005). Aflatoxins (AFs) are produced by *Aspergillus flavus*, *Aspergillus ochraceus*, and *Aspergillus parasiticus*. Contamination of AFs occurs at any stage from field to storage, whenever environmental conditions are conducive for fungi. The fungi are generally regarded as storage fungi, which grow under conditions of relatively high moisture/humidity. It causes severe liver damage and both liver and intestinal cancer in humans.

Generally, milled rice contains low levels of AFs, but parboiled rice and paddy harvested in rainy season contains high AF levels. Storage insects like rice weevil, lesser grain borer, khapra beetle, etc. also encourage AFs in paddy/rice. According to PFA Rules, 1955, the AFs in rice should not be more than 30 µg/kg. AFs belong to a group of difuranocumarinic derivatives structurally related and are produced meanly by fungi of genus *Aspergillus* spp. Its production depends on many factors such as substrate, temperature, pH, relative humidity, and the presence of other fungi (Fig. 10.1). It has been identified 18 types of AFs; the most frequent in foods are B1, B2, G1, G2, M1, and M2 (Bhatnagar et al., 2002). There is an imperative need to gather information on the microbial pathogens involved in various postharvest diseases, conditions favoring disease development, and methods of developing effective systems of disease management.

10.4.1 BIOSYNTHESIS OF AFLATOXIN BY FUNGUS

The biosynthetic pathway in *A. flavus* consists of approximately 23 enzymatic reactions and at least 15 intermediates (reviewed in Bhatnagar et al., 2006b; Brase et al., 2009) encoded by 25 identified genes clustered within a 70-kb DNA region on chromosome III (Bhatnagar et al., 2006a; Cary & Calvo, 2008; Cary & Ehrlich, 2006; Smith et al., 2007). The initial substrate acetate is used to generate polyketides with the first stable pathway intermediate being anthraquinone norsolorinic acid (NOR) (Bennett et al., 1997). This is followed by anthraquinones, xanthones, and ultimately AFs synthesis (Yu et al., 2004a,b). Few regulators of this process have been identified (Cary

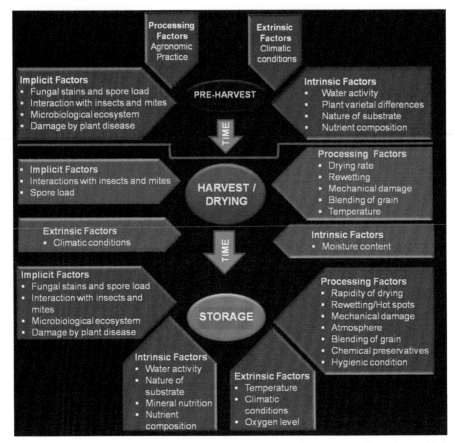

FIGURE 10.1   Interaction between intrinsic and extrinsic factors in the food chain which influences mold spoilage and mycotoxin production in stored commodities (based on Magan et al. 2004).

& Calvo, 2008), and a general model based on *Aspergillus* has recently been reviewed by Georgianna and Payne (2009). In addition to pathway-specific regulators, production of AFs is also under the control of a number of global regulatory networks that respond to environmental and nutritional cues. These include responses to nutritional factors such as carbon and nitrogen sources and environmental factors such as pH, light, oxidative stress, and temperature. Nitrogen source plays an important role in AF biosynthesis

(Bhatnagar et al., 1986). AF production, in general, is greatest in acidic medium and tends to decrease as the pH of the medium increases (Keller et al., 1997). Response to changes in pH is regulated by the globally acting transcription factor PacC, which is posttranslationally modified by a pH sensing protease (Tilburn et al., 1995). PacC binding sites indentified in the promoters of AFs biosynthetic genes could be involved in negative regulation of AFs biosynthesis during growth at alkaline pH (Ehrlich et al., 2002).

10.4.2 GENETICS OF AFLATOXIN

Cloning of genes involved in AF biosynthesis is the key to understanding the molecular biology of the pathway (Trail et al., 1995). There are 21 enzymatic steps required for AF biosynthesis and the genes for these enzymes have been cloned (Bhatnagar et al., 2003). Molecular research has targeted the genetics, biosynthesis, and regulation of AF formation in *A. flavus* and *A. parasiticus*. AFs are biosynthesized by a type II polyketide synthase, and it has been known for a long time that the first stable step in the biosynthetic pathway is the NOR, an anthraquinone (Bennett et al., 1997). A complex series of post-polyketide synthase steps follow, yielding a series of increasingly toxigenic anthraquinone and difurocoumarin metabolites (Trail et al., 1995). Sterigmatocystin (ST) is a late metabolite in the AF pathway and is also produced as a final biosynthetic product by a number of species. It is now known that ST and AFs share almost identical biochemical pathways (Bhatnagar et al., 2003). AF was one of the first fungal secondary metabolites shown to have all its biosynthetic genes organized within a DNA cluster (Fig. 10.2). These genes, along with the pathway specific regulatory genes *aflR* and *aflS*, reside within a 70-kb DNA cluster near the telomere of chromosome 3 (Sweeney et al., 1999; Georgianna & Payne, 2009). Research on *A. flavus*, *A. parasiticus*, and *A. nidulans* has led to our current understanding of the enzymatic steps in the AF biosynthetic pathway, as well as the genetic organization of the biosynthetic cluster. Two genes, *aflR* and *aflS*, located divergently adjacent to each other within the AF cluster are involved in the regulation of AF/ST-gene expression. The gene *aflR* encodes a sequencespecific DNA-binding binuclear zinc cluster (Zn(II)2Cys6) protein, required for transcriptional activation of most, if not all, of the structural genes (Georgianna & Payne, 2009).

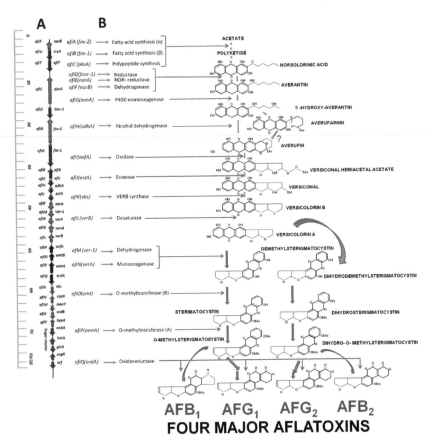

FIGURE 10.2   The gene cluster responsible for aflatoxins biosynthesis in A. flavus and A. parasiticus. (A) Clustered genes (arrows indicate the direction of gene transcription) and (B) the AF biosynthetic pathway (based on Yu et al., 2004b).

10.4.3   DETECTION OF MYCOTOXINS IN FOOD

Aflatoxigenic fungi can contaminate food commodities, including cereals, peanuts, spices, and figs. Foods and feeds are especially susceptible to colonization by aflatoxigenic *Aspergillus* species in warm climates, where they may produce AFs at several stages in the food chain, that is either at preharvest, processing, transportation, or storage (Ellis et al., 1991). The level of mold infestation and identification of the governing species are important indicators of raw material quality and predictors of the potential risk of mycotoxin occurrence (Shapira et al., 1996). Traditional methods for the identification and detection of these fungi in foods include culture in different media and morphological studies. This approach, however, is time-consuming, laborious, and requires special facilities and mycological expertise (Edwards, 2004). Moreover, these methods have a low degree of sensitivity and do not allow the specification of mycotoxigenic species (Zhao et al., 2001). Polymerase chain reaction (PCR)-based methods that target DNA are considered a good alternative for rapid diagnosis due to their high specificity and sensitivity and have been used for the detection of aflatoxigenic strains of *A. flavus* and *A. parasiticus* (Somashekar et al., 2004). However, as yet, none of these methods can reliably differentiate *A. flavus* from other species of the *A. flavus* group. In particular, *A. flavus* and *A. parasiticus* have different toxigenic profiles, *A. flavus* produces AF B1 (M1), B2, cyclopiazonic acid, aflatrem, 3-nitropropionic acid, ST, verdsicolorin A, and aspetoxin, whereas *A. parasiticus* produced AF B1 (M1), B2, G1, G2, and versicolorin A. Another important fact is that *A. flavus* and *A. fumigatus* are responsible for 90% of the aspergillosis in human beings (Gonzalez-Salgado et al., 2008). It is evident that one fundamental solution to the problem of mycotoxins in food would be to ensure that no contamination of edible crops occurred during harvesting and storage. It is equally clear, however, that such a solution is virtually unattainable, and hence that the presence of mycotoxins in food will have to be accommodated. Three approaches to the problem are most widely encountered; one involves physicochemical methods of analysis, other relies on biological assays, and another one is microscopic examination. The former approach has found most widespread acceptance for routine purposes, but some authorities feel that a chemical diagnosis should be supported with some form of demonstration that the detected material is, in fact, biologically toxic. The validity of this requirement is open to debate, but, for specific legal purposes, it may well become obligatory (Robinson, 1975).

10.5 APPROACHES TO POSTHARVEST CONTROL STRATEGIES FOR STORAGE FUNGI

It is well established that mycotoxin contamination of agricultural product can occur in the field as well as during storage (Wilkinson, 1999). Since phytopathogenic fungi such as *Fusarium* and *Alternaria* spp. can produce mycotoxins before or immediately after postharvesting, several strategies have been developed including biological and cultural control practices to help mycotoxin contamination occurring in this way.

10.5.1 IMPROVING OF DRYING AND STORAGE CONDITIONS

In cereals, mycotoxigenic fungal growth can arise in storage as a result of moisture variability within the grain itself or as a result of moisture migration results from the cooling of grains located near the interface with the wall of the storage container/silo (Topal et al., 1999). Thus, control of adequate aeration and periodical monitoring of the moisture content of silos plays an important role in the restriction of mycotoxin contamination during the storage period (Heathcote & Hibbert, 1978). The moisture level in stored crops is one of the most critical factors in the growth of mycotoxigenic molds and in mycotoxin production (Abramson, 1998) and is one of the main reasons for mycotoxin problems in grain produced in developing countries. Cereal grains are particularly susceptible to grow by Aspergilli in storage environments. The main toxigenic species are *A. flavus* and *A. parasiticus* for AFs, and *Penicillium verrucosum* is the main producers in cereals for ochratoxin A (OTA) (Lund & Frisvad, 2003), while *A. ochraceus* is typically associated with coffee, grapes, and species, AFs can be produced at aw values ranging from 0.95 to 0.99 with a minimum aw value of 0.82 for *A. flavus*, while the minimum aw for OTA production is 0.80 (Sweeney & Dobson, 1998). It has been reported that *A. flavus* will not invade grain and oilseeds when their moisture contents are in equilibrium with a relative humidity of 70% or less. The moisture content of wheat at this relative humidity is about 15%, and around 14% for maize, but it is lower for seeds containing more oil, approximately 7% and 10% for peanuts and cottonseeds, respectively (Heathcote & Hibbert, 1978), while *A. parasiticus* has been reported to produce AFs at 14% moisture content in wheat grains after 3 months of storage (Atalla et al., 2003). The second critical factor influencing postharvest mold growth and mycotoxin production is temperature. Both the main AF producing Aspergillus strains *A. flavus* and *A. parasiticus* can grow in the temperature range from 10–12°C to 42–43°C, with an optimum in the 32–33°C range, with several studies highlighting the relatively high incidence of mycotoxins such as AFs and ochratoxins in foods and feeds in tropical and subtropicals regions (Soufleros et al., 2003). The control of temperature of the stored grain at several fixed time intervals during storage may be important in determining mold growth. A temperature rise of 2–3°C may indicate mold growth or insect infestation. Until recently, little if any work has been carried out on monitoring how spoilage fungi interact with each other in the stored grain ecosystem, and the effect that this has on mycotoxin production. Magan et al. (2003) have shown that the system is in a state of dynamic flux with niche overlap altering in direct response to temperature and aw levels. It appears that the fungi present tended to occupy separate niches, based on resources utilization, and this tendency increased with drier conditions. Initially, *A. flavus* and other *Aspergillus* spp. were considered exclusively storage fungi, and AF contamination was believed to be primarily a storage problem. This is very severe in many rural areas that lack of infrastructure for drying and other appropriate storage conditions. Usually, corncobs are harvested at moisture contents that vary between 25% and 30% and are dried under the sunlight to reach 12–14% moisture content. Research has been conducted to determine the optimum temperature and moisture content of grains during storage to prevent *Aspergillus* spp. growth and AF production. In maize inoculated with *A. flavus* and stored at 27°C for 30 days with varying moisture contents, an association between moisture content and AF levels was established. At 16% moisture, AF levels reach 116 μg/kg, while a 22% moisture 2166 μg/kg AF levels were obtained (Moreno-Martínez et al., 2000). In this same study, the authors tested the protective effects of propionic acid salts (6.5–12.5 l/t) on fungal growth and AF production. All grains treated with ammonium, calcium, or sodium propionates yielded very low *A. flavus* growth and AF levels (2–5.6 μg/kg) at all moisture contents. It is well established that rapid crop drying may be useful in controlling AF contamination in storage and that in addition that crops containing different moisture values are not stored together. It is also well established that mold invasion is facilitated as a result of increased moisture levels of stored commodities. Moisture abuse can even occur in crops with very low moisture content. Another factor to bear in mind is the fact that if fungal growth does occur in storage, moisture will be released during metabolism, which will be released during metabolism, leading to the growth of other fungal species and to the production of mycotoxins such as OTA.

10.5.2 BIOLOGICAL CONTROL OF AFLATOXINS

The first approach which we will discuss is the biological control, which focuses in the use of living organisms to control pests (insects, weeds, diseases, and disease vectors) in agriculture. The objective of the biologic control is to stimulate the colonization of antagonist organism on plant surfaces to reduce the inoculum of the pathogens (FAO, 2004). Different organisms, including bacteria, yeasts, and nontoxigenic *Aspergillus* fungi, have been tested for their ability in the control of AF contamination in laboratory experiments (Yin et al., 2008, 2009), the same effect was observed in strains of *Bacillus subtilis* and *P. solanacearum* isolated from the nonrhizophere of rice soil were also able to inhibit AF accumulation (Nesci et al., 2005). In other experiments, it is showed that *B. subtilis* prevented AF contamination in rice in field tests when ears were inoculated with the bacterium 48 h before inoculation with *A. flavus* (Cuero et al., 1991). However, no reduction in AF occurred when bacteria were inoculated 48 h after inoculation with *A. flavus*. *B. subtilis* (NK-330) did not inhibit AF contamination in peanuts, when it was applied to pods prior to warehouse storage for 56 days (Smith et al., 1990). Saprophytic yeasts isolated from fruits of almond, pistachio, and walnut trees inhibited AF production by *A. flavus* in vitro (Hua et al., 1999; Masoud & Kaltoft, 2006). A strain of *Candida krusei* and a strain of *Pichia anomala* reduced AFs production by 96% and 99%, respectively, in a Petri-dish assay. Efforts are underway to apply these yeasts to almond and pistachio orchards to determine their potential for AF reduction under crop production conditions (Hua, 2002). Although they were considered to be potential biocontrol agents for management of AFs, further field experiments are necessary to test their efficacies in reducing AFs contamination under field conditions (Yin et al., 2008). Alternatively, a limited number of biocompetitive microorganisms have been shown for the management of *Fusarium* infections. Antagonistic bacteria and yeasts may also lead to reductions in preharvest mycotoxin contamination. For instance, *B. subtilis* has been shown to reduce mycotoxin contamination by *F. verticilloides* during the endophytic growth phase. Similarly, antagonistic yeasts such as *Cryptococcus nodaensis* have also been shown to inhibit various *Fusarium* species (Cleveland et al., 2003). Recent glasshouse studies by Diamond and Cooke (2003) involving the pre-inoculation of wheat ears at anthesis, with the two nonhost pathogens, *Phoma betae* and *Pytium ultimum*, showed a reduction in disease development and severity caused by *F. culmorum*, *F. avenaceum*, *F. poae*, and *M. nivale*. *A. flavus* is not considered to be an aggressive invader of preharvest corn ear tissue. Garman and Jewett (1914) reported that in years with high insect populations, the incidence of moldy ears in field rice increased. Efforts to determine the specific role of insects in the *A. flavus* infection process increased dramatically when AF was recognized as a health concern, leading to recognition that panicle feeding insects, for example, shoot borer, *Ostrinia nubilalis*, Aphids, fall armyworm, *S. frugiperda* can increase AF levels in preharvest rice (Catangui & Berg, 2006). The difficulty in establishing the relationship between insect damage and AF incidence is in part due to *A. flavus* ability to colonize panicle, infect seeds, and produce AFs in developing ears under insect-free conditions (Jones et al., 1980), and in part due to unknown factors that result in conflicting information (Abbas et al., 2009). Because the relationship between insect damage to shoot borer and AF is heavily influenced by environmental conditions, success in managing AF contamination via insect control has been highly variable. The greatest success to date regarding biological control of AFs contamination in the field has been achieved through competitive exclusion by applying on aflatoxigenic strains of *A. flavus* and *A. parasiticus* to soil of developing crops. These strains are typically referred to as atoxigenic or nontoxigenic, but those designations are often used with reference to production of AFs only (Dorner, 2004). According to Yin et al. (2008), the use of nontoxigenic *Aspergillus* strains is a strategy based on the application of nontoxigenic strains to competitively exclude naturally toxigenic strains in the same niche and compete for crop substrates. Thus, for competitive exclusion to be effective, the biocontrol nontoxigenic strains must be predominant in the agricultural environments when the crops are susceptible to be infected by the toxigenic strains (Cole & Cotty, 1990; Cotty, 1994; Dorner, 2004). For this to work, the applied strains must occupy the same niche as the naturally occurring toxigenic strains and compete for crop substrates (Dorner, 2004). Two primary factors exist that determine the effectiveness of this strategy. First, the applied strain(s) must be truly competitive and dominant relative to the toxigenic strains that are already present.

Second, the formulation used to apply the competing strain(s) must be effective in delivering the necessary quantity of conidia to achieve a competitive advantage. In addition, the timing of that application is crucial for ensuring that the necessary competitive level is present when the threat of crop infection is greatest (Cotty, 1989; Dorner, 2004). Should be noted, that not only species of *Aspergillus* used for biological control are capable of producing AFs, but also a variety of other toxins and toxic precursors to AFs including cyclopiazonic acid, ST and related compounds, and the versicolorins (Cole & Cox, 1981). In the research realized by Cotty (1994) in greenhouse, demonstrated the ability of seven non-aflatoxigenic strains of *A. flavus* to reduce AFs contamination of cottonseed when were co-inoculated with toxigenic strains. Six of these strains show significantly reduced the amount of AFs produced in cottonseed by the toxigenic strain. Strain 36 (AF36) produced the largest reduction in AF under these conditions and it was biological control of Aflatoxin Contamination of Crops 429 subsequently shown to reduce AF contamination of cottonseed in the field, when applied on colonized wheat seed (Cotty, 1994). This strain has been registered on cotton for control of AF contamination of cottonseed in Arizona, USA. It is also on a schedule for registration on pistachio in California. Additionally, this biocontrol agent was also tested for control of AF in corn (Cotty, 1996).

When corn ears were either co-inoculated with AF36 and a toxigenic strain of *A. flavus* or inoculated with AF36 at 24 h prior to inoculation with the toxigenic strain, subsequent AFs concentrations were significantly reduced, compared to inoculation with the toxigenic strain alone (Brown et al., 1991).

10.5.3   RESISTANT VARIETIES AND TRANSGENIC

Research has demonstrated that insecticides cannot be applied economically to control corn insects well enough to reduce AF to acceptable levels. The most successful approach has been the use of corn resistant to panicle-feeding insects. Several authors have shown that *Bacillus thuringensis* (*Bt*)-transformed rice hybrids, which are resistant to panicle-feeding insects, reduce AF contamination of the grain. The adoption of *Bt* rice hybrids has given producers crop with increased insect resistance, however these hybrids may only reduce AF contamination under certain circumstances. However, commercial production of these genetically modified hybrids is not allowed in some nations. Several sources of natural resistance to insects have been identified, and crosses between insect- and AF-resistant lines have shown potential to increase resistance to both insect damage and AF contamination (Williams et al., 2002). Ideally, management of AF contamination should begin with the employment of resistant genotypes as has been demonstrated by several US breeding programs. In Mexico, the wide genetic diversity of rice has not been fully exploited to identify resistance to AF contamination in breeding programs, thus impeding the reduction of AF levels in the field. Additional complications come from the fact that transgenic rice expressing insecticidal protein or any other trait to reduce AF is not viable in Mexico due to a government prohibition on the use of genetically modified rice (Plasencia, 2004). Four major genetically controlled components for which variability exist appear to be involved in determining the fate of *A. flavus*-grain interaction: (1) resistance to the infection process, (2) resistance to toxin production, (3) plant resistance to insect damage, and (4) tolerance to environmental stress (Widstrom, 1987). The latter two components have an indirect influence since their effects only reduce AF contamination but do not prevent it. Although differences among genotypes have been found, heritability of the trait appears to be low, and the genotype/environment interaction may often mask true differences among genotypes (Plasencia, 2004). There are many new and exciting preharvest prevention strategies being explored that involve new biotechnologies. These new approaches involve the design and production of plants that reduce the incidence of fungal infection, restrict the growth of toxigenic fungi, or prevent toxic accumulation. Biocontrols using nontoxigenic biocompetitive agents are also a potentially useful strategy in corn. However, the possibility of recombination with toxigenic strains is a concern (Abdel-Wahhab & Kholif, 2008). The differences between crop species appear to differ between countries. This is probably due to the differences in genetic pool within each country's breeding program and the different environmental and agronomic conditions in which crops are cultivated (Edwards, 2004). With respect to genetic resistance to *Aspergillus* infection and subsequent AF production, since the early 1970s, much work has been done to identify genetically resistant crop genotypes in both laboratory and field-based experiments to help control of aflatoxigenic mold growth and AF biosynthesis (D'Mello et al., 1998). This has led to the identification of a number of well-characterized sources of both resistance of *A. flavus* infection and to AF production. These include kernel proteins such as a 14-kDa trypsin-inhibiting protein and others including globulin 1 and 2 and a 22-kDa zeamatin protein (Chen et al., 2001). Although the role of insects in fostering *Aspergillus* colonization of rice seeds is well documented, there is little evidence that transgenic corn expressing insecticidal proteins has a significant effect on reducing AF contamination. Cry-type proteins constitute a family of insecticidal proteins from *B. thuringensis*, whose genes have been incorporated into several crops to confer protection against insect pests. In rice, several hybrids expressing distinct Cry-type proteins have been developed and widely used in the United States, Canada, Argentina, and other maize-producing countries (Plasencia, 2004). The distribution of AF in agricultural commodities has been fairly well characterized because of its importance to food supply. However, little is known on the occurrence and fate of AF in soil. Radiological assays conducted to assess the fate of AF B1 (AFB1) in soil indicated that a low level of mineralization of AFB1 to CO_2 was observed, with less than 1–8% mineralized in 120 days (Angle, 1986). Not surprisingly, several microorganisms have the potential to degrade AFs, especially bacteria, for example, *Flavobacterium* and *Mycobacterium* (Hormisch et al., 2004). In addition, *A. flavus* also is capable of degrading AFs during later stages of mycelial growth in pure culture (Huyhn & Lloyd, 1984). In recent years, molecular techniques have increased the possibilities to characterize soil microbial ecology. While molecular methods have been extensively used for studying soil bacteria, these techniques have been applied to studying soil fungi, such as the biological control agents *Colletotrichum coccodes* (Dauch et al., 2003), Trichoderma (Weaver et al., 2005), and mycorrizal fungi (Ma et al., 2005). Amplification of specific DNA fragments using PCR and specific gene probes is extremely sensitive and has the potential to detect the presence of *A. flavus* in agricultural commodities (Manonmani et al., 2005). Since all of the genes involved in the AF biosynthesis pathway have been identified and cloned (Yu et al., 2004a,b), and the entire genome of *A. flavus* sequenced (Payne et al., 2006), molecular methods for the detection of *Aspergillus* should be fairly readily adapted by using biosynthetic pathway genes as probes, as evidenced by the recent work differentiating toxigenic and atoxigenic *A. flavus*-utilizing AF gene expression using the reverse transcription polymerase chain reaction (RT-PCR) (Degola et al., 2007). Application of these molecular techniques to *A. flavus* soil ecology should greatly enhance our understanding of this fungus. *A. flavus* is commonly considered a saprophytic fungus; however, its ability to colonize growing crops and inflict economic damage clearly shows that it can and does function as an opportunistic pathogen. Despite the elucidation of many aspects influencing *A. flavus* ability to colonize crops and accumulate AFs, its activity and potential to produce AFs in soil and in crop residues has remained unexplored (Accinelli et al., 2008). One interesting approach is the engineering of cereal plants to catabolize fumonisins *in situ*. Typically, these approaches require considerable research and development but have the potential of ultimately producing low cost and effective solutions to the mycotoxin problem in rice and other cereals. Thus, this level of prevention is the most important and effective plan for reducing fungal growth and mycotoxin production.

10.6   CONCLUSION

In India, rice grains are stored in traditional household grain storage system as well as large-scale centralized system managed by two main government agencies—Food Corporation of India and Central Ware Housing Corporation. As the basic nature of food is to remain diverse, decentralized small-scale storage system are generally managed and monitored properly leading to better food quality and safety. Above research findings reveal great concern for industrial food products based on wheat grains where stored grain pest and fungi infestation cannot be determined, thus making the quality control process extremely difficult. Hence, in the light of above limited but significant research findings in depth analysis of grains infested in a natural way under field conditions especially in centralized system needs to be carried out. One of the major needs in the use of conventional and transgenic resistance for control of storage insect pests is the identification of resistant germplasm and specific mechanisms of resistance to storage pests. Rapid tests for identifying resistance factors would speed development of new varieties. Characterization of resistance levels in currently used commercial varieties may help to identify mechanisms of resistance that are compatible with agronomic traits and provide an immediate management tool for producers through selection of resistant varieties for planting.

Development of transgenic crops requires screening for proteins and other compounds that are toxic to stored-product insects. Currently, we rely on screening programs targeted for pests of crops in the field. This might include screening cereals for inhibitors of enzymes required for insect digestion and other physiological processes, or screening for naturally occurring substances in the plant or harvested grain that interfere with normal insect development. Genes for toxins produced by plants and other organisms are also potential candidates for incorporation into transgenic crops.

KEYWORDS

- biotic factors
- storage
- postharvest loss
- biochemical changes
- grain quality
- storage pest

REFERENCES

Abate-Zeru, S. The Potential for Natural Products as Grain Protectants against Stored Product Insects, MS Thesis, Okla. State Univ., Stillwater, 2001; p 68.

Abbas, H. K.; Wilkinson, J. R.; Zablotowics, R. M.; Accinelli, C.; Abel, C. A.; Bruns, H. A.; Weaver, M. Ecology of *Aspergillus flavus*, Regulation of Aflatoxin Production, and Management Strategies to Reduce Aflatoxin Contamination of Corn. *Toxin Rev.* **2009**, *28*, 142–153.

Abdel-Razek, A. S. Comparative Study on the Effect of Two *Bacillus thuringiensis* Strains of the Same Serotype on Three Coleopteran Pests of Stored Wheat. *J. Egypt. Soc. Parasitol.* **2002**, *32*, 415–424.

Abdel-Wahhab, M. A.; Kholif, A. M. Mycotoxins in Animal Feeds and Prevention Strategies: A Review. *Asian J. Animal Sci.* **2008**, *2*, 7–25.

Abramson, D. Mycotoxin Formation and Environmental Factors. In: *Mycotoxins in Agriculture and Food Safety*; Sinha, K. K., Bhatnagar, D., Eds.; Marcel Dekker, Inc.: New York, **1998**; pp 255–277.

Accinelli, C.; Abbas, H. K.; Zablotowicz, R. M.; Wilkinson, J. R. *Aspergillus flavus* Aflatoxin Occurrence and Expression of Aflatoxin Biosynthesis Genes in Soil. *Can. J. Microbiol.* **2008**, *54*, 371–379.

Agunbiade, T. A.; Nwilene, F. E.; Onasanya, A.; Semon, M.; Togola, A.; Tamo, M.; Falola, O. O. Resistance Status of Upland NERICA Rice Varieties to Termite Damage in Northcentral Nigeria. *J. App. Sci.* **2009**, *9*, 3864–3869.

Ahmad, M. *Grain Storage Management Newsletter*; University of Agriculture: Faisalabad, **1995**, p 1.

Angle, J. S. Aflatoxin Decomposition in Various Soils. *J. Environ. Sci. Health B* **1986**, *21*, 277–288.

Arthur, F. H. Aerosol Distribution and Efficacy in a Commercial Food Warehouse. *Insect Sci.* **2008**, 15, 133–140.

Atalla, M. M.; Hassanein, N. M.; El-Beith, A. A.; Youssef, Y. A. Mycotoxin Production in Wheat Grains by Different Aspergilli in Relation to Different Relative Humidities and Storage Periods. *Nahrung* **2003**, *47*, 6–10.

Barber, S. Milled Rice and Changes during Ageing. In: *Rice Chemistry and Technology*, 1st ed.; Houston, D. F., Ed.; Am. Assoc. Cereal Chem.: St Paul, MN, 1972; pp 215–263.

Benhalima, H.; Choudhary, M. Q.; Millis, K. A.; Price, N. Phosphine Resistance in Stored-product Insect Collected from Various Grain Storage Facilities. *J. Stored Product Res.* **2004**, *40*(3), 241–249.

Bennett, J. W.; Chang, P. K.; Bhatnagar, D. One Gene to Whole Pathway: The Role of Norsolorinic Acid in Aflatoxin Research. *Adv. Appl. Microbiol.* **1997**, *45*, 1–15.

Bhatnagar, D.; Cary, J. W.; Ehrlich, K. C.; Yu, J.; Cleveland, T. E. Understanding the Genetics of Regulation of Aflatoxin Production and *Aspergillus flavus* Development. *Mycopathologia* **2006b**, *162*, 155–166.

Bhatnagar, D.; Ehrlich, K. C.; Yu, J.; Cleveland, T. E. Molecular Genetic Analysis and Regulation of Aflatoxin Biosynthesis. *Appl. Microbiol. Biotechnol.* **2003**, *61*, 83–93.

Bhatnagar, D.; Proctor, R.; Payne, G. A.; Wilkinson, J.; Yu, J.; Cleveland, T. E.; Nierman, W. C. Genomics of Mycotoxigenic Fungi. In: *The Mycotoxigenic Fact Book (Food and Feed Topics)*; Barug, D., Bhatnagar, D., van Egmond, H. P., van der Kamp, J. W., van Osenbruggen, W. A., Visconti, A., Eds.; Wagningen Academic Publishers: Wagningen, The Netherlands, **2006a**; pp 157–178.

Bhatnagar, D.; Yu, J.; Ehrlich, K. C. Toxins in Filamentus Fungi. In: *Fungal allergy and Pathogenicity*; Breitenbach, M., Crameri, R., Lehrer, S. B., Eds.; Chem. Immunol.: Basel, Karger, 2002, pp 167–206.

Bhatnagar, R. K.; Ahmad, S. K.; Mukerji, G. Nitrogen Metabolism in *Aspergillus parasiticus* NRRL 3240 and *A. flavus* NRRL 3537 in Relation to Aflatoxin Production. *J. Appl. Bacteriol.* **1986**, *60*, 203–211.

Bodnaryk, R. P.; Fields, P. G.; Xie, Y.; Fulcher, K. A. US Patent No. 5,955,082, 1999.

Brase, S.; Encinas, A.; Keck, J.; Nising, C. F. Chemistry and Biology of Mycotoxins and Related Fungal Metabolites. *Chem. Rev.* **2009**, *109*, 3903–3990.

Brown, R. L.; Cotty, P. J.; Cleveland, T. E. Reduction in Aflatoxin Content of Maize by Atoxigenic Strains of *Aspergillus flavus*. *J. Food Prot.* **1991**, *54*(8), 623–626.

Campos-Figueroa, M. Attract-and-kill Methods for Control of Indianmeal Moth, *Plodia interpunctella* (Hübner) (Lepidoptera: Pyralidae), and Comparisons with Other Pheromone-based Control Methods, Ph. D. Thesis. Okla. State Univ, 2008; p 114.

Cary, J. W.; Calvo, A. M. Regulation of *Aspergillus* mycotoxin Biosynthesis. *Toxin Rev.* **2008**, *27*, 347–370.

Cary, J. W.; Ehrlich, K. Aflatoxigenicity in *Aspergillus*: Molecular Genetics, Phylogenetic Relationships and Evolutionary Implications. *Mycopathologia* **2006**, *162*, 167–177.

Catangui, M. A.; Berg, R. K. Western Beat Cutworm, *Striacosta albicosta* (Smith) (Lepidoptera: Noctuidae), as a Potential Pest of Transgenic Cry1Ab *Bacillus thuringensis* Corn Hybrids in South Dakota. *Environ Entomol.* **2006**, *35*, 1439–1452.

Chen, H.; Tang, W.; Xu, C. G.; Li, X. H.; Lin, Y. J.; Zhang, Q. F. Transgenic *indica* Rice Plants Harboring a Synthetic cry2A* Gene of *Bacillus thuringiensis* Exhibit Enhanced Resistance against Lepidopteran Rice Pests. *Theor. Appl. Genet.* **2005**, *111*, 1330–1337.

Chen, M.; Shelton, A.; Ye, G. Y. Insect-resistant Genetically Modified Rice in China: From Research to Commercialization. *Ann. Rev. Entomol.* **2011**, *56*, 81–101.

Chen, Z. Y.; Brown, R. L.; Cleveland, T. E.; Damann, K. E.; Russin, J. S. Comparison of Constitutive and Inducible Maize Kernel Proteins of Genotypes Resistant or Susceptible to Aflatoxin Production. *J. Food Prot.* **2001**, *64*, 1785–1792.

Cleveland, T.; Dowd, P. F.; Desjardins, A. E.; Bhatnagar, D.; Cotty, P. J. United States Department of Agriculture—Agricultural Research Service Research on Preharvest Prevention of Mycotoxins and Mycotoxigenic Fungi in US Crops. *Pest Manage. Sci.* **2003**, *59*, 629–642.

Cole, R. J.; Cotty, P. J. Biocontrol of Aflatoxin Production by Using Biocompetitive Agents. In: *A Perspective on Aflatoxin in Field Crops and Animal Food Products in the United States: A Symposium: ARS-83*; Robens, J., Huff, W., Richard, J., Eds.; US Department of Agriculture, Agricultural Research Service: Washington, DC, 1990; pp 62–66.

Cole, R. J.; Cox, R. H. *Handbook of Toxic Fungal Metabolites*. Academic Press: New York, 1981; p 937.

Cotty, P. J. Influence of Field Application of an Atoxigenic Strain of *Aspergillus flavus* on the Population of *A. flavus* Infecting Cotton Balls and on the Aflatoxin Content of Cottonseed. *Phytopathology* **1994**, *84*, 1270–1277.

Cotty, P. J. Aflatoxin Contamination of Commercial Cottonseed Caused by the S strain of *Aspergillus flavus*. *Phytopathology* **1996**, 86, S71.

Cotty, P. J. Virulence and Cultural Characteristics of Two *Aspergillus flavus* Strains Pathogenic on Cotton. *Phytopathology* **1989**, *79*, 808–814.

Cuero, R. G.; Duffus, E.; Osuji, G.; Pettit, R. Aflatoxin Control in Preharvest Maize: Effects of Chitosan and Two Microbial Agents. *J. Agric. Sci.* **1991**, *117*, 165–169.

D'Mello, J. P. F.; McDonald, A. M. C.; Postel, D.; Dijksma, W. T. P.; Dujardin, A.; Placinta, C. M. Pesticide Use and Mycotoxin Production in *Fusarium* and *Aspergillus phytopathogenes*. *Eur. J. Plant Pathol.* **1998**, *104*, 741–751.

Daglish, G. J.; Head, M. B.; Hughes, P. B. Field Evaluation of Spinosad as a Grain Protectant for Stored Wheat in Australia: Efficacy Against *Rhyzopertha dominica* (F.) and Fate of Residues in Whole Wheat and Milling Fractions. *Aust. J. Entomol.* **2008**, *47*, 70–74.

Dauch, A. L.; Watson, A. K.; Jabaji-Hare, S. H. Detection of the Biological Control Agent *Colletotrichum coccodes* (183088) from the Target Weed Velvetleaf and Soil by Strain Specific PCR Markers. *J. Microbiol. Methods* **2003**, *55*, 51–64.

Degola, F.; Berni, E.; Dall'Asta, C.; Spotti, E.; Marchelli, R.; Ferrero, I.; Restivo, F. M. A Multiplex RT-PCR Approach to Detect Aflatoxigenic Strains of *Aspergillus flavus*. *J. Appl. Microbiol.* **2007**, *103*, 409–417.

Diamond, H.; Cooke, B. M. Preliminary Studies on Biological Control of the Fusarium Ear Blight Complex of Wheat. *Crop Prot.* **2003**, *22*, 99–107.

Dorner, J. W. Biological Control of Aflatoxin Contamination of Crops. *J. Toxicol. Toxicol. Rev.* **2004**, *23*, 425–450.

Dubey, N. K.; Srivastava, B.; Kumar, A. Current Status of Plant Products as Botanical Pesticides in Storage Pest Management. *J. Biopest.* **2008**, *1*(2), 182–186.

Edwards, S. G. Influence of Agricultural Practices on Fusarium Infection of Cereals and Subsequent Contamination of Grain by Trichothecene Mycotoxins. *Toxicol. Lett.* **2004**, *153*, 29–35.

Ehrlich, K. C.; Montalbano, B. G.; Cary, J. W.; Cotty, P. J. Promoter Elements in the Aflatoxin Pathway Polyketide Synthase Gene. *Biochim. Biophys. Acta* **2002**, *1576*, 171–175.

Ellis, W. O.; Smith, J. P.; Simpson, B. K. Aflatoxin in Food: Occurrence, Biosynthesis, Effects on Organisms, Detection, and Methods of Control. *Crit. Rev. Food Sci. Nutr.* **1991**, *30*, 403–439.

EPA. Spinosad: Pesticide Tolerance. *Fed. Regist.* **2005**, *70*, 1349–1357.

Fields, P. G. Effect of *Pisum sativum* Fractions on the Mortality and Progeny Production of Nine Stored-grain Beetles. *J. Stored Prod. Res.* **2006**, *42*, 86–96.

Flexner, J. L.; Belnavis. D. L. Microbial Insecticides. In: *Biological and Biotechnological Control of Insect Pests. Agriculture and Environment Series*; Rechcigl, J. E., Rechcigl, N. A., Eds.; CRC Press LLC: Boca Raton, FL, 1998; pp 35–62.

FAO (Food and Agriculture Organization). Manual Tecnico: Manejo Integrado de Enfermedades en Cultivos Hidroponicos. Oficina Regional para America Latina y el Caribe, **2004**.

Fujimoto, H.; Itoh, K.; Yamamoto, M.; Kayozuka, J.; Shimamoto, K. Insect Resistant Rice Generated by a Modified Delta Endotoxin Genes of *Bacillus thuringiensis*. *Biotechol.* **1993**, *11*, 1151–1155.

Garman, H.; Jewett, H. H. The Life-history and Habits of the Corn-ear Worm (*Chloridae obsoleta*). *Kentucky Agric. Exp. Stat. Bull.* **1914**, *187*, 388–392.

Gautam, A. K.; Gupta, H.; Soni, Y. Screening of Fungi and Mycotoxins Associated with Stored Rice Grains in Himachal Pradesh. *Int. J. Theor. Appl. Sci.* **2012**, *4* (2), 128–123.

Georgianna, D. R.; Payne, G. A. Genetic Regulation of Aflatoxin Biosynthesis: From Gene to Genome. *Fungal Genet Biol.* **2009**, *46*, 113–125.

Gonzalez-Salgado, A.; González-Jaen, T.; Vazquez, C.; Patiro, B. Highly Sensitive PCR-based Detection Method Specific for *Aspergillus flavus* in Wheat Flour. *Food Addit. Contam., A: Chem. Anal. Control Expos. Risk Assess.* **2008**, *25*, 758–764.

Gosal, S. S.; Wani, S. H.; Kang, M. S. Biotechnology and Drought Tolerance. *J. Crop Imp.* **2009**, *23*, 19–54.

Harrison, R. L.; Bonning. B. C. Genetic Engineering of Biocontrol Agents for Insects. In: *Biological and Biotechnological Control of Insect. Agriculture and Environment Series*; Rechcigl, J. E., Rechcigl, N. A., Eds.; CRC Press LLC: Boca Raton, FL, 1998; pp 243–280.

Hashmi, S.; Hashmi, G.; Glazer, I.; Gaugler, R. Thermal Response of *Heterorhabditis bacteriophora* Transformed with the *Caenorhabditis elegans* hsp 70 Encoding Gene. *J. Exp. Zool.* **1998**, *281*, 164.

Heathcote, J. G.; Hibbert, J. R. *Aflatoxin Chemical and Biological Aspects*. Elsevier Scientific Publishing Company: Amsterdam, 1978.

Hermann, R.; Moskowitz, H.; Zlotkin, E.; Hammock. B. D. Positive Cooperativity among Insecticidal Scorpion Toxins. *Toxicon* **1995**, *33*, 1099–1102.

Hormisch, D.; Brost, I.; Kohring, G. W.; Gifthorn, F.; Krooppenstedt, E.; Farber, P.; Holzapfel, W. H. *Mycobacterium fluoranthenivorans* sp. *nov.*, a Fluoranthene and Aflatoxin B1 Degrading Bacterium from Contaminated Soil of a Former Coal Gas Plant. *Syst. Appl. Microbiol.* **2004**, *27*, 653–660.

Hou, X.; Fields, P. G. Granary Trial of Protein-enriched Pea Flour for the Control of Three Stored Product Insects in Barley. *J. Econ. Entomol.* **2003**, *96*, 1005–1015.

Hua, S. S.; Baker, T.; Flores-Espiritu, M. Interactions of Saprophytic Yeasts with a Nor Mutant of *Aspergillus flavus*. *Appl. Environ. Microbiol.* **1999**, *65*, 2738–2740.

Hua, S. S. Biological Control of Aflatoxin in Almond and Pistachio by Preharvest Yeast Application in Orchards. In: Special Issue: Aflatoxin/Fumonisin Elimination and Fungal Genomics Workshops. Phoenix, Arizona, October 23–26, 2001. *Mycopathologia* **2002**, *155*, 65.

Huyhn, V. L.; Lloyd, A. B. Synthesis and Degradation of Aflatoxins by *Aspergillius parasiticus*. I. Synthesis of aflatoxin B1 by Young Mycelium and its Subsequent Degradation in Aging Mycelium. *Aust. J. Biol. Sci.* **1984**, *37*, 37–43.

Jauhar, P. P. Modern Biotechnology as an Integral Supplement to Conventional Plant Breeding: The Prospects and Challenges. *Crop Sci.* **2006**, *46*, 1841–1859.

Jones, R. K.; Duncan, H. E.; Payne, G. A.; Leonard, K. J. Factors Influencing Infection by *Aspergillus flavus* in Silk-inoculated Corn. *Plant Dis.* **1980**, *64*, 859–863.

Kapoor, N.; Arya, A.; Siddiqui, M. A.; Kumar, H.; Amir, A. Physiology and Biochemical Changes during Seed Deterioration in Aged Seeds of Rice (*Oryza sativa* L.). *Am. J. Plant Physiol.* **2011**, *6*(1), 28–35.

Kavallieratos, N. G.; Athanassiou, C. G.; Michalaki, M. P.; Batta, Y. A.; Rigatos, H. A.; Pashalidou, F. G.; Balotis, G. N.; Tomanovic, Z.; Vayias, B. Effect of the Combined Use of *Metarhizium anisopliae* (Metschinkoff) Sorokin and Diatomaceous Earth for the Control of Three Stored-product Beetle Species. *Crop Prot.* **2006**, *25*, 1087–1094.

Kavallieratos, N. G.; Athanassiou, C. G.; Saitanis, C. J.; Kontodimas, D. C.; Roussos, A. N.; Tsoutsa, M. S.; Anastassopoulou, U. A. Effect of Two Azadirachtin Formulations against Adults of *Sitophilus oryzae* and *Tribolium confusum* on Different Grain Commodities. *J. Food Prot.* **2007**, *70*, 1627–1632.

Keller, N. P.; Nesbitt, C.; Sarr, B.; Phillips, T. D.; Burow, G. B. pH Regulation of Sterigmatocystin and Aflatoxin Biosynthesis in *Aspergillus* spp. *Phytopathology* **1997**, *87*, 643–648.

Koul, O.; Isman, M. B.; Ketkar, C. M. Properties and Uses of Neem, *Azadirachta indica. Can. J. Bot.* **1990**, *68*, 1–11.

Kumar, N.; Bhatt, R. P. Transgenics: An Emerging Approach for Cold Tolerance to Enhance Vegetables Production in High Altitude Areas. *Indian J. Crop Sci.* **2006**, *1*, 8–12.

Lacey, L. A.; Goettel. S. M. Current Development in Microbial Control of Insect Pests and Prospects for the Early 21st Century. *Entomophaga* **1995**, *40*, 3–27.

Lund, F.; Frisvad, J. C. *Penicillium verrucosum* in Wheat and Barley Indicates Presence of Ochratoxin A. *J. Appl. Microbiol.* **2003**, *95*, 1117–1123.

Ma, W. K.; Sicilliano, S. D.; Germida, J. J. A PCR-DGGE Method for Detecting Arbuscular mycorrizal Fungi in Cultivated Soil. *Soil Biol. Biochem.* **2005**, *37*, 1589–1597.

Magan, N.; Hope, R.; Cairns, V.; Aldred, D. Post-harvest Fungal Ecology: Impact of Fungal Growth and Mycotoxin Accumulation in Stored Grain. *Eur. J. Plant Pathol.* **2003**, *109*, 723–730.

Magan,, N.; Sanchis, V.; Aldred, D. Role of Spoilage Fungi in Seed Deterioration. In: *Fungal Biotechnology in Agricultural, Food and Environmental Applications*; Aurora, D. K., Ed.; Marcel Dekker: New York, 2004; pp 311–332.

Manonmani, H. K.; Anand, S.; Chandrashekar, A.; Rati, E. R. Detection of Atoxigenic Fungi in Selected Food Commodities by PCR. *Process Biochem.* 2005, *40*, 2859–2864.

Mason, L. J.; Obermeyer, J. Stored Grain Insect Pest Management. *Stored Product Pests*, Purdue University, 2010; pp 1–5.

Masoud, W.; Kaltoft, C. H. The Effects of Yeasts Involved in the Fermentation of *Coffea arabica* in East Africa on Growth and Ochratoxin A (OTA) Production by *Aspergillus ochraceus*. *Int. J. Food Microbiol.* 2006, *106*(2), 229–234.

Miller, L. K. Genetically Engineered Insect Virus Pesticides: Present and Future. *J. Invertebr. Pathol.* 1995, *65*, 211–216.

Moreno-Martínez, E.; Vázquez-Badillo, M.; Facio-Parra, F. Use of Propionic Acid Salts to Inhibit Aflatoxin Production in Stored Grains of Maize. *Agrociencia* 2000, *34*(2), 477–484.

Narayanasamy, P. Immunology in Plant Health and Its Impact on Food Safety. Haworth Press: New York, 2005.

Nesci, A. V.; Bluma, R. V.; Etcheverry, M. G. In Vitro Selection of Maize Rhizobacteria to Study Potential Biological Control of *Aspergillus* Section *Flavi* and aflatoxins Production. *Eur. J. Plant Pathol.* 2005, *113*(2), 159–171.

Nwilene, F. E.; Okhidievbie, O.; Agunbiade, T. A.; Traore, A. K.; Gaston, L. N.; Togola, M. A.; Youm, O. Antixenosis Component of Rice Resistance to African Rice Gall Midge, *Orseolia oryzivora*. *Int. Rice Res. Notes* 2009, *34*, 1–5.

Nwilene, F. E.; Williams, C. T.; Ukwungwu, M. N.; Dakouo, D.; Nacro, S.; Hamadoun, A.; Kamara, S. I.; Okhidievbie, O.; Abamu, F. J.; Adam, A. Reactions of Differential Rice Genotypes to African Rice Gall Midge in West Africa. *Int. J. Pest Manage.* 2002, *48*(3), 195–201.

Nwilene, F. E.; Nacro, S.; Tamo, M.; Menozzi, P.; Heinrichs, E. A.; Hamadoun, A.; Dakouo, D.; Adda, C.; Togola, A. Managing Insect Pests of Rice in Africa. In: *Realizing Africa's Rice Promise*; Wopereis, M. C. S., Ed.; Francis CAB International, 2013; pp 229–240.

O'Reilly, D. R.; Miller, L. K. Improvement of a Baculovirus Pesticide by Deletion of the *EGT* Gene. *Biotechnology* 1991, *9*, 1086–1089.

Ortiz, R. Critical Role of Plant Biotechnology for the Genetic Improvement of Food Crops: Perspectives for the Next Millennium. *Electron. J. Biotechnol.* 1998, *1*(3), 1–8.

Payne, G. A.; Nierman, W. C.; Wortman, J. R.; Pritchard, B. L.; Brown, D.; Dean, R. A. Whole Genome Comparison of *Aspergillus flavus* and *A. oryzae*. *Med. Mycol.* 2006, *44*, 9–11.

Pimentel, D. Amounts of Pesticides Reaching Target Hosts: Environmental Impacts and Ethics. *J. Agric. Environ. Ethics* 1995, *8*, 17–29.

Plasencia, J. Aflatoxins in Maize: A Mexican Perspective. *J. Toxicol.* 2004, *23*, 155–177.

Poinar, G. O. J. Genetic Engineering of Nematodes for Pest Control. In: *Biotechnology for Biological Control of Pests and Vectors*; Maramorosch, K., Ed.; CRC Press: Boca Raton, FL, 1991; pp 77–93.

Prakash, A.; Rao, J. Insect Pest Management in Stored Rice Ecosystem. In: *Stored Grain Ecosystem*; Jayas, D. G., White, N. D. G., Muir, W. E., Eds.; Marcel Dekker, Inc.: New York, Basel, Hong Kong, 1995; pp 709–736.

Rajandran, S.; Sriranjini, V. Plant Products as Fumigants for Stored-product Insect Control. *J. Stored Prod. Res.* 2008, *44*, 126–135.

Reed, C. R. Managing Stored Grain to Preserve Quality and Value. AACC International: Saint Paul, MN, 2010.

Riudavets, J.; Gabarra, R.; Pons, M. J.; Messeguer, J. Effect of Transgenic Bt Rice on the Survival of Three Nontarget Stored Product Insect Pests. *Environ. Entomol.* **2006**, *35*, 1432–1438.

Robinson, J. Ethics and Transgenic Crops: A Review. *Electron. J. Biotechnol.* **1999**, *2*, 71–81.

Robinson, R. K. The Detection of Mycotoxins in Food. *Int. J. Environ. Stud.* **1975**, *8*, 199–202.

Rushton, P. J.; Gurr, S. J. Engineering Plants with Increased Disease Resistance: What are We Going to Express? *Trends Biotechnol.* **2005**, *23*, 275–282.

Sanghera, G. S.; Kashyap, P. L.; Wani, S. H.; Singh, G. QTL Herald: A Revival of Biometrical Genetics. In: *Genetic Engineering: A New Hope for Crop Production and Improvement*; Malik, C. P., Ed., 2010b.

Sanghera, G. S.; Wani, S. H.; Gill, M. S.; Kashyap, P. L.; Gosal, S. S. RNA Interference: Its Concept and Application in Crop Plants. In: *Biotechnology Cracking New Pastures*; Malik, C. P., Verma, A., Eds.; M. D. Publishers: New Delhi, 2010a; pp 33–78.

Shapira, R.; Paster, N.; Eyal, O.; Menasherov, M.; Mett, A.; Salomon, R. Detection of Afla-toxigenic Molds in Grains by PCR. *Appl. Environ. Microbiol.* **1996**, *62*, 3270–3273.

Shelar, V. R.; Shaikh, R. S.; Nikam, A. S. Soybean Seed Quality during Storage: A Review. *Agric. Rev.* **2008**, *29*(2), 125–131.

Silvie, P.; Togola, A.; Adda, C.; Nwilene, F.; Ravaomanarivo, L.; Randriamanantsoa, R.; Menozzi, P. Le riz *Bt* dans le contexte de la riziculture en Afrique et a Madagascar. *Cahier Agricultures* (numero spécial riz), 2012.

Singh, S. R.; Jackai, L. E. N.; Dos-Santos, J. H. R.; Adalla, C. B. Insect Pests of Cowpea. In *Insect Pests of Tropical Food Legumes*; Singh, S. R., Ed.; Wiley: Chichester, 1990; pp 43–89.

Smith, C. A.; Woloshuk, C. P.; Robertson, D.; Payne, G. A. Silencing of the Aflatoxin Gene Cluster in a Diploid Strain of *Aspergillus flavus* is Suppressed by Ectopic *afl*R Expression. *Genetics* **2007**, *176*, 2077–2086.

Smith, J. S.; Dorner, J. W.; Cole, R. J. Testing *Bacillus subtilis* as a Possible Aflatoxin Inhib-itor in Stored Farmers Stock Peanuts. *Proc. Am. Peanut Res. Educ. Soc.* **1990**, *22*, 35.

Somashekar, D.; Rati, E. R.; Chandrashekar, A. PCR-restriction Fragment Length Analysis of aflR Gene for Differentiation and Detection of *Aspergillus flavus* and *Aspergillus para-siticus* in Maize. *Int. J. Food Microbiol.* **2004**, *93*, 101–107.

Soufleros, E. H.; Tricard, C.; Bouloumpasi, E. C. Occurrence of Ochratoxin A in Greek Wines. *J. Sci. Food Agric.* **2003**, *83*(3), 173–179.

St. Leger, R. J.; Joshi, L.; Bidochka, M. J.; Roberts, D. W. Construction of an Improved Mycoinsecticide Overexpressing of Toxin Protease. *Proc. Natl. Acad. Sci. U. S. A.* **1996**, *93*, 6349.

Subramanyam, B. H.; Toews, M. D.; Ileleji, K. E.; Maier, D. E.; Thompson, G. D.; Pitts, T. J. Evaluation of Spinosad as a Grain Protectant on Three Kansas Farms. *Crop Prot.* **2007**, *26*, 1021–1030.

Suzuki, Y.; Ise, K.; Li, C.; Honda, I.; Iwai. Y.; Matsukura, U. Volatile Components in Stored Rice (*Oryza sativa* L.) of Varieties with and without Lipoxygenase-3 in Seeds. *J. Agric. Food Chem.* **1999**, *47*, 1119–1124.

Sweeney, M. J.; Dobson, A. D. W. Mycotoxin Production by *Aspergillus*, *Fusarium* and *Peni-cillium* Species. *Int. J. Food Microbiol.* **1998**, *43*, 141–158.

Sweeney, M. J.; Dobson, A. D. W. Molecular Biology of Micotoxin Biosynthesis. *FEMS Microb. Lett.* **1999**, *175*, 149–163.

Thompson, G. D.; Dutton, R.; Sparks, T. C. Spinosad—A Case Study: An Example from a Natural Products Discovery Programme. *Pest Manage. Sci.* **2000**, *56*, 696–702.

Throne, J. E.; Lord, J. C. Control of Sawtoothed Grain Beetles (Coleoptera: Silvanidae) in Stored Oats by using an Entomopathogenic Fungus in Conjunction with Seed Resistance. *J. Econ. Entomol.* **2004**, *97*, 1765–1771.

Tilburn, J.; Sarkar, S.; Widdick, D. A.; Espeso, E. A.; Orejas, M.; Mungroo, J.; Penalva, M. A.; Arst, H. N. The *Aspergillus* PacC Zinc Finger Transcription Factor Mediates Regulation of Both Acid and Alkaline Expressed Genes by Ambient pH. *EMBO J.* **1995**, *14*, 779–790.

Toews, M. D.; Subramanyam, B. Survival of Stored-product Insect Natural Enemies in Spinosad-treated Wheat. *J. Econ. Entomol.* **2004**, *97*, 1174–1180.

Toews, M. D.; Campbell, J. F.; Arthur, F. H. Temporal Dynamics and Response to Fogging or Fumigation of Stored-product Coleoptera in a Grain Processing Facility. *J. Stored Prod. Res.* **2006**, *42*, 480–498.

Togola, A.; Nwilene, F. E.; Chougourou, D. C.; Agunbiade, T. Presence, Populations and Damage of the Angoumois Grain Moth, *Sitotroga cerealella* (Olivier) (Lepidoptera, Gelechiidae), on Rice Stocks in Benin. *Cahiers Agric.* **2010**, *19*(3), 205–209.

Topal, S.; Aran, N.; Pembezi, C. Turkiye'nin tarmsal mikroflorasinin mikotoksin profilleri. *Gida Dergisi* **1999**, *24*, 129–137.

Trail, F.; Mahanti, N.; Linz, J. Molecular Biology of Aflatoxins Biosynthesis. *Microbiology* **1995**, *141*, 755–765.

Treacy, M. F. Recombinant baculoviruses. In: *Biopesticides Use and Delivery. Methods in Biotechnology No. 5*; Hall, F. R., Menn, J. J., Eds.; Humana Press Inc.: Totowa, NJ, **1999**; pp 321–340.

Vail, P. V.; Tebbets, J. S.; Cowan, D. C.; Jenner, K. E. US Patent No. 5,023,182, 1991.

Van Rensburg, J. B. J. First Report of Field Resistance by the Stem Borer, *Busseola fusca* (Fuller) to *Bt*-transgenic Maize. *South Afr. J. Plant Soil.* **2007**, *24*, 147–151.

Walker, D. J. Farrell, G. *Food Storage Manual*. Natural Resources Institute/World Food Programme: Chatham, UK/Rome, Italy, **2003**.

Waller, J. M. Postharvest Diseases. In: *Plant Pathologist's Pocketbook*; Waller, J. M., Lenne, J. M., Waller, S. J., Eds.; CAB International: Wallingford, UK, 2002; pp 39–54.

Wani, S. H.; Sanghera, G. S. Genetic Engineering for Viral Disease Management in Plants. *Not. Sci. Biol.* **2010**, *2*(1), 20–28.

Wani, S. H. Inducing Fungus Resistance in Plants through Biotechnology. *Not. Sci. Biol.* **2010**, *2*(2), 14–21.

Weaver, M. A.; Vadenyapina, E.; Kenerley, C. M. Fitness, Persistence, and Responsiveness of an Engineered Strain *Thrichoderma virens* in Soil Mesocosms. *Appl. Soil Ecol.* **2005**, *29*, 125–134.

Widstrom, N. W. Breeding Strategies to Control Aflatoxin Contamination of Maize through Host Plant Resistance. In: *Aflatoxin in Maize: A Proceedings of the Workshop*; Zuber, M. S., Lillehoj, E. B., Renfro, B. L., Eds.; CIMMYT: Mexico, 1987; pp 212–220.

Wilkinson, J. M. Silage and Animal Health. *Nat. Toxins* **1999**, *7*, 221–232.

Williams, W. P.; Buckley, P, M.; Windham, G. L. Southwestern Corn Borer (Lepidoptera: Crambidae) Damage and Aflatoxin Accumulation in Maize. *J. Econ. Entomol.* **2002**, *95*, 1049–1053.

World Bank Report, Post-harvest Management—Fights Hunger with FAO, India Grains, March **2002**, *4*(3): 20–22, **1999**.

Yankanchi, S. R.; Gadache, A. H. Grain Protectant Efficacy of Certain Plant Extracts against Rice Weevil, *Sitophilus oryzae* L. (Coleoptera: Curculionidae). *J. Biopest.* **2010,** *3*(2), 511–513.

Yin, Y.; Lou, T.; Jiang, J.; Yan, L.; Michailides, T. J.; Ma, Z. Molecular Characterization of Toxigenic and Atoxigenic *Aspergillus flavus* Isolates Collected from Soil in Various Agroecosystems in China. *Appl. Microbiol.* **2009,** *107*, 1857–1865.

Yin, Y.; Yan, L.; Jiang, J. Biological Control of Aflatoxin Contamination of Crops. *J. Zhejiang Univ. Sci. B* **2008,** *9*(10), 787–792.

Youm, O.; Vayssieres, J. F.; Togola, A.; Robertson, S. P.; Nwilene, F. E. International Trade and Exotic Pests: The Risks for Biodiversity and African Economies. *Outlook Agric.* **2011,** *40*(1), 59–70.

Yu, J.; Bhatnagar, D.; Cleveland, T. E. Completed Sequence of Aflatoxin Pathway Gene Cluster in *Aspergillus parasiticus*. *FEBS Lett.* **2004a,** *564*, 126–130.

Yu, J.; Chang, P. K.; Ehrlich, K. C.; Cary, J. W.; Bhatnagar, D.; Cleveland, T. E. Clustered Pathway Genes in Aflatoxin Biosynthesis. *Appl. Environ. Microbiol.* **2004b,** *70*, 1253–1262.

Zhao, J.; Kong, F.; Li, R.; Wang, X.; Wan, Z.; Wang, D. Identification of *Aspergillus fumigatus* and Related Species by Nested PCR Targeting Ribosomal DNA Internal Transcribed Spacer Regions. *J. Clin. Microbiol.* **2001,** *39*, 2261–2266.

CHAPTER 11

MOLECULAR APPROACHES FOR MULTIPLE RESISTANCE GENES STACKING FOR MANAGEMENT OF MAJOR RICE DISEASES

B. S. KHARAYAT[1], S. YADAV[2], R. SRIVASTAVA[3], and U. M. SINGH[4*]

[1]*Division of Plant Pathology, ICAR—Indian Agricultural Research Institute, Pusa, New Delhi 110012, India*

[2]*Plant Biotechnology Division, Forest Research Institute, Dehradun, Uttarakhand, India*

[3]*Division of Molecular and Life Sciences, College of Science and Technology, Hanyang University, Ansan, Republic of Korea*

[4]*International Rice Research Institute, ICRISAT, Patancheru 502324, Tealangana State, India*

Corresponding author. E-mail: uma.singh@irri.org

CONTENTS

ABSTRACT

Breeders have successfully developed several rice cultivars lines resistant to diseases by integrating *R*-genes. However, durable, long-lasting resistance in many cases has been difficult to achieve as pathogens quickly evolve and develop counter resistance genes that circumvent the host cultivars resistance. Thus, there is need to speed up the breeding technologies with the help of new molecular tools. Nowadays, molecular breeding is used as genetic manipulation carried out at DNA levels to improve characters of interest in plants, including molecular marker-assisted selection and genomic selection. The rapid improvement in the field of molecular biology in the last few decades is very applicable to plant breeder for rice improvement. In India and the Philippines, bacterial blight resistance alleles of genes *Xa4*, *xa5*, *xa13*, *Xa21*, etc. have been successfully introgressed into several rice and hybrid rice varieties.

11.1 INTRODUCTION

Rice (*Oryza sativa* L.) is a primary staple food crop for billions of people worldwide including India. The world population is increasing very rapidly, simultaneously leading to reduction in cultivable land for rice, decreasing water level, emergence of devastating diseases and pests. Climate change is thus becoming major issue that must be addressed by researchers to develop sustainable crop varieties with resistance to biotic and abiotic stresses (Hasan et al., 2015). To ensure global food security for the continuous population growth, it is essential to manage the diseases that cause economic as well as quality losses in rice. A rising global population requires increased crop production but some reports suggests that the rate of increase in crop yield is currently declining and hence a major focus of plant breeding efforts should be on traits related to yield, stability, and sustainability. To keep the pace with increasing demand for food grains, there is a need to enhance the production at least by 1.5–2.0 M t annually in the next 10–15 years, with the back drop of declining and deteriorating resources and without adversely affecting the environment. Losses caused by plant diseases that manifest during pre- and postharvest treatment inevitably contribute to these deficiencies, especially in developing countries. Plant protection in general and the protection of crops against plant diseases in particular have an obvious role to play in meeting the growing demand for food quality and quantity. In the eighties, a plague of crops in different parts of the world caused losses to half the harvests and it is still wreaking havoc in those regions (Ragimekula et al., 2013). The yield of cultivated plants is threatened by competition and destruction from pathogens, especially when grown in large-scale monocultures or with heavy fertilizer applications. In most cases, information on the magnitude of losses caused by diseases in plants is, however, limited. Nevertheless, it is estimated that 30–40% of harvests is lost each year throughout the production chain. In addition to yield losses caused by diseases, these new elements of complexity also result in postharvest quality losses and accumulation of toxins during and after the cropping season.

For the management of diseases in rice, several approaches are followed including cultural, biocontrol, chemical, and host resistance. Due to low efficacy of bioagents and ecological hazards and development of resistance among pathogens against chemical pesticides, host resistance is considered safest, economical, and most feasible method for the management of plant diseases. In traditional breeding methods, recurrent backcrossing is commonly employed to transfer alleles at one or more loci from a donor to an elite variety (Reyes-Valdés, 2000). The expected recurrent parent (RP) genome recovery would be 99.2% by six backcrosses, which is most similar to improved variety and very cumbersome and time-consuming process also (Babu et al., 2004). Host plant resistance is an important tool for rice disease control and has played a key role in sustaining rice productivity, especially in tropical Asia (Lang et al., 2008). Single gene approach has been shown very effective in the management of plant diseases but often faces problem of breakdown of resistance. This breakdown of resistance is due to high selection pressure posed by the vertical or single gene-based resistance on pathogen which leads to development of new matching types with high virulence. Single-gene- or major-gene-based resistance is also vulnerable to boom and bust cycles in crops. There is broad agreement that combining genes for resistance (gene pyramids) is a useful approach for increasing durability, with many known successes (Mundt, 2014; Singh et al., 2014).

Breeders often spot this breakdown in resistance and hurriedly integrate a newly found effective *R*-gene into their populations. In mean time, the new *R*-gene loses its effectiveness and the boom-bust and induced coevolution between crop and pathogen continues. Plant breeders have two options to increase the durability of their resistant cultivars. The first is high-dose/refuge or multiline strategy (Pink, 2002; Rausher, 2001). A multiline or refuge will reduce the selection intensity against the susceptible genotypes by providing an acceptable host for the pathogen. The selection intensities are decreased and the number of generations necessary for the new allele to become predominant increases dramatically. The multiline strategy requires that some level of disease is acceptable and that the pathogen reproduces by sexual means. Also, multilines may not hold a necessary uniformity that many cropping systems require, which make them not feasible to practical utility. The second option for durable resistance is gene pyramiding. In theory, pyramiding several "undefeated" *R*-genes into a single cultivar will provide a more durable resistance as several mutations would need to take place, one at each of the pathogen's corresponding *Avr*-loci (McDowell & Woffenden, 2003; Pink, 2002). Advances in genomics have demonstrated that a considerable proportion, 1–2% of a plant's genome is devoted to resistance genes or genes with similar properties that could conceivably confer resistance to a pathogen possessing a complementary avirulence gene. Such genes are often clustered or occur in tandem repeats, suggesting that resistance genes with different specificities arise by gene duplication followed by intragenic and intergenic recombination, gene conversion and diversifying selection (Michelmore & Meyers, 1998). Maintenance of this genetic flux is crucial to the survival of the plants. Although conventional breeding has had a significant impact on improving resistance cultivars, the time-consuming process of making crosses and backcrosses, and the selection of the desired resistant progeny make it difficult to react adequately to the evolution of new virulent pathogens.

11.2   RICE DISEASES: A POTENTIAL THREAT TO RICE PRODUCTION IN INDIA

Rice is attacked by various insect pest and pathogens. The major pathogen groups attacking rice crop include fungi, bacteria, virus, and nematodes. Depending upon the environmental conditions, these pathogens cause epiphytotics in different proportions from time to time. With preference to geographical and climatic conditions, these pathogens cause different levels of diseases and accordingly have significant economic importance. Following are the major and emerging diseases of rice in India as well as other parts of the world.

11.2.1   BLAST

Blast disease of rice, caused by the fungus *Magnaporthe oryzae*, is one of the most devastating diseases of rice worldwide, including India (Khang et al., 2008; Liu et al., 2010). The fungus attacks all foliar parts of the rice plants. Depending on the site of symptom rice blast is referred as leaf blast, collar blast, node blast, and neck blast (Fig. 11.1). On the leaves, symptoms first appear as small bluish flecks, about 1–3 mm in diameter. The lesions of this disease on leaf blade are elliptical or spindle-shaped with brown borders and gray centers. Under favorable environmental conditions, lesions enlarge and coalesce eventually killing the leaves. Collar blast of rice occurs when the pathogen infects the collar region that can kill the entire leaf blade. The pathogen also infects the node of the stem that turns blackish and breaks easily; this condition is called node blast. Neck of the panicle can also be infected. Infected neck is girdled by a grayish-brown lesion that makes panicle fall over when infection is severe. The neck becomes shriveled and covered with a gray fluffy mycelium. The affected plants can be very easily identified by examining the bluish patch on the neck or stem. The pathogen also causes brown lesions on the branches of panicles and on the spikelet. If the infection has occurred much before the grain formation the later are not filled and the panicle remains erect (Elazegui & Islam, 2003; Singh, 2005).

FIGURE 11.1   Diagnostic symptoms of blast disease of rice. (a) Leaf blast phase: the lesions on leaf blade are elliptical or spindle shaped with brown borders and gray centers, (b) collar blast phase: pathogen infects the collar that can kill the entire leaf blade, (c) node blast phase: pathogen also infects the node of the stem that turns blackish and breaks easily, and (d) panicle blast phase: infected neck is girdled by a grayish-brown lesion that makes panicle fall over when infection is severe. (adopted and modified from Rice Knowledge Portal Management, HYPERLINK "http://www.rkmp.co.in" www.rkmp.co.in)

11.2.2   SHEATH BLIGHT

Sheath blight is caused by *Rhizoctonia solani*. This is also very devastating disease in India. It causes severe loses approximately 40–45% under favorable environmental conditions (Ou, 1985). It occurs in various states in the country including Uttarakhand, Utter Pradesh, Orissa, and Tamil Nadu. The lesions are usually observed on the leaf sheaths although leaf blades may also be affected. The initial lesions are small, ellipsoid, or ovoid, and greenish-gray (Fig. 11.2) and usually develop near the water line in lowland fields. Under favorable conditions, they enlarge and may coalesce forming bigger lesions with irregular outline and grayish-white center with dark brown borders. The presence of several large spots on a leaf sheath usually causes the death of the whole leaf. Instead of spores, the rice sheath blight (ShB) fungus produces sclerotia measuring usually 1–3 mm in diameter and relatively spherical (Mew & Rosales, 1984; Sato et al., 2004; Yadav et al., 2015).

FIGURE 11.2 Diagnostic symptoms of sheath blight of rice. Lesions with irregular outline and grayish-white center with dark brown borders (photograph: B. S. Kharayat).

11.2.3 BROWN SPOT

Brown spot disease of rice, caused by the fungus *Bipolaris oryzae*, is also one of the most devastating diseases of rice India (Khalili et al., 2012; Singh et al., 2005). The symptoms of the disease appear on the coleotile, the leaves, sheath, and also the glumes. On seedlings, the fungus produces small, circular, brown lesions, which may girdle the coleoptile and cause distortion of the primary and secondary leaves. The spots vary in size and shape from minute dots to circular, eye-shaped or oval lesions measuring 1–14 × 0.5–3 mm. They are distinct and isolated, usually fairly scattered over the leaf surface. The smaller spots are dark brown or purplish-brown. The larger spots are dark brown at the edge but they may be pale yellow, dirty white, brown, or grey in the center (Fig. 11.3). Sometimes the spots are surrounded by a yellowish halo and may coalesce and become irregular in shape. On the leaves of older plants, the fungus produces circular to oval lesions that have a light-brown to gray center surrounded by a reddish-brown margin The fungus may also infect the glumes, causing dark brown to black oval spots and may also infect the grain, causing a black discoloration (Baranwal et al., 2013; Elazegui & Islam, 2003; Singh, 2005).

FIGURE 11.3 Diagnostic symptoms of brown spot disease on rice leaf and grains. Circular to oval lesions that have a light brown to gray center surrounded by a reddish-brown margin (photograph: B. S. Kharayat).

11.2.4 *FALSE SMUT OF RICE*

Earlier this disease was used to consider of minor importance in India and other countries, now it has taken very serious attention of plant pathologist, breeder, scientist, and farmers. This disease is caused by fungus *Ustilaginoidea virens* (Cooke) Takahashi (Dodan & Singh, 1996). The fungus transforms individual grains of the panicle into greenish spore balls that have velvety appearance (Fig. 11.4). The spore balls are small at first and visible in between glumes, grow gradually to reach 1 cm or more in diameter and enclose the floral parts. They are covered with a membrane that bursts as a result of further growth. The color of the ball becomes orange and later yellowish green or greenish black (Elazegui & Islam, 2003; Singh, 2005; Sunder et al., 2010).

FIGURE 11.4 Diagnostic symptoms of false smut on rice grains. Fungus transformed individual grains of the panicle into greenish spore ball (Photograph: B. S. Kharayat).

11.2.5 BACTERIAL BLIGHT

Bacterial leaf blight of rice caused by *Xanthomonas oryzae* pv. *oryzae* is one of the most serious constraints of rice production worldwide (Sundaram et al., 2014). Leaf blight phase is most commonly seen. This phase is characterized by linear yellow to straw colored strips with wavy margins, generally on both edges of the leaf, rarely on one edge (Fig. 11.5a and b). These stripes usually start from the tip and extend downward. This is followed by drying and twisting of the leaf tip and rapid extension of marginal blight lengthwise and crosswise to cover large areas of the leaf. In severely diseased fields, grains may also be infected (Fig. 11.5a). In the tropics, infection may also cause withering of leaves or entire young plants (referred as kresek) and production of pale yellow leaves at a later stage of the growth. In dry weather, opaque and turbid drops of bacterial ooze which dry into yellowish beads can be seen on the leaf surface (Fig. 11.5c). These drops are washed down by rains (Chen et al., 2002; Swings et al., 1990).

FIGURE 11.5 Diagnostic symptoms of bacterial blight infected rice plants: (a) field view, (b) infection on single leaf showing typical wavy margins, and (c) bacterial beads on rice leaf (photograph: B. S. Kharayat).

11.2.6 *PANICLE BLIGHT OR GLUM BLOTCH DISEASE OF RICE*

It is an emerging bacterial disease of rice in India. It is caused by *Burkholderia glumae*. Bacterial panicle blight (BPB) emerged recently in epidemic proportions in rice-growing areas of Northern India including Uttar Pradesh, Haryana, and Delhi (Mondal et al., 2015). BPB is destructive on emerging panicles and leaf sheaths of basmati and non-basmati rice varieties. The diseased panicle bears light to dark brown, partially or fully discolored glumes. Under severe conditions, grain filling in the diseased panicles is affected resulting in chaffy grains. Sheath and leaf lesions appear as irregular elongated brown patches (Mondal et al., 2015). The disease is first detected as a light-to-medium brown discoloration of the lower third to half of the hulls (Fig. 11.6). The typical panicle symptoms of this disease are straw-colored panicles containing florets with a darker base and a reddish-brown line across the floret between darker straw-colored and light straw-colored areas (Nandakumar et al., 2009). The stem below the infected grain remains green. The grain discoloration and the green stems are the key diagnostic characteristics. BPB tends to develop in circular patterns in the field, with the most severely infected panicles in the center, remaining upright due to sterility or no grain in ears.

FIGURE 11.6   The diagnostic symptoms of panicle blight of rice. The typical symptoms of this disease are straw-colored panicles containing florets with a darker base and a reddish-brown line across the floret between darker straw-colored and straw-colored areas (photograph: B. S. Kharayat).

11.2.7 TUNGRO DISEASE

In 1979, tungro was found to be associated with two viruses: an RNA virus, *Rice tungro spherical virus*, and a DNA/RNA virus, *Rice tungro bacilliform virus* (Azzam & Chancellor, 2002). Plants affected by tungro exhibit stunting and reduced tillering. Yellowing of the tips of young leaves is the first symptom to appear, and this may include interveinal chlorosis or chlorotic mottling (Azzam & Chancellor, 2002). The panicles are small and not completely exerted and bear mostly sterile or partially filled grains often covered with dark brown specks. Tungro are transmitted by the green leafhoppers (Dahal et al., 1990; Santa-Cruz et al., 2003).

11.3 MODERN MOLECULAR APPROACHES IN THE DEVELOPMENT OF DURABLE RESISTANCE

The modern molecular techniques make it possible to use markers and probes to track the introgression of several *R*-genes into a single cultivar from various sources during a crossing program. In spite of optimism on conventional breeding for continued yield improvement, new technologies such as DNA markers serve as a new tool to detect the presence of allelic variation in the genes underlying the economic traits. By reducing the reliance on laborious and fallible screening procedures, DNA markers have enormous potential to improve the efficiency and precision of conventional plant breeding via marker-assisted selection (MAS).

11.3.1 MOLECULAR MARKERS

Molecular markers have become important tools for genetic analysis and crop improvement. DNA-markers, being neutral and unlimited in number, have allowed scanning of the whole genome and assigning landmarks in high density on every chromosome in many plant species, which makes them fit for indirect selection. Rapid development in marker technology make it not limited to, restriction fragment length polymorphism (RFLP), random amplified polymorphic DNA, amplified fragment length polymorphism, inter simple sequence repeat (ISSR), but to microsatellites or simple sequence repeat (SSR), cleaved amplified polymorphic sequence (CAPS), expressed sequence tag, diversity arrays technology, and single nucleotide polymorphism have been used in several crops (Doveri et al., 2008). Each and every marker system has their own advantages and disadvantages (Singh et al., 2014). For MAS work, various factors to be considered in selecting one or more of these marker systems (Panigrahi, 2011; Semagn et al., 2006). Five common feature of a suitable molecular marker are (1) must be polymorphic, (2) codominant inheritance, (3) should be reproducible, (4) easy and cheap to detect, and (5) randomly and frequently distributed throughout the genome.

11.3.2 GENE PYRAMIDING

The most important applications of DNA markers to plant breeding is marker-assisted gene pyramiding (MAGP). Gene pyramiding has been applied to enhance resistance to disease and insects by combining two or more than two genes at a time. Several pyramided lines have been developed in rice against bacterial blight and blast (Huang et al., 1997; Joseph et al., 2004; Luo et al., 2012; Singh et al., 2001; Sundaram et al., 2008). The success in pyramiding qualitative gene and quantitative trait loci (QTLs) for resistance to stripe rust in barley has been reported (Castro et al., 2003). In this case, markers allow selecting for QTL-allele-linked markers that have the same phenotypic effect. Pyramiding of multiple genes or QTLs is recommended as a potential strategy for improvement of quantitatively inherited traits in plant breeding (Richardson et al., 2006). The cumulative effects of multiple-QTL pyramiding have been proven in crop species like rice, wheat, barley, and soybean (Jiang et al., 2007a,b; Joseph et al., 2004; Li et al., 2010; Richardson et al., 2006; Sundaram et al., 2008; Wang et al., 2012).

The process of simultaneously combining multiple genes/QTLs together into a single genotype is known as pyramiding. Through conventional breeding it is possible but extremely difficult or impossible at early generations. Using conventional phenotypic selection, individual plants must be phenotypically screened for all traits tested. Therefore, it may be very difficult to assess plants for traits with destructive bioassays or from certain population types (e.g., F_2). DNA markers may facilitate selection because DNA marker for multiple specific genes/QTLs can be tested using a single DNA sample (at any growth stage) without phenotyping and are nondestructive. The most widespread application for pyramiding has been for combining multiple disease resistance genes. MAS might even be the only practical method to pyramid disease resistance genes that have similar phenotypic effects, especially where one gene masks the presence of other genes (Sanchez et al., 2000; Walker et al., 2002). When applying such strategies in practical breeding it has to be taken into account the fact that the pyramiding has to be repeated after each crossing, because the pyramided resistance genes are segregating in the progeny (Werner et al., 2005).

11.4 MARKER-ASSISTED GENE STACKING SCHEMES

Different approaches have been used for pyramiding of multiple genes/ QTLs: multiple-parent crossing or complex crossing, backcrossing, and recurrent selection (Scheme 11.1–3). The suitability of breeding scheme for MAGP depends on the number of genes/QTLs required for improvement of traits, the number of parents that contain the required genes/QTLs, the heritability of traits of interest, and other factors (e.g., marker-gene association, expected duration to complete the plan and relative cost). For MAGP, in general, there may be three strategies or breeding schemes: stepwise, simultaneous/synchronized, and convergent backcrossing or transfer.

Supposing one cultivar W is superior in comprehensive performance but lack of a trait of interest, and four different genes/QTLs contributing to the trait have been identified in four germplasm lines (e.g., P1, P2, P3, and P4). Three MAGP schemes (adapted from Jiang, 2013) for pyramiding the genes/ QTLs can be described as follow:

In the stepwise backcrossing, different four target genes/QTLs are transferred into the RP W. In the one step of backcrossing program, one gene/ QTL is targeted and selected. The selected QTLs further targeted followed by next step of backcrossing for another gene/QTL, until all target genes/ QTLs have been introgressed into the RP. The advantage of stepwise backcrossing is that gene pyramiding is more precise and easier to implement as it involves only one gene/QTL at one time and thus the population size and genotyping amount will be small for the easy handling. The improved RP may be released before the final step as long as the integrated genes/ QTLs (e.g., two or three) meet the requirement. The disadvantage of stepwise backcrossing is that it takes a longer time to complete the breeding program (Scheme 11.1).

SCHEME 11.1   Stepwise backcrossing for improvement of rice (adapted from Jiang, 2013).

The second important backcross breeding program commonly used is known as simultaneous or synchronized backcrossing. In this backcross breeding program, RP W is first crossed to each of four donor parents to produce four single-cross F_1s. After crossing, two of the four single-cross F_1s are again crossed with each other to produce two double-cross F_1s, and these two double-cross F_1s are crossed again to produce a hybrid integrating all four target genes/QTLs in heterozygous state. Further, the hybrid and/ or progeny with heterozygous markers for all four target genes/QTLs is subsequently crossed back to the RP, RP W until a satisfactory recovery of the RP genome occurred, and finalized by one generation of selfing. The advantage of this back cross breeding method is that it takes the shortest time to complete the breeding program. However, in the backcrossing all target genes/QTLs are involved at the same time, and thus it requires a large population and more genotyping, which is tough to handle (Scheme 11.2).

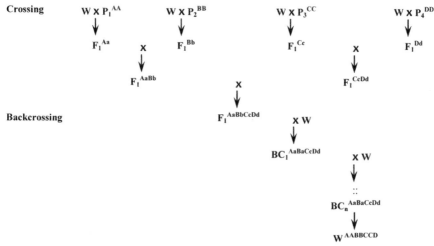

SCHEME 11.2 Simultaneous/synchronized backcrossing (adapted from Jiang, 2013).

The third backcross breeding program, convergent backcrossing is a method of the combining the advantages of stepwise and synchronized backcrossing. First, the four target gene/QTLs are transferred separately from the donors into the RP W by single crossing followed by the backcrossing based on markers linked to the target genes/QTLs, to produce four improved lines (WAA, WBB, WCC, and WDD). Further, two of the improved lines are crossed with each other and the two hybrids are then intercrossed to integrate all four genes/QTLs together and develop the final improved line with all four genes/QTLs pyramided (i.e., WAABBCCDD). Presently, convergent backcrossing is more acceptable backcross breeding program, because in this scheme time of crossing were reduced (compared to stepwise transfer) but gene fixation and/or pyramiding is also more easily assured (compared to simultaneous transfer). Thus, this breeding program is important for the stacking of more than one genes (Scheme 11.3).

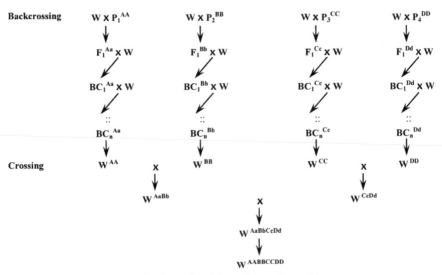

SCHEME 11.3   Convergent backcrossing (adapted from Jiang, 2013).

Theoretical issues and efficiency of molecular assisted back cross (MABC) for the different gene pyramiding have been investigated by the different computer simulations (Ribaut et al., 2002; Servin et al., 2004; Ye & Smith, 2008). There are several practical application of MABC to gene pyramiding has been reported in many crops, including rice, wheat, barley, cotton, soybean, common bean, and pea, especially for developing durable resistance to stresses in crops. However, there is very restricted information accessible about the release of commercial cultivars resulted from this strategy. Somers et al. (2005) used a molecular breeding strategy to initiate multiple pest-resistance genes into Canadian wheat by using high throughput SSR genotyping and half-seed analysis to process backcrossing and selection for six FHB resistance QTLs, plus orange blossom wheat midge resistance gene *Sm1* and leaf rust resistance gene *Lr21*. They also used 45–76 SSR markers to perform background selection in backcrossing populations to increase speed the restoration of the RP genetic background in the breeding program. The above-discussed strategy resulted in 87% fixation of the elite genetic background at the BC_2F_1 on average and successfully introduced all (up to 4) of the chromosome segments containing FHB, *Sm1* and *Lr21* resistance genes in four separate crosses (Somers et al., 2005). Joshi and Nayak (2010) and Xu (2010) recently reviewed the techniques and practical cases in marker-based gene pyramiding in the breeding program.

By using molecular markers to select or pyramid for multiple genes/QTLs is more complex and less proven. Recurrent selection is broadly regarded as an useful strategy for the advancement of polygenic traits. However, the usefulness and efficiency of selection are not so reasonable in various cases because phenotypic selection is exceedingly dependent upon environments and genotypic selection takes extensive time (2–3 crop seasons at least for one cycle of selection). Marker-assisted recurrent selection (MARS) is a method which permits performing genotypic selection and intercrossing in the similar crop season for one cycle of selection. Therefore, MARS could improve the efficiency of recurrent selection and speed up the advancement of the procedure (Jiang et al., 2007a), principally helps in integrating multiple favorable genes/QTLs from different resource through recurrent selection based on a multiple-parental population. For complex traits like grain yield, biotic and abiotic resistance, MARS has been proposed for "forward breeding" of resident genes and pyramiding multiple QTLs (Crosbie et al., 2006; Eathington, 2005; Ragot et al., 1995; Ribaut et al., 2000, 2010). Ribaut et al. (2010), also defined the MARS is a recurrent selection method using molecular markers for the classification and selection of multiple genomic regions gripped in the appearance of complex traits to accumulate the best-performing genotype within a single or across related populations.

11.5 ADVANTAGES OF MAS OVER CONVENTIONAL METHODS

In addition to the cost and time savings, for a number of breeding scenarios, MAS methods are expected to proffer momentous advantages compared with conventional selection methods.

11.5.1 GENE STACKING FOR A SINGLE TRAIT

MAS allows breeders to recognize the presence of multiple genes/alleles connected to a single trait, when the alleles do not bring to bear individually detectable outcomes on the expression of the trait, for example, when one gene presents resistance to a specific disease, breeders would be powerless to use traditional phenotypic screening to add an additional gene to the same cultivar in order to augment the durability of resistance. In such cases, MAS would be the only practicable choice, provided markers are available for such genes.

11.5.2 EARLY DETECTION

MAS allows the detection of desirable alleles/ traits in the seedling stage itself well before the trait is expressed phenotypically in the older age of crop. This advantage can be particularly very important in slow growing and long duration crops, which save the time and money.

11.5.3 RECESSIVE GENES

MAS allows breeders to recognize heterozygous plants that carry a recessive allele of concern whose presence cannot be detected phenotypically in the population. In traditional breeding approaches, an extra step of selfing is required to detect phenotypes associated with recessive genes in the breeding program.

11.5.4 HERITABILITY OF TRAITS

MAS is mainly helpful in selection for traits with low heritability up to a point, expands from MAS increase with decreasing heritability.

11.5.5 SEASONAL CONSIDERATIONS

MAS proposes the potential of time savings compared with conventional selection when it is essential to monitor for traits whose appearance depends on seasonal parameters. By using molecular markers, at any time of the year, plant breeders can monitor for the presence of an allele connected with traits that are expressed only during certain growing seasons of the crop. For example, CIMMYT's wheat breeding station in northern Mexico is usually used for the screening of segregating germplasm for leaf rust resistance. However, expression of leaf rust is not uniform in all growing seasons. When there are seasons with small expression of leaf rust, markers, if available, can be a precious alternative as a tool for monitoring leaf rust gene.

11.5.6 GEOGRAPHICAL CONSIDERATIONS

MAS is essential to screen for the traits whose expression depends on geographical contemplations. Using molecular markers, plant breeders in one location can screen for the presence of an allele(s) associated with characters expressed only in other locations.

11.5.7 MULTIPLE GENES, MULTIPLE TRAITS

MAS also offers potential discounts when there is a need to select for multiple traits simultaneously. With conventional techniques, it is often compulsory to carry out separate trials for the screening of individual traits.

11.5.8 BIOLOGICAL SECURITY CONSIDERATIONS

MAS provides impending advantages over selection based on the use of potentially harmful biological agents (e.g., artificial viral infections or artificial infestations with pathogens), which may need specific security measures at time of screening.

11.6 VARIETY RELEASED/DOCUMENTED THROUGH MAS

Rice is one of the smallest genome-containing cereal crop among the cultivated cereal and also most extensively studied species among cereals. To date, MAS in rice breeding has mainly been utilized for the pyramiding of biotic resistance (disease resistances), that is, bacterial blight and blast. The pyramided bacterial blight resistance genes, $Xa4 + xa5 + Xa21$, articulated strong resistance to the different virulent bacterial blight isolates and compared with individual resistance genes that are moderately to completely susceptible (Jeung et al., 2006). The resistance genes $xa5$, $xa13$, and $Xa21$ have been also pyramided into an indica rice cultivar (PR106) using MAS that expressed well-built resistance to the different bacterial blight races of India (Singh et al., 2001). Hittalmani et al. (2000) also pyramided three major genes ($Pi1$, $Piz-5$, and $Pita$) using RFLP markers from three different parents for rice blast into a single rice cultivar Co-39. Two commercially cultivated rice cultivars (Angke and Conde) were released in 2002 for cultivation in Indonesia by the MAS breeding program. They hold gene pyramids $Xa4+xa5$ and $Xa4 + Xa7$, respectively (Bustamam et al., 2002). In the Philippines, two rice cultivars (NSIC Rc142 and NSIC Rc154) have the gene combination $Xa4 + xa5 + Xa21$ successfully released. These genes have been integrated into the susceptible mega rice cultivar IR64 genetic background using MAS (Toenniessen et al., 2003), and in China, the photo-sensitive genic male sterile line 3418s (Luo et al., 2003), restorer lines R8006 and R1176 (Cao et al., 2003), and Kang 4183 (Luo et al., 2005) were successfully adopted with an elevated level of resistance to bacterial blight by using the bacterial blight resistant gene *xa21*.

Marker-assisted backcross breeding (MABB) tied with phenotypic selection for agronomic, grain, and cooking quality traits has been used to incorporate BB-resistance genes *Xa13* and *Xa21* into "Pusa Basmati 1" (Joseph et al., 2004). One of the improved lines was released in India as "Improved Pusa Basmati 1" for commercial cultivation in 2007 (Gopal-akrishnan et al., 2008), and this is one of the first product of MAS-breeding programs to be used in India. However, the susceptibility of "Improved Pusa Basmati 1" and other Basmati rice varieties to rice blast and ShB diseases remain a major concern in the different rice-growing region of India. Soon after, Singh et al. (2012) identified a blast resistance gene *Pi54* and ShB resistance QTL-qSBR11-1 in a rice cultivar "Tetep" to Improved Pusa Basmati 1 through MAS and the improved lines have desirable Basmati grain and cooking quality characteristics, in tandem with inbuilt resis-tance to bacterial blight, blast and ShB, and yield on par with "Improved Pusa Basmati 1." The improved multiple biotic stress-resistant rice lines will now be assessed under multilocation trials for the releasing to farmers as improved Basmati cultivars. There are other achievements in rice blast resistant breeding program which include the applications of the blast resis-tant genes, such as the *Pid1*, *Pib*, and *Pita* pyramided to G46B (Chen et al., 2004), the *Pi2* introduced into Zhenshan97B (Chen et al., 2004) and the *Pi1*, *Pi2*, and *Pi33* introgressed to Jin23B (Chen et al., 2008). Parallel to the above efforts, the resistance breeding team at Directorate of Rice Research, Hyderabad, India has introgressed three different bacterial blight resistance genes *Xa21*, *xa13*, and *xa5* into the elite rice variety cultivar Samba Mahsuri through marker-assisted breeding program (Sundaram et al., 2008). A three-gene pyramid rice line, RPBio-226 (IET 19046) was identified to acquire high-yield, good level, and broad-spectrum bacterial blight resistance and excellent grain quality. Recently, this line has been released for commercial cultivation as a new variety "Improved Samba Mahsuri." Further, a sister line of Improved Samba Mahsuri, RPBio-210 (IET 19045), which has elevated level of bacterial blight resistance, high yield and also good grain quality has been freshly registered to the National Bureau of Plant Genetic Resources, New Delhi, India as a novel germplasm (Sundaram et al., 2010). Recently, Shanti et al. (2010) also introgressed four different *Xa4*, *xa13*, *xa5*, and *Xa21* genes into the hybrid rice parental lines KMR3, PRR78, IR58025B, Pusa 6B, and the popular cv. Mahsuri, whereas Zhan et al. (2012) developed an elite restorer line R8012 carrying multiple genes (*Pi25/Xa21/xa13/xa5*) through MAS, in which all the resistance genes can confer resistance to BB and blast.

The performance of the bacterial blight resistant accession of Pusa RH10 produced by intercrossing the improved parental lines showed superiority over the original Pusa RH10 (Basavaraj et al., 2010). Importantly, we now have BB-resistant Basmati breeding lines in the genetic background of Pusa Basmati-1 (Joseph et al., 2004), Pusa RH10 (an aromatic hybrid, Basavaraj et al., 2010), and a traditional Basmati, Type-3 (Rajpurohit et al., 2011). Pandey et al. (2013) also improved the two traditional BB-susceptible Basmati varieties (Taraori Basmati and Basmati 386), through the approach of limited marker-assisted backcrossing for introgression of two major BB resistance genes, *Xa21* and *Xa13*, together with phenotype-based selection for improvement of their plant type and yield. The different available, demonstrated rice varieties developed by the utility of molecular markers in improvement of biotic stress resistance of rice are discussed in Table 11.1.

11.7   CONCLUSION AND FUTURE PROSPECTS

Incorporation of multiple resistance genes into single cultivars through backcrossing for rice diseases management is exciting but very exhausting, challenging, and time-taking task for the rice breeder. Modern molecular approaches like MAS can solve the problem of many rice breeders. Gene pyramiding is emphasized to obtain many complex biochemical pathways in plants for crop improvement and durable resistance. Although the development of matching type or more fit pathogen race against a specific resistance gene is considered very slow in case of multiple gene resistance strategy or against stacked genes. However, a potential risk of the development of super-race is still into consideration among scientific communities when a gene pyramiding strategy is used for the management of plant pathogens. Therefore, the development of pyramided lines and their long-term field efficacy that can assure a sustainable plant disease management is still a question mark for future.

TABLE 11.1 Marker-assisted Selection Study for the Improvement of Rice.

S. No.	Target trait	Gene(s)/QTL(s)	Type/name of marker(s) used	Reference	Remarks
1	BB resistance	Xa21	STS (pTA248)	Ronald et al. (1992)	MAS applied for MABB
2	BB resistance	Xa4, xa5, and Xa10	Gene linked RFLP and RAPD markers	Yoshimura et al. (1995)	MAS applied for gene pyramiding
3	BB resistance	Xa4, xa5, xa13, and Xa21	STS for Xa4 CAPS for xa5 (RG556 + DraI), CAPS for xa13 (RG136 + HinfI), STS for Xa21 (pTA248)	Huang et al. (1997)	MAS applied for gene pyramiding
4	BB resistance	Xa21	STS (pTA248)	Reddy et al. (1997)	MAS applied for early generation selection for BB resistance
5	BB resistance	Xa21	STS (pTA248)	Chen et al. (2000)	MAS applied for MABB
6	BB resistance	xa5, xa13, and Xa21	CAPS for xa5 (RG556 + DraI) CAPS for xa13 (RG136 + HinfI), STS for Xa21 (pTA248)	Sanchez et al. (2000)	MAS applied for gene pyramiding
7	Blast resistance	Pi1, Piz-5, Pi2, Pita	RFLP markers for Pi1, Pi2, and Pita and a PCR-based SAP marker for Piz-5	Hittalmani et al. (2000)	MAS applied for gene pyramiding (target variety: C039)
8	BB resistance	xa5, xa13, and Xa21	CAPS for xa5 (RG556 + DraI), CAPS for xa13 (RG136 + HinfI), STS for Xa21 (pTA248)	Singh et al. (2001)	MAS applied for MABB (target variety: PR106)
9	BB resistance	xa5, xa13, and Xa21	CAPS for xa5 (RG556 + DraI), CAPS for xa13 (RG136 + HinfI), STS for Xa21 (pTA248)	Davierwala et al. (2001)	MAS applied for gene pyramiding
10	BB resistance + blast resistance	Xa21 and Piz	STS for Piz, transgene specific marker for Xa21	Narayanan et al. (2002)	MAS applied for pyramiding of target traits. Xa21 gene originally introduced into donor lines through genetic engineering (target variety: IR50)

TABLE 11.1 (Continued)

S. No.	Target trait	Gene(s)/QTL(s)	Type/name of marker(s) used	Reference	Remarks
11	Blast resistance	Pi1	SSR and ISSR markers	Liu et al. (2002)	MAS applied for backcross breeding (target variety: Zhenshan 97A)
12	BB resistance	xa5	CAPS (RG556 + DraI)	Toenniessen et al. (2003)	MAS applied for MABB
13	BB resistance	Xa4, xa5, and Xa21	STS for Xa4 CAPS for xa5 (RG556 + DraI) STS for Xa21 (pTA248)	Leung et al. (2004)	MAS applied for gene pyramiding
14	BB	Xa7 and Xa21	STS for Xa7	Zhang et al. (2006)	MAS applied for gene pyramiding
15	BB resistance	xa5, xa13, and Xa21	CAPS for xa5 (RG556+DraI) CAPS for xa13 (RG136 + HinfI) STS for Xa21 (pTA248)	Sundaram et al. (2008)	MAS applied for backcross breeding (target variety: Samba Mahsuri)
16	BB resistance + Grain quality	xa13 and Xa21	CAPS for xa13 (RG136+HinfI) STS for Xa21 (pTA248)	Joseph et al. (2004), Gopalakrishnan et al. (2008)	MAS applied for backcross breeding (target variety: Pusa Basmati 1)
17	BB resistance	Xa4, X17, and Xa21	STS for Xa4 and Xa7, STS for Xa21 (pTA248)	Perez et al. (2008)	BB resistance
18	BB resistance	Xa4, xa5, xa13, and Xa21	STS for Xa4 CAPS for xa5 (RG556 + DraI), CAPS for xa13 (RG136 + HinfI), STS for Xa21 (pTA248)	AICRIP Progress Report. Vol. 1 (2000–2008)	MAS applied for gene pyramiding (target varieties: Swarna and IR64, some pre-breeding lines in the genetic background of Lalat and Tapaswini possessing BB resistance also developed by CRRI and nominated for AICRIP trials)
19	BB resistance	xa5 and xa13	CAPS for xa13 (RG136 + HinfI), STS for Xa21 (pTA248)	Sundaram et al. (2009)	MAS applied for backcross breeding (target variety: Triguna)

TABLE 11.1 (Continued)

S. No.	Target trait	Gene(s)/QTL(s)	Type/name of marker(s) used	Reference	Remarks
22	Blast resistance	Pi-9(t)	pB8	Wen and Gao (2011)	Introgressed the broad-spectrum blast resistant gene Pi-9(t) from the donor parent P2 into hybrid restorer Luhui17 by using MAS technique
21	BB resistance	Xa4, xa5, xa13, and Xa21	STS for Xa4, CAPS for xa5 (RG556 + DraI), CAPS for xa13 (RG136 + HinfI), STS for Xa21 (pTA248)	Shanti et al.(2010)	MAS applied for pyramiding the BB resistance genes into the hybrid rice parental lines KMR3, PRR78, IR58025B, Pusa 6B and the popular cv. Mahsuri
22	BB resistance + Blast resistance	xa5, xa13, Xa21, and Pi25	CAPS for xa5 (RG556 + DraI), CAPS for xa13 (RG136 + HinfI), STS for Xa21 (pTA248), and STS for Pi25 (SA7)	Zhan et al. (2012)	MAS applied for pyramiding multiple genes (Pi25/Xa21/xa13/xa5) in to elite restorer line R8012 and its hybrid (Zhong 9A/R8012) playing a vital role in securing rice production in China
23	BB resistance + blast resistance + ShB	xa13, Xa21, Pi54, and qSBR11-1	CAPS for xa13 (RG136 + HinfI), STS for Xa21 (pTA248), SSR for Pi54 (RM206) SSR for qSBR11-1 (flanking markers RM224 and RM7443)	Singh et al. (2012)	The rice cultivar "Improved Pusa Basmati 1" (carrying the BB resistance genes xa13 and Xa21) was used as the recurrent parent and cultivar "Tetep" (carrying the blast resistance gene Pi54 and ShB resistance QTL, qSBR11-1) was the donor
24	BB resistance	xa13 and Xa21	CAPS for xa13 (RG136 + HinfI), STS for Xa21 (pTA248)	Pandey et al. (2013)	Improved the two traditional BB-susceptible Basmati varieties (Taraori Basmati and Basmati 386)

TABLE 11.1 (Continued)

S. No.	Target trait	Gene(s)/QTL(s)	Type/name of marker(s) used	Reference	Remarks
25	BB resistance + Blast resistance	*Xa21* and *Pi54*	STS for *Xa21* (pTA248), SSR for *Pi54* (RM206)	Hari et al. (2013)	Introgression of bacterial blight and blast resistance into IR58025B, an elite maintainer line of rice
26	BB resistance	*Xa4, xa5, xa13,* and *Xa21*	STS for *Xa4*, CAPS for *xa5* (RG556 + DraI), CAPS for *xa13* (RG136 + HinfI), STS for *Xa21* (pTA248)	Dokku et al. (2013)	Three resistance genes, that is, *xa5, xa13,* and *Xa21* were transferred from IRBB 60 through MABC to supplement the *Xa4* gene present in Tapaswini, an elite cultivar having a wide coverage
27	Blast resistance	*Pi-b* and *Pi-kh*	RM208 for Pi-b RM206 for Pi-kh	Tanweer et al. (2015)	*Pi-b* and *Pi-kh* genes transferred from the Pongsu Seribu 2 variety into popular Malaysian rice variety MR219
28	Blast resistance	*Pi-7(t), Pi-d(t)1, Pir2-3(t),* and *qLN2*	RM263 for Pi-7(t), MR5961 for Pi-d(t)1, Pir2-3(t), and qLN2	Hasan et al. (2015)	Pongsu Seribu 1 rice variety used as donor of the blast resistance *Pi-7(t), Pi-d(t)1* and *Pir2-3(t)* genes and *qLN2* QTL into high-yielding Malaysian rice variety, MR263

BB, bacterial blight.
Source: Adapted and modified from Collard and Mackill (2008).

KEYWORDS

- **molecular marker**
- **pyramiding**
- **biotic stress**
- **marker-assisted selection**
- **genomics**

REFERENCES

Azzam, O.; Chancellor, T. C. B. The Biology, Epidemiology and Management of Rice Tungro Disease in Asia. *Plant Dis.* **2002,** *86,* 88–100.

Babu, R.; Nair, S. K.; Prasanna, B. M.; Gupta, H. S. Integrating Marker Assisted Selection in Crop Breeding—Prospects and Challenges. *Curr. Sci.* **2004,** *87,* 607–619.

Baranwal, M. K.; Kotasthane, A.; Magculia, N.; Mukherjee, P. K.; Savary, S.; Sharma, A. K.; Singh, H. B.; Singh, U. S.; Sparks, A. H.; Variar, M.; Zaidi, N. A Review on Crop Losses, Epidemiology and Disease Management of Rice Brown Spot to Identify Research Priorities and Knowledge Gaps. *Eur. J. Plant Pathol.* **2013,** *136,* 443–457.

Basavaraj, S. H.; Singh, V. K.; Singh, A.; Singh, A.; Singh, A.; Yadav, S.; Ellur, R. K.; Singh, D.; Gopala Krishnan, S.; Nagarajan, M.; Mohapatra, T.; Prabhu, K. V.; Singh, A. K. Marker-assisted Improvement of Bacterial Blight Resistance in Parental Lines of Pusa RH10, a Superfine Grain Aromatic Rice Hybrid. *Mol. Breed.* **2010,** *2,* 293–305.

Bustamam, M.; Tabien, R. E.; Suwarno, A.; Abalos, M. C.; Kadir, T. S.; Ona, I.; Bernardo, M.; Veracruz, C. M.; Leung, H. Asian Rice Biotechnology Network: Improving Popular Cultivars through Marker-assisted Backcrossing by the NARES. In: Poster Presented at the International Rice Congress, September 16–20, 2002, Beijing, China, 2002.

Cao, L. Y.; Zhuang, J. Y.; Zhan, X. D.; Zheng, K. L.; Cheng, S. H. Hybrid Rice Resistant to Bacterial Blight Developed by Marker Assisted Selection. *Chin. J. Rice Sci.* **2003,** *17*(2), 184–186.

Castro, A. J.; Capettini, F.; Corey, A. E.; Filichkina, T.; Hayes, P. M.; Kleinhofs, A.; Kudrna, D.; Richardson, K.; Sandoval-Islas, S.; Rossi, C.; Vivar, H. Mapping and Pyramiding of Qualitative and Quantitative Resistance to Stripe Rust in Barley. *Theor. Appl. Genet.* **2003,** *107,* 922–930.

Chen, H.; Chen, Z.; Ni, S.; Zuo, S.; Pan, X.; Zhu, X. Pyramiding Three Genes with Resistance to Blast by Marker Assisted Selection to Improve Rice Blast Resistance of Jin23B. *Chin. J. Rice Sci.* **2008,** *22,* 23–27.

Chen, H.; Wang, S.; Zhang, Q. New Gene for Bacterial Blight Resistance in Rice Located on Chromosome 12 Identified from Minghui 63, an Elite Restorer Line. *Phytopathology* **2002,** *92,* 750–754.

Chen, S.; Lin, X. H.; Xu, C. G.; Zhang, Q. F. Improvement of Bacterial Blight Resistance of Minghui 63 an Elite Restorer Line of Hybrid Rice, by Molecular Marker-assisted Selection. *Crop Sci.* **2000,** *40,* 239–244.

Chen, X. W.; Li, S. G.; Ma, Y. Q.; Li, H. Y.; Zhou, K. D.; Zhu, L. H. Marker-assisted Selection and Pyramiding for Three Blast Resistance Genes, *Pi-d (t) 1*, *Pi-b*, *Pi-ta2*, in Rice. *Chin. J. Biotechnol.* **2004**, *20*, 708–714.

Collard, B. C. Y.; Mackill, D. J. Marker-assisted Selection: An Approach for Precision Plant Breeding in the Twenty-first Century. *Philos. Trans. R. Soc. Lond. B: Biol. Sci.* **2008**, *363*, 557–572.

Crosbie, T. M.; Eathington, S. R.; Johnson, G. R.; Edwards, M.; Reiter, R.; Stark, S.; Mohanty, R. G.; Oyervides, M.; Buehler, R. E.; Walker, A. K.; Dobert, R.; Delannay, X.; Pershing, J. C. ; Hall, M. A.; Lamkey, K. R. Plant Breeding: Past, Present and Future. In: *Plant Breeding: The Arnel R. Hallauer International Symposium*; Lamkey, K. R., Lee, M., Eds.; Blackwell Publishing: Oxford, UK, 2006; pp 3–50.

Dahal, G.; Hibino, H.; Cabunagan, R. C.; Tiongco, E. R.; Flores, Z. M.; Aguiero, V. M. Changes in Cultivar Reaction to Tungro Due to Changes in "virulence" of the Leafhopper Vector. *Phytopathology* **1990**, *80*, 659–665.

Davierwala, A. P.; Reddy, A. P. K.; Lagu, M. D.; Ranjekar, P. K.; Gupta, V. S. Marker Assisted Selection of Bacterial Blight Resistance Genes in Rice. *Biochem. Genet.* **2001**, *39*, 261–278.

Directorate of Rice Research. Progress Reports of Hybrid Rice Project and AICRIP (various issues). DRR: Hyderabad, 2000–2008.

Dodan, D. S.; Singh, R. *False Smut of Rice: Present Status*; Agricultural Reviews—Agricultural Research Communications Centre India, 1996; pp 227–240.

Dokku, P.; Das, K. M.; Rao, G. J. N. Pyramiding of Four Resistance Genes of Bacterial Blight in Tapaswini, an Elite Rice Cultivar, through Marker-assisted Selection. *Euphytica* **2013**, *192*, 87–96.

Doveri, S.; Maheswaran, M.; Powell, W. Molecular Markers—History, Features and Applications. In: *Principles and Practices of Plant Genomics*; Kole, C., Abbott, A. G., Eds.; Vol 1: Genome Mapping. Science Publishers: Enfield, Jersey, Plymouth, 2008.

Eathington, S. R. Practical Applications of Molecular Technology in the Development of Commercial Maize Hybrids. In: Proceedings of the 60th Annual Corn and Sorghum Seed Research Conferences, American Seed Trade Association, Washington, DC, **2005**.

Elazegui, F.; Islam, Z. *Diagnosis of Common Diseases of Rice*. IRRI, 2003; p 25.

Gopalakrishnan, S.; Sharma, R. K.; Rajkumar, K. A.; Joseph, M.; Singh, V. P.; Singh, A. K.; Bhat, K. V.; Singh, N. K.; Mohapatra, T. Integrating Marker Assisted Background Analysis with Foreground Selection for Identification of Superior Bacterial Blight Resistant Recombinants in Basmati Rice. *Plant Breed.* **2008**, *127*, 131–139.

Hari, Y.; Srinivasarao, K.; Viraktamath, B. C.; Hariprasad, A. S.; Laha, G. S.; Ilyas Ahmed, M.; Natarajkumar, P.; Sujatha, K.; Srinivas Prasad, M.; Pandey, M.; Ramesha, M. S; Neeraja, C. N.; Balachandran, S. M.; Shobharani, N.; Kemparaju, B.; Madhanmohan, K.; Sama, V. S. A. K.; Hajira, S. K.; Baachiranjeevi, C. H.; Pranathi, K.; Ashok Reddy, G.; Madhav, M. S.; Sundaram, R. M. Marker-assisted Introgression of Bacterial Blight and Blast Resistance into IR 58025B, an Elite Maintainer Line of Rice. *Plant Breed.* **2013**, *132*, 586–594.

Hasan, M. M.; Rafii, M. Y.; Ismail, M. R.; Mahmood, M.; Rahim, H. A.; Alam, M. A.; Askani, S.; Malek, M. A.; Latif, M. A. Marker-assisted Backcrossing: A Useful Method for Rice Improvement. *Biotechnol. Biotechnol. Equip.* **2015**, *29*, 237–254.

Hittalmani, S.; Parco, A.; Mew, T. V.; Zeigler, R. S.; Huang, N. Fine Mapping and DNA Marker-assisted Pyramiding of the Three Major Genes for Blast Resistance in Rice. *Theor. Appl. Genet.* **2000**, *100*, 1121–1128.

Huang, N.; Angeles, E. R.; Domingo, J.; Magpantay, G.; Singh, S;, Zhang, G.; Kumaravad-ivel, N.; Bennett, J.; Khush, G. S. Pyramiding of Bacterial Blight Resistance Genes in Rice: Marker Assisted Selection using RFLP and PCR. *Theor. Appl. Genet.* **1997,** *95,* 313–320.

Jeung, J. U.; Heu, S. G.; Shin, M. S.; Veracruz, C. M.; Jena, K. K. Dynamics of *Xanthomonas oryzae* pv. Oryzae Populations in Korea and their Relations to Known Bacterial Blight Resistance Genes. *Phytopathology* **2006,** *96,* 867–875.

Jiang, G. L.; Dong, Y.; Shi, J.; Ward, R. W. QTL Analysis of Resistance to Fusarium Head Blight in the Novel Wheat Germplasm CJ 9306. II. Resistance to Deoxy-nivalenol Accu-mulation and Grain Yield Loss. *Theor. Appl. Genet.* **2007b,** *115,* 1043–1052.

Jiang, G. L.; Shi, J.; Ward, R. W. QTL Analysis of Resistance to Fusarium Head Blight in the Novel Wheat Germplasm CJ 9306. I. Resistance to Fungal Spread. *Theor. Appl. Genet.* **2007a,** *116,* 3–13.

Jiang, G. Molecular Markers and Marker-Assisted Breeding in Plants. 2013. DOI:10.5772/52583.

Joseph, M.; Gopalakrishnan, S.; Sharma, R. K.; Singh, A. K.; Singh, V. P.; Singh, N. K. Mohapatra, T. Combining Bacterial Blight Resistance and Basmati Quality Characteristics by Phenotypic and Molecular Marker Assisted Selection in Rice. *Mol. Breed.* **2004,** *13,* 377–387.

Joshi, R. K.; Nayak, S. Gene Pyramiding—A Broad Spectrum Technique for Developing Durable Stress Resistance in Crops. *Biotechol. Mol. Biol. Rev.* **2010,** *5,* 51–60.

Khalili, E.; Sadravi, M.; Naeimi, S.; Khosravi, V. Biological Control of Rice Brown Spot with Native Isolates of Three *Trichoderma* Species. *Braz. J. Microbiol.* **2012,** *43,* 297–305.

Khang, C. H.; Park, S. Y.; Lee, Y. H.; Valent, B.; Kang, S. Genome Organization and Evolu-tion of the AVR-Pita Avirulence Gene Family in the *Magnaporthe grisea* Species Complex. *Mol. Plant Microb. Interact.* **2008,** *21*(5), 658–670.

Lang, N.; Buu, B. C.; Ismail, A. Molecular Mapping and Marker-assisted Selection for Major-gene Traits in Rice (*Oryza sativa* L.). *Omonrice* **2008,** *16,* 50–56.

Leung, H.; Wu, J.; Liu, B.; Bustaman, M.; Sridhar, R.; Singh, K.; Redona, E.; Quang, V. D.; Zheng, K.; Bernardo, M.; Wang, G.; Leach, J.; Choi, I. R.; Vera Cruz, C. Sustainable Disease Resistance in Rice: Current and Future Strategies. In: New directions for a diverse planet. Proceedings of the 4th International Crop Science Congress, 26 Sept. 1–Oct., 2004, Brisbane, Australia. Published on CDROM, 2004. www.cropscience.org.au.

Li, X.; Han, Y.; Teng, W.; Zhang, S.; Yu, K.; Poysa, V.; Anderson, T.; Ding, J.; Li, W. Pyra-mided QTL Underlying Tolerance to *Phytophthora* Root Rot in Mega-environment from Soybean Cultivar 'Conrad' and 'Hefeng 25'. *Theor. Appl. Genet.* **2010,** *121,* 651–658.

Liu, D. W.; Oard, S. V.; Oard, J. H. High Transgene Expression Levels in Sugarcane (*Saccharum officinarurn* L) Driven by Rice Ubiquitin Promoter RUBQ2. *Plant Sci.* **2003,** *165,* 743–750.

Liu, G.; Lu, G.; Zeng, L.; Wang, G. L. Two Broadspectrum Blast Resistance Genes, Pi9(t) and Pi2(t), are Physically Linked on Rice Chromosome 6. *Mol. Genet. Genom.* **2002,** *267,* 472–480.

Liu, J.; Wang, X.; Mitchell, T.; Hu, Y.; Liu, X., Dai, L.; Wang, G. L. Recent Progress and Understanding of the Molecular Mechanisms of the Rice–*Magnaporthe oryzae* Interaction. *Mol. Plant Pathol.* **2010,** *11*(3), 419–427.

Luo, Y. C.; Wang, S. H.; Li, C. Q.; Wang, D. Z.; Wu, S.; Du, S. Y.; Breeding of the Photo-period-sensitive Genetic Male Sterile Line '3418S' Resistant to Bacterial Bright in Rice by Molecular Marker Assisted Selection Rice. *Acta Agron. Sin.* **2003,** *29*(3), 402–407.

Luo, Y. C.; Wang, S. H.; Li, C. Q.; Wu, S.; Wang, D.; Du, S. Improvement of Bacterial Blight Resistance by Molecular Marker Assisted Selection in a Wide Compatibility Restorer Line of Hybrid Rice. *Chin. J. Rice Sci.* **2005,** *19*(1), 36–40.

Luo, Y. C.; Wu, S.; Wang, S. H.; Li, C. Q.; Zhang, D. P.; Zhang, Q.; Zhao, K. J.; Wang, C. L.; Wang, D. Z.; Du, S. Y.; Wang, W. X. Pyramiding Two Bacterial Blight Resistance Genes into a CMS Line R106A in Rice. *Sci. Agric. Sin.* **2005,** *38*, 2157–2164.

McDowell, J. M.; Woffenden, B. J. Plant Disease Resistance Genes: Recent Insights and Potential Applications. *Trend. Biotechnol.* **2003,** *21*, 178–183.

Mew, T. W.; Rosales, A. M. *Relationship of Soil Microorganisms to Rice Sheath Blight Development in Irrigated and Dryland Rice Cultures*; Technical Bulletin ASPAC Food and Fertilizer Technology Center: Taipei City, Taiwan, 1984; p 11.

Michelmore, R. W.; Meyers, B. C. Clusters of Resistance Genes in Plants Evolve by Divergent Selection and a Birth-and-death Process. *Genome Res.* **1998,** *8*, 1113–1130.

Mondal, K. K.; Mani, C.; Verma, G. Emergence of Bacterial Panicle Blight Caused by *Burkholderia glumae* in North India. *Plant Dis.* **2015,** *99*(9), 1268.

Mundt, C. C. Durable Resistance: A Key to Sustainable Management of Pathogens and Pests. *Inf. Genet. Evol.* **2014,** *27*, 446–455.

Nandakumar, R.; Shahjahan, A. K. M.; Yuan, X. L.; Dickstein, E. R.; Groth, D. E.; Clark, C. A.; R. Cartwright, D.; Rush, M. C. *Burkholderia glumae* and *B. gladioli* cause Bacterial Panicle Blight in Rice in the Southern United States. *Plant Dis.* **2009,** *93*, 896–905.

Narayanan, N. N.; Baisakh, N.; Vera Cruz, N.; Gnananmanickam, S. S.; Datta, K.; Datta, S. K. Molecular Breeding for the Development of Blast and Bacterial Blight Resistance in Rice cv. IR50. *Crop Sci.* **2002,** 2072–2079.

Ou, S. H. *Rice Disease*, 2nd ed. Commonwealth Mycological Institute Pub: Kew, Surrey, 1985.

Pandey, M. K.; Shobha Rani, N.; Sundaram, R. M.; Laha, G. S.; Madhav, M. S.; Srinivasa Rao, K.; Injey, S.; Yadla, H.; Varaprasad, G. S.; Subba Rao, L. V.; Kota, S.; Sivaranjani, A. K. P.; Viraktamath, B. C. Improvement of Two Traditional Basmati Rice Varieties for Bacterial Blight Resistance and Plant Stature through Morphological and Marker-assisted Selection. *Mol. Breed.* **2013,** *31*, 239–246.

Panigrahi, J. Molecular Mapping and Map Based Cloning of Genes in Plants. In *Plant Biotechnology and Transgenics*; Thangadurai, D., Ed.; Bentham Science Publishers: USA. 2011.

Perez, L. M.; Redona, E. D.; Mendioro, M. S.; Vera Cruz, C. M.; Leung, H. Introgression of *Xa4, Xa7* and *Xa21* for Resistance to Bacterial Blight in Thermo sensitive Genetic Male Sterile Rice (*Oryza sativa* L.) for the Development of Two-line Hybrids. *Euphytica* **2008,** *164*, 627–636.

Pink, D. A. C. Strategies Using Genes for Non-durable Disease Resistance. *Euphytica* **2002,** *124*, 227–236.

Ragimekula, N.; Varadarajula, N. N.; Mallapuram, S. P.; Gangimeni, G.; Reddy, R. K.; Kondreddy, H. R. Marker Assisted Selection in Disease Resistance Breeding. *J. Plant Breed. Genet.* **2013,** *1*(2), 90–109.

Ragot, M.; Biasiolli, M.; Dekbut, M. F.; Dell'orco, A.; Margarini, L.; Thevenin, P.; Vernoy, J.; Vivant, J.; Zimmermann, R.; Gay, G. Marker-assisted Backcrossing: A Practical Example. In: *Les Colloques, No. 72*; Berville, A., Tersac, M., Eds.; INRA: Paris, 1995, pp 45–46.

Rajpurohit, D.; Kumar, R.; Kumar, M.; Paul, P.; Awasthi, A.; Basha, P. O.; Puri, A.; Jhang, T.; Singh, K.; Dhaliwal, H. S. Pyramiding of Two Bacterial Blight Resistance and a Semi-dwarfing Gene in Type 3 Basmati using Marker-assisted Selection. *Euphytica* **2011,** *178*, 111–126.

Rausher, M. Co-evolution and Plant Resistance to Natural Enemies. *Nature* **2001**, *411*, 857–864.

Reddy, J. N.; Baroidan, M. R.; Bernardo, M. A.; George, M. L. C.; Sridhar, R. Application of Marker Assisted Selection in Rice for Bacterial Blight Resistance Gene, Xa-21. *Curr. Sci.* **1997**, *73*, 873–875.

Reyes-Valdés, M. H. A Model for Marker-based Selection in Gene Introgression Breeding Programs. *Crop Sci.* **2000**, *40*, 91–98.

Ribaut, J. M.; de Vicente, M. C.; Delannay, X. Molecular Breeding in Developing Countries: Challenges and Perspectives. *Curr. Opin. Plant Biol.* **2010**, *13*, 1–6.

Ribaut, J. M;, Edmeades, G.; Perotti, E.; Hoisington, D. QTL Analysis, MAS Results and Perspectives for Drought-tolerance Improvement in Tropical Maize. In: *Molecular Approaches for the Genetic Improvement of Cereals for Stable Production in Water-limited Environments*; Ribaut, J. M., Poland, D., Eds.; CIMMYT: Mexico, 2000; pp 131–136.

Richardson, K. L.; Vales, M. I.; Kling, J. G.; Mundt, C. C.; Hayes, P. M. Pyramiding and Dissecting Disease Resistance QTL to Barley Stripe Rust. *Theor. Appl. Genet.* **2006**, *113*, 485–495.

Ronald, P. C.; Albano, B.; Tabien, R.; Abenes, M. L. P.; Wu, K. S.; McCouch, S. R.; Tanksley, S. D. Genetic and Physical Analysis of the Rice Bacterial Blight Disease Resistance Locus Xa-21. *Mol. Gen. Genet.* **1992**, *236*, 113–120.

Sanchez, A. C.; Brar, D. S.; Huang, N.; Li, Z.; Khush, G. S Sequence Tagged Site Marker-assisted Selection for Three Bacterial Blight Resistance Genes in Rice. *Crop Sci.* **2000**, *40*, 792–797.

Santa-Cruz, F. C.; Hull, R.; Azzam, O. Changes in Level of Virus Accumulation and Incidence of Infection are Critical in the Characterization of Rice Tungro Bacilliform Virus (RTBV) Resistance in Rice. *Arch. Virol.* **2003**, *148*, 1465–1483.

Sato, H.; Ideta, O.; Ando, I.; Kunihiro, Y.; Hirabayashi, H.; Iwano, M.; Miyasaka, A.; Nemoto, H.; Imbe, T. Mapping QTLs for Sheath Blight Resistance in the Rice Line WSS2. *Breed. Sci.* **2004**, *54*, 265–271.

Semagn, K.; Bjornstad, A.; Ndjiondjop, M. N. An Overview of Molecular Marker Methods for Plant. *Afr. J. Biotechnol.* **2006**, *25*, 2540–2569.

Servin, B.; Martin, O. C.; Mezard, M.; Hospital, F. Toward a Theory of Marker-assisted Gene Pyramiding. *Genetics* **2004**, *168*, 513–523.

Shanti, M. L.; Shenoy, V. V.; Lalitha Devi, G.; Mohan Kumar, V.; Premalatha, P.; Naveen Kumar, G.; Shashidhar, H. E.; Zehr, U. B.; Freeman, W. H. Marker-assisted Breeding for Resistance to Bacterial Leaf Blight in Popular Cultivar and Parental Lines of Hybrid Rice. *J. Plant Pathol.* **2010**, *92*(2), 495–501.

Singh, R. S. *Plant Diseases*, 8th ed.; Oxford and IBH Publishing Co. Pvt. Ltd., 2005; p 720.

Singh, R.; Dabur, K. R.; Malik, R. K.; Long-term Response of Zero-tillage: Soil Fungi, Nematodes and Diseases of Rice–Wheat System. *Technical Bulletin (7)*; CCS HAU Rice Research Station/Department of Nematology and Directorate of Extension Education, CCS HAU: Kaul/Hisar, 2005; pp 16.

Singh, S.; Sidhu, J. S.; Huang, N.; Vikal, Y.; Li, Z.; Brar, D. S.; Dhaliwal, H. S.; Khush, G. S. Pyramiding Three Bacterial Blight Resistance Genes (xa5, xa13 and Xa21) using Marker-assisted Selection into Indica Rice Cultivar PR106. *Theor. Appl. Genet.* **2001**, *102*, 1011–1015.

Singh, U. M.; Tiwari, G.; Babu, B. K.; Srivastava, R Marker-assisted Breeding Approaches for Enhancing Stress Tolerance in Crops in Changing Climate Scenarios. In *Climate Change Effect on Crop Productivity*; 2014; p 397.

Singh, V. K.; Singh, A.; Singh, S. P.; Ellur, R. K.; Singh, D.; Gopalakrishnan, S.; Nagarajan, M.; Vinod, K. K.; Singh, U. D.; Rathore, R.; Prasanthi, S. K.; Agrawal, P. K.; Bhatt, J. C.; Mohapatra, T.; Prabhu, K. V.; Singh, A. K. Incorporation of Blast Resistance Gene in Elite Basmati Rice Restorer Line PRR78, Using Marker Assisted Selection. *Field Crop Res.* **2012**, *128*, 8–16.

Somers, D. J.; Thomas, J.; DePauw, R.; Fox, S.; Humphreys, G.; Fedak, G. Assembling Complex Genotypes to Resist *Fusarium* in Wheat (*Triticum aestivum* L.). *Theor. Appl. Genet.* **2005**, *111*, 1623–1631.

Sundaram, R. M.; Chatterjee, S.; Oliva, R.; Laha, G. S.; Cruz, L. J. E.; Sonti, R. V. Update on Bacterial Blight of Rice: Fourth International Conference on Bacterial Blight. *Rice* **2014**, *7*, 12.

Sundaram, R. M.; Priya, M. R. V.; Laha, G. S.; Shobha Rani, N.; Srinivasa Rao, P.; Balachandran, S. M.; Ashok Reddy, G.; Sarma, N. P.; Sonti, R. V. Introduction of Bacterial Blight Resistance into Triguna, a High Yielding, Mid-early Duration Rice Variety by Molecular Marker Assisted Breeding. *Biotechnol. J.* **2009**, *4*, 400–407.

Sundaram, R. M.; Vishnupriya, M. R.; Biradar, S. K.; Laha, G. S.; Reddy, G. A.; Shoba Rani, N.; Sarma, N. P.; Sonti, R. V. Marker Assisted Introgression of Bacterial Blight Resistance in Samba Mahsuri, an Elite Indica Rice Variety. *Euphytica* **2008**, *160*, 411–422.

Sundaram, R. M.; Vishnupriya, M. R.; Shobha Rani, N.; Laha, G. S.; Viraktamath, B. C.; Balachandran, S. M.; Sarma, N. P.; Mishra, B.; Ashok Reddy, G.; Sonti, R. V. RPBio-189 (IET19045) (IC569676; INGR09070), a Paddy (Oryza sativa) Germplasm with High Bacterial Blight Resistance, Yield and Fine-grain Type. *Ind. J. Plant Genet. Res.* **2010**, *23*, 327–328.

Sunder, S.; Singh, R.; Dodan, D. S. Evaluation of Fungicides, Botanicals and Non-conventional Chemicals against Brown Spot of Rice. *Ind. Phytopathol.* **2010**, *63*, 192–194.

Swings, J.; Van Den Mooter, M.; Vauterin, L.; Hoste, B.; Gillis, M.; Mew, T. W.; Kersters, K. Reclassification of the Causal Agents of Bacterial Blight *Xanthomonas campestris* pv. *oryzae* and Bacterial Leaf Streak *Xanthomonas campestris* pv. *oryzicola* of Rice as Pathovars of *Xanthomonas oryzae* New Species Ex Ishiyama 1922. Revived Name. *Int. J. Syst. Bacteriol.* **1990**, *40*, 309–311.

Tanweer, F. A.; Rafii, M. Y.; Sijam, K.; Rahim, H. A.; Ahmed, F.; Latif, M. A. Current Advance Methods for the Identification of Blast Resistance Genes in Rice. *Crit. Rev. Biol.* **2015**, *338*, 321–334.

Toenniessen, G. H.; O'Toole, J. C.; DeVries. J. Advances in Plant Biotechnology and its Adoption in Developing Countries. *Curr. Opin. Plant Biol.* **2003**, *6*, 191–198.

Walker, D.; Roger Boerma, H.; All, J.; Parrott, W. Combining cry1Ac with QTL Alleles from PI 229358 to Improve Soybean Resistance to Lepidopteran Pests. *Mol. Breed.* **2002**, *9*, 43–51.

Wang, X.; Jiang, G.L.; Green, M.; Scott, R. A.; Hyten, D. L.; Cregan, P. B. Quantitative Trait Locus Analysis of Saturated Fatty Acids in a Population of Recombinant Inbred Lines of Soybean. *Mol. Breed.* **2012**, *12*, 9704–9710.

Wen, S.; Gao, B. Introgressing Blast Resistant Gene Pi-9(t) into Elite Rice Restorer Luhui17 by Marker Assisted Selection. *Rice Genom. Genet.* **2011**, *2*(4), 31–36.

Werner, K.; Friedt, W.; Ordon, F. Strategies for Pyramiding Resistance Genes against the Barley Yellow Bosaic Virus Complex (BaMMV, BaYMV, BaYMV-2). *Mol. Breed.* **2005**, *16*, 45–55.

Xu, Y. *Molecular Plant Breeding.* CAB International, 2010.

Yadav, S.; Anuradha, G.; Kumar, K. K.; Vemireddy, L. R.; Sudhakar, R.; Donempudi, K.; Venkata, D.; Jabeen, F.; Narasimhan, Y. K.; Marathi, B.; Siddiq, E. A. Identification of QTLs and Possible Candidate Genes Conferring Sheath Blight Resistance in Rice (*Oryza sativa* L.). *Spr. Plus* 2015, *4*, 175.

Ye, G.; Smith, K. F. Marker-assisted Gene Pyramiding for Inbred Line Development: Basic Principles and Practical Guidelines. *Int. J. Plant Breed.* **2008,** 2, 1–10.

Yoshimura, S.; Yoshimura, A.; Nelson, R. J.; Mew, T. W.; Iwata, N. Tagging Xa-1, the Bacterial Blight Resistance Gene in Rice, by using RAPD Markers. *Jpn. J. Breed.* **1995,** *45*, 81–85.

Zhan, X.; Zhou, H. P.; Chai, R. Y.; Zhuang, J. Y.; Cheng, S. H.; Cao, L. Y. Breeding of R8012, a Rice Restorer Line Resistant to Blast and Bacterial Blight Through Marker-assisted Selection. *Rice Sci.* **2012,** *19*(1), 29–35.

Zhang, J.; Li, X.; Jiang, G.; Xu, Y.; He, Y. Pyramiding of Xa7 and Xa21 for the Improvement of Disease Resistance to Bacterial Blight in Hybrid Rice. *Plant Breed.* **2006,** *125*(6), 600–605.

CHAPTER 12

BIOTIC STRESS MANAGEMENT IN RICE THROUGH RNA INTERFERENCE

SARITA[1,2*], GEETA[3], SUNIL KUMAR YADAV[1,2], and DEEPTI SRIVASTAVA[1]

[1]*Plant Molecular Biology and Genetic Engineering Division, CSIR-National Botanical Research Institute, Lucknow 226001, India*

[2]*Academy of Scientific and Innovative Research, New Delhi 110022, India*

[3]*Govt. P. G. College Noida, Gautam Buddhnagar, Sec-39, Noida, Uttar Pradesh, India*

Corresponding author. E-mail: ssarita07@gmail.com

CONTENTS

ABSTRACT

Rice is one of the most important staple crops, second largest produced cereal, feeding more than half of the world's population. Rice is subjected to important diseases which dramatically reduce crop yields up to 30%. Biotic factors that damages paddy crop are virus, bacteria, fungi, nematode, and insect pests. There are several methods of rice disease management but all existing methods have some limitations. RNA interference (RNAi)-induced gene silencing emerged as an effective tool to engineer resistant plants to different types of biotic stresses. This chapter emphasizes on different types of biotic factors that negatively affects the rice production. The available methods of biotic stress management are discussed along with their limitations. RNAi-mediated management of rice diseases has some advantages over existing disease management programs. This approach can also be used as a potential tool for disease management of other commercially important crops which are highly susceptible to various types of biotic stresses.

12.1   INTRODUCTION

Rice (*Oryza sativa* L.) belongs to family Poaceae with more than 40,000 cultivated varieties. Rice feeds more than half of the world's population and is grown on every continent apart from Antarctica. Rice is the second largest produced cereal and will be the main daily staple food of about 4.6 billion people by 2025. Increasing population and economic development have been posing a growing pressure for increase in food production (Zhang, 2007). According to FAO forecasts estimate, the food production must be double by 2050. So to increase rice production appears crucial for human life. Due to vast area of growth, rice is constantly exposed to interaction with various organisms from insects to bacteria and subjected to important diseases. A recent study estimated an annual loss between 120 and 200 million tons of grain due to insects, diseases, and weeds in rice fields in tropical Asia (Willocquet et al., 2004). Biotic factors that damages paddy crop are virus, bacteria, fungi, nematodes, and insect pests. The details of damage by different biotic factors are discussed in very brief below.

12.2 VIRUS

Viral diseases constitute a serious threat to increase rice production in Southeast Asian countries. The most commonly encountered symptoms are abnormal growth of the plant and changes of color. The teratological symptoms are various degrees of stunting, increased, or reduced number of tillers; twisting, crinkling, or rolling of leaves; formation of galls on leaves and culms; and necrotic lesions on culms. In general, the changes of color on leaves of infected rice plants vary either from green to yellow to white/orange. Plant viruses are transmitted by mechanical means, insects, mites, nematodes, fungi, dodders, pollen, seed, grafting, budding, vegetative propagation, or soil (Sasaya, 2015).

12.3 BACTERIA

Bacterial diseases of rice include bacterial blight, leaf streak, foot rot, grain rot, sheath brown rot, and pecky rice. Bacterial blight is one of the most important diseases of rice, found in tropical and temperate regions. It causes greatest economic impact in Asia, known to result in significant crop loss, ranging up to 60% of potential yield or several billions in direct economic terms. Symptoms includes water-soaked stripes on leaf blades, yellow or white stripes on leaf blades, leaves appearing grayish in color, plants wilting and rolling up, leaves turning yellow, stunted plants, plant death, youngest leaf on plant turning yellow.

Symptoms of bacterial leaf streak begin with fine translucent streak between veins. As the disease progresses, the streaks become yellowish gray, the lesion coalesce, then eventually turn brown to grayish white causing the leaves to die (Agrios, 2005).

12.4 FUNGI

Significant yield losses due to fungal attacks occur in most of the agricultural and horticultural species. More than 70% of all major crop diseases are caused by fungi (Agrios, 2005). Crops of all kinds often suffer heavy losses. Fungal diseases are rated either the most important or second most important factor contributing to yield losses in rice (Lee et al., 2007). The most important fungal diseases of rice are "blast," "heliminthosporiose," "stem rot," and "foot rot;" of these, "blast" disease is more destructive and widespread.

12.5 NEMATODES

Plant–parasitic nematodes feed and reproduce on living plants and are capable of active migration in the rhizosphere, on aerial plant parts, and/or inside the plant (Soriano et al., 2004). More than 150 species of plant–parasitic nematodes have been associated with rice. A parasitic relationship with rice, however, has been demonstrated for only a few of these nematode species. They are disseminated by wind, irrigation and flood water, tools, machinery, animals, and infected plant propagation materials. Estimated annual losses due to plant–parasitic nematodes on rice yield worldwide range from 10% to 25%. The lower estimate corresponds to an annual monetary value of US$16 billion (Bridge et al., 2005).

12.6 INSECT PESTS

Most of the world's rice production is from irrigated and rain-fed lowland rice fields where insect pests are constraints. The warm and humid environment in which rice is grown is conducive to the proliferation of insects. Heavily fertilized, high-tillering modern varieties, and the practice of multicropping rice throughout the year favor the buildup of pest populations. The rice plant is subject to attack by more than 100 species of insects; 20 of them can cause economic damage. Together they infest all parts of the plant at all growth stages, and a few transmit viral diseases. The most commonly cited crop loss data from rice are those of Cramer (1967), who estimated worldwide losses in rice production due to insect damage to be about 34.4%. The major insect pests that cause significant yield losses are leafhoppers and planthoppers. Brown plant hopper (BPH) causes the most serious damage to the rice crop globally among all rice pests. They cause direct damage as well as transmit viruses. Other insects are stem borers, and a group of defoliator species. As in many other agroecosystems, the rice agroecosystem has a few primary pests that may actually limit production under certain conditions. Chemical methods such as insecticide to control the insect have been negatively affecting the quality of edible crops like rice. Therefore, it is important to develop rice cultivars with adequate levels of resistance to the striped insect pests.

12.7 RESPONSE TO BIOTIC STRESS IN RICE

The pathogen attack leads to two types of stress response in rice. First response triggered by membrane-bound receptors and second by effector molecules which regulates transcription of various stress related genes by binding to intracellular receptors. The expression patterns of genes in response to a range of biotic stresses showed similarities in the transcriptome responses to biotic stress including infection with bacterium (Narsai et al., 2013), fungus (Marcel et al., 2010; Ribot et al., 2008), parasite (Swarbrick, 2008), and virus (GSE11025). In response to all four biotic stresses there is significant downregulation of nucleotide metabolism functions and an upregulation of genes encoding calcium signaling functions, miscellaneous transport functions, biotic and abiotic stress response functions, glutathione-S-transferases, and both WRKY and NAC transcription factors (Fig. 12.1).

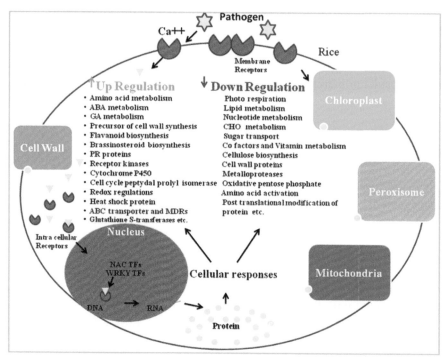

FIGURE 12.1 Cellular response to different biotic factors in rice (based on Narsai et al., 2013).

Components of calcium signaling are involved in general stress recognition process, including in response to biotic stresses (Fig. 12.1). In addition to calcium signaling, a conserved upregulation of specific NAC and WRKY TFs were also reported across all the biotic stresses (Fig. 12.1). The importance of transcriptional regulation in response to infection is most notably evidenced by the significant alteration in resistance or stress response, studied in rice plants by knockout/overexpression of specific NAC and WRKY transcription factors. These NAC and WRKY transcription factors clearly appear to not only be regulated at the transcript level in response to infection, but are also directly involved in regulating transcript abundance, given their role as transcription factors. Furthermore, apart from the genes encoding, glutathione-S-transferases, genes in all of the other functional categories (Fig. 12.1) contained significant enrichment of putative NAC-binding elements in the promoter regions of these genes, which suggests that these genes, including the WRKY TFs, may be regulated by NAC TFs (Fig. 12.1). Welner et al. (2012) experimentally confirmed the relationship between a NAC and WRKY TFs and suggested a conserved and related role for these TFs. Narsai et al. (2013) also found that five NAC transcription factors that were differentially expressed in response to Xoo infection have shown results in increased stress tolerance.

12.8   DIFFERENT APPROACHES OF BIOTIC STRESS MANAGEMENT IN RICE

There are several approaches for the biological, biochemical, and molecular management of biotic stress in rice.

12.8.1   USE OF ANTIBIOTICS

Antibiotics are used to cure the bacterial and fungal diseases in rice. Seeds usually treated with 0.1 g streptocycline and 0.1 g copper sulfate or 0.3 g agrimycin-100 and 0.1 g copper oxychloride in 1 L of water for 20 min for prevention of bacterial diseases. Foliar spray of 0.05 g streptocycline and 0.05 g copper sulfate are also used to cure bacterial diseases, but their inappropriate exposure resulted in development of multidrug resistant strains. The first antibiotic that was found to inhibit the growth of rice blast fungus on rice leaves was "cephalothecin," produced by a species of Cephalothecium (Yoshii, 1949). Following this, "antiblastin" (Suzuki,

1954), "antimycin-A" (Harada, 1955), "blastmycin" (Watanabe et al., 1957), and "blasticidin-A" (Fukunaga et al., 1968) were found and tested but due to their chemical instability and toxicity to fish. In 1955, a new systemic antibiotic, Blasticidin S was developed from *Streptomyces griseo chromogenes*. It was found to be effective mainly in postinfectional control, did not works as protectant and was highly toxic to plants and mammals (Ou, 1985). Shortly after the discovery of blasticidin S, a new antibiotic Kasugamycin, produced by *Streptomyces kasugaensis* was discovered. It gave excellent control of rice blast and had very less toxicity to mammals and rice plant (Okamoto, 1972). In around 1970, in the areas where the antibiotics have been used extensively and exclusively for blast control, population of *P. oryzae* began to show resistance to antibiotic compounds (Uesugi, 1978). However, after halting the use of antibiotics in the areas with resistant populations of *P. oryzae*, the population of resistant types reduced to nearly zero and later the use of antibiotics in some areas success-fully resumed (Uesugi, 1978).

12.8.2   USE OF ANTIVIRAL COMPOUNDS

A few compounds, such as 2-thiouracil and 8-azaquanine, used as foliar spray, as solution for watering plants, or for dipping plants have been reported to prevent virus infection or to suppress symptom development and to diminish the virus concentration, but treated plants revert to the original condition after the treatment has ceased. None of these compounds has yet had any commercial application for controlling virus disease in the field nor have they been tried with rice plants infected with virus. The major difficulty is that to be effective the compound must inhibit virus infection and multi-plication without damaging the plant. Virus multiplication is so intimately bound up with cell processes that any compound blocking virus synthesis is likely to have damaging effects on the plant (Waziril, 2015).

12.8.3   USE OF FUNGICIDES

Chemicals, mainly fungicides are the most frequently and widely used method of plant disease management worldwide. For rice blast, most aggressive and successful chemical control program in world has been shown by Japan. The copper fungicides were first effectively used in Japan but as they are highly phytotoxic, a more attractive alternative was sought.

Subsequently, copper fungicides were used in mixture with phenylemercuric acetate (PMA) which was more effective than copper alone in rice blast control and were less toxic to the rice plant. Later, mixture of PMA and slaked lime provides much more effective control of rice blast and was less toxic and cheap, hence used extensively. However, these fungicides are toxic to mammals and are cause severe environmental pollutants (Ou, 1985). Then, the organophosphorus fungicides were introduced to control blast in Japan, but in the late 1970s, the reports of resistance in *P. oryzae* to these compounds started emerging. The phosphorothiolate fungicides, including iprobenfos and edifenphos, were introduced for rice blast control in 1963. Iprobenfos and isoprothiolane have systemic action and are used mainly as granules for application on the surface of paddy water (soil application). Copper fungicides were found effective for rice blast control in India as well, but it was seen that high-yielding varieties were copper-shy; hence, the emphasis was shifted to another group of fungicides, namely, dithiocarbamate and edifenphose but they were having shorter residual activity. So in 1974–1975, the first-generation systemic fungicides benomyl, carbendazim, and others were evaluated and found effective. Following these, many systemic fungicides with different mode of action, like antimitotic compounds, melanin inhibitors, ergosterol biosynthesis inhibitor, and other organic compounds were discovered for rice blast control (Siddiq, 1996). In a chemical scheduling trial, Bavistin 1 g/L spray at tillering + Hinosan 1 g/L at heading and after flowering provided the best yield increase. Tricyclazole and Pyroquilon fungicides as seed dressers have been found effective to provide protection to seed up to 8 weeks after sowing. Some of the recently developed chemicals for blast control are Carpropamid (1999, melanin biosynthesis inhibitor), Fenoxanil (2002, melanin biosynthesis inhibitor), and Tiadinil (2004, plant activator). In the most recent field, evaluation of commercial fungicidal formulations Rabicide (tetrachlorophthalide), Nativo (tebuconazole + trifloxystobin), and Score (difenoconazole) which are found most effective (Katagiri and Uesugi, 1978; Usman et al., 2009) reported the frequency of emergence of resistant mutants of *P. oryzae* against different chemicals.

12.8.4 USE OF INSECTICIDES

Biotic stress-induced damages can be reduced using appropriate cultural practices and pesticides. Most commonly used pesticides in rice are as follows in the different diseases (Table 12.1).

TABLE 12.1 Some Commonly Used Pesticides in Rice.

Organophosphorous	
Phenthoate 500 g/L EC	RWM, RSB, RB
Chlorpyrifos 200 g/L EC	CW, RSC, RLF, RSB
Chlorpyrifos 400 g/L EC	CW, RSC, RLF, RSB
Dimethoate 400 g/L EC	TH
Quinalphos 250 g/L EC	TH, RSB, RB, RFC
Diazinon 5% GR	RWM, RGM, RSB
Diazinon 500 g/L EC	RWM, TH, RB, RFC
Phenthoate 55% DP	RB

Adapted and modified from Rola and Pingali (1993).

However, reported development of pesticide resistant strains of blast fungus together with the withdrawal of many products from the market because of the ever tougher regulatory frameworks on pest management chemicals pose a major economic risk and emphasize the real need for alternative strategies (Barker et al., 1985).

12.8.5 MOLECULAR BREEDING

Many morphological, anatomical, physiological, and biochemical factors have been reported to be associated with resistance, each controlled by different sets of genes. Molecular breeding used for development of resistant varieties by changes in the genetic background of promising lines through the introduction of a new resistance genes. Plant breeders' and geneticists' attempts to produce new varieties that better tolerate pathogen attack have not resulted in any satisfactory cultivars yet, due to prolonged time consuming process. Consequently, there exists a high demand for novel efficient methods for controlling plant diseases, as well as for producing plants of interest with increased resistance to biotic stress (Collard & Mackill, 2008; Hasan et al., 2015).

12.8.6 TRANSGENIC APPROACH

The ability to maintain or increase rice production in a cost-effective manner will rely on developing varieties that can be productive in response to a variety of abiotic or biotic stresses. The use of biotechnological approaches to develop crops resistant to a given stress imposition often takes a single gene approach where stress-induced genes encoding proteins, often transcription factors, are over expressed in transgenic plants, resulting in greater tolerance to a given stress imposition. While this is promising, there are several barriers translating laboratory based experiments to field situations, including the use of model plants compared to crop plants and the impositions of single stress compared to multiple stresses. Ban on field trials of transgenic crops in many developing countries like India also discourages the researchers and scientific community to follow this approach (Jayaraman & Jia, 2012).

12.8.7   RNA INTERFERENCE AND ITS SCOPE IN BIOTIC STRESS MANAGEMENT IN RICE

RNA interference (RNAi) is a potential gene regulatory approach in functional genomics that helps in crop improvement significantly by downregulation of gene expression in a highly precise manner without affecting the expression of other genes. RNAi approaches have also been used effectively to knockout the expressions of genes and to understand their biological functions (Anandalakshmi, 2013). Additional research in this field has been focused on a number of areas including siRNA, microRNAs, hairpin RNA, and promoter methylation (Fig. 12.2).

During RNAi, long double-stranded RNA (dsRNA) is diced into small fragments ~21 nucleotides long by an enzyme called "Dicer." These small fragments are called small-interfering RNAs (siRNA). siRNA binds to proteins from a special family referred as the Argonaute proteins and forms RNA-induced silencing complex (RISC). After binding to an Argonaute protein, one strand of the dsRNA is removed (guide strand), leaving the remaining strand (active strand) available to bind to messenger RNA target sequences by base pairing (A binds U, G binds C, and vice versa). After that, the RISC can either cleave the target messenger RNA, destroying it, or recruit accessory factors to regulate the target sequence in other ways. MicroRNAs represent a natural form of developmentally important siRNAs. Like siRNAs, microRNAs are made by Dicer, but microRNAs are derived from single-stranded RNAs that fold back on themselves to generate small regions of double-stranded RNA so-called stem-loops instead of the long double-stranded RNA that produces siRNAs. miRNAs can guide Argonaute proteins to repress messenger RNAs that match the microRNA incompletely, allowing one miRNA to regulate hundreds of genes. Small RNA can also lead to RNA-induced transcriptional silencing by promotion of methylation at promoter region (Hammond et al., 2000).

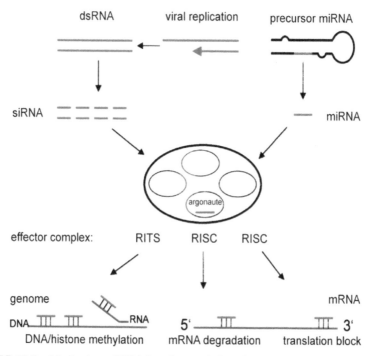

FIGURE 12.2 Mechanism of RNA interference (adapted and modified from Hammond et al., 2000).

Plant pathologists and plant biotechnologists have adopted different approaches to develop pathogen-resistant genotypes, but in last decade, RNAi-induced gene silencing emerged as an effective tool to engineer pathogen-resistant plants and battling some of the most notoriously challenging diseases caused by viruses, bacteria, fungi (Wani et al., 2010) and insect pests (Mao et al., 2007). RNAi has also been identified as a natural mechanism for regulation of gene expression in all higher organisms from plants to humans and promises greater accuracy and precision to plant improvement. Furthermore, the ability to turn off the expression of a single gene makes RNAi an appealing therapeutic approach to treat infectious diseases or genetic disorders that resulted from the inappropriate and undesirable activity of a gene, as in many cancers and neurodegenerative diseases. Argonaute proteins bind many naturally occurring small RNAs to defend against molecular parasites such as transposable elements, viral genetic elements to maintain chromosome structure and stability, and regulate developmental timing and differentiation. Keeping beside the significance of RNAi in the genome integrity maintenance as well as growth and development, RNAi-induced gene silencing is vital in plant's stress management. Modifying the genes by the interference of small RNAs is one of the ways through which plants react to the environmental stresses. Hence, investigating the role of small RNAs in regulating gene expression assists the researchers to explore the potentiality of RNAi in biotic stress management. This novel approach opens new avenues for crop improvement by developing disease resistant or biotic stress tolerant and high-yielding elite varieties (Navarro et al., 2006).

The application of tissue specific inducible gene silencing in combination with the use of appropriate promoters to silence several genes simultaneously will result in protection of crops against destructive pathogens. Manipulating new RNAi pathways, which generate small RNA molecules to amend gene expression in crops, can produce new quality traits and having better potentiality of protection against abiotic and biotic stresses. This RNAi approach has opened new avenues in the development of eco-friendly techniques for plant improvement as specific genes responsible for stress can be suppressed. Nutritional quality improvement, change in morphology and enhanced secondary metabolite production are some of the other advantages of RNAi technology (Navarro et al., 2006).

12.8.7.1 METHODS FOR RNAI INDUCTION IN PLANTS

In RNAi technology of disease management, one of the biggest challenges is the delivery of the active molecules that can induce the RNAi pathway in plants. A number of methods for delivery of dsRNA or siRNA into different cells and tissues used by researchers including transformation with dsRNA-forming vectors for selected gene(s) by *Agrobacterium*-mediated transformations (Chuang & Meyerowtiz, 2000; Waterhouse et al., 2001); delivery cognate dsRNA of uidA GUS (β-glucuronidase) and TaGLP2a:GFP (green fluorescent protein) reporter genes into single epidermal cells of maize, barley, and wheat by particle bombardment (Schweizer et al., 2000), introducing a tobacco rattle virus-based vector in tomato plants by infiltration (Liu et al., 2002); delivery of dsRNA into tobacco suspension cells by cationic oligopeptide polyarginine–siRNA complex; infecting plants with viral vectors that produce dsRNA and delivery of siRNA into cultured plant cells of rice, cotton, and slash pine for gene silencing by nanosense pulsed laser-induced stress wave (Tang et al., 2006). Wani and Sanghera (2010) suggested that among the approaches of delivery of dsRNA to plant cells, the most reliable and common are agro-infiltration, micro-bombardment and virus-induced gene silencing. Xue et al. (2015) proposed a simple and convenient protocol for dsRNA synthesis, their delivery by microinjection and RNAi verification in rice planthoppers.

12.8.7.2 RNAI AND BIOTIC STRESS MANAGEMENT IN RICE

Rice is not only one of the most important cereal crops in the world, but also is a model monocot plant for biological research (Devos & Gale, 2000). The rice genome is one-sixth the size of the maize genome and 40 times smaller than the wheat genome (IRGSP 2005). The small genome size has enriched a set of resources available for molecular biological studies (Jung et al., 2008). As rice is both a model and crop plant, it offers a system to directly determine the effect of stress on growth and yield in specific field varieties, even if under laboratory conditions. The ability to maintain or increase rice production in a cost-effective manner will rely on developing varieties that can be productive in response to a variety of abiotic or biotic stresses. Since the recent sequencing of the rice genome, the functional identification of rice genes has become increasingly important. Various tagged lines have been generated; however, the number of tagged genes available is not sufficient for extensive study of gene function. To help identify the functions of genes in rice, Miki and Shimamoto (2004) developed a gateway vector, pANDA, for RNAi of rice genes and can be used for *Agrobacterium* transformation of rice. In the construct, hairpin RNA derived from a given gene is transcribed from a strong maize ubiquitin promoter, and an intron is placed 5' upstream of inverted repeats to enhance RNA expression. They also analyzed rice genes using this vector and showed that mRNA expression in transgenic plants was suppressed more than 90%. They detected siRNA indicative of RNA silencing in each silenced plant. They also developed a similar vector, pANDAmini, for direct transfer into leaf cells or protoplasts which can be used for transient suppression of gene function in rice. These vectors should help identify the functions of rice genes which play major role in different biotic stress responses and complement existing methods for functional genomics of rice.

12.8.7.3   RNA INTERFERENCE IN RICE FOR VIRAL DISEASES MANAGEMENT

RNAi technology employs as a natural antiviral defense mechanism to cause resistance against viral diseases by virus-induced gene silencing. RNA-silencing hosts target protein translation and process the virus-mediated dsRNA, which results by pathogen replication into vsiRNAs (virus-mediated siRNAs). The vsiRNAs then target and suppress gene expression and protein translation in the virus genes. For stabilization of the defense system, virus encodes "viral suppressor of RNA silencing protein" that has been identified and isolated from various plant virus. Scorza et al. (2001) revealed for the first time RNAi role for virus resistance in woody perennial species and produced *Plum pox virus* (PPV) resistant plants containing the PPV coat protein gene. Plants can also control viral diseases by RNAi and reveal resistance when having proper antisense or hairpin RNAi constructs. Homology-dependent selective degradation of RNA, RNAi, is involved in several biological phenomena, including adaptive defense against viruses in plants. RNAi-mediated resistance has been demonstrated to be effective to viral diseases in plants. At least 20 viruses can infect rice, and about 16 of them may seriously affect rice yield. The distribution of each virus is generally restricted to only one of the continents in which rice is grown, for example, Rice tungro viruses and Rice stripe virus in Asia, Rice hoja blanca virus in South America, and Rice yellow mottle virus in Africa (Fargette et al., 2006). Two strategies, called protein-mediated and RNA-mediated resistance, have been the underlying principles to develop successful transgenic viral resistance in rice (Sanford & Johnston, 1985). Both strategies depend on the concept of "pathogen-derived resistance," wherein virus-encoded proteins or RNA are used to interfere with crucial steps in the infection cycle.

Rice tungro is a viral disease seriously affecting rice production in South and Southeast Asia. Tungro is caused by the simultaneous infection in rice of Rice tungro bacilliform virus (RTBV), a double-stranded DNA virus, and Rice tungro spherical virus, a single-stranded RNA virus. Tyagi et al. (2008) developed transgenic rice plants expressing DNA encoding ORF IV of RTBV, both in sense and in antisense orientation, resulting in the formation of dsRNA. Specific degradation of the transgene transcripts and the accumulation of small RNA were observed in transgenic plants. It was also found that different resistance responses against RTBV occurred in the transgenic plants expressing dsRNA. In RTBV-O-Ds1 line, there was an initial rapid buildup of RTBV levels, followed by a sharp reduction, resulting in approximately 50-fold lower viral titers. In RTBV-ODs2 line, RTBV DNA levels gradually rose from an initial low to almost 60% of that of the control at 40 days after inoculation. Line RTBV-ODs1 showed symptoms of tungro similar to the untransformed control lines, whereas line RTBV-ODs2 showed extremely mild symptoms.

More than 40 viral suppressors have been identified in plant viruses (Ruiz & Voinnet, 2007). Results from some of the well-studied virus suppressors indicated that they interfere with systemic signaling for silencing. *Pns10* encoded by *Rice dwarf virus* (RDV) suppressed local and systemic S-PTGS but not IR-PTGS suggesting that *Pns10* also targets an upstream step of dsRNA formation in the silencing pathway (Cao et al., 2005). The nonstructural protein Pns12 of RDV is one of the early proteins expressed in cultured insect cells, and it is one of 12 proteins that initiate the formation of the viroplasm, the putative site of viral replication. Shimizu et al. (2007) introduced Pns12- and Pns4-specific RNAi constructs into rice plants and found that the resultant transgenic plants accumulated specific siRNAs. Progenies of the transgenic rice plants with Pns12-specific RNAi constructs were strongly resistant to RDV infection (Table 12.2).

These studies clearly demonstrate that RNAi-mediated resistance is a practical and effective way to control viral infection in crop plants. Moreover, ribozyme, a class of small catalytic RNA molecules that possess sequence-specific RNA cleavage activity, was also demonstrated to be useful in engineering viral resistant transgenic plants. Transgenic rice plants expressing a ribozyme gene with 350 nucleotide hybridizing arms directed against RDV S5 mRNA displayed high resistance or delayed and attenuated disease symptoms (Han et al., 2000).

12.8.7.4   RNA INTERFERENCE IN RICE FOR BACTERIAL DISEASE MANAGEMENT

Gnanamanickam et al. (1999) obtained multiple resistance in the elite indica rice variety IR50 for two important diseases of rice – blast and blight. They developed blast resistant lines of IR50 introgressed with *Pi2* gene for blast resistance as the starting material for genetic transformation with bacterial blight resistance gene *Xa21*. This approach combines both the traditional breeding and the transgenic approach for deriving elite rice cultivars with multiple resistance to pathogens.

Rice WRKY45 is transcriptional activator acting through W-boxes. Epistasis analysis suggested that *WRKY45* acts in the SA signaling pathway independently of OsNPR1, a rice ortholog of *Arabidopsis* NPR1), which

TABLE 12.2 RNAi Technology against Pathogens/Insect-pests of Rice.

Pathogen/Insect-pests	Gene target	Objective	Reference
RTBV	ORF IV of *RTBV*	RNA-interference in rice against Rice *tungro bacilliform* virus results in its decreased accumulation in inoculated rice plants	Tyagi et al. (2008)
Rice dwarf virus	*Pns10*	Identification of an RNA silencing suppressors from a plant double-stranded RNA virus	Cao et al. (2005)
Rice dwarf virus	*Pns4, Pns12*	The repression of cell wall- and plastid-related genes and the induction of defense-related genes in rice plants infected with Rice dwarf virus	Shimizu et al. (2007)
Magnaporthe oryzae	*WRKY45*	Rice WRKY45 plays a crucial role in benzothiadiazole-inducible resistance to fungal blast and bacterial blight diseases disease resistance	Takatsuji et al. (2007)
Magnaporthe grisea and Xan-thomonas oryzae pv. oryzae	*OsSSI2*	Suppression of the rice fatty-acid desaturase gene OsSSI2 enhances resistance to blast and leaf blight diseases in rice	Jiang et al. (2009)
Xanthomonas oryzae pv. oryzae	*Os-11N3*	Rice *xa13* recessive resistance to bacterial blight is defeated by induction of the disease susceptibility gene Os-11N3	Antony et al. (2010)
Xanthomonas oryzae pv. oryzae	*OsPAD4*	Rice OsPAD4 functions differently from Arabidopsis At-PAD4 in host–pathogen interactions	Ke et al. (2014)
Magnaporthe grisea	*eGFP*	RNA silencing in the pathogenic fungus *Magnaporthe oryzae*	Kadotani et al. (2003)
Magnaporthe oryzae	*LRR-RLKs, OsBRR1*	A Putative leucine-rich repeat receptor kinase, OsBRR1, is involved in rice blast resistance	Peng et al. (2009)
Magnaporthe grisea	*OsACDR1*	Rice OsACDR1 (*Oryza sativa* accelerated cell death and re-sistance 1) is a potential positive regulator of fungal disease resistance	Kim et al. (2009)

TABLE 12.2 (Continued)

Pathogen/Insect-pests	Gene target	Objective	Reference
Cochliobolus miyabeanus (Bipolaris oryzae)	OsMPK5	Abscisic acid-induced resistance against the brown spot pathogen Cochliobolus miyabeanus in rice involves MAP kinase-mediated repression of ethylene signaling	Vleesschauwer et al. (2010)
Magnaporthe oryzae	OsNPR1	Involvement of OsNPR1/NH1 in rice basal resistance to blast fungus Magnaporthe oryzae	Feng et al. (2011)
Magnaporthe oryzae	OsRap2.6	OsRap2.6 transcription factor contributes to rice innate immunity through its interaction with RACK1	Wamaitha et al. (2012)
Magnaporthe oryzae	OsERF922	The rice ERF transcription factor OsERF922 negatively regulates resistance to Magnaporthe oryzae and salt tolerance	Liu et al. (2012)
Magnaporthe oryzae	OsACS2	The role of ethylene biosynthesis and signaling in resistance to fungal diseases in rice	Helliwell (2013)
Magnaporthe oryzae	WRKY42	The WRKY45-2 WRKY13 WRKY42 transcriptional regulatory cascade is required for rice resistance to fungal pathogen	Cheng et al. (2015)
BPH	Ra	Identification and characterization of Bph14, a gene conferring resistance to brown plant hopper in rice	Du et al. (2009)
Nilaparvata lugens	NlHT1 Nlcar and Nltry	Strategies for developing Green Super Rice.	Zha et al. (2011)
Nilaparvata lugens	NlEcR-a, NlEcR-b, NlEcR-c	The insect ecdysone receptor is a good potential target for RNAi-based pest control	Yu et al. (2014)
Scirpophaga incertulas	Cyp18A1, Ces	New insights into an RNAi approach for plant defense against piercing–sucking and stem-borer insect pests	Li et al. (2015)

RTBV, rice tungro bacilliform virus; RACK1, receptor for activated kinase-C 1. Adapted and modified from Tan and Yin (2004) and Younis et al. (2014).

distinguishes *WRKY45* from any known *Arabidopsis* WRKY transcription factors. *WRKY45* regulates two defense-related genes, encoding a gluta-thione-S-transferase and a cytochrome P450. Takatsuji et al. (2007) knock-down *WRKY45* by RNAi in rice and found that knockdown lines compromised BTH inducible resistance to blast disease, so they suggested that *WRKY45* is essential for benzothiadiazole (BTH)-inducible defense responses and play a crucial role in fungal blast and bacterial blight diseases. They also suggested that some environmental factor(s) triggered defense reactions by acting downstream of WRKY45 transcription. The high degree of multi-disease resistance accompanied by minor growth retardation due to over expression of WRKY45 makes this gene a promising tool for developing practically useful multidisease resistant rice by a transgenic approach. Fatty acids and their derivatives play important signaling roles in plant defense responses. It has been shown that suppressing a gene for stearoyl acyl carrier protein fatty-acid desaturase enhances the resistance of Arabidopsis (SSI2) and soybean to multiple pathogens. Jiang et al. (2009) performed functional analyses of a rice homolog of SSI2 (OsSSI2) in disease resistance of rice plants. A transposon insertion mutation (Osssi2-Tos17) and RNAi-mediated knockdown of OsSSI2 (OsSSI2-kd) reduced the oleic acid (18:1) level and increased that of stearic acid (18:0), indicating that OsSSI2 is responsible for fatty-acid desaturase activity. These plants displayed spontaneous lesion formation in leaf blades, retarded growth, slight increase in endogenous free salicylic acid (SA) levels, and SA/BTH-specific inducible genes, including WRKY45, a key regulator of SA/BTH-induced resistance, in rice. More-over, the OsSSI2-kd plants showed markedly enhanced resistance to the blast fungus *Magnaporthe grisea* and leaf-blight bacteria *Xanthomonas oryzae* pv. *oryzae* (Xoo). Their results suggest that OsSSI2 is involved in the negative regulation of defense responses in rice and induction of SA-respon-sive genes, including WRKY45, is likely responsible for enhanced disease resistance in OsSSI2-kd rice plants (Table 12.2).

The rice (*O. sativa*) gene xa13 is a recessive resistance allele of Os-8N3, a member of the NODULIN3 (N3) gene family, located on rice chromo-some 8. Os-8N3 is susceptibility (S) gene for *Xoo*, the causal agent of bacte-rial blight, and the recessive allele is defeated by strains of the pathogen producing any one of the type III effectors AvrXa7, PthXo2, or PthXo3, which are all members of the transcription activator-like (TAL) effector family. Both AvrXa7 and PthXo3 induce the expression of a second member of the N3 gene family, named Os-11N3. Antony et al. (2010) performed insertion mutagenesis and RNA-mediated silencing of Os-11N3 which resulted in plants with loss of susceptibility specifically to strains of *Xoo* dependent on AvrXa7 or PthXo3 for virulence. They showed that AvrXa7 drives expression of Os-11N3 and that AvrXa7 interacts and binds specifically to an effector binding element within the Os-11N3 promoter, lending support to the predictive models for TAL effector binding specificity. The result indicates that variations in the TAL effector repetitive domains are driven by selection to overcome both dominant and recessive forms of resistance to bacterial blight in rice. The finding that Os-8N3 and Os-11N3 encode closely related proteins also provides evidence that N3 proteins have a specific function in facilitating bacterial blight disease.

The extensively studied Arabidopsis phytoalexin deficient 4 (AtPAD4) gene plays an important role in Arabidopsis disease resistance; however, the function of its sequence ortholog in rice is unknown. Ke et al. (2014) studied function of ortholog of AtPAD4 in rice (OsPAD4) and found functionally different from AtPAD4 in host–pathogen interactions. OsPAD4 encodes a plasma membrane protein but that of AtPAD4 encodes a cytoplasmic and nuclear protein. Suppression of OsPAD4 by RNAi increased rice susceptibility to the biotrophic pathogen Xoo, which causes bacteria blight disease in local tissue. OsPAD4-RNAi plants also show compromised wound-induced systemic resistance to Xoo. The increased susceptibility to Xoo was associated with reduced accumulation of jasmonic acid (JA) and phytoalexin momilactone A (MOA). Exogenous application of JA complemented the phenotype of OsPAD4-RNAi plants in response to Xoo. Their study suggested that OsPAD4 functions differently than AtPAD4 in response to pathogen infection. OsPAD4 plays an important role in wound-induced systemic resistance and involved in defense signaling against Xoo is JA-dependent, whereas AtPAD4 mediates systemic acquired resistance and involved in defense signaling against biotrophic pathogens is SA-dependent. OsPAD4 is required for the accumulation of terpenoid-type phytoalexin MOA in rice-bacterium interactions, but AtPAD4-mediated resistance is associated with the accumulation of indole-type phytoalexin camalexin.

12.8.7.5 RNA INTERFERENCE IN FUNGAL DISEASE MANAGEMENT

Systematic silencing of *M. grisea*, a causal organism of rice blast was carried out by Kadotani et al. (2003) by using the enhanced green florescent protein gene as a model. To assess the ability of RNA species induce silencing in fungus, plasmid construct expressing sense, antisense, and hairpin RNA were introduced into an e*GFP*-expressing transformants. The fluorescence of *eGFP* in the transformants was silenced much more efficiently by hairpin

RNA of *eGFP* than by other RNA species. In the silenced transformants, the accumulation of *eGFP* mRNA was drastically reduced, but not methylation of coding or promoter regions was involved. The siRNA molecules of 19–23 nucleotides were observed in both sense and antisense strands of *eGFP* gene. RNAi strategy was also used to specifically knockdown 59 individual rice genes encoding putative LRR-RLKs, and a novel rice blast resistance-related gene (designated as OsBRR1) was identified by screening T_0 RNAi population using a weakly virulent isolate of *Magnaporthe oryzae*, Ken 54-04. Wild-type plants (*O. sativa* L. cv. "Nipponbare") showed intermediate resistance to Ken 54-04, while OsBRR1 suppression plants were susceptible to Ken 54-04 (Peng et al., 2009). Furthermore, OsBRR1-overexpressing plants exhibited enhanced resistance to some virulent isolates (97-27-2, 99-31-1, and zhong 10-8-14). OsBRR1 expression was low in leaves and undetectable in roots under normal growth conditions, while its transcript was significantly induced in leaves infected with the blast fungus (Ken 54-04) and was moderately affected by ABA, JA, and SA treatment. Over expression or RNAi suppression of OsBRR1 did not causes visible developmental changes in rice plants.

Rice-accelerated cell death and resistance 1 (OsACDR1) encodes a putative Raf-like mitogen-activated protein kinase kinase kinase (MAPKKK). Kim et al. (2009) reported upregulation of the OsACDR1 transcript by a range of environmental stimuli involved in eliciting defense-related pathways. They also apply biochemical, gain, and loss-of-function approaches to characterize OsACDR1 function in rice. The OsACDR1 protein showed autophosphorylation and possessed kinase activity. Rice plants overexpressing OsACDR1 exhibited spontaneous hypersensitive response (HR)-like lesions on leaves, upregulation of defense-related marker genes, and accumulation of phenolic compounds and secondary metabolites (phytoalexins). The transgenic plants also acquired enhanced resistance to a fungal pathogen (*M. grisea*) and showed inhibition of appressorial penetration on the leaf surface. In contrast, loss-of-function and RNA-silenced OsACDR1 rice mutant plants showed downregulation of defense-related marker genes expressions and susceptibility to *M. grisea*. Furthermore, transient expression of an OsACDR1:GFP fusion protein in rice protoplast and onion epidermal cells revealed its localization to the nucleus. They suggested that OsACDR1 plays an important role in the positive regulation of disease resistance in rice (Table 12.2).

Vleesschauwer et al. (2010) demonstrated that exogenously administered ABA enhances basal resistance of rice (*O. sativa*) against the brown spot-causing ascomycete *Cochliobolus miyabeanus* anamorph form *Bipolaris oryzae*. Microscopic analysis of early infection events in control and ABA treated plants revealed that ABA-inducible resistance (ABA-IR) is based on restriction of fungal progression in the mesophyll. They also showed that ABA-IR does not rely on boosted expression of SA, JA, or callose dependent resistance mechanisms but, requires a functional Ga-protein. In addition, several lines of evidence are suggested that ABA drives its positive effect on brown spot resistance through antagonistic cross talk with the ethylene (ET) response pathway. Exogenous ethephon application enhances susceptibility, whereas genetic disruption of ET signaling renders plants less vulnerable to *C. miyabeanus* attack, thereby inducing a level of resistance similar to that observed on ABA-treated wild-type plants. Moreover, ABA treatment alleviates *C. miyabeanus*-induced activation of the ET reporter gene EBP89, while repression of pathogen-triggered EBP89 transcription via RNAi-mediated knockdown of OsMPK5, an ABA-primed mitogen-activated protein kinase gene, compromises ABA-IR. Collectively, these data favor a model whereby exogenous ABA enhances resistance against *C. miyabeanus* at least in part by suppressing pathogen-induced ET action in an OsMPK5-dependent manner.

M. oryzae, another species of blast fungus, is a major threat to worldwide rice production. Plant basal resistance is activated by virulent pathogens in susceptible host plants. OsNPR1/NH1, a rice homolog of NPR1 that is the key regulator of systemic acquired resistance in *Arabidopsis thaliana*, was shown to be involved in the resistance of rice to bacterial blight disease caused by *Xoo* and BTH-induced blast resistance. However, the role of OsNPR1/NH1 in rice basal resistance to blast fungus *M. oryzae* remains uncertain. Feng et al. (2011) isolated the OsNPR1 gene and identified from rice cultivar Gui99. They developed transgenic Gui99 rice plants harboring OsNPR1-RNAi, and the OsNPR1-RNAi plants were significantly more susceptible to *M. oryzae* infection. Northern hybridization analysis showed that the expression of pathogenesis-related (PR) genes, such as PR-1a, PBZ1, CHI, GLU, and PAL, was significantly suppressed in the OsNPR1-RNAi plants. Consistently, overexpression of OsNPR1 in rice cultivars Gui99 and TP309 conferred significantly enhanced resistance to *M. oryzae* and increased expression of the above-mentioned PR genes. These results revealed that OsNPR1 is involved in rice basal resistance to the blast pathogen *M. oryzae* and suggested the role of OsNPR1 in rice disease resistance.

Wamaitha et al. (2012) reported that transcription factor OsRap2.6 interacts with another transcription factor RACK1A. They showed 94% similarity between the OsRap2.6 AP2 domain and Arabidopsis Rap2.6

(AtRap2.6). They did bimolecular fluorescence complementation (BiFC) assays in rice protoplasts using tagged OsRap2.6 and RACK1A with the C-terminal and N-terminal fragments of Venus (Vc/Vn) indicated that OsRap2.6 and RACK1A interacted and localized in the nucleus and the cytoplasm. Moreover, OsRap2.6 and OsMAPK3/6 interacted in the nucleus and the cytoplasm. Expression of defense genes PAL1 and PBZ1 as well as OsRap2.6 was induced after chitin treatment. Disease resistance analysis using OsRap2.6 RNAi and overexpressing (Ox) plants infected with the rice blast fungus indicated that OsRap2.6 RNAi plants were highly susceptible, whereas OsRap2.6 Ox plants had an increased resistance to the compatible blast fungus and suggested the role of OsRap2.6 in rice innate immunity through its interaction with RACK1A in compatible manner.

Rice OsERF922, encoding an APETELA2/ET response factor (AP2/ERF) type transcription factor, is rapidly and strongly induced by abscisic acid (ABA) and salt treatments, as well as by both virulent and avirulent pathogenic strains of *M. oryzae*, the causal agent of rice blast disease. OsERF922 is localized to the nucleus, binds specifically to the GCC box sequence, and acts as a transcriptional activator in plant cells (Liu et al., 2012) knockdown OsERF922 by using RNAi and found enhanced resistance against *M. oryzae*. The resistance of the RNAi plants was associated with increased expression of PR, PAL, and the other genes encoding phytoalexin biosynthetic enzymes. They also developed OsERF922-overexpressing plants and found reduced expression of these defense-related genes and enhanced susceptibility to *M. oryzae*. In addition, the OsERF922-overexpressing lines exhibited decreased tolerance to salt stress with an increased $Na+/K+$ ratio in the shoots. The ABA levels were found increased in the over expressing lines and decreased in the RNAi plants. Expression of the ABA biosynthesis-related genes, 9-cis-epoxycarotenoid dioxygenase (NCED) 3 and 4, was upregulated in the OsERF922-overexpressing plants, and NCED4 was down regulated in the RNAi lines. They suggested that OsERF922 is involved into the cross-talk between biotic and abiotic stress-signaling networks through modulation of the ABA levels.

Helliwell (2013) used the RNAi method, to generate transgenic rice lines with suppressed expression of the single *M. oryzae* inducible-OsACS2 gene. The OsACS2-RI lines exhibited reduced basal levels of ET, and reduced expression of ET-signaling genes. A series of elicitation and inoculation studies were performed with the use of these ET-deficient OsACS2-RI lines, in conjunction with previously developed ET-signaling mutants OsEIN2a+b-RI and OsEIL1-RI lines. He found that both ET-deficient and ET-insensitive lines showed an attenuation of mechanisms involved in basal resistance, including decreased expression of a chitin-binding receptor, lower and/or delayed expression of genes involved in basal resistance and phytoalexin biosynthesis, a reduction in early HR-like cell death, reduced callose deposition, and increased susceptibility to multiple isolates of *M. oryzae*. These results suggest that both ET production and signaling are involved in basal resistance. A second study involved generation of transgenic lines with inducible overexpression of OsACS2 (OsACS2-OX). In contrast to the ET-deficient and insensitive lines, the OsACS2-OX lines showed increased resistance to both a moderately and a highly virulent isolate of *M. oryzae*, as well as *R. solani*. These results suggest that increasing endogenous levels of ET may enhance host resistance to a broad spectrum of pathogens and pathogen races. A third study performed using a negative regulator of ET production, OsEOL1 (ET-overproducer like 1) revealed more information about regulation in ET production involved in ET-mediated resistance. OsEOL1 is a BTB (broad-spectrum/tamtrac/bric a brac) domain-containing protein, and a putative E3 ligase component. Expression of OsEOL1 increased at 48 h post-inoculation, occurring simultaneously to a decrease in ET production in rice. Transgenic rice lines with stress-inducible over expression of *OsEOL1* showed increased resistance to spray-inoculated, but not punch-inoculated isolates of *M. oryzae*. This suggests that OsEOL1 is a positive regulator of resistance to *M. oryzae*, depending on the time of induction. The results of these experiments suggest that ET production and signaling are necessary components of basal resistance; however, ET-mediated resistance is dependent.

At least 11 rice WRKY transcription factors have been reported to regulate rice response to *M. oryzae* either positively or negatively. However, the relationships of these WRKYs in the rice defense signaling pathway against *M. oryzae* are unknown. Many studies have revealed that rice WRKY13 (as a transcriptional repressor) and WRKY45-2 enhance resistance to *M. oryzae*. Cheng et al. (2015) suggested that rice WRKY42, functioning as a transcriptional repressor, suppresses resistance to *M. oryzae*. WRKY42–RNAi and WRKY42-overexpressing (oe) plants showed increased resistance and susceptibility to *M. oryzae*, accompanied by increased or reduced JA content, respectively, compared to wild-type plants. JA pretreatment enhanced the resistance of WRKY42-oe plants to *M. oryzae*. WRKY13 directly suppressed WRKY42. WRKY45-2, functioning as a transcriptional activator and directly activated WRKY13. In addition, WRKY13 directly suppressed WRKY45-2 by feedback regulation. The WRKY13-RNAi WRKY45-2-oe and WRKY13-oe WRKY42-oe double transgenic lines showed increased susceptibility to *M. oryzae* compared with WRKY45-2-oe and WRKY13-oe plants, respectively. These results suggest that the three WRKYs form a sequential transcriptional regulatory cascade. WRKY42 may negatively regulate rice response to *M. oryzae* by suppressing JA signaling-related genes, and WRKY45-2 transcriptionally activates WRKY13, whose encoding protein in turn transcriptionally suppresses WRKY42 to regulate rice resistance to *M. oryzae* (Cheng et al., 2015).

Bipolaris oryzae is the causal agent of rice brown spot disease and is responsible for significant economic losses. *B. oryzae* is classified in the subdivision Deuteromycotina (imperfect fungi), class Deuteromycetes, order Moniliales, and family Dematiaceae and is the causal agent of brown spot disease of rice. It affects the quality and the number of grains per panicle and reduces the kernel weight (Mew & Gonzales, 2002). ABA, a plant hormone is involved in many plant processes and regulates gene expression during adaptive responses to various environmental signals. Although role of ABA in abiotic stress is well established but nowadays emerging evidence indicates that ABA is also significantly involved in the regulation and integration of pathogen defense response. Vleesschauwer et al. (2010) demonstrated that exogenously administered ABA enhances basal resistance of rice (*O. sativa*) against the brown spot-causing ascomycete *C. miyabeanus* anamorph form *B. oryzae*. Microscopic analysis of early infection events in control and ABA treated plants revealed that ABA-inducible resistance (ABA-IR) is based on restriction of fungal progression in the mesophyll. They also showed that ABA-IR does not rely on boosted expression of SA-, JA-, or callose dependent resistance mechanisms but requires a functional Ga-protein. In addition, several lines of evidence are suggested that ABA drives its positive effect on brown spot resistance through antagonistic cross talk with the ET-response pathway. Exogenous ethephon application enhances susceptibility, whereas genetic disruption of ET signaling renders plants less vulnerable to *C. miyabeanus* attack, thereby inducing a level of resistance similar to that observed on ABA-treated wild-type plants. Moreover, ABA treatment alleviates *C. miyabeanus*-induced activation of the ET reporter gene EBP89, while derepression of pathogen-triggered EBP89 transcription via RNAi-mediated knockdown of OsMPK5, an ABA-primed mitogen-activated protein kinase gene, compromises ABA-IR. Collectively, these data favor a model whereby exogenous ABA enhances resistance against *C. miyabeanus* at least in part by suppressing pathogen-induced ET action in an OsMPK5-dependent manner (Vleesschauwer et al., 2010).

12.8.7.6   RNA INTERFERENCE IN RICE FOR INSECT PEST MANAGEMENT

The observation that RNAi could also be effective in reducing gene expression, measured by mRNA level, when fed to insects (Turner et al., 2006) has led to two recent articles in which transgenic plants producing dsRNAs are shown to exhibit partial resistance to insect pests. Transgenic maize producing dsRNA directed against V-type ATPase of corn rootworm showed suppression of mRNA in the insect and reduction in feeding damage compared to controls (Baum et al., 2007). Similarly, transgenic tobacco and Arabidopsis plant material expressing dsRNA directed against a detoxification enzyme (Cyt P450 gene CYP6AE14) for gossypol in cotton bollworm caused the insect to become more sensitive to gossypol in the diet (Mao et al., 2007). This approach holds great promise for future development of insect resistant rice varieties.

Mao et al. (2007) reported a new strategy about plant-mediated herbivorous insects RNAi, which describes the suppression of a critical insect-gene through insect feeding on plant engineered to develop a specific dsRNA that can prompt dissection of gene functions in these insects. Du et al. (2009) cloned *Bph14*, a gene conferring resistance to BPH at seedling and maturity stages of the rice plant, using a map-base cloning approach. They suggested that *Bph14* encodes a coiled-coil, nucleotide-binding, and leucine-rich repeat (CC-NB-LRR) protein. Sequence comparison indicates that *Bph14* carries a unique LRR domain that might function in recognition of the BPH insect invasion and activating the defense response. *Bph14* is predominantly expressed in vascular bundles, the site of BPH feeding. Expression of *Bph14* activates the SA signaling pathway and induces callose deposition in phloem cells and trypsin inhibitor production after plant hopper infestation, thus reducing the feeding, growth rate, and longevity of the BPH insects. They also used RNAi to suppress the expression of *Ra* in the RI35 rice plants. The RNAi-transgenic lines were susceptible and were killed by the BPHs in the tests. So they concluded that *Ra* confers the resistance phenotype and is the *Bph14* gene.

To analyze the potential of exploiting RNAi-mediated effects in this insect, Zha et al. (2011) identified genes (Nlsid-1 and Nlaub) encoding proteins that might be involved in the RNAi pathway in *N. lugens*. Both genes are expressed ubiquitously in nymphs and adult insects. Three genes (the hexose transporter gene NlHT1, the carboxypeptidase gene Nlcar and the trypsin-like serine protease gene Nltry) that are highly expressed in the *N. lugens* midgut were isolated and used to develop dsRNA constructs for transforming rice. RNA-blot analysis of transgenic lines showed that the dsRNAs were transcribed and some of them were processed to siRNAs. When nymphs were fed on rice plants expressing dsRNA, levels of transcripts of the targeted genes in the mid gut were reduced (as high as 73% for the Nltry gene); however, no lethal phenotype was observed after feeding dsRNAs.

Yu et al. (2014) reported that the insect ecdysone receptor (EcR) is a good potential target for RNAi-based pest control in the brown planthopper *Nilaparvata lugens*, a serious insect pest of rice plants by using a 360-bp fragment (*NlEcR-c*) that is common between *NlEcR-A* and *NlEcR-B* for feeding RNAi experiments. They found significantly decreased the relative mRNA expression levels of *NlEcR* compared with those in the ds*GFP* control. Feeding RNAi also resulted in a significant reduction in the number of offspring per pair of *N. lugens*. They developed transgenic rice line expressing *NlEcR* dsRNA by *Agrobacterium*-mediated transformation. The results of qRT-PCR showed that the total copy number of the target gene in all transgenic rice lines was 2. Northern-blot analysis showed that the small RNA of the hairpin ds*NlEcR-c* was successfully expressed in the transgenic rice lines. After feeding newly hatched nymphs of *N. lugens* on the transgenic rice lines, effectiveness of RNAi was observed. The *NlEcR* expression levels in all lines examined were decreased significantly compared with the control. In all lines, the survival rate of the nymphs was nearly 90%, and the average number of offspring per pair in the treated groups was significantly less than that observed in the control, with a decrease of 44.18–66.27%. Their findings support the RNAi-based pest control strategy and its importance for the management of rice insect pests.

Haichao (2015) immersed rice roots for twenty-four hours in fluorescent-labeled dsEYFP (enhanced yellow fluorescent protein) and observed fluorescence in the rice sheath, stem, and in planthoppers feeding on the rice. The expression levels of Ago and Dicer in rice and planthoppers were induced by dsEYFP. When rice roots were soaked in dsActin, their growth was also significantly suppressed. Rice/Maize planthoppers or Asian corn borers fed on rice crop irrigated with a solution containing the dsRNA of an insect target gene increased the insect's mortality rate significantly. They demonstrated that dsRNAs can be absorbed by crop roots, trigger plant and insect RNAi and enhance piercing–sucking and stem-borer insect mortality rates and also confirmed that dsRNA was stable under outdoor conditions. These results indicate that the root dsRNA soaking can be used as a bioinsecticide strategy during crop irrigation (Table 12.2).

12.9   LIMITATIONS OF RNAI TECHNOLOGY

RNAi has great potential for use in insect pest control and numerous studies indicate that target gene silencing by RNAi could lead to insect death. This phenomenon has been considered as an effective strategy for insect pest control, and it is termed RNAi-mediated crop protection. Insect dsRNA expression in transgenic crops can increase plant resistance to biotic stress; however, creating transgenic crops to defend against every insect pest is impractical. There are many limitations using RNAi-based technology for pest control, with the effectiveness of target gene selection and reliable dsRNA delivery being two of the major challenges. These significant challenges must be overcome before RNAi-based pest control can become a reality. One challenge is the proper selection of a good target gene for RNAi. With respect to target gene selection, at present, the use of homologous genes and genome-scale high-throughput screening are the main strategies adopted by researchers. Once the target gene is identified, dsRNA can be delivered by microinjection or by feeding as a dietary component. However, microinjection, which is the most common method, can only be used in laboratory experiments. Expression of dsRNAs directed against insect genes in transgenic plants and spraying dsRNA reagents have been shown to induce RNAi effects on target insects. Hence, RNAi-mediated crop protection has been considered as a potential new-generation technology for pest control, or as a complementary method of existing pest control strategies; however, further development to improve the efficacy of protection and range of species affected is necessary.

12.10   CONCLUSION AND FUTURE PERSPECTIVES

Biotic stress disproportionately affects farm productivity around the world with immense annual yield losses. For these reasons, control of microbial pathogens continues to be an agronomic and scientific challenge, and innovative and ground-breaking strategies are required to meet the requirement of a growing population. It has been suggested that biotechnology can contribute to the agronomic improvement of rice, when used in combination with traditional or in conventional breeding methods, which will make it possible to achieve the required increase in crop production and protection against various pathogens of rice. Recent work suggested that novel RNAi-based plant protection strategies may provide new opportunities for improving the world's food supplies and thus can have a huge impact on world's economy.

A great number of basic research studies have enabled the rapid increase of knowledge in dsRNA-mediated silencing of target genes. Whereas the first investigations focused on the use of model organisms, it is now becoming possible to apply this knowledge toward modifying specific traits in agriculturally relevant crop plants. In addition to metabolic engineering and host–plant-induced gene silencing mediated enhancement of disease resistance, RNAi strategies may be used to improve food safety by controlling the growth of phytopathogenic, mycotoxin-producing fungi. Before these exciting possibilities can be realized, some significant challenges remain. Are all detoxifying genes equally amenable to this approach? Will this technology extend to other pathogen groups? Will ingestible RNAi be circumvented by sequence polymorphisms in other organisms? And will the successes seen in the lab translate into effective pathogen control in the field? Despite these questions, there is no doubt that researchers and farmers have grounds to look forward to a new era in biotic stress management in rice.

KEYWORDS

- rice
- biotic stress
- down regulation
- transcription factor
- RNAi

REFERENCES

Agrios; George, N. *Plant Pathology*, 5th ed., Elsevier Academic Press: Burlington, MA, 2005.

Anandalakshmi, R. Application of RNAi for Engineering Disease Tolerance in Crops with Special Reference to Horticultural Crops. *J. Mycol. Plant Pathol.* **2013,** *43,* 111.

Antony, G.; Zhou, J.; Huang, S.; Li, T.; Liu, B.; White, F.; Yang, B. Rice *xa13* Recessive Resistance to Bacterial Blight is Defeated by Induction of the Disease Susceptibility Gene Os-11N3. *Plant Cell* **2010,** *22,* 3864–3876.

Barker, R; Herdt, R. W.; Rose, B. The Rice Economy of Asia. Resources for the Future with IRRI: Washington, D. C., and Los Baños, Laguna, Philippines, 1985.

Baum, J. A.; Bogaert, T.; Clinton, W; Heck, G. R.; Feldmann, P.; Ilagan, O.; Johnson, S.; Plaetinck, G.; Munyikwa, T.; Pleau, M. Control of Coleopteran Insect Pests through RNA Interference. *Nat. Biotechnol.* **2007,** *25,* 1322–1326.

Bridge, J.; Plowright, R. A. Peng, D. Nematode Parasites of Rice. In: *Plant–Parasitic Nematodes in Subtropical and Tropical Agriculture*; Luc, M., Sikora, R. A., Bridge, J., Eds.; CAB International: Wallingford, **2005**; pp 87–130.

Cao, X.; Zhaou, P.; Zhang, X.; Zhu, S.; Zhong, X.; Xiao, Q; Ding, B; Li, Y. Identification of an RNA Silencing Suppressors from a Plant Double Stranded RNA Virus. *J. Virol.* **2005**, *79*, 13018–13027.

Cheng, H.; Liu; H.; Deng, Y.; Xiao, J.; Li, X.; Wang, S. The WRKY45-2 WRKY13 WRKY42 Transcriptional Regulatory Cascade Is Required for Rice Resistance to Fungal Pathogen. *Plant Physiol.* **2015**, *167*, 1087–1099.

Chuang, C. F.; Meyerowitz, E. M. Specific and Heritable Genetic Interference by Double-stranded RNA in *Arabidopsis thaliana*. *Proc. Nat. Acad. Sci. U. S. A.* **2000**, *97*, 4985–4990.

Collard, B. C.; Mackill, D. J. Marker-assisted Selection: An Approach for Precision Plant Breeding in the Twenty-first Century. *Philos. Trans. Res. Soc. B* **2008**, *363*, 557–572.

Cramer, P. The Stroop Effect in preschool Aged Children: A Preliminary Study. *J. Genet. Psychol.* **1967**, *111*, 9–12.

Devos, K. M.; Gale, M. D. Genome Relationships: The Grass Model in Current Research. *Plant Cell* **2000**, *12*, 637–646.

Du, B.; Zhang, W.; Liu, B.; Hu, J.; Wei, Z.; Shi, Z.; He, R.; Zhu, L.; Chen, R.; Han, B.; He, G. Identification and Characterization of *Bph14*, a Gene Conferring Resistance to Brown Planthopper in Rice. *Proc. Natl. Acad. Sci. U. S. A.* **2009**, *106*, 22163–22168.

Fargette, D.; Ghesquiere, A.; Albar, L.; Thresh, J. M. Virus Resistance in Rice. In: *Natural Resistance Mechanism of Plants to Viruses*; Loebenstein, G., Carr, J. P., Eds.; Springer: Dordrecht, The Netherlands, 2006; pp 431–446.

Feng, J. X.; Cao, L.; Li, J.; Duan, C. J.; Luo, X. M.; Le, N.; Wei, H.; Liang, S.; Chu, C.; Pan; Tang, J. L. Involvement of OsNPR1/NH1 in Rice Basal Resistance to Blast Fungus *Magnaporthe oryzae*. *Eur. J. Plant Pathol.* **2011**, *11*, 9801–9807.

Fukunaga, K.; Misato, T.; Ishii, I.; Asakawa, M.; Katagiri, M. Research and Development of Antibiotics for Rice Blast Control. *Bull. Nat. Inst. Agric. Sci. Tokyo* **1968**, *22*, 1–94.

Gnanamanickam, S. S.; Priyadarisini, V. B.; Narayanan, N. N.; Vasudevan, P.; Kavitha, S. An Overview of Bacterial Blight Disease of Rice and Strategies for Its Management. *Curr. Sci.* **1999**, *77*, 1435–1443.

Haichao, L. New insights into an RNAi Approach for Plant Defence against Piercing–Sucking and Stem-borer Insect Pests. *Plant Cell Environ.* **2015**, *38*, 2277–2285.

Hammond, S.; Bernstein, E.; Beach, D.; Hannon, G. An RNA-directed Nuclease Mediates Post-transcriptional Gene Silencing in Drosophila Cells. *Nature* **2000**, *404*, 293–296.

Han, S.; Wu, Z.; Yang, H.; Wang, R.; Yie, Y.; Xie, L.; Tien, P. Ribozyme-mediated Resistance to Rice Dwarf Virus and the Transgene Silencing in the Progeny of Transgenic Rice Plants. *Transgen. Res.* **2000**, *9*, 195–203.

Harada, Y. Studies on a New Antibiotic for Rice Blast Control. Lecture Given at the Annual Meeting of the Agricultural Chemical Society of Japan, 1955.

Hasan, M.; Rafi, M. Y.; Ismail, M. R.; Mahmood, M.; Rahim, H. A.; Alam, M. Marker Assisted Backcrossing: A Useful Method for Rice Improvement. *Biotechnol. Biotechnol. Equip.* **2015**, *29*, 237–254.

Helliwell, E. The Role of Ethylene Biosynthesis and Signaling in Resistance to Fungal Diseases in Rice. Dissertation the Pennsylvania State University, **2013**, *139*, 709–975.

Jayaraman, K.; Jia, H. GM Phobia Spreads in South Asia. *Nat. Biotechnol.* **2012**, *30*, 1017–1019.

Jiang, C. J.; Shimono, M.; Maeda, S.; Inoue, H.; Mori, M.; Hasegawa, M.; Sugano, S.; Takat-suji, H. Suppression of the Rice Fatty-acid Desaturase Gene OsSSI2 Enhances Resistance to Blast and Leaf Blight Diseases in Rice. *Mol. Plant Microbe Interact.* **2009**, *7*, 820–829.

Jung, K. H.; An, G.; Ronald, P. C. Towards a Better Bowl of Rice: Assigning Function to Tens of Thousands of Rice Genes. *Nat. Rev. Genet.* **2008**, *9*, 91–101.

Kadotani, N.; Nakayashiki, H.; Tosa, H.; Mayama, S. RNA Silencing in the Pathogenic Fungus *Magnaporthe oryzae*. *Mol. Plant Microbe Interact.* **2003**, *16*, 769–776.

Katagiri, M.; Uesugi, Y. In Vitro Selection of Mutants of *Pyricularia oryzae* Resistant to Fungicides [rice]. *Ann. Phytopathol. Soc. Jpn.* **1978**, *44*, 448.

Ke, Y.; Liu, H.; Li, X.; Xiao, J.; Wang, S. Rice OsPAD4 Functions Differently from Arabidopsis AtPAD4 in Host–Pathogen Interactions. *Plant J.* **2014**, *78*, 619–631.

Kim, J. A.; Cho, K.; Singh, R.; Jung, Y. H.; Jeong, S. H.; Kim, S. H.; Lee, J. E.; Cho, Y. S.; Agrawal, G. K; Rakwal, R.; Tamogami, S.; Kersten, B.; Jeon, J. S.; An. G.; Jwa, N. S. Rice OsACDR1 (*Oryza sativa* Accelerated Cell Death and Resistance 1) is a Potential Positive Regulator of Fungal Disease Resistance. *Mol. Cells* **2009**, *28*, 431–439.

Lee, J. W.; Gwark, K. S.; Park, J. Y.; Park, M. L.; Choi, D. H.; Kwon, M.; Choi, I. G. Biological Pretreatment of Softwood *Pinus densiflons* by Three White Rot Fungi. *J. Microbiol.* **2007**, *45*, 485–491.

Li, R.; Zhang, J; Li, J.; Zhou, G.; Wang, Q.; Bian, W.; Erb, M; Lou, Y. Prioritizing Plant Defence Over Growth through WRKY Regulation Facilitates Infestation by Non-target Herbivores. *Elife* **2015**, *4*, e04805.

Liu, D.; Chen, X.; Liu, J.; Ye, J.; Guo, Z. The Rice ERF Transcription Factor OsERF922 Negatively Regulates Resistance to *Magnaporthe oryzae* and Salt Tolerance. *J. Exp. Bot.* **2012**, *63*, 3899–3912.

Liu, Y.; Schiff, M.; Marathe, R.; Dinesh, K. S. P. Tobacco *Rar1*, *EDS1* and *NPR1/NIM1* Like Genes are Required for *N*-mediated Resistance to Tobacco Mosaic Virus. *Plant J.* **2002**, *30*, 415–429.

Mao, Y. B.; Cai, W. J.; Wang, J. W.; Hong, G. J.; Tao, X. Y.; Wang, L. J.; Huang, Y.P.; Chen, X. Y. Silencing a Cotton Bollworm *P450* Monoxygenase Gene by Plant-mediated RNAi Impairs Larval Tolerance of Gossypol. *Nat. Biotechnol.* **2007**, *25*, 1307–1313.

Marcel, S.; Sawers, R.; Oakeley, E.; Angliker, H.; Paszkowski, U. Tissue-adapted Invasion Strategies of the Rice Blast Fungus *Magnaporthe oryzae*. *Plant Cell* **2010**, *22*, 3177–3187.

Mew, T. W.; Gonzales, P. A Handbook of Rice Seed-born Fungi. Science Publishers, Inc.: Los Banos, Philippines: International Rice Research Institute (IRRI) and Enfield, NH, **2003**; pp 83.

Miki, D.; Shimamoto, K. Simple RNAi Vectors for Stable and Transient Suppression of Gene Function in Rice. *Plant Cell Physiol.* **2004**, *45*, 490–495.

Narsai, R.; Wang, C.; Chen, J.; Wu, J.; Shou, H.; Whelan, J. Antagonistic, Overlapping and Distinct Responses to Biotic Stress in Rice (*Oryza sativa*) and Interactions with Abiotic Stress. *BMC Genom.* **2013**, *14*, 93.

Navarro, L.; Dunoyer, P.; Jay, F.; Arnold, B.; Dharmasiri, N.; Estelle, M. A Plant miRNA Contributes to Antibacterial Resistance by Repressing Auxin Signaling. *Science* **2006**, *312*, 436–439.

Okamoto, M. On the Characteristics of Kasumin, Antibiotic Fungicide. *Jpn. Pesticide Inform.* **1972**, *10*, 66–69.

Ou, S. H. *Rice Diseases*. International Rice Research Institute: Manila, Philippines, 1985.

Peng, H.; Zhang, Q, Li.; Y.; Lei, C.; Zhai, Y.; Sun, X.; Sun, D.; Sun, Y.; Lu, T. A Putative Leucine-rich Repeat Receptor Kinase, OsBRR1, Is Involved in Rice Blast Resistance. *Planta* **2009**, *230*, 377–385.

Ribot, C.; Hirsch, J.; Balzergue, S.; Tharreau, D.; Nottéghem, J. L.; Lebrun, M. H.; Morel, J. B. Susceptibility of Rice to the Blast Fungus, *Magnaporthe grisea. J. Plant Physiol.* **2008**, *165*, 114–124.

Rola, A. C.; Pingali, P. L. *Pesticides, Rice Productivity, and Farmers Health*. International Rice Research Institute, Manila/Word Resources Institute, 1993; pp 1–6.

Ruiz, F.; Voinnet, O. Roles of Plant Small RNAs in Biotic Stress Responses. *Ann. Rev. Plant Biol.* **2008**, *60*, 485–510.

Sanford, J. C.; Johnston, S. A. The Concept of Parasite-derived Resistance-deriving Resistance Genes from the Parasite's Own Genome. *J. Theor. Biol.* **1985**, *113*, 395–405.

Sasaya, T. Detection Methods for Rice Viruses by a Reverse-transcription Loop-mediated Isothermal Amplification (RT-LAMP). *Methods Mol. Biol.* **2015**, *1236*, 49–59.

Schweizer, P.; Pokorny, J.; Lefert, P. S.; Dudle, R. Double-stranded RNA Interferes with Gene Function at the Single-cell Level in Cereals. *Plant J.* **2000**, *24*, 895–903.

Scorza, R.; Callahn, A.; Levy, L.; Damsteegt, V.; Webb, K.; Ravelonandro, M. Post-transcriptional Gene Silencing in Plum Pox Virus Resistant European Plum Containing the Plum Pox Potyvirus Coat Protein Gene. *Transgenic Res.* **2001**, *10*, 201–209.

Shimizu, T.; Satoh, K.; Kikuchi, S.; Omura, T. The Repression of Cell Wall- and Plastid-related Genes and the Induction of Defense-related Genes in Rice Plants Infected with Rice Dwarf Virus. Mol. *Plant Microbe Interact.* **2007**, *20*, 247–254.

Siddiq, E. A. Rice. In: *50 Years of Crop Science Research in India*; Paroda, R. S., Chadha, K. L., Eds.; ICAR (Indian Council of Agricultural Research): India, 1996.

Soriano, I. R.; Riley, I. T.; Potter, M. J.; Bowers, W. S. Phytoecdysteroids: A Novel Defense against Plant Parasitic Nematodes. *J. Chem. Ecol.* **2004**, 30, 1885–1889.

Suzuki, H. Studies on Antiblastin (I–IV). *Ann. Phytopathol. Soc. Jpn.* **1954**, *18*, 138.

Swarbrick, P. J. Global patterns of gene expression in rice cultivars undergoing a susceptible or resistant interaction with the parasitic plant *Striga hermonthica. New Phytol.* **2008**, *179*, 515–529.

Takatsuji, H.; Shimono, M.; Sugano, S.; Nakayama, A.; Jiang, C. J.; Hayashi, N. Rice WRKY45 Plays a Crucial Role in Benzothiadiazole-inducible Resistance to Fungal Blast and Bacterial Blight Diseases Disease Resistance. Research Unit. *Res. Highl.* **2007**, 7–8.

Tan, F. L.; Yin, J. Q. RNAi, a New Therapeutic Strategy against Viral Infection. *Cell Res.* **2004**, *14*, 460–466.

Tang, W.; Weidner, D. A.; Hu, B. Y.; Newton, R. J.; Hu, X. Efficient Delivery of Small Interfering RNA to Plant Cells by a Nanosecond Pulsed Laser-induced Wave for Post Transcriptional Gene Silencing. *Plant Sci.* **2006**, *171*, 375–81.

Turner, C. T.; Davy, M. W.; MacDiarmid, R. M.; Plummer, K. M.; Birch, N. P.; Newcomb, R. D. RNA Interference in the Light Brown Apple Moth, *Epiphyas postvittana* (Walker) Induced by Double-stranded RNA Feeding. *Insect. Mol. Biol.* 2006, *15*, 383–391.

Tyagi, H.; Rajasubramaniam, S.; Rajam, M. V.; Dasgupta, I. RNA-interference in Rice against Rice *Tungro Bacilliform* Virus Results in Its Decreased Accumulation in Inoculated Rice Plants. *Transgen. Res.* **2008**, *17*, 897–904.

Uesugi, Y. Resistance of Phytopathogenic Fungi to Fungicides. *Jpn. Pestic. Inform. (Jpn.)* **1978**, *35*, 5–9.

Usman, G. M.; Wakil, W.; Sahi, S. T.; Saleem, I. Y. Influence of Various Fungicides on the Management of Rice Blast Disease. *Mycopathology* **2009**, *7*, 29–34.

Vleesschauwer, D. D.; Yang, Y.; Cruz, C. V.; Hofte, M. Abscisic Acid-induced Resistance against the Brown Spot Pathogen *Cochliobolus miyabeanus* in Rice Involves MAP Kinase-mediated Repression of Ethylene Signaling. *Plant Physiol.* **2010,** *152,* 2036–2052.

Wamaitha, M. J.; Yamamoto, R.; Wong, H. L.; Kawasaki, T.; Kawano, Y.; Shimamoto, K. OsRap2.6 Transcription Factor Contributes to Rice Innate Immunity Through its Interaction with Receptor for Activated Kinase-C 1 (RACK1). *Rice* **2012,** *5,* 35.

Wani, S. H.; Sanghera, G. S. Genetic Engineering for Viral Disease Management in Plants. *Nat. Sci. Biol.* **2010,** *2,* 20–28.

Wani, S. H.; Sanghera, G. H.; Singh, N. B. Biotechnology and Plant Disease Control-Role of RNA Interference. *Am. J. Plant Sci.* **2010,** *1,* 55–68.

Watanabe, K.; Tanaka, T.; Fukuhara, K.; Miyairi, N.; Yonehara, H.; Umezawa, H. Blastmycin, a New Antibiotic from *Streptomyces* sp. *J. Antibiot.* **1957,** *10*(1), 39–45.

Waterhouse, P. M.; Wang, M. B.; Lough, T. Gene Silencing as an Adaptive Defence against Viruses. *Nature* **2001,** *411,* 834–842.

Waziril, H. M. Plants as Antiviral Agents. *J. Plant Pathol Microbes* **2015,** *6,* 2–5.

Welner, D. H.; Lindemose, S.; Grossmann, J. G.; Mollegaard, N. E.; Olsen, A. N.; Helgstrand, C.; Skriver, K.; Lo Leggio, L. DNA Binding by the Plant-specific NAC Transcription Factors in Crystal and Solution: a Firm Link to WRKY and GCM Transcription Factors. *Biochem. J.* **2012,** *444,* 395–404.

Willocquet, L.; Elazegui, F. A.; Castilla, N.; Fernandez, L.; Fischer, K. S.; Peng, S.; Teng, P. S.; Srivastava, R. K.; Singh, H. M.; Zhu, D.; Savary, S. Research Priorities for Rice Disease and Pest Management in Tropical Asia: A Simulation Analysis of Yield Losses and Management Efficiencies. *Phytopathology* **2004,** *94,* 672–682.

Xue, J.; Ye, Y. X.; Jiang, Y. Q.; Zhuo, J. C.; Huang, H. J.; Cheng, R. L.; Xu, H. J.; Zhang, C. X. *Efficient RNAi of Rice Planthoppers Using Microinjection.* 2015. Available at: http://www.nature.com/protocolexchange/protocols/3639.

Yoshii, K. Studies on Cephalothecium as a Means of Artificial Immunization of Agricultural Crops. *Ibid.* **1949,** *13,* 37–40.

Younis, A.; Siddique, M. I.; Kim, C. K.; Lim, K. B. RNA Interference (RNAi) Induced Gene Silencing: A Promising Approach of Hi-tech Plant Breeding. *Int. J. Biol. Sci.* **2014,** *10*(10), 1150–1158.

Yu, R.; Xu, X.; Liang, Y.; Tian, H.; Pan, Z.; Jin, S.; Wang, N.; Zhang, W. The Insect Ecdysone Receptor is a Good Potential Target for RNAi-based Pest Control. *Int. J. Biol. Sci.* **2014,** *10,* 1171.

Zha, W.; Peng, X.; Chen, R.; Du, B.; Zhu, L. Knockdown of Midgutgenes by dsRNA-Transgenic Plant-mediated RNA Interference in the Hemipteran Insect *Nilaparvata lugens.* *PLoS One* **2011,** *6,* e20504.

Zhang, Q. Strategies for Developing Green Super Rice. *Proc. Natl. Acad. Sci. U.S.A.* **2007,** *104,* 16402–16409.

INDEX

Printed and bound by CPI Group (UK) Ltd, Croydon, CR0 4YY

23/10/2024

01777701-0014